SOLID STATE PHYSICS LITERATURE GUIDES
Volume 11

CRYSTAL GROWTH
BIBLIOGRAPHY
SUPPLEMENT

Solid State Physics Literature Guides

Prepared under the auspices of the Research Materials Information Center *
Oak Ridge National Laboratory

General Editor: T. F. Connolly†

Solid State Division
Oak Ridge National Laboratory‡
Oak Ridge, Tennessee

Volume 1: Ferroelectric Materials and Ferroelectricity—1970

Volume 2: Semiconductors—Preparation, Crystal Growth, and Selected Properties—1972

Volume 3: Groups IV, V, and VI Transition Metals and Compounds— Preparation and Properties—1972

Volume 4: Electrical Properties of Solids—Surface Preparation and Methods of Measurement—1972

Volume 5: Bibliography of Magnetic Materials and Tabulation of Magnetic Transition Temperatures—1972

Volume 6: Ferroelectrics Literature Index—1974

Volume 7: Scattering of Thermal Neutrons—A Bibliography—1974

Volume 8: Bibliography of Microwave Optical Technology—1976

Volume 9: Laser Window and Mirror Materials —1977

Volume 10: Crystal Growth Bibliography—1979

Volume 11: Crystal Growth Bibliography—Supplement—1981

* The Research Materials Information Center has been discontinued.
†Deceased.
‡Oak Ridge National Laboratory is operated by Union Carbide Corporation for the U.S. Department of Energy.

SOLID STATE PHYSICS LITERATURE GUIDES
Volume 11

CRYSTAL GROWTH BIBLIOGRAPHY

SUPPLEMENT

Compiled by

A. M. Keesee, T. F. Connolly,

and

G. C. Battle, Jr.

Research Materials Information Center
Solid State Division
Oak Ridge National Laboratory
Oak Ridge, Tennessee

With a Foreword by

Lynn A. Boatner

SPRINGER SCIENCE+BUSINESS MEDIA, LLC

Library of Congress Cataloging in Publication Data

Keesee, Anne M.
  Crystal growth bibliography.

  (Solid state physics literature guides ; v. 11)
  Includes index.
  1. Crystals—Growth—Bibliography. I. Connolly, T. F. II. Battle, G. C. III. Title. IV.
Series.
Z7144.S58S65 [Z5524.C8] [QD921]   [016.548'5]   81-19923
ISBN 978-1-4615-9620-2   ISBN 978-1-4615-9618-9 (eBook)
DOI 10.1007/978-1-4615-9618-9   016.5304'1s   AACR2

FOREWORD

Man's first experience with the ordered state of matter to which we now apply the generic term "crystals" came about when he found specimens of some of the natural crystalline mineral substances that are relatively common in the surface and near-surface areas of the earth's crust. His first widespread use of these natural materials in which their crystalline nature was of importance was undoubtedly in fabricating jewelry and otherwise adorning his weapons, tools, and household items. Both the Old and New Testaments of the Bible document the use of crystalline gems, and the Romans are credited with first employing diamonds—a metastable crystalline form of carbon—in jewelry. Various civilizations appear to have ascribed magical powers to some natural crystals, and they are known to have been widely accepted in Europe as having medicinal properties during the Middle Ages. Given early man's appreciation of the symmetry and beauty of natural crystals, it is not surprising that his earliest interest in working with these materials appears to have been directed toward techniques for duplicating or manufacturing these substances that were so highly valued as gems. Although the exact beginning of the science that we now know as "crystal growth" cannot be precisely specified, we do know that Robert Boyle had attempted to grow crystals that could be used as gems prior to 1672. Much later, in 1873, M. A. Gaudin was able to grow single crystals of ruby (i.e., doped Al_2O_3). Gaudin's work was followed by the work of Edmond Fremy and culminated in the development, in 1891, of the flame-fusion crystal growth process by Fremy's student Auguste Verneuil. This process is still widely used today in the growth of single crystals of sapphire and other refractory oxides.

Although the earliest developments in crystal growth came about as a result of man's efforts to duplicate natural materials that were employed as gemstones, it is interesting to note that similar activities today represent only a small fraction of the total monetary and human investment in the growth of single crystals. Modern scientific developments have revealed that the unique electronic, chemical, and physical properties of single crystals can be exploited in making thousands of useful devices. In fact, the impact of electronic devices based on single crystals of materials such as silicon, germanium, and gallium arsenide has been so great that some have called these developments a second industrial revolution or, in this case, an electronic revolution. The use of single crystals in current technology is, of course, not limited solely to electronic devices such as rectifiers, integrated circuits, detectors, etc., but includes applications in optical components, magnetic devices, lasers, and even mechanical applications in bearings, strain gauges, and turbine blades. The ongoing development of optical techniques to duplicate the functions of electronic circuitry in data processing and storage or communications and control applications promises to increase further the importance of the growth of high-quality single crystals.

Considering the accepted critical nature of the science of crystal growth in modern technology, the importance of a volume such as the present one in which a large body of bibliographic material is collected and organized in a systematic manner (i.e., with an extremely useful permuted title index) is apparent. The current volume, in which over 1100 references are compiled, is a supplement to an earlier compilation of over 5000 references on crystal growth that was published as Volumes 10A and 10B of the series *Solid State Physics Literature Guides.* The necessity of updating this earlier compilation is, in itself, testimony

to the increasing pace of development in the science of the growth of single crystals and to the importance of this state of matter to modern life.

L. A. Boatner
Crystal Growth and Characterization Group
Oak Ridge National Laboratory

INTRODUCTION

Coverage

This supplement (over 1100 references) to the 1979 *Crystal Growth Bibliography* (1972—1977 — over 5000 references) covers the period 1978—1979. The bibliography is restricted to the crystal growth of inorganic materials and was largely drawn from the literature collection of the Research Materials Information Center, which is now closed. It includes theoretical, review, and experiment (or "recipe") papers, technical reports, and books.

The coverage of epitaxy presented a problem, since authors do not always make it clear whether or not the epitaxial growth described resulted in single or polycrystalline structures. Papers are of course included where single crystallinity was claimed or illustrated by a definite electron diffraction pattern. Stated attempts to grow single crystals, even when failures, are included. As for the many where a decision could not be made, exclusion was the general rule. Theoretical and review papers are included.

Two books, of the many good books on crystal growth, are essential complements to this bibliography:

> *The Chemistry of Imperfect Crystals,* 2nd Revised Edition.
> Volume 1, Preparation, Purification, Crystal Growth and Phase Theory
> Kröger, F. A.
> North-Holland Publishing Company, Amsterdam—London; American Elsevier Publishing Company, Inc., New York (1973)
>
> (Includes an extensive tabulation of crystals grown by a variety of methods, with over 1100 references for the table alone.)
>
> *Crystal Growth*
> Wilke, K.-T.
> VEB Deutscher Verlag der Wissenschaften, Berlin (1973)
>
> (Has very comprehensive separate tables of crystals grown, methods, crucibles, preparation and purification of starting materials. Lists suppliers of materials and crystal-growing equipment. Its 177 pages of references are supplied as a separate brochure. This book is in German, but the tabulations make it usable without knowledge of the language.)

Permuted Title Index

The keywords (in the center column) of the computer-ordered permuted title index are in strict alphabetic order — which means, for example, that BaO appears before BP. Once recognized, this should cause no difficulty. A preliminary study of the index for the few other computer-imposed idiosyncrasies, produced by the numerical ordering of subscripts, will greatly facilitate its use.

Author Index

Russian Surnames. Because of variations in transliteration of Russian characters, an author's surname in the index may be spelled as you expect to find it, in one or more ways which you do not expect, or both. The most frequent variations are:

(1) Omission of an apostrophe after a consonant, usually within a surname, but sometimes after the last letter. Sometimes j or y is substituted for the apostrophe. Note that the apostrophe is indexed before the letter a.

(2) Interchangeable use of i, j, and y, usually after a vowel, but sometimes after a consonant; either within or at the end of the name.

(3) Substitution of y (sometimes i) for ii, at the end of the name, after a consonant.

Less frequent variations are: substitution of j for zh; interchangeable use of ks and x, v and w, tk and tik.

Initials. Differences in practices among journals, editors, abstractors, and compilers cause widespread problems with initials:

(1) An author's name may appear with one or two initials, or both ways.

(2) The initials of a hyphenated given name may appear with or without a hyphen, or both ways.

Less frequent variations are: with or without Jr. or III; two or three initials; one or three initials.

The Russian initials Yu and Ya may appear as Y, Eh may appear as E, and V and B are sometimes used interchangeably.

CONTENTS

<1>
Rate Equations for Dislocation-Free and Dislocation-Assisted Growth of Gallium (Ga, theory)
Abbaschian, G.J.; Ravitz, S.F.
J. Cryst. Growth 44, 453-66 (1978)
1978
429

<2>
Growth Morphology and Crystal Habit of Struvite Crystals (MgNH4PO4.6H2O) (solution, theory)
Abbona, F.; Boistelle, R.
J. Cryst. Growth 46, 339-54 (1979)
1979
443

<3>
New Gel Method for Growing Large Needles and Single Crystals of Lead Chloride (PbCl)
Abdulkhadar, M.; Ittyachen, M.A.
J. Cryst. Growth 48, 149-54 (1980)
1980
454

<4>
Effect of Crucible Shape on a Crystallization Front (Bridgman, Kinetics, theory)
Abduragimov, G.A.; Vishryakov, E.M.; Kurbanov, K.R.
Inorg. Mater. 14, 1371-73 (1978)
1978
439

<5>
Magnetic and Mossbauer Studies of FeNi2BO5 and FeNiBO4 (ferrites, vapor transport)
Abe, M.; Kaneta, K.; Gomi, M.; Nomura, S.
Mat. Res. Bull. 14, 519-26 (1979)
1979
442

<6>
Mossbauer Study of FeMo2S4 (vapor transport)
Abe, M.; Kaneta, K.; Uchino, K.
J. Phys. Soc. Japan 44, 1739-40 (1978)
1978
428

<7>
Congruent (Diffusionless) Vapor Transport (theory, kinetics)
Abernathy, J.R.; Greenwell, D.W.; Rosenberger, F.
J. Cryst. Growth 47, 145-54 (1979)
1979
443

<8>
Electronic Bandgap Measurements of SnS2 Polytypes (vapor transport)
Acharya, S.; Srivastava, O.N.
Phys. Stat. Sol. (a) 56, K1-K4 (1979)
1979
455

<9>
Epitaxial Growth of Potassium Lithium Niobate Single-Crystal Films on Potassium Bismuth Niobate
Single Crystals by RF Sputtering (KxLi1-xNbO3)
Adachi, M.; Hori, M.; Shiosaki, T.; Kawabata, A.
Japan. J. Appl. Phys. 17, 2053-54 (1978)
1978
429

<10>

<10>
Elastic and Piezoelectric Properties of Potassium Lithium Niobate (KLN) Crystals
(K2.89Li1.55Nb5.11O15, Czochralski)
Adachi, M.; Kawabata, A.
Japan. J. Appl. Phys. 17, 1969-73 (1978)
1978
429

<11>
Epitaxial Growth of Potassium Lithium Niobate Single-Crystal Films on Potassium Bismuth Niobate
Single Crystals by the FGM Technique (KxLi1-xNbO3)
Adachi, M.; Shiosaki, T.; Kawabata, A.
Japan. J. Appl. Phys. 18, 193-4 (1979)
1979
445

<12>
Growth of Tungstenite Single Crystals by Direct Vapour Transport Method (WS2, sublimation)
Agarwal, M.K.; Reddy, K.W.; Patel, H.B.
J. Cryst. Growth 46, 139-42 (1979)
1979
430

<13>
Growth Conditions and Crystal Structure Parameters of Layer Compounds in the Series $Mo1-xWxSe2$
(vapor transport)
Agarwal, M.K.; Wani, P.A.
Mat. Res. Bull. 14, 825-30 (1979)
1979
439

<14>
Single Crystal Growth of Stannous Selenide (SnSe, Bridgman)
Agnihotri, O.P.; Jain, A.K.; Gupta, B.K.
J. Cryst. Growth 46, 491-94 (1979)
1979
443

<15>
Positron Annihilation in ReO3 (vapor transport)
Akahane, T.; Chiba, T.; Tsuda, N.
J. Phys. Soc. Japan 46, 815-21 (1979)
1979
453

<16>
Acoustooptic Characteristics of LiBi(MoO4)2 (Czochralski)
Akimov, S.V.; Dudnik, E.F.; Stolpakova, T.M.; Dovchenko, G.V.
Sov. Phys. Solid State 20, 547-48 (1978)
1978
427

<17>
Epitaxial Stratified Growth (kinetics, theory)
Aleksandrov, L.N.
Kristall Tech. 13, 765-71 (1978)
1978
432

<18>
Computer Model of Autoepitaxial Growth of Ge Films (theory)
Aleksandrov, I.N.; Loginova, R.V.; Gaiduk, E.A.
Inorg. Mater. 14, 769-71 (1978)
1978
436

<19>
The Mechanism of Silicon Epitaxial Layer Growth from Ion-Molecular Beams (Si)
Aleksandrov, L.N.; Lutovich, A.S.; Belorusets, E.D.
Phys. Stat. Sol. 54(a), 463-69 (1979)
1979
457

<20>
Synthesis and Crystal Growth of Refractory Materials by RF Melting in a Cold Container (review,
alpha-Al2O3, Al2O3:Cr2O3)
Aleksandrov, V.I.; Osiko, V.V.; Prokhorov, A.M.; Tatarintsev, V.M.
pp. 421-80 in Current Topics in Materials Science, Vol. 1, E. Kaldis (ed.), North Holland
Publishing Company (1978)
1978
452

<21>
Crystal Structure of beta-KEr3F10 (melt, slow cooling)
Aleonard, S.; Guitel, J.C.; Roux, M.T.
J. Solid State Chem. 24, 332-44 (1978) (in French)
1978
418

<22>
Growth and X-Ray Analysis of Single Crystals of EuSb4S7 and EuSb2Se4 (vapor transport)
Aliev, O.M.; Rustamov, E.G.; Guseinov, G.G.; Guseinov, M.S.
Inorg. Mager. 14, 1052-53 (1978)
1978
436

<23>
Influence of Isotopic Composition on the Density and Molecular Volume of Solid Crystalline
Lithium Nitrate Hydrates (solution, LiNO3.3H2O, LiNO3.3D2O)
Alimova, I.A.; Ryskin, G.Ya.
Sov. Phys. Solid State 20, 191-93 (1978)
1978
436

<24>
Li3N: A Promising Li Ionic Conductor (electrolytic deposition, review)
Alpen, U.v.
J. Solid State Chem. 29, 379-92 (1979)
1979
448

<25>
Telluride Halides of Copper with High Ionic Partial Conductivity. III. On the Phase Transition of
CuTeBr and the Structure of the High Temperature Phase (Bridgman)
Alpen, U.v.; Fenner, J.; Predel, B.; Rabenau, A.; Schluckebier, G.
Z. Anorg. Allg. Chem. 438, 5-14 (1978)
1978
421

<26>
Electrocrystallization. Fundamental Aspects (review, theory)
Amblard, J.; Maurin, G.; Wiart, R.
Paris, Tech. Ing. D906, 67-89 (1978) (in French)
1978
451-A

<27>
Doping of Silicon Crystals with Phosphorus During Growth from the Gas Phase (Si)
Amner, S.A.; Karelin, B.V.
Inorg. Mater. 15, 575-77 (1979)
1979
457

<28>

<28>
Influence of Environmental Saturation and Electric Field on Growth and Evaporation of Epitaxial
Ice Crystals (H2O, vapor transport, sublimation, evaporation)
Anderson, B.J.; Hallett, J.
J. Cryst. Growth 46, 427-44 (1979)
1979
443

<29>
The Crystal Structure of Cu4(PO4)2O (flux)
Anderson, J.B.; Shoemaker, G.L.; Kostiner, E.
J. Solid State Chem. 25, 49-57 (1978)
1978
420

<30>
Properties of Fe2P Crystals Prepared from a Liquid Copper Medium
Andersson, Y.; Rundqvist, S.; Beckman, O.; Lundgren, L.; Nordblad, P.
Phys. Stat. Sol. (a) 49, K153-56 (1978)
1978
438

<31>
On the Resistivity of the Magnetically Ordered Ground State of the Intermediate Valence Compound
TmSe (slow cooling)
Andres, K.; Walsh, W.M.,Jr.; Darack, S.; Rupp, L.W.,Jr.; Longinotti, L.D.
Solid State Commun. 27, 825-28 (1978)
1978
430

<32>
Growth of Large Crystals of Tetramethyltetrathiafulvalenium Tetracyanoquinodimethanide
(TMTTF-TCNQ) (solution)
Anzai, H.
J. Cryst. Growth 47, 733-35 (1979)
1979
454

<33>
Orientation and Temperature Dependence of the Flow Stress in the Intermetallic Compound Ni3Ge
Single Crystals (Bridgman)
Aoki, K.; Izumi, O.
J. Mat. Sci. 13, 2313-2C (1978)
1978
429

<34>
Magneto-Seebeck Effect of Pb-Doped Bi-Sb (Bridgman)
Aono, T.
Japan. J. Appl. Phys. 17, 843-49 (1978)
1978
429

<35>
Conditions of LPE Growth for Lattice Matched GaInAsP/InP DH Lasers with (100) Substrate in the
Range of 1.2-1.5 micrometers
Arai, S.; Itaya, Y.; Suematsu, Y.; Kishino, K.; Katayama, S.
Japan. J. Appl. Phys. 17, 2067-68 (1978)
1978
429

<36>
Crystal Structure of RbEu3F10 (melt)
Arbus, A.; Fournier, M.-T.; Picaud, B.; Boulon, G.; Vedrine, A.
J. Solid State Chem. 31, 11-21 (1980) (in French)
1980
455

<37>
The Systems RbF-EuF3 and CsF-EuF3. Study of the Phases RbLnF4 and CsLnF4 (melt, RbEu2F7)
Arbus, A.; Picaud, B.; Fournier, M.T.; Vedrine, A.; Cousseins, J.C.
Mat. Res. Bull. 13, 33-41 (1978)
1978
424

<38>
Solution Growth of RbAg4I5 Crystals
Arend, H.; Huber, W.; Freudenreich, W.; Surbeck, H.
J. Cryst. Growth 46, 286-88 (1979)
1979
443

<39>
Layer Perovskites of the (CnH2n+1NH3)2MX4 and NH3(CH2)mNH3MX4 Families with M = Cd, Cu, Fe, Mn or
Pd and X = Cl or Br: Importance, Solubilities and Simple Growth Techniques (solution)
Arend, H.; Huber, W.; Mischgofsky, F.H.; Richter-Van Leeuwen, G.K.
J. Cryst. Growth 43, 213-23 (1978)
1978
416

<40>
Preparation of Pure Hydroxyapatite Single Crystals by Hydrothermal Recrystallization (Ca5(PO4)3OH)
Arends, J.; Schuthof, J.; van der Linden, W.H.; Bennema, P.; van den Berg, P.J.
J. Cryst. Growth 46, 213-20 (1979)
1979
442

<41>
Some Aspects of the Epitaxial Vapour Growth of Semiconductors: Elements, III-V Compounds and
Alloys (Ge, Si, GaAs, BE)
Arizumi, T.
pp. 343-420 in Current Topics in Materials Science, Volume 1, E. Kaldis (ed.), North Holland
Publishing Company (1978)
1978
452

<42>
The Vapour Transport of NbB2 and TaB2 (kinetics, theory)
Armas, B.; Jeffes, J.H.E.; Hocking, M.G.
J. Cryst. Growth 44, 605-12 (1978)
1978
427

<43>
Spherical Monocrystals for X-Ray Work Obtained by Plasma Remelting of Alloy Powder (Cu9Al4)
Arnberg, L.; Westman, S.
J. Appl. Cryst. 11, 148-50 (1978)
1978
425

<44>
Growing Optical films of Aluminum Yttrium Garnet (RE substituted Y3Al5O12, flux)
Arsen'ev, P.A.; Bagdasarov, Kh.S.; Fenin, V.V.
Sov. Phys. Cryst. 23, 376-77 (1978)
1978
434

<45>

<45>
Growth of ZnSxSe1-x Layers on ZnS and ZnSe Substrates by a Solid-State-Diffusion Technique
Asami, S.; Ebina, A.; Takahashi, T.
Japan. J. Appl. Phys. 17, 779-85 (1978)
1978
429

<46>
Structure, Spectroscopy, and Stimulated Emission of Crystals of Yttrium Holmium Aluminum Garnets
((Y1-xHox)3Al5O12, Czochralski)
Ashurov, M.Kh.; Voron'kc, Yu.K.; Zharikov, E.V.; Kaminskii, A.A.; Osiko, V.V.; Sobol', A.A.;
Timoshechkin, M.I.; Fedorov, V.A.; Shabaltai, A.A.
Inorg. Mater. 15, 979 (1979)
1979
457

<47>
Liquidus Measurements in the Pb-Sn-Te System (Pb1-xSnxTe, PbTe, flux)
Astles, M.G.; Hatto, P.; Crocker, A.J.
J. Cryst. Growth 47, 379-83 (1979)
1979
454

<48>
Influence of Density of Liquid Phase in the Y3Fe5O12-BaO.B2O3 System on Convection of a Solution
with a Temperature Difference
Avvakumova, L.A.; Bodyachevskii, S.V.
Inorg. Mater. 15, 662-64 (1979)
1979
457

<49>
Growth of a Crystal Needle of Si2ON2 by Nitridization in the Si-SiO2 System in Compacted Powders
Azuma, M.; Yamada, T.; Hayashi, H.
J. Ceram. Soc. Japan 86, 20-27 (1978) (in Japanese)
1978
451-A

<50>
Properties of CuCr2Se4 Crystals Obtained from Solutions in Melts
Babitsyna, A.A.; Chernitsyna, M.A.; Tsurkan, V.V.; Emel'yanova, T.A.; Kalinnikov, V.T.; Veselago,
V.G.
Inorg. Mater. 15, 333-35 (1979)
1979
457

<51>
Crystal Growth of CuCr2Se4
Babitsyna, A.A.; Emel'yanova, T.A.; Chernitsyna, M.A.; Kalinnikov, V.T.
Zh. Neorg. Khim., SSSR 23, 267-68 (1978) (in Russian)
1978
431-A

<52>
Crystallization of WO2 from the Gas Phase (vapor transport)
Babushkin, A.V.; Klinkova, L.A.; Skrebkova, E.D.
Inorg. Mater. 13, 1690-92 (1978)
1978
419

<53>

<53>
The Growth of In-Rich Bulk Single Crystals of GaxIn1-xPyAs1-y Via the Gradient Freeze Method
(slow cooling, melt, doped)
Bachmann, K.J.; Thiel, F.A.; Ferris, S.
J.Cryst. Growth 43, 752-55 (1978)
1978
432

<54>
Melt and Solution Growth of Bulk Single Crystals of Quaternary III-V Alloys (review, kinetics)
Bachmann, K.J.; Thiel, F.A.; Schreiber, H.,Jr.
Prog. Crystal Growth Charact. 2, 171-206 (1979)
1979
453

<55>
Semiconductor Surface and Crystal Physics Studied by MBE (GaAs, Ga1-xAlxAs, review of MBE process)
Bachrach, R.Z.
Prog. Cryst. Growth. Charact. 2, 115-44 (1979)
1979
453

<56>
Growth and Phase Transitions in Single Crystals of (KxNa1-x)NbO3 (flux)
Badurski, M.; Stroz, K.
J. Cryst. Growth 46, 274-76 (1979)
1979
443

<57>
Analytical Study of the Thermal Transfer Process in an Installation for Preparation of Single
Crystals by the Method of Directed Vertical Crystallization (theory, kinetics)
Bagdasarov, Kh.S.; Goryainov, L.A.
Fiz. Khim. Obrabot. Mater. No. 1, 73-78 (1978) (in Russian)
1978
451-A

<58>
Crystal Structure of the Phosphate of Trivalent Manganese Mn(PO3)3 (flux)
Bagieu-Beucher, M.
Acta Cryst. B34, 1443-46 (1978)
1978
421

<59>
A Fluctuation Theory of Normal Crystal Growth for Metallic One-Component Systems. Two Cases of
Modelling the Melt Growth Interface Motion
Baikov, Yu.A.; Zelenev, Yu.V.; Haubenreisser, W.
Phys. Stat. Sol. 55(a), 123-35 (1979)
1979
457

<60>
Reaction of Germanium (111) Single Crystals with GeBr4 Vapor (Ge, epitaxy, vapor transport,
kinetics, theory)
Balkanov, M.R.; Repinskii, S.M.; Kisel', Yu.I.; Popov, V.P.
Inorg. Mater. 14, 124-26 (1978)
1978
425

<61>

<61>
Observations of the Movement of a Molten Zone in Germanium Under the Influence of the Peltier
Effect (Ge)
Balooch, M.; Cabiri, A.E.
J. Cryst. Growth 43, 277-79 (1978)
1978
416

<62>
Novel Reactor for High Volume Low-Cost Silicon Epitaxy (Si, vapor transport, kinetics)
Ban, V.S.
J. Cryst. Growth 45, 97-107 (1978)
1978
430

<63>
Mass Spectrometric Study on Chemical Transport of VO2
Bando, Y.; Kyoto, M.; Takada, T.; Muranaka, S.
J. Cryst. Growth 45, 20-24 (1978)
1978
430

<64>
Semiconductor Single Crystals Obtained by the Traveling Solvent Method from (Pb0.8Sn0.2)Te
Bansaragtschin, B.; Link, R.; Lehmann, G.
Kristall Tech. 13, 269-79 (1978) (in German)
1978
451-A

<65>
Low Temperature Conductivity of Gd2(SO4)38H2O (crystal) and Dy2Ti2O7 (powder) as a Function of
Magnetic Field (solution)
Barclay, J.A.; Paterson, L.; Bingham, D.; Moze, O.
Cryogenics 18, 535-37 (1978)
1978
438

<66>
Crystal Growth (Proceedings of the Third International Summer School on Crystal Growth, 1977)
(review)
Bardsley, W.(ed.); Hurle, D.T.J.(ed.); Mullin, J.B.(ed.)
North Holland Series in Crystal Growth, Vol. 2, North-Holland, Amsterdam (1979)
1979
448

<67>
The Growth Figures of MnAs (whiskers, theory)
Barner, K.; Berg, H.
J. Cryst. Growth 46, 763-70 (1979)
1979
445

<68>
Dislocation-Free and Low-Dislocation Quartz Prepared by Hydrothermal Crystallization (SiO2)
Barns, R.L.; Freeland, P.E.; Kolb, E.D.; Laudise, R.A.; Patel, J.R.
J. Cryst Growth 43, 676-86 (1978)
1978
432

<69>
Some Aspects of Polytypism in Crystals (SiC, ZnS, CdI2, mica, theory, review)
Baronnet, A.
Prog. Cryst. Growth. Charact. 1, 151-211 (1978)
1978
424

<70>
Solubility of Phlogopite in Basic Aqueous Solutions Under Hydrothermal Conditions (mica,
KMg3AlSi3O10(OH)2)
Baronnet, A.; Amouric, M.; Chabot, B.; Corny, P.
J. Cryst. Growth 43, 255-63 (1978)
1978
416

<71>
Preparation of Barium Lead Hexa-Aluminate Single Crystal Layers by the Liquid Phase Epitaxy
Technique (flux, (Ba,Pb)Al12O19)
Bartels, G.; Mateika, D.; Robertson, J.M.
J. Crystal Growth 47, 414-20 (1979)
1979
454

<72>
A New Substrate Holder for Liquid Phase Epitaxy
Bartels, G.; Passig, G.
J. Cryst. Growth 44, 363-64 (1978)
1978
426

<73>
Growth of Films of CdS Evaporated in Vacuum onto Silicon (vapor transport, electron beam melting,
kinetics, theory)
Barter, B.H.; Boswarva, I.M.; Holt, D.B.; Jeffes, J.H.E.; Steyn, J.B.
J. Cryst. Growth 47, 623-34 (1979)
1979
454

<74>
The Effects of Growth Speed on the Compositional Variations in Crystals of Cadmium Mercury
Telluride (Cd1-xHgxTe, slow cooling)
Bartlett, B.E.; Capper, P.; Harris, J.E.; Quelch, M.J.T.
J. Cryst.Growth 46, 623-29 (1979)
1979
443

<75>
A Study of Casting in the CdxHg1-xTe System
Bartlett, B.E.; Capper, P.; Harris, J.E.; Quelch, M.J.T.
J. Cryst. Growth 47, 341-50 (1979)
1979
454

<76>
The Synthesis of the First Stage Graphite Salt C8(+)AsF6(-) and Its Relationship to the First
Stage Graphite/AsF5 Intercalate (C8AsF6)
Bartlett, N.; McQuillan, B.; Robertson, A.S.
Mat. Res. Bull. 13, 1259-64 (1978)
1978
430

<77>
Crystallographic and Magnetic Structure of RbFeCl3.2D2O and CsFeCl3.2D2O (solution)
Basten, J.A.J.; van Vlimmeren, Q.A.G.; de Jonge, W.J.M.
Phys. Rev. B 18, 2179-84 (1978)
1978
430

<78>

<78>
Crystallography of Aligred Fe-Al Eutectoid (FeAl, FeAl2)
Bastin, G.F.; Van Loo, F.J.J.; Vrolijk, J.W.G.A.; Wolff, L.R.
J. Cryst. Growth 43, 745-51 (1978)
1978
432

<79>
Magnetic Mixed Valent TmSe (melt)
Batlogg, B.; Ott, H.R.; Kaldis, E.; Thoni, W.; Wachter, P.
Phys. Rev. B 19, 247-59 (1979)
1979
445

<80>
Structure of New Rhenium Apatites Containing Large Anions: Ba10(ReO5)6Br2 and Ba10(ReO5)6I2
(vapor transport)
Baud, G.; Besse, J.-P.; Sueur, G.; Chevalier, R.
Mat. Res. Bull. 14, 675-82 (1979) (in French)
1979
442

<81>
Crystal Structure of La3ReO8
Baud, G.; Besse, J.P.; Chevalier, R.; Gasperin, M.
J. Solid State Chem. 29, 267-72 (1979)
1979
448

<82>
Growth of ZnSiP2 from Gas Phase Under Non-Stoichiometric Conditions -- Microanalytical Evaluation
cf Composition (vapor transport)
Baum, H.; Winkler, K.
Kristall Tech. 13, 645-55 (1978) (in German)
1978
440

<83>
The Crystal Structure of the Tetragermanate K2Ba[Ge4O9]2
Baumgartner, C.; Vollenkle, H.
Monatshefte Chem. 109, 1145-53 (1978)
1978
432

<84>
Development of Depressions and Voids During LPE Growth of GaAs (film)
Bauser, E.
Appl. Phys. 15, 243-52 (1978)
1978
415

<85>
Kinetic Studies of the Intercalation of 2H-NbSe2 with Hydrazine by Transmittance Measurements
(theory)
Beal, A.R.; Acrivos, J.V.
Phil. Mag. B 37, 409-21 (1978)
1978
425

<86>
Growth and Physical Properties of ZnSiP2 (vapor transport, solution)
Becherer, H.; Buhrig, E.; Hein, K.; John, H.; Kirsten, P.; Schneider, H.A.; Siegel, W.; Winkler,
K.
Kristall Tech. 13, 1053-57 (1978)
1978
432

11

<87>

<87>
A New Organic Low Temperature Conductor: HMTSF-TNAP (solution, organics)
Bechgaard, K.; Jacobsen, C.S.; Andersen, N.H.
Solid State Commun. 25, 875-79 (1978)
1978
419

<88>
New Mixed Halide Compounds MFX of Divalent Rare Earths (M = Sm, Eu, Tm, and Yb - X = Cl, Br, and
I) (melt, SmFCl, SmFBr, SmFI, EuFCl, EuFBr, EuFI, TmFCl, TmFBr, TmFI, YbFCl, YbFBr, YbFI)
Beck, H.P.
J. Solid State Chem. 23, 213-17 (1978)
1978
409

<89>
Liquid Phase Epitaxy of II-IV-V2, Chalcopyrite Compounds (ZnSnAs2, solution)
Bedair, S.M.; Littlejohn, M.A.
J. Electrochem. Soc. 125, 952-56 (1978)
1978
426

<90>
Connections Between Thermodynamic Quantities and Vacancy and Diffusion Characteristics in Binary
Metallic Solid Solutions (theory)
Beke, D.L.; Godeny, I.; Kedves, F.J.; Erdelyi, G.
J. Phys. Chem. Solids 40, 543-55 (1979)
1979
445

<91>
Vapour Growth in a Microgravity Environment (review)
Belouet, C.
Thin Solid Films 58, 1-8 (1979)
1979
445

<92>
X-Ray Topographic Study of Growth Sector Strains in KDP Single Crystals (solution)
Belouet, C.; Stacy, W.T.
J. Cryst. Growth 44, 315-19 (1978)
1978
426

<93>
Some Problems of the Crystallization Kinetics and Growth Mechanism of Y3Al5O12 Crystals on
Oriented Seeds in Melt-Solutions (flux)
Belyaev, L.M.; Bykov, A.B.; Timofeeva, V.A.
Sov. Phys. Cryst. 23, 215-17 (1978)
1978
434

<94>
The Mechanism of Growth of KDP Crystals in the Presence of Impurities (solution)
Belyustin, A.V.; Kolina, A.V.
Sov. Phys. Cryst. 23, 126-27 (1978)
1978
440

<95>
Crystal Growth and Phase Transitions in the NbxCr1-xO2 (x greater than 0.5) System (vapor
transport)
Ben-Dor, L.; Shimony, Y.
J. Cryst. Growth 43, 1-4 (1978)
1978
417

<96>

<96>
Wolfram (W)
Benesovsky, F.
Gmelin Handbuch der Anorganischen Chemie, Main Series, 8th edition, Suppl. Volume A1 (1979)
1979
439

<97>
Crystallization During Polymerization of Lithium Dihydrogen Phosphate. II. Crystal Growth by
Dimer Addition (LiPO3)
Benkhoucha, R.; Wunderlich, B.
Z. Anorg. Allg. Chem. 444, 267-76 (1978)
1978
438

<98>
Transmission Electron Microscopy Observations of Precipitates in ZnTe: Shape of the Tellurium
Solidus Line (Bridgman)
Bensahel, D.; Dupuy, M.; Pfister, J.C.
J. Cryst. Growth 47, 727-32 (1979)
1979
454

<99>
GaSb and InSb Crystals Grown by a Vertical and Horizontal Travelling Heater Method
Benz, K.W.; Muller, G.
J. Cryst. Growth 46, 35-42 (1979)
1979
430

<100>
Crystal Growth of the Para- and Antiferroelectric Phases of Deuterium Ammonium Dihydrogen
Arsenate (DADA) (NH4(1-x)D4xH2(1-x)D2xAsO4, solution)
Berdowski, J.; Opilski, A.
J. Cryst. Growth 43, 381-84 (1978)
1978
417

<101>
Liquidus of the Ternary System Pb-Tl-Te
Berg, L.G.; Latypov, Z.M.
Inorg. Mater. 13, 1290-93 (1978)
1978
419

<102>
Structure of La2Fe1.76S5 (melt)
Besrest, F.; Collin, G.
J. Solid State Chem. 24, 301-9 (1978) (in French)
1978
418

<103>
Crystal Structure of Double Oxides of Rhenium. II. The Lanthanum-Rhenium Oxide La6Re4O18 (melt)
Besse, J.-P.; Baud, G.; Chevalier, R.; Gasperin, M.
Acta Cryst. B34, 3532-35 (1978)
1978
433

<104>
High Pressure Synthesis and Structure of a New Pyrochlore Containing Rhenium: Ca1-xRe2O6(OH)2x
(x = 0.3)
Besse, J.P.; Baud, G.; Chevalier, R.; Joubert, J.C.
Mat. Res. Bull. 13, 217-20 (1978)
1978
424

\<105\>
Interaction Between Successive Monatomic Step Trains of Different Density on Evaporating Crystal
Surfaces (NaCl, theory)
Bethge, H.; Hcche, H.; Katzer, D.; Keller, K.W.; Bennema, P.; van der Hoek, B.
J. Cryst. Growth 48, 9-18 (1980)
1980
454

\<106\>
Magnetic and Neutron Scattering Experiments on the Antiferromagnetic Layer- Type Compounds
K2Mn1-xMxF4 (M + Fe, Co) (K2Mn1-xFexF4, slow cooling)
Bevaart, L.; Frikkee, E.; Lebesque, J.V.; de Jongh, L.J.
Phys. Rev. B 18, 3376-92 (1978)
1978
429

\<107\>
Growth of Nickel-Zinc Ferrite Single Crystals from BaO-B2O3-ZnO-NiO-Fe2O3 Melts with a
Nonstoichiometric Content of Ferrite-Forming Oxides (NixZn1-xFe2O4, flux)
Beznaternykh, L.N.; Masbchenko, V.G.
Inorg. Mater. 15, 81-85 (1979)
1979
445

\<108\>
Polar Twins in Lithium Metasilicate and Metagermanate (Li2SiO3, Li2GeO3) Single Crystals
(Bridgman)
Bhalla, A.S.; Cross, L.E.; Newnham, R.E.
J. Cryst. Growth 46, 262-64 (1979)
1979
443

\<109\>
Crystal Growth of Antimony Sulphur Iodide (SbSi, flux)
Bhalla, A.S.; Spear, K.E.; Cross, L.E.
Mat. Res. Bull. 14, 423-29 (1979)
1979
442

\<110\>
Growth & Etching of InBi Single Crystals (melt, zone melting)
Bhatt, V.P.; Desai, C.F.
Indian J. Pure Appl. Phys. 16, 960-62 (1978)
1978
444

\<111\>
Growth and Cleavage of Se-Te Alloy Crystals (Bridgman)
Bhatt, V.P.; Trivedi, S.B.
J. Cryst. Growth 44, 262-63 (1978)
1978
424

\<112\>
Growth of Hollow Se-Te Whisker Crystals (theory)
Bhatt, V.P.; Trivedi, S.B.
Kristall Tech. 13, 1435-38 (1978)
1978
444

<113>

<113>
Irreversible Thermodynamics of the Cellular Growth of a Dilute Binary Alloy Single Crystal
(Fe-Ni, eutectics, theory, dendrites)
Billia, B.; Capella, L.
J. Cryst. Growth 44, 235-40 (1978)
1978
424

<114>
Growth and Characterization of Undoped ZnSe Epitaxial Layers Obtained by Organometallic Chemical
Vapour Deposition
Blanconnier, P.; Cerclet, M.; Henoc, P.; Jean-Louis, A.M.
Thin Solid Films 55, 375-86 (1978)
1978
447

<115>
The Growth of Single Crystals of Some Organic Compounds by the Czochralski Technique and the
Assessment of their Perfection (organics)
Bleay, J.; Hooper, R.M.; Narang, R.S.; Sherwood, J.N.
J. Cryst. Growth 43, 589-96 (1978)
1978
419

<116>
Growth of Si Strips by Stepanov's Method
Bletskan, N.I.; Buzynin, A.N.; Zaichko, V.V.; Selin, V.V.
Inorg. Mater. 14, 466-68 (1978)
1978
430

<117>
Problems in the Melt- and Vapor Growth of Silicon for Integrated Circuits and Solar Cells (Si,
review)
Bloem, J.
J. Solid State Chem. 27, 19-27 (1979)
1979
448

<118>
Mechanisms of the Chemical Vapour Deposition of Silicon (Si)
Bloem, J.; Giling, L.J.
pp. 147-342 in Current Topics in Materials Science, Volume 1, E. Kaldis (ed.), North Holland
Publishing Company (1978)
1978
452

<119>
Preparation and Properties of Single Crystals of Solid Solutions GaxIn1-xP (GaP, InP)
Bodnar', I.V.; Makovetskaya, L.A.; Smirnova, G.F.
Inorg. Mater. 14, 1067-70 (1978)
1978
439

<120>
Crystal Structure of Ba2Fe6O11 (flux)
Boivin, J.-C.; Thomas, D.; Pouillard, G.; Perrot, P.
J. Solid State Chem. 29, 101-08 (1979) (in French)
1979
448

<121>

<121>
Growth of Bi2(SeO4)3 Single Crystals in Silica Gels
Boncheva-Mladenova, Z.; Dishovsky, N.
J. Cryst. Growth 47, 82-84 (1979)
1979
443

<122>
Growth of Single Crystals of Ag2SeO4 in Silica Gels
Boncheva-Mladenova, Z.; Dishovsky, N.
J. Cryst. Growth 47, 467-68 (1979)
1979
454

<123>
Monte Carlo Simulation of the Crystal-Melt Interface of a Lennard-Jones Substance (theory)
Bonissent, A.; Gauthier, E.; Finney, J.L.
Phil. Mag. 39B, 49-59 (1979)
1979
444

<124>
Preparation by Vapor Transport of Monocrystals of AlF3, FeF3, and TiF3
Bonnamy, C.; Launay, J.-C.; Pouchard, M.
Rev. Chim. Miner. 15, 178-84 (1978) (in French)
1978
437

<125>
Reproducible Preparation of Twin-Free InP Crystals Using the LEC Technique (Czochralski)
Bonner, W.A.
Mat. Res. Bull. 15, 63-72 (1980)
1980
454

<126>
Nature, Origin and Effect of Dislocations in Epitaxial Semiconductor Layers (III-V, review)
Booker, G.R.; Titchmarsh, J.M.; Fletcher, J.; Darby, D.B.; Hockly, M.; Al-Jassim, M.
J. Cryst. Growth 45, 407-25 (1978)
1978
430

<127>
Chemical Synthesis and Crystal Growth of Laser Quality Praseodymium Pentaphosphate (PrP5O14)
Borkowski, B.; Grzesiak, E.; Kaczmarek, F.; Katuski, Z.; Karolczak, J.; Szymanski, M.
J. Cryst. Growth 44, 320-24 (1978)
1978
426

<128>
The Effect of the Solidification Front Shape on the Temperature Distribution in the Crystal
(NaCl, theory, melt, Czochralski)
Borodin, V.A.; Davidova, L.B.; Erofeev, V.N.; Shdanov, A.V.; Startsev, S.A.; Tatarchenko, V.A.
J. Cryst. Growth 46, 757-62 (1979)
1979
445

<129>
On the Ordering of Fe Atoms in FexNbS2 (vapor transport)
Boswell, F.W.; Prodan, A.; Vaughan, W.R.; Corbett, J.M.
Phys. Stat. Scl. 45(a), 469-81 (1978)
1978
420

<130>

<130>
Synthesis and Structure of Cs2S2 (hydrothermal)
Bottcher, P.
J. Less-Common Metals 63, 99-103 (1979)
1979
447

<131>
The Compound Zn4(P2S6)3 (melt)
Bouchetiere, M.; Toffoli, P.; Khodadad, P.
C.R. Acad. Sci. 286C, 79-81 (1978) (in French)
1978
415

<132>
Crystal Structure of Zn4(P2S6)3 (melt)
Bouchetiere, M.; Toffoli, P.; Khodadad, P.; Rodier, N.
Acta Cryst. B34, 384-87 (1978)
1978
418

<133>
Crystal Growth and EPR Study of Cr+-Doped MgTe Single Crystals (sublimation)
Boulanger, D.; Martin, G.S.
Phys. Stat. Sol. 85(b), 59-601 (1978)
1978
419

<134>
Solvent Effects and the Growth Kinetics of Ionic Crystals (CaSO4.2H2O, solution, theory)
Bourne, J.R.; Davey, R.J.
J. Cryst. Growth 44, 613-14 (1978)
1978
427

<135>
The First Oxomanganate(II): Na14Mn2O9 = Na14[MnO4]2O
Brachtel, G.; Hoppe, R.
Z. Anorg. Allg. Chem. 438, 97-104 (1978)
1978
421

<136>
Local Selective Homoepitaxy of Silicon at Reduced Temperatures Using a Silicon-Iodine Transport System (Si)
Braun, P.D.; Kosak, W.
J. Cryst. Growth 45, 118-25 (1978)
1978
430

<137>
Czochralski Crystal Growth of Rare Earth Tetraborides (YB4, TbB4, ErB4)
Bressel, B.; Chevalier, B.; Etourneau, J.; Hagenmuller, P.
J. Cryst. Growth 47, 429-33 (1979)
1979
454

<138>
The Growth of Single Crystals of Langbeinites Rb2Cd2(SO4)3, Tl2Cd2(SO4)3, K2Co2(SO4)3, and their Phase Transitions (solution)
Brezina, B.; Fouskova, A.
Kristall Tech. 13, 623-29 (1978)
1978
440

<139>
Crystal Growth from Liquids at High Temperatures (review, melt, flux, Czochralski, Kyropoulos,
float zone melting, Bridgman)
Brice, J.C.
Prog. Cryst. Growth Charact. 1, 256-88 (1978)
1978
426

<140>
A Molecular Dynamics Study of Surface Melting (theory, vapor transport, kinetics)
Broughton, J.Q.; Woodcock, L.V.
J. Phys. C: Solid State Phys. 11, 2743-62 (1978)
1978
435

<141>
Crystal Growth and Characterization of Lithium Metasilicate (Li2SiO3, Czochralski)
Brun, M.K.; Bhalla, A.S.; Spear, K.E.; Cross, L.E.; Berger, R.S.
J. Cryst. Growth 47, 335-40 (1979)
1979
454

<142>
Crystal Structure of Cu4(PO4)2O (flux)
Brunel-Laugt, M.; Durif, A.; Guitel, J.C.
J. Solid State Chem. 25, 39-47 (1978) (in French)
1978
420

<143>
Gadolinium Gallium Garnet (Gd3Ga5O12, review)
Bruni, F.J.
in Crystals for Magnetic Applications, Vol. 1, edited by C. J. M. Rooijmans, Springer-Verlag, New
York (1978)
1978
431-A

<144>
Investigation of the Mechanism and Kinetics of Growth of LPE GaAs (theory, epitaxy)
Bryskiewicz, T.
J. Cryst. Growth 43, 101-14 (1978)
1978
417

<145>
The Crystal Structures of Two Modifications of HfNi3 (melt)
Bsenko, L.
Acta Cryst. B34, 3201-04 (1978)
1978
433

<146>
The Crystal Structure of Hf8Ni21 (melt)
Bsenko, L.
Acta Cryst. B34, 3204-7 (1978)
1978
433

<147>
The Crystal Structure of Hf3Ni7 (melt)
Bsenko, L.
Acta Cryst. B34, 3207-10 (1978)
1978
433

<148>

<148>
Sublimation Growth and X-Ray Topographic Characterization of CdTe Single Crystals
Buck, P.; Nitsche, R.
J. Cryst. Growth 48, 29-33 (1980)
1980
454

<149>
A Theoretical Study of Temperature Distributions During Czochralski Crystal Growth (Cu, Si)
Buckley-Golder, I.M.; Humphreys, C.J.
Phil. Mag. 39A, 41-57 (1979)
1979
444

<150>
Simple Pressurized Chambers for Liquid Encapsulated Czochralski Crystal Growth (InP, CuInS2)
Buehler, E.
J. Cryst. Growth 43, 584-88 (1978)
1978
419

<151>
Preparation and Physical Properties of Boracite Crystals (Chemical transport, N3B7O13X)
Bugakov, V.I.; Orlovskii, V.P.; Belyaevskii, T.V.; Bobrov, Yu.A.; Egorov, V.L.; Morozov, N.N.;
Pakhcmcv, V.I.
Inorg. Mater. 15, 816-20 (1979)
1979
457

<152>
Integrated Optical Devices Fabricated by MBE (GaAs, Ga1-xAlxAs, review, film)
Burnham, R.D.; Scifres, D.R.
Prog. Crystal Growth Charact. 2, 95-113 (1979)
1979
453

<153>
Americium Ditelluride (AmTe2-x
Burns, J.H.; Damien, D.; Haire, R.G.
Acta Cryst. B35, 143-44 (1979)
1979
433

<154>
Magnetization and Neutron Studies of UTe and USb0.8Te0.2
Busch, G.; Vogt, O.; Delapalme, A.; Lander, G.H.
J. Phys. C: Solid State Phys. 12, 1391 (1979)
1979
450

<155>
Ferroelectric (Pb,Ba)5Ge3O11 Crystals (flux)
Bush, A.A.; Venevtsev, Yu.N.
Inorg. Mater. 14, 759-61 (1978)
1978
432

<156>
The Thermal Oxidation of GaAs
Butcher, D.N.; Sealy, B.J.
J. Phys. D: Appl. Phys. 11, 1451-56 (1978)
1978
428

<157>
Growth of Single Crystals of Cinnabar (alpha-HgS) at Rather Low Temperature (solution, doped)
Butti, C.; Masse, G.
Kristall Tech. 14, 5-8 (1979)
1979
452

<158>
An X-Ray Topographic Assessment of Cadmium Mercury Telluride (CdxHg1-xTe, Bridgman)
Bye, K.I.
J. Mat. Sci. 14, 619-25 (1979)
1979
446

<159>
Phase Equilibria in Solvent-Garnet Systems (flux, Y3Al5O12, Y3Fe5O12)
Bykov, A.B.; Timofeeva, V.A.
Sov. Phys. Cryst. 23, 96-99 (1978)
1978
440

<160>
Single Crystal and Film Growth of Actinide Pnictides by Chemical Vapour Transport (review)
Calestani, G.; Spirlet, J.C.; Muller, W.
J. Phys. (Paris) 40, Suppl. 4, 106-8 (1979)
1979
448-T

<161>
Preparation and Single-Crystal Growth of Protactinium Arsenides by Chemical Vapour Transport
(PaAs2, Pa3As4, PaAs)
Calestani, G.; Spirlet, J.C.; Rebizant, J.; Muller, W.
J. Less-Common Metals 66, 207-212 (1979)
1979
456

<162>
A Simple Method for Growing V3Si Single Crystals (electron beam melting)
Callaghan, T.; Schwanebeck, J.; Toth, L.; Dayan, M.; Goldman, A.M.
J. Appl. Phys. 49, 2523-25 (1978)
1978
419

<163>
In-Situ Observation of Solidification Interfaces with a Very High Voltage Electron Microscope
Camel, D.; Lemaignan, C.; Pelissier, J.
J. Cryst. Growth 47, 317-19 (1979)
1979
443

<164>
An Improved System for the Bridgman Growth of Crystals with Toxic and/or Highly Volatile
Components (equipment)
Capper, P.; Harris, J.E.; Nicholson, D.; Cole, D.
J. Cryst. Growth 46, 575-81 (1979)
1979
443

<165>
On the Cooling Rates of Large-Diameter Silicon Crystals (Si, theory, kinetics, Czochralski)
Capper, P.; Wilkes, J.G.
Appl. Phys. Lett. 32, 187-89 (1978)
1978
411

<166>

<166>
Evaluation of Supersaturation in Crystal Growth from Solution (theory, organics)
Cardew, P.T.; Davey, R.J.; Garside, J.
J. Cryst. Growth 46, 534-38 (1979)
1979
443

<167>
Melt Growth of Nd:Y3Al5O12(Nd:YAG) Using the Heat Exchange Method (HEM) (Y3Al5O12, melt, slow cooling)
Caslavsky, J.I.; Viechnicki, D.
J. Cryst. Growth 46, 601-606 (1979)
1979
443

<168>
Percolation Theory and Crystallization (entropy, specific heat, melting, theory)
Chaban, I.A.
Sov. Phys. Solid State 20, 863 (1978)
1978
430

<169>
Synthesis of Rare-Earth Carbonates Under Hydrothermal Conditions (YOHCO3, GdOHCC3, ErOHCO3, Y2O(OH)2CO3, Er2O2CO3)
Chai, B.H.T.; Mroczkowski, S.
J. Cryst. Growth 44, 84-97 (1978)
1978
432

<170>
On the Incorporation of Metallic Impurities in Synthetic Quartz Single Crystals (SiO2, solution)
Chakraborty, D.
J. Cryst. Growth 44, 599-603 (1978)
1978
427

<171>
Low-Temperature Specific Heat and Magnetic Susceptibility of Nonmetallic Vanadium Bronze (NaxV2O5, KxV2O5, AgxV2O5, theory)
Chakraverty, B.K.; Sienko, M.J.; Bonnerot, J.
Phys. Rev. B 17, 3781-89 (1978)
1978
420

<172>
On the Growth of NH4Cl Dendrites at Very Low Supersaturation (solution)
Chan, S.-K.; Reimer, H.-H.; Kahlweit, M.
J. Cryst. Growth 43, 229-34 (1978)
1978
416

<173>
Computer Simulation of Convection in Floating Zone Melting. I. Pure Rotation Driven Flows (Si, Al2O3, kinetics, float zone melting, theory)
Chang, C.E.
J. Cryst. Growth 44, 168-77 (1978)
1978
424

<174>
Computer Simulation of Convection in Floating Zone Melting. II. Combined Free and Rotation Driven
Flows (kinetics, theory, electron beam melting)
Chang, C.E.
J. Cryst. Growth 44, 178-86 (1978)
1978
424

<175>
Thermocapillary Convection in Floating Zone Melting: Influence of Zone Geometry and Prandtl
Number at Zero Gravity (Si, Al2O3, theory, float zone melting, kinetics, Si)
Chang, C.E.; Wilcox, W.F.; Lefever, R.A.
Mat. Res. Bull. 527-36 (1979)
1979
442

<176>
Semiconductor Superlattices by MBE and Their Characterization (GaAs, Ga1-xAlxAs, In1-xGaxAs,
GaSb1-yAsy)
Chang, L.L.; Esaki, L.
Prog. Cryst. Growth Charact. 2, 3-14 (1979)
1979
453

<177>
Silicide Formation by High-Dose Si+-Ion Implantation of Pd (electron beam melting, evaporation)
Chapman, G.E.; Lau, S.S.; Matteson, S.; Mayer, J.W.
J. Appl. Phys. 50, 6321-27 (1979)
1979
454

<178>
The Crystal Structure and Twinning Behavior of Ferric Molybdate, Fe2(MoO4)3 (hydrothermal)
Chen, H.-Y.
Mat. Res. Bull. 14, 1583-90 (1979)
1979
454

<179>
Convection Phenomena During the Growth of Sodium Chlorate Crystals from Solution (NaClO3)
Chen, F.S.; Shlichta, P.J.; Wilcox, W.R.; Lefever, R.A.
J. Cryst. Growth 47, 43-60 (1979)
1979
443

<180>
Crystal Growth and Crystal Chemistry of NiAs-Type Compounds: MnSb, CoSb, and NiSb (Czochralski)
Chen, T.; Mikkelsen, J.C.,Jr.; Charlan, G.B.
J. Cryst. Growth 43, 5-12 (1978)
1978
417

<181>
Kinetics of Crystallization of Alloys of Multiple Constituents (theory)
Cherepanova, T.A.
Dokl. Akad. Nauk SSSR 238, 162-65 (1978) (in Russian)
1978
451-A

<182>
Fast Growth of Ordered AB Crystals: A Monte Carlo Simulation for Ionic Crystal Growth (theory)
Cherepanova, T.A.; van der Eerden, J.P.; Bennema, P.
J. Cryst. Growth 44, 537-44 (1978)
1978
427

<183>

<183>
Growth cf Crystals, Volume 11 (review, theory)
Chernov, A.A. (ed.)
Consultants Bureau, New York (1979) (Translated from the Russian)
1979
442-TC

<184>
Theoretical Analysis of Equilibrium Adsorption Layers in CVD Systems (Si-H-Cl, Ga-As-H-Cl) (GaAs,
theory, epitaxy)
Chernov, A.A.; Rusaikin, M.P.
J. Cryst. Growth 45, 73-81 (1978)
1978
430

<185>
Preparation and Investigation of Crystals of Solid Solutions of the System GaAs-ZnSe (Bridgman)
Chernyshov, A.I.; Zeleva, G.M.; Kirovskaya, I.A.
Inorg. Mater. 14, 800-802 (1978)
1978
436

<186>
New Ternary Mc(II)-Compcunds InxMo15Se19 Containing Mo6Se8 and Mo9Se11 Units
Chevrel, R.; Sergent, M.; Seeber, B.; Fischer, O.; Gruttner, A.; Yvon, K.
Mat. Res. Bull. 14, 567-77 (1979)
1979
442

<187>
Growth of Crystalline Slabs of Layered InSe by the Czochralski Method
Chevy, A.; Gouskov, A.; Besson, J.M.
J. Cryst. Growth 43, 756-59 (1978)
1978
432

<188>
LPE Growth of Quaternary GayIn1-yAsxP1-x (film)
Chiao, S.H.; Moon, R.L.
Prog. Cryst. Growth. Charact. 2, 261-68 (1979)
1979
000

<189>
Lasing and Fluorescence in K5NdLi2F10 (flux)
Chinn, S.R.; Hong, H.Y-F.; Bayard, M.; Lempicki, A.; McCollum, B.
Report ESD-TR-78-245, M.I.T., Lincoln Laboratories, 1978, pp. 9-11
1978
428

<190>
Single-Crystal-Aluminum Schottky-Barrier Diodes Prepared by Molecular-Beam Epitaxy (MBE) on GaAs
(Al, films)
Cho, A.Y.; Dernier, P.D.
J. Appl. Phys. 49, 3328-32 (1978)
1978
420

<191>
Germanium-Iodine Equilibrium from 700 to 1000 K (Ge, vapor transport, theory, thermodynamics)
Choukrcun, S.; Launay, J.C.; Pouchard, M.; Hagenmuller, P.
J. Cryst. Growth 43, 597-606 (1978)
1978
419

<192>
Crystal Growth of Niobium Nitride and Niobium Carbide Nitride (NbN, NbN-NbC, NbxNyCz, zone melting)
Christensen, A.N.; Rusche, C.
J. Cryst. Growth 44, 383-86 (1978)
1978
429

<193>
The Affinity of Crystal Growth and Dissolution in Aqueous Solution with Special Reference to Calcium Sulphate Dihydrate (CaSO4.2H2O, kinetics, theory)
Christoffersen, J.; Christoffersen, M.R.; Van Rosmalen, G.M.; Marchee, W.G.J.
J. Cryst. Growth 47, 607-12 (1979)
1979
454

<194>
Crystalline Defects in Solid Phase Epitaxy Si Films Deposited at Elevated Temperatures (vapor transport)
Christou, A.; Davey, J.E.; Tseng, W.
Appl. Phys. Lett. 32, 683-85 (1978)
1978
419

<195>
Solar Cells from Zone-Refined Metallurgical Silicon (Si, Czochralski, zone melting, vapor transport)
Chu, T.L.; Chu, S.S.; Kelm, R.W.,Jr.
J. Electrochem. Soc. 125, 595-97 (1978)
1978
425

<196>
Purification and Characterization of Metallurgical Silicon (Si, melt, Czochralski)
Chu, T.L.; van der Leeden, G.A.; Yoo, H.I.
J. Electrochem. Soc. 125, 661-65 (1978)
1978
425

<197>
The Praseodymium-Gallium System from 0 to 50 At% Gallium (melt, Pr2Ga, Pr5Ga3, PrGA)
Cirafici, S.; Franceschi, E.
J. Less-Common Metals 66, 137-43 (1979)
1979
449

<198>
Crystallochemical Study of the System Ni3Pb2S2-Ni3Pb2Se2 and Note on the Shandite Structure (Ni3Pb2Se2, vapor transport)
Clauss, A.; Warasteh, M.; Weber, K.
Neu. Jb. Mineral., Monatsh. 6, 256-68 (1978) (in German)
1978
451-A

<199>
Control of Substrate Degradation in InP LPE Growth with PH3 Partial Pressure (flux)
Clawson, A.R.; Lum, W.Y.; McWilliams, G.E.
J. Cryst. Growth 46, 300-303 (1979)
1979
443

<200>

<200>
Preparation and Characterization of a Europium(II) Bromide-Chloride Phase EuBr1.5Cl0.5 (melt, slow cooling, hygroscopic)
Clink, B.L.; Eick, H.A.
J. Solid State Chem. 28, 321-28 (1979)
1979
448

<201>
Theoretical Solute Redistribution During a Modified Form of Zone Refining-Cascade Purification (metals)
Clyne, T.W.
J. Cryst. Growth 47, 85-92 (1979)
1979
443

<202>
A Complexity in the Solidification Behaviour of Molten Y3Al5O12 (Czochralski)
Cockayne, B.; Lent, B.
J. Cryst. Growth 46, 371-78 (1979)
1979
443

<203>
Single Crystal Growth of 12CaO.7Al2O3 (Czochralski)
Cockayne, B.; Lent, B.
J. Cryst. Growth 46, 467-73 (1979)
1979
443

<204>
Electrocrystallization of NbO and Hollow NbO2 Needles from Molten Salts (flux)
Cohen, U.
J. Cryst. Growth 46, 147-50 (1979)
1979
430

<205>
Hydrothermal Synthesis of Several Mixed Oxides of the A(6+)B2(3+)O6 Type Under Very High Pressure (UCr2O6, UFe2O6, UMn2O6, UMnNiO6, MoCr2O6)
Collomb, A.; Capponi, J.J.; Gondrand, M.; Joubert, J.C.
J. Solid State Chem. 23, 315-19 (1978)
1978
415

<206>
Lateral Solute Segregation During Unidirectional Solidification of a Binary Alloy with a Curved Solid-Liquid Interface (Ge:Ga, Bridgman, theory)
Coriell, S.R.; Sekerka, R.F.
J. Cryst. Growth 46, 479-82 (1979)
1979
443

<207>
Characterization of (001) Tilt Boundaries Produced by Epitaxial Growth on Bicrystalline Substrates of NaCl (Au, Ag, vapor transport)
Cosandey, F.; Komem, Y.; Bauer, C.L.
Phys. Stat. Sol. 48(a), 555-63 (1978)
1978
426

<208>
La8Ru4O21: A Mixed-Valence Ternary Ruthenium Oxide of a New Hexagonal Structure Type (flux)
Cotton, F.A.; Rice, C.E.
J. Solid State Chem. 24, 359-65 (1978)
1978
418

<209>
The Crystal Structure of La3Ru3O11: A New Cubic KSbO3 Derivative Oxide with No Metal-Metal
Bonding (flux)
Cotton, F.A.; Rice, C.E.
J. Solid State Chem. 25, 137-42 (1978)
1978
422

<210>
Observation by Chemical Attack and X-Ray Topography of Cellular Structures of Monocrystals of
Dilute Binary Metallic Alloys (Cu - 0.5%Si, Cu1-xSbx - x = 0.0015-0.01)
Coulet, A.L.; Billia, B.; Capella, L.
J. Cryst. Growth 47, 469-72 (1979)
1979
454

<211>
Low Temperature CVD Garnet Growth (Er3Fe5O12, (Eu,Y)3Fe5O12 and (Eu,Yb)3Fe5O12, vapor transport,
epitaxy, films)
Cowher, M.E.; Sedgwick, T.O.
J. Cryst. Growth 46, 399-402 (1979)
1979
443

<212>
Properties of GaN Grown on Sapphire Substrates (epitaxy)
Crouch, R.K.; Debnam, W.J.; Fripp, A.L.
J. Mat. Sci. 13, 2358-64 (1978)
1978
429

<213>
Growth of Single Crystals of Beta Rhombohedral Boron by a Modified Verneuil Method (B)
Cueilleron, J.; Viala, J.C.
J. Cryst. Growth 43, 250-54 (1978) (in French)
1978
416

<214>
Growth of Single Crystals of Zinc Selenide from the Vapour Phase (ZnSe, vapor transport,
sublimation)
Cutter, J.R.; Woods, J.
J. Cryst. Growth 47, 405-13 (1979)
1979
454

<215>
Specific Heat of V6O11 Between 0.4 and 50 K (vapor transport, theory)
Dad Khattak, G.; Keesom, P.H.; Faile, S.P.
Solid State Commun. 26, 441-44 (1978)
1978
432

<216>

<216>
Thin Film Growth of (MnxZn1-x)FeO4 by Liquid Phase Epitaxy on Bridgman Grown Zn2TiO4 Substrates (flux)
Damen, J.P.M.; Robertson, J.M.; Huyberts, M.A.H.
J. Cryst. Growth 47, 486-92 (1979)
1979
453

<217>
Preparation of V3Ga Single Crystals by Iodine Transport
Das, B.N.; Ayers, J.D.
J. Cryst. Growth 43, 397-99 (1978)
1978
417

<218>
Impurity Effects on Low-Angle Boundary Formation in Silicon Single Crystals (Si, Czochralski)
Dashevsky, M.Ya.; Eidenson, A.M.
Kristall Tech. 14, 29-36 (1979)
1979
452

<219>
The Role of Dislocations in the Growth of Ammonium Dihydrogen Phosphate Crystals from Aqueous Solution (ADP, NH4H2PO4)
Davey, R.J.; Ristic, R.I.; Zizic, B.
J. Cryst. Growth 47, 1-4 (1979)
1979
443

<220>
Electrical and Optical Properties of Copper Indium Ditelluride Crystals Grown from Near-Stoichiometric Compositions (doped, zone melting)
Davis, J.G.; Bridenbaugh, P.M.; Wagner, S.
J. Electron. Mater. 7, 39-45 (1978)
1978
410

<221>
The Adsorption of Zr onto W(100) Surfaces
Davis, P.R.
Surface Science 91, 385-399 (1980)
1980
456

<222>
An Improved Method of Stirring for LPE Growth (Sm,Y)3(Fe,Ga)5O12 (flux, theory, film)
de Brouckere, L.
J. Cryst. Growth 43, 734-38 (1978)
1978
432

<223>
The Effect of Doping on Microdefect Formation in As-Grown Dislocation-Free Czochralski Silicon Crystals
de Kock, A.J.R.; Stacy, W.T.; van de Wijgert, W.M.
Appl. Phys. Lett. 34, 611-13 (1979)
1979
441

<224>
Conditions for Stable Growth of Epitaxial GaP Layers by Molten Salt Electrodeposition (flux)
De Mattei, R.C.; Elwell, D.; Feigelson, R.S.
J. Cryst. Growth 44, 545-52 (1978)
1978
427

<225>
Growth Rate Limitations in Electrochemical Crystallization (metals, Nb, NaxWO3, Czochralski, electrolytic deposition)
De Mattei, R.C.; Feigelson, R.S.
J. Cryst. Growth 44, 115-20 (1978)
1978
424

<226>
Neutron-Scattering Study of the Incommensurate Phase Transition of Rb2ZnBr4 (theory, solution)
de Pater, C.J.; van Dijk, C.
Phys. Rev. B 18, 1281-93 (1978)
1978
436

<227>
Growth and Some Properties of Cerium Sulphate Enneahydrate Single Crystals (Ce2(SO4)3.9H2O, solution)
de Saja, A.; Pastor, J.M.; Rull, F.; de Saja, J.A.
Kristall Tech. 13, 909-14 (1978)
1978
432

<228>
Preparation Method of High Purity Aluminum Single Crystals with Low Dislocation Density (Al, zone melting, theory)
Deguchi, Y.; Kamigaki, N.; Kashiwaya, K.; Kino, T.
Japan. J. Appl. Phys. 17, 611-16 (1978)
1978
417

<229>
Thermal and Dielectric Properties of LiKSO4 and LiCsSO4
Delfino, M.; Iolacono, G.M.; Smith, W.A.; Shaulov, A.; Tsuo, Y.H.; Bell, M.I.
J. Solid State Chem. 31, 131-134 (1980)
1980
455

<230>
Hydrothermal Crystallization of Sodium Zirconogermanates (Na2ZrGeO5. Na2ZrGe2O7, Na4Zr2Ge3O12, Na3HZr[GeO4]2)
Dem'yanets, L.N.; Nosyrev, N.A.
Sov. Phys. Cryst. 23, 333-35 (1978)
1978
434

<231>
Kinetics of Gallium Phosphide Synthesis by Chemical Gas Transport Reactions (GaP)
Dement'ev, Yu.S.; Sokolov, E.B.; Fedorov, V.A.; Il'in, A.G.; Kotova, Yu.A.
Inorg. Mater. 14, 796-800 (1978)
1978
436

<232>
Hydrothermal Crystallization of Magnetic Oxides (review)
Demianets, L.N.
in Crystals for Magnetic Applications, Vol. 1, C.J.M. Rooijmans (ed.), Springer-Verlag, New York (1978)
1978
431-A

<233>

<233>
X-Ray Topographic Study of Growth Defects in Hydrothermal Na2CoGeO4 Single Crystals in Relation to Their Growth Conditions
Demianets, L.N.; Duderov, N.G.; Lobachev, A.N.; Lefaucheux, F.; Robert, M.C.; Authier, A.
J. Cryst. Growth 44, 570-80 (1978)
1978
427

<234>
Hydrothermal Synthesis of Crystals (review)
Demianets, L.N.; Lobachev, A.N.
Kristall Tech. 14, 509-25 (1979)
1979
452

<235>
About SnF2 Stannous Flouride. I. Crystallochemistry of alpha-SnF2
Denes, G.; Pannetier, J.; Lucas, J.; Le Marouille, J.Y.
J. Solid State Chem. 30, 335-343 (1979)
1979
455

<236>
An Application of the Flux Function Method to a Five-Component System: The Chemical Vapour Transport of Ni-Cl Boracite (no materials grown - Ni3B7O13Cl theory)
Depmeier, W.; Schmid, H.; Nolang, B.I.; Richardson, M.W.
J. Cryst. Growth 46, 716-21 (1979)
1979
443

<237>
Flux Growth and Characterization of Fluorite Crystals (CaF2)
Desai, C.C.; John, V.
J. Cryst. Growth 44, 625-28 (1978)
1978
427

<238>
Flux Growth of Single Magnetic Sublattice Antiferromagnetic Garnets Ca3Fe2Ge3O12 and Ca3Mn2Ge3O12
Desvignes, J.M.; Le Gall, H.
Mat. Res. Bull. 13, 141-46 (1978)
1978
424

<239>
Growth of Oriented Bicrystals of Nickel Oxide (NiO, arc image, float zone, Verneuil)
Dhalenne, G.; Revcolevschi, A.; Gervais, A.
J. Cryst. Growth 44, 297-305 (1978)
1978
426

<240>
Preparation and Basic Physical Properties of BiTeI Single Crystals (Bridgman)
Dich, N.T.; Losiak, P.; Horak, J.
Czech. J. Phys. B28, 1297-1303 (1978)
1978
433

<241>
The Crystal Structure of Triclinic WO3 (sublimation)
Diehl, R.; Brandt, G.; Salje, E.
Acta Cryst. B34, 1105-11 (1978)
1978
421

<242>
Seeded Growth of Large Single Crystals of CdS from the Vapor Phase
Dierssen, G.H.; Gabor, T.
J. Cryst. Growth 43, 572-76 (1978)
1978
419

<243>
On the Growth Mechanism of Sodalite Single Crystals Grown by the Hydrothermal Method on Single
Crystal Seeds Coated with Informative Interfacial Layers (Na4Al3Si3O12Cl)
Distler, G.I.; Kobzareva, S.A.; Lobachev, A.N.; Melnikov, O.K.; Triodina, N.S.
Kristall Tech. 13, 1025-34 (1978) (in Russian)
1978
432

<244>
Growth of Homogeneous Mixed Crystals with a Large Segregation Tendency by the Bridgman Method -
Pseudobinary CdxHg1-xTe
Dittmar, G.
Kristall Tech. 13, 639-43 (1978)
1978
440

<245>
Spectroscopic Properties of Er3Al5-xGaxO12 Films Obtained by Liquid-Phase Epitaxy
Dmitruk, M.V.; Zhekov, V.I.; Prokhorov, A.M.; Timoshechkin, M.I.
Inorg. Mater. 15, 976-79 (1979)
1979
457

<246>
Profile-Sapphire and Its Structural Perfection (alpha-Al2O3, melt)
Dobrovinskaya, E.R.; Litvinov, L.A.; Pishchik, V.V.
Kristall Tech. 13, 289-92 (1978)
1978
451-A

<247>
Twinning and Crystal Structure of Mn0.75Ga2.17S4 (flux)
Dogguy-Smiri, L.; Dung, N.-H.; Pardo, M.-P.
Mat. Res. Bull. 13, 661-65 (1978)
1978
430

<248>
Periodic Doping Structure in GaAs (GaAs, Ga1-xAlxAs, molecular beam epitaxy, review)
Dohler, G.H.; Ploog, K.
Prog. Cryst. Growth. Charact. 2, 145-68 (1979)
1979
453

<249>
Improvement of Crystal Composition in Ga1-xAlxAs LPE Layers Grown under Conditions of Constant
Cooling Rate (melt, theory)
Doi, A.; Hirao, M.; Ito, R.
Japan. J. Appl. Phys. 17, 503-7 (1978)
1978
428

<250>
Infra-Red Transmitting Materials. Part 1. Crystalline Materials (review, theory)
Donald, I.W.; McMillan, P.W.
J. Mat. Sci. 13, 1151-76 (1978)
1978
420

<251>

<251>
Growth Kinetics and Structure of Epitaxial Garnet Layers as Functions of the Mismatch of Crystal Lattice Parameters((Y, Ei, Sm, Ca)3(Fe, Ge)5 O12, flux)
Dorfman, V.F.; Petrushinina, S.A.; Shupegin, M.L.
Thin Solid Films 62, 157-164 (1979)
1979
458

<252>
Low Temperature Method for the Growth of Lithium Nitride Single Crystals (Li3N)
Down, M.G.; Pulham, R.J.
J. Cryst. Growth 47, 133-34 (1979)
1979
443

<253>
Crystal Growth of Some Alkaline-Earth Rare-Earth Pentaoxometallates (flux, Sr2LaAlO5, Sr2PrAlO5, Sr2NdAlO5, Sr2SmAlO5, Sr2EuAlO5, Sr2GdAlO5, Sr2TbAlO5, Sr2NdAlO5, Sr2NdFeO5, Sr2SmFeO5, Sr2EuFeO5, Sr2GdFeO5)
Drofenik, M.; Golic, L.; Kolar, D.
J. Cryst. Growth 47, 735-42 (1979)
1979
454

<254>
Dichlorosilane as Silicon Source for Epitaxial Films on Insulators (Si on sapphire, vapor transport)
Druminski, M.; Nagel, E.; Wieczorek, C.
Siemens Forsch. Entwickl. Ber. 7, 63-70 (1978)
1978
454

<255>
Kinetics of Silicon Growth Under Low Hydrogen Pressure (vapor transport, Si)
Duchemin, M.J.; Bonnet, M.M.; Koelsch, M.F.
J. Electrochem. Soc. 125, 637-44 (1978)
1978
425

<256>
Na2CoGeO4 Single Crystal Growth and Study of Their Piezoelectric and Elastic Properties (hydrothermal)
Duderov, N.G.; Demianets, L.N.; Lobachev, A.N.; Pisarevsky, Yu.V.; Sil'vestrova, I.M.
J. Cryst. Growth 44, 483-91 (1978)
1978
429

<257>
Crystallization of Sodium Perborate from Aqueous Solutions. I. Nucleation Rates in Pure Solution in the Presence of a Surfactant (NaBO2.H2O2.3H2O, kinetics, theory)
Dugua, J.; Simon, B.
J. Cryst. Growth 44, 265-79 (1978)
1978
426

<258>
Crystallization of Sodium Perborate from Aqueous Solutions. II. Growth Kinetics of Different Faces in Pure Solution and in the Presence of a Surfactant (NaBO2.H2O2.3H2O)
Dugua, J.; Simon, B.
J. Cryst. Growth 44, 280-86 (1978)
1978
426

<259>

<259>
Crystal Structure of Larthanum Polysulfide LaS2
Dugue, J.; Carre, D.; Guittard, M.
Acta Cryst. B34, 403-6 (1978)) (in French)
1978
418

<260>
Structure of Silver Dichromate: Ag2Cr2O7 (solution)
Durif, A.; Averbuch-Pouchot, M.T.
Acta Cryst. B34, 3335-37 (1978)
1978
433

<261>
Crystal Structure of the Mercury Nitro-Phosphate Hg4PO4NO3.H2O (melt)
Durif, A.; Tordjman, I.; Masse, R.; Guitel, J.-C.
J. Solid State Chem. 24, 101-5 (1978) (in French)
1978
416

<262>
Dislocation-Free Czochralski Growth of (110) Silicon Crystals (Si)
Dyer, L.D.
J. Cryst. Growth 47, 533-40 (1979)
1979
453

<263>
Growth of Carbon Tetrabromide Crystals for Optical Studies (CBr4, vapor transport)
Ebisuzaki, Y.
J. Cryst. Growth 43, 64E-50 (1978)
1978
419

<264>
Photoconduction of Lead Trititanate (PbTi3O7, flux)
Ehara, S.; Muramatsu, K.; Hattori, T.; Nashio, T.; Shimazu, M.
Japan. J. Appl. Phys. 17, 1153 (1978)
1978
428

<265>
Solubility Curves in High-Temperature Melts for the Growth of Single Crystals of Rare Earth
Vanadates and Phosphates (YPO4, YVO4, REPO4, flux)
Eigermann, W.; Muller-Vogt, G.; Wendl, W.
Phys. Stat. Sol. (a) 49, 145-48 (1978)
1978
436

<266>
The Physical Properties of TGS Single Crystals, Grown from Aqueous TGS Solutions Containing
Aniline
Eisner, J.
Ferroelectrics 17, 575-78 (1978)
1978
415

<267>
A Series of Lead Tungsten Bronzes (PbxWO3)
Ekstrom, T.; Tilley, R.J.D.
J. Solid State Chem. 24, 209-18 (1978)
1978
417

<268>

<268>
Effect of Some Thermal Parameters on the Directional Solidification Process (theory, eutectics,
Al-Al3Ni)
El-Mahallawy, N.A.; Farag, M.M.
J. Cryst. Growth 44, 251-58 (1978)
1978
424

<269>
Growth and Electrical Properties of Sputter-Deposited Single-Crystal GaSb Films on GaAs Substrates
Eltoukhy, A.H.; Greene, J.E.
J. Appl. Phys. 50, 6396-6405 (1979)
1979
454

<270>
Priorities in the Initial Use of Spacelab for Crystal Growth (vapor transport)
Elwell, D.
Proc. Royal Soc. London A 361, 151-56 (1978)
1978
423

<271>
Man-Made Gemstones (review)
Elwell, D.
Halsted Press, New York (1979)
1979
451-A

<272>
The Role of B2O3 in PbO/PbF2/B2O3 Fluxes (Y3Al15O12, theory)
Elwell, D.; Coe, I.M.
J. Cryst. Growth 44, 553-60 (1978)
1978
427

<273>
Specific Heat of a Quasi-Two-Dimensional Antiferromagnet Ni(OH)2, hydrothermal)
Enoki, T.; Tsujikawa, I.
Japan. J. Appl. Phys. 45, 1515-19 (1978)
1978
429

<274>
Single-Crystal Elastic Constants of Al2Cu (Bridgman, arc melting)
Eshelman, F.R.; Smith, J.F.
J. Appl. Phys. 49, 3284-88 (1978)
1978
420

<275>
Preferred Growth Directions and Perfection of Single Crystals of Germanium (Ge, theory, kinetics)
Esin, V.O.; Krivonosova, A.S.
Inorg. Mater. 14, 461-65 (1978)
1978
430

<276>
A Double-Ellipsoid Mirror Furnace for Zone Crystallization Experiments in Spacelab (CdTe, zone
melting, sublimation)
Eyer, A.; Nitsche, R.; Zimmermann, H.
J. Cryst. Growth 47, 219-29 (1979)
1979
443

<277>

<277>
Solution Growth and Some Properties of GaP Bulk Crystals
Fabig, B.; Hildisch, L.
Acta Phys. Acad. Sci. Hung. 44, 5-11 (1978)
1978
432

<278>
Optical Properties of Zn3P2 (evaporation)
Fagen, E.A.
J. Appl. Phys. 50, 6505-15 (1979)
1979
454

<279>
Modified Chemical Vapor Growth of Cinnabar (HgS) and GaP in Closed Systems
Faile, S.P.
J. Cryst. Growth 43, 129-32 (1978)
1978
417

<280>
Growth of Uranium Dioxide Crystals Using Tellurium Tetrachloride and Argon (UO2, vapor transport)
Faile, S.P.
J. Cryst. Growth 43, 133-34 (1978)
1978
417

<281>
The Modified Entrainment Method and Its Application to the Study of Heterogeneous Reactions
(GaAs, vapor transport, theory)
Faktor, M.M.; Garrett, I.; Lyons, M.H.
J. Cryst. Growth 46, 21-30 (1979)
1979
430

<282>
The Crystal Structure of CuTeO4 (hydrothermal)
Falck, L.; Lindqvist, O.; Mark, W.; Philippot, E.; Moret, J.
Acta Cryst. B34, 1450-53 (1978)
1978
421

<283>
The Verneuil Process (review)
Falckenberg, R.
pp. 109-184 in Crystal Growth, Vol. 2, C.H.L. Goodman (ed.), Plenum Press, New York (1978)
1978
429-T

<284>
Powder Feeder for Crystals of Large Diameter Grown by the Verneuil Technique (MgAl2O4,
alpha-Al2O3)
Falckenberg, R.
Kristall Tech. 13, 747-52 (1978)
1978
440

<285>
Hydronium Beta" Alumina: A Fast Proton Conductor (0.84Na2O.0.84MgO.5Al2O3, theory)
Farrington, G.C.; Briant, J.L.
Mat. Res. Bull. 13, 763-73 (1978)
1978
430

<285>

<286>
Ionic Conductivity in N3O+ Beta Alumina (solution)
Farrington, G.C.; Briant, J.L.; Breiter, M.W.; Roth, W.L.
J. Solid State Chem. 24, 311-19 (1978)
1978
418

<287>
A High Temperature High Purity Source for Metal Beam Epitaxy (Ag, film)
Farrow, R.F.C.; Williams, G.M.
Thin Solid Films 55, 303-15 (1978)
1978
447

<288>
Electron Energy-Loss Spectroscopy of TiS2 (vapor transport)
Feldkamp, L.A.; Shinozaki, S.S.; Kukkonen, C.A.; Paile, S.P.
Phys. Rev. B 19, 2291-94 (1979)
1979
447

<289>
Analysis of the Static Method of the Growth of Cadmium Sulphide Crystals with Large Temperature
Gradients (CdS, kinetics, theory, vapor transport)
Ferianc, M.
Kristall Tech. 13, 891-97 (1978)
1978
432

<290>
The Binary Systems Cobalt-Gallium and Nickel-Gallium Compared (CoGa3, NiGa4, flux)
Feschotte, P.; Eggimann, P.
J. Less-Common Metals 63, 15-30 (1979)
1979
447

<291>
Single Crystals with Compositions in the Range TiB1.89 - TiB1.96 Prepared by Chemical Transport
(TiB2)
Feurer, R.; Constant, G.
J. Less-Common Metals 67, 107-114 (1979)
1979
458

<292>
A Possible Method for the Growth of Homogeneous Mercury Cadmium Telluride Single Crystals
(Hg1-xCdxTe, Bridgman)
Fiorito, G.; Gasparrini, G.; Passoni, D.
J. Electrochem. Soc. 125, 315-17 (1978)
1978
411

<293>
Magnetic, Transport, and Thermal Properties of Ferromagnetic EuB6 (review)
Fisk, Z.; Johnston, D.C.; Cornut, B.; von Molnar, S.; Oseroff, S.; Calvo, R.
J. Appl. Phys. 50, 1911-34 (1979)
1979
439

<294>
Change in the Energy Band Structure of Eulytine on Transition to the Amorphous State
(Czochralski, Bi4Si3O12)
Flerova, S.A.; Bochkova, T.M.
Sov. Phys. Solid State 20, 703-4 (1978)
1978
430

35

<295>
On the Preparation of Pure, Doped and Reduced KNbO3 Single Crystals (Czochralski)
Fluckiger, U.; Arend, H.
J. Cryst. Growth 43, 406-16 (1978)
1978
419

<296>
Mechanism of Habit Change of ADP Crystals by Fe3+, Based on Mossbauer Studies (NH4H2PO4, solution)
Fontcuberta, J.; Rodriguez, R.; Tejada, J.
J. Cryst. Growth 44, 593-98 (1978)
1978
427

<297>
Solid State Reactions to CdTeMoO6 and Its Structural Characterization (slow cooling)
Forzatti, P.; Tieghi, G.
J. Solid State Chem. 25, 387-90 (1978)
1978
428

<298>
High-Efficiency p+-n-n+ Back-Surface-Field Silicon Solar Cells (Si)
Fossum, J.G.; Burgess, E.L.
Appl. Phys. Lett. 33, 238-40 (1978)
1978
429

<299>
Vapor Phase Transport and Crystal Growth of Molybdenum Trioxide and Molybdenum Ditelluride
(MoO3-x, MoTe2)
Fourcaudot, G.; Gourmala, M.; Mercier, J.
J. Cryst. Growth 46, 132-35 (1979)
1979
430

<300>
Crystal Structure of the Zinc Iodide ZnI2 (vapor transport)
Fourcroy, P.H.; Carre, D.; Rivet, J.
Acta Cryst. B34, 3160-62 (1978)
1978
433

<301>
Surface Processes Controlling the Growth of GaxIn1-xAs and GaxIn1-xP Alloy Films by MBE
(kinetics)
Foxon, C.T.; Joyce, B.A.
J.Cryst. Growth 44, 75-83 (1978)
1978
432

<302>
Growth Morphology of Weddellite (CaC2O4.2XH2O, gel)
Franchini-Angela, M.; Aquilano, D.
J. Cryst. Growth 47, 719-26 (1979)
1979
454

<303>
Performance of a New High-Intensity Silicon Solar Cell (Si, float zone melting)
Frank, R.I.; Kaplow, R.
Appl. Phys. Lett. 34, 65-67 (1979)
1979
441

<304>

<304>
The Morphology of Hydrothermally Grown Strontium Paracelsian (Sr[Al2Si2O8]
Franke, W.; Ghobarkar, E.; Heimann, R.B.
J. Cryst. Growth 46, 474-78 (1979)
1979
443

<305>
Epitaxial Growth of Single Crystal Cd1-xZnxS Layers on (111) GaAs Substrates Using the
Close-Spaced Geometry (vapor transport)
Franzosi, P.; Ghezzi, C.; Gombia, E.
J. Cryst. Growth 44, 306-14 (1978)
1978
426

<306>
Preparation of Films of IV-VI Compounds in a Quasi-Sealed Volume (SnTe, PbTe, PbSe, evaporation)
Frenk, D.M.; Avgustimov, V.L.; Grushin, A.I.; Solonichnyi, Ya.V.; Shperun, V.M.; Maslyak, N.T.
Fiz. Khim. Obrabot. Mater. No 3, 104-6 (1978) (in Russian)
1978
451-A

<307>
Elastic Constants of Niobium-Zirconium, Hafnium, and Tungsten Alloys (Nb-Zr, Nb-Hf, electron beam
melting, zone melting)
Frey, M.L.; Lonnee, J.E.; Shannette, G.W.
J. Appl. Phys. 49, 4406-10 (1978)
1978
436

<308>
Liquid-Phase Epitaxy of ZnSe from Zn-Ga Solution (film)
Fujita, S.; Mimoto, H.; Noguchi, T.
J. Cryst. Growth 45, 281-86 (1978)
1978
430

<309>
Growth of Cubic ZnS, ZnSe and ZnSxSe1-x Single Crystals by Iodine Transport
Fujita, S.; Mimoto, H.; Takebe, H.; Noguchi, T.
J. Cryst. Growth 47, 326-34 (1979)
1979
454

<310>
High Temperature Form of Pb2WO5 and Transformation Phenomena to Its Low Form (Czochralski, flux)
Fujita, T.; Huamatsu, K.
Mat. Res. Bull. 14, 5-12 (1979)
1979
433

<311>
Growth of LiTaO3 Single Crystals for Saw Device Applications (Czochralski)
Fukuda, T.; Matsumura, S.; Hirano, H.; Ito, T.
J. Cryst. Growth 46, 179-84 (1979)
1979
442

<312>
Organometallic VPE Growth of InAs
Fukui, T.; Horikoshi, Y.
Japan. J. Appl. Phys. 18, 2157-58 (1979)
1979
454

<313>

<313>
New Gaseous Impurity and Its Effect on the Purity of LPE GaAs (vapor transport)
Fukui, T.; Kobayashi, T.
J. Cryst. Growth 45, 243-47 (1978)
1978
430

<314>
On the Growth Mechanism of Polycrystalline Snow Crystals with a Specific Grain Boundary (H2O)
Furukawa, Y.; Kobayashi, T.
J. Cryst. Growth 45, 57-65 (1978)
1978
430

<315>
Raman Scattering Studies in the Solid Electrolytes of the RbAg4I5 Family (KAg4I5, NH4Ag4I5, solution)
Gallagher, D.A.; Klein, M.V.
Phys. Rev. B 19, 4282-91 (1979)
1979
443

<316>
The Crystal Structure of Te3Nb2O11
Galy, J.; Lindqvist, O.
J. Solid State Chem. 27, 279-86 (1979)
1979
448

<317>
A New Structural Family M2"M'Ta5O15. Crystal Structure of CaTlTa5O15 (flux)
Ganne, M.; Dion, M.; Verbaere, A.; Tournoux, M.
J. Solid State Chem. 29, 9-13 (1979) (in French)
1979
448

<318>
Controlled Growth of alpha-Fe Single Crystal Whiskers (hydrogen reduction of ferrous halides, float zone melting)
Gardner, R.N.
J. Cryst. Growth 43, 425-32 (1978)
1978
419

<319>
Flux Growth of Some Fluoride Crystals Under Reducing Conditions. IV (ZrFeF6, ZrCrF6, ZrNiF6, ZrCoF6, ZrMnF6, ZrVF6, RbTiF4, RbVF4, RbVF3, RbV2F6, CsVF4, KFe2F6, KMnF3, RbMnF3)
Garrard, B.J.; Wanklyn, B.M.
J. Cryst. Growth 47, 159-63 (1979)
1979
443

<320>
Viscosity Measurements in the System Dy2O3-K2O-MoO3 (DyKMo2O8, melt)
Garrard, B.J.; Yanagisawa, K.; Wanklyn, B.M.
Mat. Res. Bull. 14, 1001-1005 (1979)
1979
443

<321>
Growth of Large Single Crystals of Heusler Alloys, Ni2MnSn and Ni2Mn1-xVxSn, for Neutron Inelastic Scattering Experiments (Czochralski, slow cooling)
Garrett, J.D.; Greedan, J.E.; Locke, K.E.; Stager, C.V.
J. Cryst. Growth 46, 463-66 (1979)
1979
443

<322>

<322>
Direct Observation of Secondary Nuclei Production (KAl(SO4)2.12H2O, MgSO4.7H2O, solution)
Garside, J.; Larson, M.A.
J. Cryst. Growth 43, 694-704 (1978)
1978
432

<323>
Crystal Growth of Some Rare Earth Trifluorides (GdF3, TbF3, DyF3, HoF3, ErF3,
Bridgman-Stockbarger, Czochralski)
Garton, G.; Walker, P.J.
Mat. Res. Bull. 13, 129-33 (1978)
1978
424

<324>
The Crystal Structure of Lithium Uranate (Li2UO4, flux)
Gebert, E.; Hoekstra, H.R.; Reis, A.H.,Jr.; Peterson, S.W.
J. Inorg. Nucl. Chem. 40, 65-68 (1978)
1978
416

<325>
Interaction Between Technological Parameters and Crystal Quality During the Crystal Growth of
Silicon Crystals, Free from Dislocations (theory)
Geil, W.; Schmugge, K.
Kristall Tech. 14, 343-50 (1979)
1979
441

<326>
Effects of Radial and Axial Temperature Gradients on Tensions in Crystals (theory, Czochralski,
melt, Si)
Geil, W.; Schmugge, K.
Kristall Tech. 13, 195-210 (1978) (in German)
1978
451-A

<327>
Growth of Phthalic Anhydride in a Closed Crystal-Vapour System (organics, vapor transport)
George, J.; Premachandran, S.K.
J. Cryst. Growth 43, 126-28 (1978)
1978
417

<328>
Growth of Single Crystals of Cubic Zinc Sulphide by Double Decomposition Reaction in Melt (ZnS,
flux)
George, V.; Patel, S.M.
J. Cryst. Growth 43, 497-500 (1978)
1978
419

<329>
Infrared Trichroism of Ferroelectric Triglycine Sulphate Thin Films Grown by a New Epitaxial
Process (TGS)
Gerbaux, X.; Hadni, A.
J. Optics 9, 57-60 (1978)
1978
421

<330>
Correlation of the Interface Reaction Constant with Elementary Surface Processes (theory)
Ghez, R.
J. Cryst. Growth 43, 618-20 (1978)
1978
419

<331>
The Crystal Structure of Synthetic Lautarite, Ca(IO3)2 (flux)
Ghose, S.; Wan, C.
Acta Cryst. B34, 84-88 (1978)
1978
415

<332>
Vapor Growth of HgCr2Se4 (epitaxy, theory)
Gibart, P.
J. Cryst. Growth 43, 21-27 (1978)
1978
417

<333>
The Liquidus Temperature and Growth-Dissolution Kinetics of Garnet Liquid Phase Epitaxy (flux, kinetics, theory)
Giess, E.A.; Faktor, M.M.; Frank, F.C.
J. Cryst. Growth 46, 62C-22 (1979)
1979
443

<334>
Kinetics of Growth of Calcium Sulfate Crystals at Heated Metal Surfaces (solution, CaSO4)
Gill, J.S.; Nancollas, G.H.
J. Cryst. Growth 48, 34-40 (1980)
1980
454

<335>
Crystal Structures of Pt4In9S17 and Pb3In6.67S13 (melt)
Ginderow, D.
Acta Cryst. B34, 1804-11 (1978)
1978
425

<336>
Synthesis and Characterization of the Ternary Compound ZnGeP2 (melt, slow cooling, high pressure, Bridgman)
Girault, B.; Gouskov, A.; Bougnot, J.
Mat. Res. Bull. 13, 457-67 (1978)
1978
424

<337>
Effect of Stirring on Crystalline Quality of Solution Grown Crystals -- Case of Potash Alum (KAl(SO4)2.12H2O)
Gits-Leon, S.; Lefaucheux, F.; Robert, M.C.
J. Cryst. Growth 44, 345-55 (1978)
1978
426

<338>
Growth of Whiskers by the Vapor-Liquid-Solid Mechanism
Givargizov, E.I.
pp. 79-146 in Current Topics in Materials Science, Volume 1, E. Kaldis (ed.), North Holland Publishing Company (1978)
1978
452

<339>

<339>
Calcium Orthovanadate, Ca3(VO4)2 - A New High-Temperature Ferroelectric (Czochralski)
Glass, A.M.; Abrahams, S.C.; Ballman, A.A.; Loiacono, G.
Ferroelectrics 17, 579-82 (1978)
1978
415

<340>
LPE Growth of Lithium Ferrite on Spinel Substrate Crystals (films, Li0.5Fe2.5O4)
Glass, H.L.; Liaw, J.H.W.
Mat. Res. Bull. 13, 353-59 (1978)
1978)
424

<341>
Crystallization of HgTe Crystals from the Vapour Phase
Golacki, Z.; Dziuba, Z.; Furmanik, Z.; Makowski, J.
J. Cryst. Growth 46, 293-96 (1979)
1979
443

<342>
Crystallization of HgCdTe Mixed Crystals from the Vapour Phase by Chemical Transport
Golacki, Z.; Makowski, J.
J. Cryst. Growth 47, 749-50 (1979)
1979
454

<343>
Heteroepitaxy of a Deposited Amorphous Germanium Layer on a Silicon Substrate by Laser Annealing
(Ge film)
Golecki, I.; Kennedy, E.F.; Lau, S.S.; Mayer, J.W.; Tseng, W.F.; Eckardt, R.C.; Wagner, R.J.
Thin Solid Films 57, L13-15 (1979)
1979
449

<344>
Growth of Zircon Single Crystals from a Solution in a Melt (ZrSiO4)
Golenko, V.P.; Matveev, S.I.
Inorg. Mater. 15, 710-11 (1979)
1979
457

<345>
Behavior of Boron in Ge Single Crystals Grown by the Czochralski Method
Goncharov, L.A.; Egorov, K.G.; Kervalishvili, P.D.; Leonov, P.A.; Orlov, P.B.; Khorvat, A.M.
Inorg. Mater. 14, 772-75 (1978)
1978
436

<346>
Behavior of Be as an Impurity During Crystallization of Ge by the Czochralski Method
Goncharov, L.A.; Kervalishvili, P.D.
Inorg. Mater. 14, 775-77 (1978)
1978
436

<347>
Morphology of Ice Droxtals Grown from Supercooled Water Droplets (H2O)
Gonda, T.; Yamazaki, T.
J. Cryst. Growth 45, 66-69 (1978)
1978
430

<348>
Crystal Growth - Theory and Techniques, Volume 2 (review, epitaxy, Verneuil, III-V compounds, Si)
Goodman, C.H.L.
Plenum Publishing Corporation, New York (1978)
1978
433-TC

<349>
Subsidiary Electrical Heating for Verneuil Furnaces in the USSR (review)
Goodman, C.H.L.
p. 185 in Crystal Growth, Vol. 2, C.H.L. Goodman (ed.), Plenum Press, New York (1978)
1978
429-T

<350>
The Thallium Tungstate Tl2W4O13: A Tunnel Structure Related to the Hexagonal Tungsten Bronze
Goreaud, M.; Labbe, Ph.; Monier, J.C.; Raveau, B.
J. Solid State Chem. 30, 311-319 (1979)
1979
455

<351>
Heat Capacity of Metals Near the Melting Point and the Vacancy Mechanism of Melting (theory,
compilation, 39 metals)
Gorecki, T.
Acta Phys. Polonica A56, 523-26 (1979)
1979
454

<352>
Epitaxial Growth of A(III)B(V) Semiconductors from Vapour Phase (review, theory)
Gorog, T.; Lendvay, E.
Acta. Phys. Sci. Hung. 44, 13-29 (1978)
1978
432

<353>
GaAs/AlAs Layered Films (epitaxy)
Gossard, A.C.
Thin Solid Films 57, 3-13 (1979)
1979
449

<354>
Liquid Phase Epitaxial Deposition of GaP on GaAs
Gottschalch, V.; Butter, E.; Jacobs, K.; Kramer, P.
J. Cryst. Growth 44, 157-62 (1978)
1978
424

<355>
On the Transition from Faceted to Non-Faceted Growth in Melt-Grown Crystals (organic)
Griffith, W.T.
J. Cryst. Growth 47, 473-75 (1979)
1979
454

<356>
The Growth of beta-Si3N4 Single Crystals
Grun, R.
J. Cryst. Growth 46, 143-46 (1978)
1978
430

<357>

<357>
Physical Properties of the Quarternary Chalcogenides Cu2(I)B(II)C(IV)X4 (BII=Zn, Mn, Fe, Co - CIV
= Si, Ge, Sn - X = S,Se) (Cu2ZnGeS4, Cu2ZnSnS4, Cu2ZnGeSe4, Cu2ZnSnSe4, Cu2MnSiS4, Cu2MnGeS4,
Cu2MnSnS4, Cu2MnGeSe4, Cu2MnSnSe4, Cu2FeGeS4, Cu2CoGeS4, vapor transport)
Guen, L.; Glaunsinger, W.S.; Wold, A.
Mat. Res. Bull. 14, 463-67 (1979)
1979
442

<358>
Limitations in Using Kilohertz Radio Frequencies for Float Zone Silicon Crystals (Si, theory)
Gupta, K.P.; Gregory, R.O.; Rossnick, M.
J. Cryst. Growth 44, 526-32 (1978)
1978
427

<359>
The Preparation of Single Crystals of the Rare Earth Borides by the Solution Method and a Study
of Their Properties (flux, LaB6, SmB6, SmB4, EuB6, YbB6)
Gurin, V.N.; Korsukova, M.M.; Nikanorov, S.P. ; Smirnov, I.A.; Stepanov, N.N.; Shul'man, S.G.
J. Less-Common Metals 67, 115-123 (1979)
1979
458

<360>
Silicon Ribbon Growth via the Ribbon-To-Ribbon (RTR) Technique: Process Update and Material
Characterization (Si, laser melting, review)
Gurtler, R.W.; Baghdadi, A.; Ellis, R.J.; Lesk, I.A.
J. Electron. Mater. 7, 441-77 (1978)
1978
434

<361>
Structure and Growth Peculiarities of Tl-Se-TlInSe2 (TlSe, zone melting, slow cooling)
Guseinov, G.D.; Guseinov, G.G.; Kerimova, E.M.; Ismailov, M.Z.; Rustamov, V.D.; Rzajeva, L.A.
Mat. Res. Bull. 13, 975-82 (1978)
1978
430

<362>
Tetragonal Distortion in Heteroepitaxial Layers: Ge On GaAs (vapor transport, film)
Hagen, W.
J. Cryst. Growth 43, 739-44 (1978)
1978
432

<363>
In Situ X-Ray Topography of Epitaxial Ge Layers During Growth
Hagen, W.; Queisser, H.J.
Appl. Phys. Lett. 32, 269-70 (1978)
1978
421

<364>
Growth and Morphology of Copper-Nickel Alloy Crystals by Hydrogen Reduction of a CuI-NiBr2
Mixture (Cu-Ni, dendrites, whiskers)
Hamamura, K.; Takenouchi, K.
J. Cryst. Growth 46, 804-6 (1979)
1979
445

<365>

<365>
Influence of the Source Material and the Source Shape on the Sublimation Properties of Cadmium,
Selenium, and Cadmium Selenide (CdSe, epitaxy, kinetics, films, theory)
Hamersky, J.
Thin Solid Films 51, 1-11 (1978)
1978
437

<366>
Electron Microscopic Characterization of the Hydrothermal Growth of Synthetic 11 Angstroms
Tobermorite (Ca6Si6O18.4H2O) Crystals
Hamid, S.A.
J. Cryst. Growth 46, 421-26 (1979)
1979
443

<367>
Measurement of Crystallization of Potassium Bromate from Its Quiescent Aqueous Solution by
Differential Scanning Calorimeter. Growth Rate of Crystallite (KBrO3)
Harano, Y.; Oota, K.
J. Chem. Eng. (Japan) 11, 119-24 (1978)
1978
451-A

<368>
Optically Pumped LPE-Grown Hg1-xCdxTe Lasers (films, flux)
Harman, T.C.
J. Electron. Mater. 8, 191-200 (1979)
1979
446

<369>
Thermal Analysis of Solidification in Web-Dendritic Ribbon Growth (Si, kinetics, theory)
Harrill, M.D.; Rhodes, C.A.; Faust, J.W.,Jr.; Hilborn, R.B.,Jr.
J. Cryst. Growth 44, 34-44 (1978)
1978
432

<370>
Skull Melter Growth of Magnetite (Fe3O4)
Harrison, H.R.; Aragon, R.
Mat. Res. Bull. 13, 1097-1104 (1978)
1978
429

<371>
Two-Phase (alpha/beta)-Brass Bicrystals Produced by a Solid-Solid Diffusion Method. II.
Diffusion-Induced Twins in the alpha-phase (theory)
Hashimoto, S.; Eberhardt, A.; Baudelet, B.
Phil. Mag. 38A, 651-71 (1978)
1978
444

<372>
Two-Phase (alpha/beta)-Brass Bicrystals Produced by a Solid-Solid Diffusion Method. I.
Morphology and Crystallographic Relationships Between and alpha- and beta-phases (Cu-Zn, theory)
Hashimoto, S.; Eberhardt, A.; Suery, M.; Baudelet, B.
Phil. Mag. 38A, 629-50 (1978)
1978
444

<373>

<373>
Magnetic Properties of Rare Earth Copper Intermetallic Compounds, RCu2. I. Heavy Rare Earth
(TbCu2, DyCu2, HoCu2, ErCu2)
Hashimoto, Y.; Fujii, H.; Fujiwara, H.; Okamoto, T.
J. Phys. Soc. Japan 47, 67-72 (1979)
1979
453

<374>
Magnetic Properties of Rare Earth Copper Intermetallic Compounds, RCu2. II. Light Rare Earth
(PrCu2, NdCu2)
Hashimoto, Y.; Fujii, H.; Fujiwara, H.; Okamoto, T.
J. Phys. Soc. Japan 47, 73-76 (1979)
1979
453

<375>
Morphology and Growth Mechanism of Vapor Grown Cd Crystals as Affected by Bi Impurity
Hasiguti, R.R.; Ishibashi, T.; Yumoto, H.
J. Cryst. Growth 45, 13-16 (1978)
1978
430

<376>
Dynamic Sublimation Technique for Growing Mercuric Iodide Crystals in Small Opened Ampoules (HgI2)
Hassan, M.A.; Pearce, G.; Edwards, J.P.M.
J. Cryst. Growth 44, 473-74 (1978)
1978
429

<377>
Crystal Structure, Growth, and Optical Properties of Alkali-Trihalogenoplumbates M(PbX3)
(CsPbCl3, CsPbBr3, CsPbI3, KPbI3, RbPbI3, vapor transport)
Haupt, H.J.; Heidrich, F.; Kunzel, H.; Mauersberger, P.
Z. Physik. Chem. 110, 63-73 (1978) (in German)
1978
444

<378>
Ferroelasticity and Phase Transformation in Rb2Hg(CN)4 Spinel (solution)
Haussuhl, S.
Acta Cryst. A34, 965-68 (1978)
1978
433

<379>
The Crystal Structure of Ba2V2O7 (melt)
Hawthorne, F.C.; Calvo, C.
J. Solid State Chem. 26, 345-55 (1978)
1978
435

<380>
Preferred Growth Direction of Pure Cu Single Crystals by the Floating-Zone Method
Hayashi, S.; Inoue, T.; Komatsu, H.
Kristall Tech. 14, K1-K2 (1979)
1979
444

<381>
Growth of Cobalt Single Crystals by the Electron Beam Floating-Zone Method (Co)
Hayashi, S.; Ono, S.; Komatsu, H.
Kristall Tech. 13, 263-67 (1978)
1978
451-A

<382>

<382>
Protein Crystallizations (solution, organics)
Weidner, E.
J. Cryst. Growth 44, 135-44 (1978)
1978
424

<383>
Growth of USb2 Single Crystals and Their Structural Perfection (vapor transport, melt, flux)
Henkie, Z.; Misiuk, A.
Kristall Tech. 14, 539-43 (1979)
1979
452

<384>
InP Growth and Properties (Czochralski, doped)
Henry, R.L.; Swiggard, E.M.
J. Electron. Mater. 7, 647-57 (1978)
1978
438

<385>
Heat-Flow-Controlled Growth During Li2B4O7 Crystallization (melt)
Herron, L.W.; Bergeron, C.G.
J. Amer. Ceram. Soc. 62, 110-11 (1979)
1979
444

<386>
Electrical Properties and Charge Imbalance for Ca,Ge-Substituted Garnet Films grown by Liquid
Phase Epitaxy from PbO-E2O3 Fluxed Melts
Hibiya, T.; Hidaka, Y.; Suzuki, K.
J. Appl. Phys. 49, 2765-69 (1978)
1978
419

<387>
Plasma Growth of Rutile Crystals and Their Photoelectronic Properties (TiO2)
Hillhouse, R.W.A.; Woods, J.
Phys. Stat. Sol. (a) 46, 163 (1978)
1978
438

<388>
(111) Cu2O Growth Modes on (111)Cu Surfaces (epitaxy)
Ho, J.H.; Vook, R.W.
J. Cryst. Growth 44, 561-69 (1978)
1978
427

<389>
Structure of Monocrystalline WV2O6 at 298 and 383 K (vapor transport)
Hodeau, J.L.; Gondrand, M.; Labeau, M.; Joubert, J.C.
Acta Cryst. B34, 3543-47 (1978)
1978
433

<390>
The Crystal Structure of V4O7 at 120 degrees K (vapor transport)
Hodeau, J.L.; Marezio, M.
J. Solid State Chem. 23, 253-63 (1978)
1978
415

<391>

<391>
Structural Aspects of the Metal-Insulator Transitions in (Ti0.9975V0.0025)407 (vapor transport)
Hodeau, J.L.; Marezio, M.
J. Solid State Chem. 29, 47-62 (1979)
1979
448

<392>
The Growth Kinetics of Urea Monocrystals from Aqueous Solution (CO(NH2)2, theory)
Hodorowicz, S.A.; Treivus, E.B.
J. Cryst. Growth 47, 573-76 (1979)
1979
453

<393>
Fast Growth in GaAs VPE at Low Temperature and High Partial Pressures (vapor transport)
Hollan, L.; Durand, J.M.
J. Cryst. Growth 46, 665-70 (1979)
1979
443

<394>
MBE Techniques for IV-VI Optoelectronic Devices (review)
Holloway, H.; Walpole, J.N.
Prog. Crystal Growth Charact. 2, 49-94 (1979)
1979
453

<395>
Fiber Optic Probe for Thermal Profiling of Liquids During Crystal Growth (solution)
Holmes, D.E.
Rev. Sci. Instrum. 50, 662-63 (1979)
1979
444

<396>
Crystal Structure and Ionic Conductivity of Li24Zn(GeO4)4 and other new Li+ Superionic Conductors (kinetics, melt)
Hong, H.Y.-P.
Mat. Res. Bull. 13, 117-24 (1978)
1978
424

<397>
Crystal Structure of K5NdLi2F10 (flux)
Hong, H.Y.-P.; McCollum, B.C.
Mat. Res. Bull. 14, 137-42 (1979)
1979
433

<398>
Advances in Preparative Chemistry of Oxides and Fluorides (review, hydrothermal, theory)
Hoppe, R.
J. Solid State Chem. 27, 99-103 (1979)
1979
448

<399>
Determination of the Lattice Constant of Epitaxial Layers of III-V Compounds (theory)
Hornstra, J.; Bartels, W.J.
J. Cryst. Growth 44, 513-17 (1978)
1978
427

<400>
Low Temperature Thermal and Magnetic Properties of CeP and CeAs (sublimation)
Hulliger, F.; Ott, H.R.
Z. Physik B 29, 47-59 (1978)
1978
415

<401>
Crystal Structure and Antiferromagnetism of EuSb2 (melt)
Hulliger, F.; Schmelczer, R.
J. Solid State Chem. 26, 389-96 (1978)
1978
435

<402>
On the Alkali Metal Tungsten Bronzes, in Particular those of Potassium, Rubidium and Cesium
(review, KxWO3, RbxWO3, CsxWO3)
Hussain, A.
Chem. Commun. No. 2, 1978, Department of Inorganic Chemistry, Arrhenius Laboratory, University of
Stockholm
1978
449

<403>
Growth of CuInS2 and Its Characterization (vapor transport)
Hwang, H.L.; Sun, C.Y.; Leu, C.Y.; Cheng, C.L.; Tu, C.C.
Rev. Phys. Appl. 13, 745-52 (1978)
1978
433

<404>
Single Crystal Growth of Akermanite (Ca2MgSi2O7) and Gehlenite (Ca2Al2SiO7) by the Floating Zone
Method
Ii, N.; Shindo, I.
J. Cryst. Growth 46, 569-74 (1979)
1979
443

<405>
Electrical Properties of TlTe Single Crystals (melt, theory)
Ikari, T.; Hashimoto, K.
Phys. Stat. Sol. (b) 86, 239-48 (1978)
1978
424

<406>
Formation of Stress and Dislocations in Crystal Growth (Si, Ge, GaAs, InAs, InSb, Al2O3, melt,
theory)
Indenbom, V.L.
Kristall Tech. 14, 493-507 (1979)
1979
452

<407>
The Growth, X-Ray Studies and Microstructure of Solution-Grown Mixed Crystals of NaCl and KCl
(NaCl-KCl eutectic)
Ingle, S.G.; Ghadekar, S.R.
J. Phys. D: Appl. Phys. 11, 913-17 (1978)
1978
420

<408>

<408>
An Evidence of Spiral Growth on the Surface of Melt-Grown KCl Crystals (Czochralski, theory)
Inoue, T.; Komatsu, H.
Japan. J. Appl. Phys. 17, 1479-82 (1978)
1978
426

<409>
Effects of Growth Rates on the Defect Generation in KCl Crystals Grown by the Czochralski Method
Inoue, T.; Komatsu, H.
Kristall Tech. 13, 1045-51 (1978)
1978
432

<410>
Electrical Properties of p- and n-Type CuInSe2 Single Crystals (melt)
Irie, T.; Endo, S.; Kimura, S.
Japan. J. Appl. Phys. 18, 1303-10 (1979)
1979
451

<411>
Crystal Growth of Tl3VS4 (Bridgman-Stockbarger)
Isaacs, T.J.; Roland, G.W.
J. Cryst. Growth 47, 712-18 (1979)
1979
454

<412>
Optical and Electrical Properties of CdGeAs2 (Bridgman)
Iseler, G.W.; Kildal, H.; Menyuk, N.
J. Electron. Mater. 7, 737-55 (1978)
1978
438

<413>
Epitaxial Growth of CrO2 on Sapphire in Air (vapor transport)
Ishibashi, S.; Namikawa, T.
Japan. J. Appl. Phys. 17, 249-50 (1978)
1978
428

<414>
Epitaxial Growth of Ferromagnetic CrO2 Films in Air
Ishibashi, S.; Namikawa, T.; Satou, M.
Mat. Res. Bull. 14, 51-57 (1979)
1979
433

<415>
Epitaxial Growth of Ferroelectric PLZT ((Pb,La)(Zr,Ti)O3) Thin Films
Ishida, M.; Tsuji, S.; Kimura, K.; Matsunami, H.; Tanaka, T.
J. Cryst. Growth 45, 393-98 (1978)
1978
430

<416>
Liquid Phase Epitaxial Growth of Zn and S Doped GaAs
Ishii, M.; Tanaka, T.; Susaki, W.
J. Cryst. Growth 46, 265-68 (1979)
1979
443

<417>

<417>
Growth Mechanism of Multi-Steps on the Surface of Vapour Grown FeCr2S4
Ishizuki, H.
Japan. J. Appl. Phys. 17, 1171-75 (1978)
1978
436

<418>
Growth of Triple Unit Cell Steps on the Surface of Vapour Grown FeCr2S4 Single Crystals. I
Ishizuki, H.; Nakada, I.
J. Cryst. Growth 44, 632-34 (1978)
1978
427

<419>
Step Growth by Elementary Steps and Multi-Steps on the Surface of Vapour Grown FeCr2S4
Ishizuki, H.; Nakada, I.
Japan. J. Appl. Phys. 17, 43-47 (1978)
1978
428

<420>
Preparation of Ge-Si Epitaxial Alloys by Sputtering
Ito, K.
J. Cryst. Growth 45, 340-45 (1978)
1978
430

<421>
Crystallography of the fcc Phase in Uni-Directionally Solidified Zn-Al Eutectic
Ito, K.; Fujimoto, K.
J. Cryst. Growth 48, 141-48 (1980)
1980
454

<422>
Growth of p-Type InP Single Crystals by the Temperature Gradient Method
Ito, K.; Ito, H.
J. Cryst. Growth 45, 246-51 (1978)
1978
430

<423>
Vapor Growth of Hexagonal GeO2 Needle Crystals
Ito, S.; Kodaira, K.; Matsushita, T.
Mat. Res. Bull. 13, 97-100 (1978)
1978
424

<424>
Vapor Growth of Zn2GeO4 Single Crystals
Ito, S.; Yoneda, N.; Shimada, S.; Tsunashima, A.; Kodaira, K.; Matsushita, T.
J. Cryst. Growth 47, 310-12 (1979)
1979
443

<425>
Birefringence in CdSiP2 (solution)
Itoh, N.; Fujinaga, T.; Nakau, T.
Japan. J. Appl. Phys. 17, 951-52 (1978)
1978
429

<426>

<426>
Crystal Growth in c Direction and Crystallographic Polarity in ZnO Crystals (vapor transport, film)
Iwanaga, H.; Shibata, N.; Nittono, O.; Kasuga, M.
J. Cryst. Growth 45, 228-32 (1978)
1978
430

<427>
Growth Mechanism of Hollow ZnO Crystals from ZnSe. II
Iwanaga, H.; Yamaguchi, T.; Shibata, N.
J. Cryst. Growth 43, 71-76 (1978)
1978
417

<428>
Crystal Growth and Sublimation in II-VI Compounds Along their Polar Axis (ZnO, CdSe, CdS, ZnS, kinetics)
Iwanaga, H.; Yoshiie, T.; Yamaguchi, T.; Shibata, N.
J. Cryst. Growth 47, 703-11 (1979)
1979
454

<429>
Direct Synthesis and Structure of Epitaxial Cadmium Sulphide Layers on Cadmium Single Crystals (CdS, evaporation)
Iwanov, D.; Nanev, C.
J. Mat. Sci. 13, 1449-54 (1978)
1978
428

<430>
Magnetic Properties of (Mn,Cr)P and (Mn,Fe)P Compounds ($Mn_{1-x}Fe_xP$, x = 0 to 0.15, theory)
Iwata, N.; Fujii, H.; Okamoto, T.
J. Phys. Soc. Japan 46, 778-83 (1979)
1979
453

<431>
Growth of Anatase (TiO2) Crystals by Chemical Transport Reactions with HBr and HCl
Izumi, F.; Kodama, H.; Ono, A.
J. Cryst. Growth 47, 139-44 (1979)
1979
443

<432>
Effect of Growth Conditions of Polytypism in Cadmium Iodide Crystals (CdI2, theory, solution, evaporation)
Jain, P.C.; Trigunayat, G.C.
J. Cryst. Growth 48, 107-13 (1980)
1980
454

<433>
Influence of the VPE Parameters on the Epitaxial Film Properties for the ($Pb_{1-x}Sn_x)_{1+y}Te_{1-y}$ System
Jakobus, T.; Bornung, J.
J. Cryst. Growth 45, 224-27 (1978)
1978
430

<434>
Raman Spectra of ZrS3
Jandl, S.; Deville Cavellin, C.; Harbec, J.Y.
Solid State Commun. 31, 351-53 (1979)
1979
445

<435>
Morphological Stability Analysis in Chemical Vapour Deposition Processes. II (theory)
Jansen, A.K.; van den Brekel, C.H.J.
J. Cryst. Growth 43, 371-77 (1978)
1978
417

<436>
Crystal Growth and the Crystal Structures of Two Modifications of Gold Monobromide, I-AuBr and
P-AuBr (vapor transport)
Janssen, E.M.W.; Wiegers, G.A.
J. Less Common Metals 57, P47-57 (1978)
1978
410

<437>
Liquid-Phase Electroepitaxy: Growth Kinetics (Peltier effect, GaAs, theory, electrolytic
deposition, kinetics)
Jastrzebski, I.; Lagowski, J.; Gatos, H.C.; Witt, A.F.
J. Appl. Phys. 40, 5909-19 (1978)
1978
434

<438>
Synthesis and Crystal Structure of Iron Polyphosphide FeP4
Jeitschko, W.; Braun, D.J.
Acta Cryst. B34, 3196-3201 (1978)
1978
433

<439>
Electrical Properties of Pb1-xCdxS Epitaxial Films (vapor transport)
Jensen, J.D.; Schoolar, R.B.
J. Electron. Mater. 7, 237-52 (1978)
1978
434

<440>
The Growth and Properties of Crystalline Rubidium and Cesium Niobates (Czochralski, Cs8Nb22O59,
Rb8Nb22O59)
Jones, G.R.; Robertson, D.S.
J. Cryst. Growth 43, 115-19 (1978)
1978
417

<441>
Growth of Chromium Silicide, Cr3Si, Crystals (float zone melting)
Jorgensen, J.-E.; Rasmussen, S.E.
J. Cryst. Growth 47, 124-26 (1979)
1979
443

<442>
Single Crystal Growth of Ferroelectric Potassium Dihydrogen Orthophosphate in Silica Gels (KDP)
Joshi, M.S.; Antony, A.V.
J. Mat. Sci. 13, 939-44 (1978)
1978
418

<443>
Nucleation in Supersaturated Potassium Dihydrogen Orthophosphate Solutions (KDP)
Joshi, M.S.; Antony, A.V.
J. Cryst. Growth 46, 7-9 (1979)
1979
430

<444>

<444>
Growth of Fibrous Sodium Aluminosilicate Crystals in Silica Gel (zeolite, Na0.5AlSi6O14.xH2O)
Joshi, M.S.; Bhoskar, B.T.
J. Cryst. Growth 47, 654-58 (1979)
1979
454

<445>
Flux Growth of Lead Neodymium Chlorapatite (Na2Nd2Pb6(PO4)6Cl, slow cooling, evaporation)
Joukoff, B.; Fadly, M.; Ostorero, J.; Makram, H.
J. Cryst. Growth 43, 81-84 (1978)
1978
417

<446>
Crystal Growth and Imperfections in Gadolinium Molybdate and Other Isotypic Rare-Earth Molybdates
(Czochralski, melt, Gd2(MoO4)3, Sm2(MoO4)3, Tb2(MoO4)3)
Joukoff, B.; Grimouille, G.
J. Cryst. Growth 43, 719-26 (1978)
1978
432

<447>
Crystal Growth and Crystallographic Data of Some Ln2(MoO4)3 Type Mixed Rare Earth Molybdates
(RE(MoO4)3, Czochralski)
Joukoff, B.; Grimouille, G.; Leroux, G.; Daguet, C.; Pougnet, A.M.
J. Cryst. Growth 46, 445-50 (1979)
1979
443

<448>
Coalescence Growth of Smoke Particles Prepared by a Gas-Evaporation Technique (MoO3, WO3, theory)
Kaito, C.
Japan. J. Appl. Phys. 17, 601-9 (1978)
1978
427

<449>
Current Topics in Materials Science, Volume 2 (Review)
Kaldis, E. (ed.)
North-Holland Publishing Company (1979)
1979
456-A

<450>
Current Topics in Materials Science, Volume 1 (review)
Kaldis, E. (ed.)
North-Holland Publishing Company (1978)
1978
452-TC

<451>
Impurity Redistribution in EFG (Si, ribbon, theory, kinetics)
Kalejs, J.P.
J. Cryst. Growth 44, 329-44 (1978)
1978
426

<452>
Aluminum Redistribution in EFG of Silicon Ribbon (Si)
Kalejs, J.P.; Freedman, G.M.; Wald, F.V.
J. Cryst. Growth 48, 74-84 (1980)
1980
454

<453>
Far Infrared Optical Study of alpha-In2Se3 Compound (melt, slow cooling)
Kambas, K.; Spyridelis, J.
Mat. Res. Bull. 13, 653-60 (1978)
1978
430

<454>
Preparation of ThSiO4 Single Crystals by a Vapour Phase Reaction
Kamegashira, N.
J. Mat. Sci. 14, 505-6 (1979)
1979
436

<455>
Growth of Single Crystals of U1-xThxO2 Solid Solutions by Chemical Transport Reactions
Kamegashira, N.; Ohta, F.; Naito, K.
J. Cryst. Growth 44, 1-4 (1978)
1978
432

<456>
Single Crystal Growth of LiInS2 (directional solidification)
Kamijoh, T.; Kuriyama, K.
J. Cryst. Growth 46, 801-803 (1979)
1979
445

<457>
Growth, Spectroscopic Investigations, and Some New Stimulated Emission Data of Gd3Ga5O12:Nd3+
Single Crystals (Nd3Ga5C12, doped, Czochralski)
Kaminskii, A.A.; Osiko, V.V.; Sarkisov, S.E.; Timoshechkin, M.I.; Zharikov, E.V.; Bohm, J.;
Reiche, P.; Schultze, D.
Phys. Stat. Sol. (a) 49, 305-11 (1978)
1978)
436

<458>
Growth, Spectroscopy, and Stimulated Emission of Cubic Bi4Ge3O12 Crystals Doped with Dy3+, Ho3+,
Er3+, Tm3+, or Yb3+ Ions (Czochralski)
Kaminskii, A.A.; Sarkisov, S.E.; Butaeva, T.I.; Denisenko, G.A.; Hermoneit, B.; Bohm, J.;
Grosskreutz, W.; Schultze, D.
Phys. Stat. Sol. (a) 56, 725-736 (1979)
1979
458

<459>
Local Modes in Rb1-cKc Alloys: A Neutron Scattering Study (Bridgman, theory)
Kamitakahara, W.A.; Copley, J.R.D.
Phys. Rev. B 18, 3772-81 (1978)
1978
434

<460>
Liquid Phase Epitaxial Growth of Lattice-Matched GaP-Ga0.5Al0.5P-GaP Layer Under Controlled
Phosphorus Vapor Pressure and Reduced Oxygen Contamination (GaP, Ga0.5Al0.5P, flux)
Kan, H.; Katsuno, H.; Sukegawa, T.
J. Cryst. Growth 46, 637-43 (1979)
1979
443

<461>

<461>
Improvements in SSD (synthesis, solute diffusion) Growth of GaP
Kaneko, K.; Ayabe, M.; Isawa, N.
Japan. J. Appl. Phys. 18, 861-68 (1979)
1979
445

<462>
Growth Rate of Iron Whiskers (Fe, vapor transport, theory)
Kaneko, T.
J. Cryst. Growth 44, 14-22 (1978)
1978
432

<463>
Electrical Conductivity and Galvanomagnetic Effects in (SN)x Single Crystal
Kaneto, K.; Yamamoto, M.; Yoshino, K.; Inuishi, Y.
J. Phys. Soc. Japan 47, 167-75 (1979)
1979
453

<464>
Uniaxial Ferromagnetism in NiZrF6.6H2O (solution)
Karnezos, M.; Meier, D.I.; Friedberg, S.A.
Phys. Rev. B 17, 4375-83 (1978)
1978
434

<465>
Effect of Pressure on Electrical Resistivity and Thermoelectric Power of FeS (Bridgman)
Karunakaran, C.; Vijayakumar, V.; Vaidya, S.N.; Kunte, N.S.; Suryanarayana, S.
Mat. Res. Bull. 15, 201-206 (1980)
1980
456

<466>
Morphology Control of NdP5O14 Single Crystals Grown from Polyphosphoric Acids (solution)
Kasano, H.; Furuhata, Y.
J. Electrochem. Soc. 126, 1567-72 (1979)
1979
451

<467>
Ultrasonic Velocities of K2PbCu(NO2)6 Near the Cooperative Jahn-Teller Phase Transitions
(solution)
Kashida, S.
J. Phys. Soc. Japan 45, 1874-79 (1978)
1978
438

<468>
Upper and Lower Critical Fields of TaS2(Pyridine)1/2 (vapor transport)
Kashihara, Y.; Nishida, A.; Yoshioka, H.
J. Phys. Soc. Japan 46, 1112-18 (1979)
1979
453

<469>
Chromium Doped GaAs Vapor Phase Epitaxy
Kato, Y.; Mori, Y.; Morizane, K.
J. Cryst. Growth 47, 12-20 (1979)
1979
443

<470>

<470>
Physico-Chemical Properties of the Non-Stoichiometric VS2 and V5S8 Phases (vapor transport)
Katsuta, H.; McLellan, R.B.; Suzuki, K.
J. Phys. Chem. Solids 4C, 1089-91 (1979)
1979
454

<471>
Crystal Growth and Characterization of In1.9As0.1Se3 (vapor transport)
Katty, A.; Castro, C.A.; Odile, J.P.; Soled, S.; Wold, A.
J. Solid State Chem. 24, 107-11 (1978)
1978
416

<472>
Deposition of Crystals from the Plasmas of ZrO2, HfO2, ThO2 and CeO2 (electrolytic deposition)
Kawabuchi, K.; Magari, S.
J. Appl. Phys. 50, 6222-29 (1979)
1979
454

<473>
Structural Transitions and Phase Composition of Thin Films of Aluminum with Silicon
Kazimirov, V.P.; Batalin, G.I.; Buzaneva, E.V.; Kaganov, V.Ya.; Trainist, T.P.; Shkavro, A.G.
Inorg. Mater. 15, 170-72 (1979)
1979
457

<474>
Structural and Magnetic Properties of "One-Dimensional" Barium Vanadium Triselenide (BaVSe3)
Kelber, J.; Reis, A.H.Jr.; Aldred, A.T.; Meuller, M.H.; Massenet, O.; DePasquali , G.; Stucky, G.
Journal of Solid State Chemistry, 30, 357-364 (1979)
1979
455

<475>
Solubility of KLa(MoO4)2 in Aqueous K2MoO4 Solutions in Hydrothermal Conditions
Kharchenko, L.Yu.; Protasova, V.I.; Klevtsov, P.V.
Inorg. Mater. 15, 1025-27 (1979)
1979
457

<476>
Linear Expansion Anisotropy of CdGa2Se4 (melt)
Khuseinov, B.; Mavlonov, Sh.; Umarov, B.S.
Inorg. Mater. 14, 675-77 (1978)
1978
432

<477>
Formation of Defects in Zinc Selenide Crystals Grown from the Melt Under Argon Pressure (ZnSe,
Bridgman)
Kikuma, I.; Furukoshi, M.
J. Cryst. Growth 44, 467-72 (1978)
1978
429

<478>
Morphological Instability Under Constitutional Supercooling During the Crystal Growth of InSb
from the Melt Under Stabilizing Thermal Gradient (Bridgman, theory)
Kim, K.M.
J. Cryst. Growth 44, 403-13, (1978)
1978
429

56

<479>

<479>
Maximum Stable Zone Length in Float-Zone Growth of Small-Diameter Sapphire and Silicon Crystals
(Al2O3, Si)
Kim, K.M.; Dreeben, A.B.; Schujko, A.
J. Appl. Phys. 50, 4472-74 (1979)
1979
442

<480>
Evaluation of Yttrium Iron Garnet Single Crystals Grown by the Floating Zone Method (Y3Fe5O12)
Kimura, S.; Shindo, I.; Kitamura, K.; Mori, Y.; Takamizawa, H.
J. Cryst. Growth 44, 621-24 (1978)
1978
427

<481>
Morphological Change from Copper/alpha-copper-Zinc Whiskers to Dendrites in Zinc Reduction Growth
(Cu, alpha-Cu-Zn)
Kishi, K.
J. Cryst. Growth 45, 517-21 (1978)
1978
430

<482>
Heteroepitaxial Growth of ZnS on GaP by the Close-Spaced Technique (kinetics)
Kitagawa, M.; Saraie, J.; Tanaka, T.
J. Cryst. Growth 45, 198-203 (1978)
1978
430

<483>
Interface Shape and Horizontal Variations of Al and Ga Contents in Substituted YIG Single
Crystals Grown by the Floating Zone Method (Y3Fe5O12, Y3Fe4GaO12, Y3Fe4AlO12)
Kitamura, K.; Ii, N.; Shindo, I.; Kimura, S.
J. Cryst. Growth 46, 277-85 (1979)
1979
443

<484>
Re-Investigation of the Re-Entrant Corner Effect in Twinned Crystals (theory)
Kitamura, M.; Hosoya, S.; Sunagawa, I.
J. Cryst. Growth 47, 93-99 (1979)
1979
443

<485>
The Influence of Impurities on the Crystallization of Supercooled Copper Melts (Cu)
Klein, F.J.; Potschke, J.
J. Cryst. Growth 46, 112-18 (1979)
1979
430

<486>
Crystal Structure of the Double Molybdate K2Ni(MoO4)2 (slow cooling, flux)
Klevtsova, R.F.; Klevtsov, P.V.
Sov. Phys. Cryst. 23, 143-46 (1978)
1978
434

<487>
Transfer of VO2 by Iodine (vapor transport, doped)
Klinkova, L.A.; Skrebkova, E.D.
Inorg. Mater. 14, 278 (1978)
1978
436

<488>
Crystals of CTA-DNA: A Growth and Morphological Study (solution)
Kliya, M.O.; Osica, V.D.
J. Cryst. Growth 47, 635-46 (1979)
1979
454

<489>
Fivefold Twinned Silicon Crystals Grown in an Al-16 wt.% Si Melt (Si, flux)
Kobayashi, K.; Hogan, L.M.
Phil. Mag. 40, 399-407 (1979)
1979
452

<490>
Crystal Growth and Assessment of SnxPb1-xTe Mixed Crystals (PbxSn1-xTe, review)
Kobayashi, K.I.I.; Kato, Y.; Komatsubara, K.F.
Prog. Cryst. Growth Charact. 1, 117-49 (1978)
1978
424

<491>
Computational Simulation of the Melt Flow During Czochralski Growth (theory)
Kobayashi, N.
J. Cryst. Growth 43, 357-63 (1978)
1978
417

<492>
Power Required to Form a Floating Zone and the Zone Shape (theory)
Kobayashi, N.
J. Cryst. Growth 43, 417-24 (1978)
1978
419

<493>
Epitaxial Relationships During the Formation of Three-Dimensional Snow Dendrites (H2O)
Kobayashi, T.; Furukawa, Y.
J. Cryst. Growth 45, 48-56 (1978)
1978
430

<494>
Chemical Vapor Deposition of Scandium Hydride (ScH2, epitaxy, film)
Kobayashi, T.; Takei, H.
J. Cryst. Growth 45, 29-36 (1978)
1978
430

<495>
Crystal Growth of Mn15Si26 (vapor transport, Bridgman)
Kojima, T.; Nishida, I.; Sakata, T.
J. Cryst. Growth 47, 589-92 (1979)
1979
453

<496>
Hydrothermal Synthesis of Aluminum Orthophosphate (AlPO4)
Kolb, E.D.; Laudise, R.A.
J. Cryst. Growth 43, 313-19 (1978)
1978
417

<497>

<497>
High Efficiency GaAs Thin Film Solar Cells by Peeled Film Technology (epitaxy)
Konagai, M.; Sugimoto, M.; Takahashi, K.
J. Cryst. Growth 45, 277-80 (1978)
1978
430

<498>
The Kinetics of Dissolution (theory)
Konak, A.R.
Kristall Tech. 13, 29-33 (1978)
1978
451-A

<499>
Electrical Resistivity in In-Heusler Cu2.0Mn0.94In1.0 Single Crystal (Bridgman)
Kondo, K.; Saito, Y.; Sato, K.; Katayama, T.
Trans. Japan. Inst. Metals 19, 301-2 (1978)
1978
435

<500>
LPE Growth of Li(Nb,Ta)C3 Solid-Solution Thin Film Waveguides on LiTaO3 Substrates (LiNb1-xTaxO3, flux)
Kondo, S.; Sugii, K.; Miyazawa, S.; Uehara, S.
J. Cryst. Growth 46, 314-22 (1979)
1979
443

<501>
Growth of Antimony on Bismuth (Sb film, epitaxy)
Kondratenko, V.V.; Fedorenko, A.I.
Sov. Phys. Cryst. 23, 86-89 (1978)
1978
440

<502>
Si-MBE: Growth and Sb Doping (Si, film)
Konig, U.; Kibbel, H.; Kasper, E.
J. Vac. Sci. Tech. 16, 985-89 (1979)
1979
452

<503>
Model of Silicon Epitaxial Growth in SiCl4-HCl-H2 System Based on Flow Graph (vapor transport, theory, Si)
Korec, J.
J. Cryst. Growth 46, 362-70 (1979)
1979
443

<504>
Flow-Graph Model of Epitaxial Growth of A(III)B(V) Compounds from the Vapour Phase (theory)
Korec, J.
J. Cryst. Growth 46, 655-64 (1979)
1979
443

<505>
Heat Capacity of EuS near the Ferromagnetic Transition (Bridgman, vapor transport, theory)
Kornblit, A.; Ahlers, G.; Buehler, E.
Phys. Rev. B 17, 282-92 (1978)
1978
415

<506>

<506>
An LPE Growth of Thick InxGa1-xAs
Kotani, T.; Yamaguchi, A.; Akita, K.; Nakai, S.
J. Cryst. Growth 43, 543-45 (1978)
1978
419

<507>
On the Dissolution of Grown Synthetic Quartz Inside the Autoclaves (SiO2, hydrothermal)
Kotru, P.N.
Kristall Tech. 13, 35-41 (1978)
1978
451-A

<508>
Flux Growth of Lanthanum Borate, LaBO3
Kotru, P.N.; Wanklyn, B.M.
J. Mat. Sci. 14, 755-57 (1979)
1979
446

<509>
Preparation and Principal Physicochemical Properties of Alkali Metal Indates and Thioindates
(LiInS2, Bridgman-Stockbarger)
Kovach, S.K.; Semrad, E.E.; Voroshilov, Yu.V.; Gerasimenko, V.S.; Slivka, V.Yu.; Stasyuk, N.P.
Inorg. Mater. 14, 1693-97 (1978)
1978
441

<510>
Defects in Single Crystals Prepared at High Temperature (theory, kinetics)
Kroger, F.A.
Ann. Chim. Fr. 4, 445-50 (1979)
1979
454

<511>
Control of the Composition of Crystals Grown in Electric and Magnetic Fields (Czochralski)
Krylov, A.S.; Romanenko, V.N.
Inorg. Mater. 14, 4-8 (1978)
1978
425

<512>
Growth and Morphology of Spherical Crystals of omega-Phase in the Cd-Hg System
Kubiak, R.
Kristall Tech. 13, K1-K3 (1978)
1978
451-A

<513>
Accurate X-ray Determination of the Lattice Parameters and the Thermal Expansion Coefficients of
VO2 near the Transition Temperature (vapor transport)
Kucharczyk, D.; Niklewski, T.
J. Appl. Cryst. 12, 370-73 (1979)
1979
452

<514>
Temperature Dependence of the Degree of Doping in Silicon Epitaxial Growth from the Gas Phase.
I. Phosphorus Content and Segregation Coefficient (Si, vapor transport, theory)
Kuhne, H.
Kristall Tech. 13, 939-46 (1978)
1978
432

<515>

<515>
Temperature Dependence of Degree of Doping in Growth of Epitaxial Silicon from the Gas Phase. II.
Phosphorus Content and Differential Molar Solution Enthalpies (Si, vapor transport, theory)
Kuhne, H.
Kristall Tech. 13, 1059-66 (1978)
1978
432

<516>
Thin Film Deposition From Beams of Ionized Atoms and Clusters (Ag on Si, epitaxy)
Kuiper, A.E.T.; Thomas, G.E.; Schouten, W.J.
J. Cryst. Growth 45, 332-33 (1978)
1978
430

<517>
Observations of Microfacets Near Irregularly Remelted Surfaces in Pulled GaSb Crystals
(Czochralski)
Kumagawa, M.
J. Cryst. Growth 44, 291-96 (1978)
1978
426

<518>
Second Harmonic Generation in a Sputtered LiNbO3 Film on MgO (epitaxy)
Kunomura, K.; Ishitani, A.; Matsubara, T.; Hayashi, I.
J. Cryst. Growth 45, 355-60 (1978)
1978
430

<519>
Growth Kinetics and Habit Modification of Barium Molybdate Single Crystals in Silica Gel (BaMoO4)
Kurien, K.V.; Ittyachen, M.A.
J. Cryst. Growth 47, 743-45 (1979)
1979
454

<520>
Single Crystal Growth of Li2ZnGe (melt)
Kuriyama, K.; Kamijoh, T.
J. Cryst. Growth 46, 151-53 (1979)
1979
430

<521>
Crystal Perfection in Czochralski Grown Nickel Single Crystals (Ni, theory)
Kuriyama, M.; Boettinger, W.J.; Burdette, H.E.
J. Cryst. Growth 43, 287-300 (1978)
1978
417

<522>
Wide Silicon Ribbon Crystals (EDFG, Si)
Kuroda, E.; Matsuda, H.; Kozuka, H.; Maki, M.
J. Cryst. Growth 43, 388-90 (1978)
1978
417

<523>
Evaluation of Temperature Distribution of Melt in Silicon Ribbon Growth (EDFG)
Kuroda, E.; Matsuda, H.; Kozuka, H.; Maki, M.
Japan. J. Appl. Phys. 18, 471-77 (1979)
1979
445

61

<524>

<524>
Growth and Characterization of Silicon Ribbon Crystals Grown with Wetting and Non-Wetting Dies
(Si, EDFG)
Kuroda, E.; Matsuda, H.; Maki, M.
Phys. Stat. Sol. (a) 48, 105-11 (1978)
1978
428

<525>
On the Effect of pH of Solution on the Kinetics of Crystallization (theory)
Kuznetsov, D.A.; Hodorowicz, S.
Kristall Tech. 13, 1413-26 (1978)
1978
444

<526>
Real Structure of Heteroepitaxial Zinc Oxide Films Grown on Sapphire (vapor transport, ZnO)
Kuznetsov, G.F.; Semiletov, S.A.; Bagamadova, A.M.
Sov. Phys. Cryst. 23, 193-96 (1978)
1978
434

<527>
Crystal Chemistry of Perovskite Phases of the Systems Th0.25NbO3-NaNbO3 (NaxTh0.25(1-x)NbO3,
melt, slow cooling)
Labeau, M.; Joubert, J.C.
J. Solid State Chem. 25, 347-53 (1978)
1978
428

<528>
Structure of the Lennard-Jones (100) Crystal-Liquid Interface (theory)
Ladd, A.J.C.; Woodcock, L.V.
J. Phys. C: Solid State Phys. 11, 3565-76 (1978)
1978
426

<529>
Surface Charge and Ledge Dynamics in Cadmium Sulfide (CdS, evaporation)
Lam, S.-T.; Munir, Z.A.
J. Cryst. Growth 47, 373-78 (1979)
1979
454

<530>
Anisotropy of Round Evaporation Spirals on Rocksalt Surfaces (NaCl, theory)
Lampert, B.; Reichelt, K.
J. Cryst. Growth 47, 77-81 (1979)
1979
443

<531>
Preparation, Phase Equilibria, and Crystal Chemistry of La, Pr, and Nd Hydroxide Bromides and
Hydroxide Iodides (Nd(OH)2Br, Pr(OH)2Br, La7(OH)18Br3, Nd7(OH)18Br3, Pr7(OH)18Br3, La7(OH)18I3,
Pr7(OH18I3, hydrothermal)
Lance-Gomez, E.T.; Haschke, J.M.
J. Solid State Chem. 23, 275-79 (1978)
1978
415

<532>

<532>
Semi-Periodic Closed Growth Lamellae on the Crystal-Melt Interface of LiCl Doped Rock Salt
Crystals (NaCl, Czochralski)
Landers, R.; Dupuis, M.M.; Grange, G.; Mutaftschiev, B.
J. Cryst. Growth 43, 655-58 (1978)
1978
432

<533>
Nitrides--Structures and Crystal Growth (review, vapor transport, flux)
Lang, J.; Laurent, Y.; Haunaye, M.; Marchand, R.
Prog. Cryst. Growth Charact. 2, 207-35 (1979)
1979
453

<534>
Evidence for a Universal Law of Dendritic Growth Rates (theory)
Langer, J.S.; Sekerka, R.F.; Fujioka, T.
J. Cryst. Growth 44, 414-18 (1978)
1978
429

<535>
Effect of the Buoyancy Parameter on Czochralski Bulk Flow in Garnet Growth (theory, kinetics)
Langlois, W.E.
J. Cryst. Growth 46, 743-46 (1979)
1979
445

<536>
Digital Simulation of Czochralski Bulk Flow in Microgravity (Al2O3, theory, kinetics)
Langlois, W.E.
J. Cryst. Growth 48, 25-28 (1980)
1980
454

<537>
Solar Furnace Annealing of Amorphous Si Layers (films)
Lau, S.S.; von Allmen, E.; Golecki, I.; Nicolet, M-A.; Kennedy, E.F.; Tseng, W.F.
Appl. Phys. Lett. 35, 327-29 (1979)
1979
452

<538>
Are Icicles Single Crystals? (H2O)
Laudise, R.A.; Barns, R.L.
J. Cryst. Growth 46, 379-86 (1979)
1979
443

<539>
Electron Microscopy Study of Some Molybdenum Oxide Crystals (MoO3, Mo4O11, microcrystals)
Lavina, A.; Aznarez, J.A.; Ortiz, C.
J. Cryst. Growth 48, 100-106 (1980)
1980
454

<540>
Epitaxial Growth of Antimony Electrodeposits on Oriented Single Crystal Surfaces of Gold and
Silver (Sb)
Lazzari, M.; Peraldo Bicelli, L.; Rivolta, B.; La Vecchia, A.
J. Mat. Sci. 13, 739-49 (1978)
1978
416

<541>

<541>
Crystal Growth by the Thermic Screen Translation (TST) Technique: A Modified Bridgman Method
(ZnF2, CoF2, BaF2, MgF2, KY3F10)
Le Gal, H.; Grange, Y.
J. Cryst. Growth 47, 445-57 (1979)
1979
454

<542>
Physical Methods Used for the Characterization of Modes of Epitaxial Growth from the Vapor Phase
(review, theory)
Le Lay, G.; Kern, R.
J. Cryst. Growth 44, 197-222 (1978)
1978
424

<543>
Crystal Structure and Physical Properties of CuO0.65VS2 (vapor transport)
Le Nagard, N.; Collin, G.; Gorochov, O.
Mat. Res. Bull. 14, 155-62 (1979) (in French)
1979
442

<544>
Preparation, Crystal Structure, and Physical Properties (transport, magnetic susceptibility, and
NMR) of the Spinel CuV2S4 (vapor transport)
Le Nagard, N.; Katty, A.; Collin, G.; Gorochov, O.; Willig, A.
J. Solid State Chem. 27, 267-77 (1979) (in French)
1979
448

<545>
Crystal Quality of Gel Grown CaHPO4.2H2O Crystals
Lefaucheux, F.; Robert, M.C.; Arend, H.
J. Cryst. Growth 47, 313-14 (1979)
1979
443

<546>
Properties of Rapidly-Grown RTR Silicon Sheet and Their Effects on Solar Cell Processing (zone
melting, Si)
Legge, R.; Baghdadi, A.; Gurtler, R.; Sopori, B.
J. Electron. Mater. 8, 675-88 (1979)
1979
442

<547>
Crystal Structure of Se2TiO6 (slow cooling)
Legros, J.-P.; Galy, J.
C.R. Acad. Sci. 286C, 705-7 (1978)
1978
427

<548>
The Growth of Nickel Single Crystals by Recrystallization Method (Ni)
Lejcek, P.; Sima, V.
Kristall Tech. 13, K38-39 (1978)
1978
451-A

<549>
Unidirectional Solidification of Irregular Silver-Germanium Eutectic Alloys (Ag-Ge)
Lemaignan, C.; Malmejac, Y.
J. Cryst. Growth 46, 771-78 (1979)
1979
445

<550>

<550>
Crystal Growth of Triple Nitrites in Gels (K2CuPb(NO2)6, Tl2CuPb(NO2)6, K2NiPb(NO2)6,
Tl2NiPb(NO2)6, Tl2NiCd(NO2)6, Tl2NiHg(NO2)6, Tl2CoPb(NO2)6, Tl2CoBa(NO2)6)
Lentz, A.; Fabian, W.
J. Cryst. Growth 47, 121-23 (1979) (in German)
1979
443

<551>
Neodymium Incorporation into YAl-Borate Crystals in Preparation from Solutions in Molten
Potassium Trimolybdate ((Y,Nd)Al3(BO3), flux)
Leonyuk, N.I.; Pashkova, A.V.; Belov, N.V.
Kristall Tech. 14, 47-50 (1979)
1979
452

<552>
Preparation and Properties of Bulk GaxIn1-xAs
Leu, Y.-T.; Thiel, F.A.; Scheiber, H.,Jr.; Rubin, J.J.; Miller, B.I.; Bachmann, K.J.
J. Electron. Mater. 8, 663-74 (1979)
1979
442

<553>
Nitrilotri(Methylenephosphonic Acid) Adsorption on Barium Sulfate Crystals and its Influence on
Crystal Growth (BaSO4, solution)
Leung, W.H.; Nancollas, G.H.
J. Cryst. Growth 44, 163-67 (1978)
1978
424

<554>
Growing Monocrystalline Bars of SiC from the Gas Phase (vapor transport)
Levin, V.I.; Tairov, Yu.M.; Travadzhyan, M.G.; Tsvetkov, V.F.; Chernov, M.A.
Inorg. Mater. 14, 830-33 (1978)
1978
436

<555>
Structural and Electrical Properties of Layered Transition Metal Selendies VxTi1-xSe2 and
TaxTi1-xSe2 (TiSe2, VSe2, TaSe2, vapor transport)
Levy, F.; Froidevaux, Y.
J. Phys. C: Solid State Phys. 12, 473-87 (1979)
1979
450

<556>
Epitaxial Growth of CdS on CuInSe2
Li, P.W.; Plovnick, R.H.
Mat. Res. Bull. 13, 791-95 (1978)
1978
430

<557>
Crystal Structure of the Indium Selenide In2Se3 (vapor transport)
Likforman, A.; Carre, D.; Hillel, R.
Acta Cryst. B34, 1-5 (1978) (in French)
1978
415

<558>
Mass Transport Processes Associated with the Epitaxial Growth of ZnS in H2: The Effects of a
Bypass Flow (theory)
Lilley, P.; Al-Saddawi, S.D.
J. Cryst. Growth 46, 1-6 (1979)
1979
430

<559>
Orientation Dependence of 1-Alanine Incorporation in TGS Crystals (solution)
Lillicrap, B.J.; Wood, J.D.C.
J. Mat. Sci. 13, 681-84 (1978)
1978
410

<560>
Growth and Electrical Properties of Deuterated TGS Produced by the Rotating Disc Technique
(solution)
Lillicrap, B.J.; Wood, J.D.C.; Wood, V.H.; Shaw, N.
J. Phys. D: Appl. Phys. 12, 633-43 (1979)
1979
450

<561>
Study of Temperature Distribution in High-Temperature Furnaces for Silicon Carbide Monocrystal
and Epitaxial Layer Growing (SiC, melt)
Lilov, S.K.; Tairov, Yu.M.; Tsevtkov, V.F.
Kristall Tech. 13, 1351-55 (1978)
1978
433

<562>
Study of Silicon Carbide Epitaxial Growth Kinetics in the SiC-C System (SiC, theory)
Lilov, S.K.; Tairov, Yu.M.; Tsvetkov, V.F.
J. Cryst. Growth 46, 269-73 (1979)
1979
443

<563>
Melting and Crystallization of Corundum (alpha-Al2O3)
Lingart, Yu.K.; Bodyachevskii, S.V.
Inorg. Mater. 14, 458-59 (1978)
1978
427

<564>
Growth of Pb1-xSnXTe Single Crystals of Graded Composition by the Travelling Solvent Method
(theory)
Link, R.; Notzel, N.; Ermisch, W.
Kristall Tech. 13, 1391-97 (1978) (in German)
1978
444

<565>
Crystallisation of Some Ferrite Materials from Lead Flux with Addition of V2O5 (Y3Fe5O12,
Ho3Fe5O12, Dy3Fe5O12, Er3Fe5O12, Ca3Fe2Ge3O12, HoFeO3, DyFeO3, ErFeO3)
Lipko, H.A.; Piekarczyk, W.
pp. 211-17 in II International Conference "Microwave Ferrites", Gabionna, Poland, October 1978
1978
441

<566>
Crystal Growth from Solution Using Cylindrical Seeds (TGS, TBFB)
Loiacono, G.M.; Osborne, W.N.
J. Cryst. Growth 43, 401-5 (1978)
1978
419

<567>
Single Crystal Growth and Properties of Deuterated Triglycine Fluoroberyllate (TGFB)
Loiacono, G.M.; Osborne, W.N.; Delfino, M.; Kostecky, G.
J. Cryst. Growth 46, 105-11 (1979)
1979
430

<568>
Hot Wall Epitaxy (II-VI compounds, IV-VI compounds, III-V compounds, evaporation, review, theory)
Lopez-Otero, A.
Thin Solid Films 49, 3-57 (1978)
1978
416

<569>
CdTe Thin Films Grown by Hot Wall Epitaxy
Lopez-Otero, A.; Huber, W.
J. Cryst. Growth 45, 214-17 (1978)
1978
430

<570>
Preparation and Crystal Structure of CsAlF4
Losch, R.; Hebecker, C.
Z. Naturforsch. 34b, 131-34 (1979) (in German)
1979
444

<571>
Physical Properties and Lithium Intercalates of CrPS4 (slow cooling)
Louisy, A.; Ouvrard, G.; Schleich, D.M.; Brec, R.
Solid State Commun. 28, 61-66 (1978)
1978
429

<572>
The Variability of the Rate of Growth of Adamantane Crystals from the Vapour Under Constant Conditions (C10016, theory, vapor transport)
Lubetkin, S.D.; Dunning, W.J.
J. Cryst. Growth 43, 77-80 (1978)
1978
417

<573>
The Preparation of Large Spherical Iron Single Crystals (Fe, strain-anneal)
Lubitz, K.; Goltz, G.
Appl. Phys. 19, 237-39 (1979)
1979
441

<574>
Temperature Gradient Measurement in Liquid Epitaxial Growth Systems (flux)
Ludington, B.W.; Immorlica, A.A.,Jr.
J. Cryst. Growth 47, 619-22 (1979)
1979
454

<575>

<575>
Localized Adsorption in One Layer on a Crystal-Solution Interface (theory)
Lundager Madsen, H.E.
J. Cryst. Growth 46, 495-503 (1979)
1979
443

<576>
Design Considerations for Molecular Beam Epitaxy Systems (review)
Luscher, P.E.; Collins, D.M.
Prog. Cryst. Growth Charact. 2, 15-32 (1979)
1979
453

<577>
Epitaxy of NdAl3(BO3)4 for Thin Film Minature Lasers (flux)
Lutz, F.; Leiss, M.; Muller, J.
J. Cryst. Growth 47, 130-32 (1979)
1979
443

<578>
Epitaxial Layers of the Laser Material Nd(Ga,Cr)3(BO3)4 (flux)
Lutz, F.; Ruppel, D.; Leiss, M.
J. Cryst. Growth 48, 41-44 (1980)
1980
454

<579>
Simultaneous Diffusion of Calcium and Strontium in KCl Single Crystals
Machida, H.; Fredericks, W.J.
J. Phys. Chem. Solids 39, 797-805 (1978)
1978
436

<580>
Synthesis and Characterization of Europium(II)-Holoborates, Eu2B509Cl and Eu2B509Br (melt)
Machida, K.; Ishino, T.; Adachi, G.; Shiokawa, J.
Mat. Res. Bull. 14, 1529-34 (1979)
1979
454

<581>
Elastic Constants and Stability of bcc In-Tl Alloys (Bridgman)
Madhava, M.R.; Saunders, G.A.
Phys. Rev. B 18, 5340-49 (1978)
1978
435

<582>
Dielectric and Optical Properties of K5Nd(MoO4)4 (Czochralski)
Maeda, M.; Sakiyama, K.; Ikeda, T.
Japan. J. Appl. Phys. 18, 25-29 (1979)
1979
445

<583>
Liquid Encapsulation Zone-Refining of PbS (zone melting, vapor transport, sublimation)
Maier, H.; Daniel, D.R.; Herkert, R.; Luck, J.
J. Mat. Sci. 13, 297-300 (1978)
1978
411

<584>

<584>
Solid Phase Epitaxial Growth of Si Through Al Film
Majni, G.; Ottaviani, G.
J. Cryst. Growth 45, 132-37 (1978)
1978
430

<585>
Growth Kinetics of (111) Si Through an Al Layer by Solid Phase Epitaxy
Majni, G.; Ottaviani, G.
J. Cryst. Growth 46, 119-24 (1979)
1979
430

<586>
Magnetic Properties of CdCr2Se4-ZnCr2Se4 Single Crystals (melt)
Makhotkin, V.E.; Veselago, V.G.; Kalinnikov, V.T.
Sov. Phys. Solid State 20, 777-79 (1978)
1978
430

<587>
Structure and Luminescence Properties of alpha-CsNd(PO3)4 Crystals
Maksimova, S.I.; Palkina, K.K.; Loshchenov, V.B.; Kuznetsov, V.G.
Inorg. Mater. 15, 760-64 (1979)
1979
457

<588>
Improving the Electrical Characteristics of CdS Single Crystals Grown from the Melt
Malikov, V.Ya.; Kostenko, V.I.; Sysoev, L.A.
Inorg. Mater. 14, 166-68 (1978)
1978
436

<589>
New Model of LPE Growth: Layer Thickness Calculation (theory, kinetics)
Malinin, A.J.; Nevsky, C.B.
J. Electron. Mater. 7, 757-74 (1978)
1978
438

<590>
New Model of LPE Growth: Growth Rate Calculation (theory)
Malinin, A.J.; Nevsky, C.B.; Khrjapov, V.T.; Minadjinov, M.S.; Noghinov, A.L.
J. Electron. Mater. 7, 775-89 (1978)
1978
438

<591>
Theory of Crystallization During Epitaxial Growth from a Solution in a Melt (GaP, flux)
Malinin, A.Yu.; Nevskii, O.B.
Inorg. Mater. 14, 1378-84 (1978)
1978
439

<592>
Hydrothermal Crystallization in the System BaO-GeO2-Na2O-H2O (barium germanates, Ba3Ge9O21.H2O,
BaGeO3.H2O, Na4Ge9O20)
Malinovskii, Yu.A.; Kuznetsov, V.A.; Pobedimskaya, E.A.; Belov, N.V.
Sov. Phys. Cryst. 23, 299-302 (1978)
1978
434

<593>
alpha-In2Te3 (melt)
Mamedov, A.S.; Mamedov, K.P.; Gasanov, G.Sh.; Bagirov, S.B.; Niftiev, G.M.
Inorg. Mater. 13, 1592-95 (1978)
1978
419

<594>
Relative Supersaturation and Supercooling in Fluxed Melt Systems (Ho2Si207, TbVO 4, DyKMo208)
Maqsood, A.; Wanklyn, B.
J. Mat. Sci. 15, 405-408 (1980)
1980
456

<595>
Flux Growth of Polymorphic Rare-Earth Disilicates, R2Si207 (R = Tm, Er, Ho, Dy) (Tm2Si207,
Er2Si207, Ho2Si207, Dy2Si207)
Maqsood, A.; Wanklyn, B.M.; Garton, G.
J. Cryst. Growth 46, 671-80 (1979)
1979
443

<596>
Ternary Lanthanoid-Transition Metal Pnictides with ThCr2Si2-Type Structure (phosphides, flux)
Marchand, R.; Jeitschko, W.
J. Solid State Chem. 24, 351-57 (1978)
1978
418

<597>
Crystal Structure of beta-Eu2(II)SiO4 (melt)
Marchand, R.; L'Haridon, P.; Laurent, Y.
J. Solid State Chem. 24, 71-76 (1978) (in French)
1978
416

<598>
The Growth of CdAs2 and ZnAs2 Single Crystals from the Vapor Phase
Marenkin, S.F.; Huseynov, B.; Shevchenko, V.Y.; Belyskiy, N.K.
J. Cryst. Growth 44, 259-61 (1978)
1978
424

<599>
Growth of CdP4 Single Crystals (evaporation, vapor transport)
Marenkin, S.F.; Samiev, S.Kh.; Shevchenko, V.Ya.
Inorg. Mater. 14, 1532-33 (1978)
1978
441

<600>
Luminescence and Bandgap Studies of High Purity MgxZn1-xTe Ternary Alloys (Bridgman)
Marine, J.; Ternisien D'Ouville, T.; Schaub, B.; Laugier, A.; Barbier, D.; Guillaume, J.C.;
Rommelaere, J.F.; Chevallier, J.
J. Electron. Mater. 7, 17-31 (1978)
1978
410

<601>
A New Method to Follow Crystal Growth by Coulter Counter (solution, organics)
Markovic, M.; Komunjer, L.
J. Cryst. Growth 46, 701-705 (1979)
1979
443

<602>

<602>
Growth Conditions for Sodium-Titanium Bronze Crystals in NaBO2 (flux, Na2Fe2Ti6O14, Na2NiTi7O16)
Marnier, G.; Cecchi, J.E.
J. Cryst. Growth 43, 153-64 (1978)
1978
416

<603>
Growth of InP Crystals by the Synthesis Solute Diffusion Method
Marshall, A.J.; Gillessen, K.
J. Cryst. Growth 43, 651-52 (1978)
1978
419

<604>
Application of a Heat Pipe to Czochralski Growth. Part I. Growth and Segregation Behavior of Ga-Doped Ge (theory)
Martin, E.P.; Witt, A.F.; Carruthers, J.R.
J. Electrochem. Soc. 126, 284-7 (1979)
1979
432

<605>
Surface Morphology of GaAs Layers Grown by Liquid-Phase Epitaxy (flux, theory, kinetics)
Maruyama, S.
Japan. J. Appl. Phys. 18, 1217-22 (1979)
1979
451

<606>
Reproductive Epitaxy
Maslov, V.N.
Inorg. Mater. 15, 281-92 (1979)
1979
457

<607>
Growth and Characterization of Transition Metal Silicides (chemical vapor transport, solution, cold crucible melting)
Mason, K.N.
Prog. Crystal Growth Charact. 2, 269-307 (1979)
1979
000

<608>
Crystal Structure of AgH2PO4. Crystallography of AgH2AsO4 (solution)
Masse, R.; Guitel, J.-C.; Durif, A.
J. Solid State Chem. 23, 369-73 (1978) (in French)
1978
415

<609>
Crystal Structure of a New Variety of Nickel Pyrophosphate: Ni2P2O7 (slow cooling)
Masse, R.; Guitel, J.C.; Durif, A.
Mat. Res. Bull. 14, 337-41 (1979) (in French)
1979
442

<610>
Magnetic and Electrical Properties of BaVS3 and BaVxTi1-xS3 (flux)
Massenet, O.; Since, J.J.; Mercier, J.; Avignon, M.; Buder, R.; Nguyen, V.D.; Kelber, J.
J. Phys. Chem. Solids 40, 573-77 (1979)
1979
445

<611>

<611>
Czochralski Growth of Barium Hexaaluminate Single Crystals (BaAl12O19)
Mateika, D.; Laudan, H.
J. Cryst. Growth 46, 85-90 (1979)
1979
430

<612>
Effect of High-Temperature Annealing in Cadmium on the Electrical Transport Properties of Single
Crystals of Cadmium Sulphide (CdS, vapor transport, theory)
Mathur, P.C.; Sethi, B.E.; Sharma, O.P.; Talwar, P.L.
J. Phys. C: Solid State Phys. 12, 2333-39 (1979)
1979
450

<613>
Vapour-Phase Epitaxial Growth of ZnS on GaP
Matsuda, N.; Akasaki, I.
J. Cryst. Growth 45, 192-97 (1978)
1978
430

<614>
X-Ray Study of LEC-Grown InP Crystals (Czochralski, theory)
Matsui, J.; Watanabe, H.; Seki, Y.
J. Cryst. Growth 46, 563-68 (1979)
1979
443

<615>
Epitaxial Growth of ZnS by Zn-S-H2 CVD Method
Matsumoto, T.; Ishida, T.
Japan. J. Appl. Phys. 17, 227-28 (1978)
1978
428

<616>
Heteroepitaxial Growth of beta-SiC on Silicon Substrate Using SiCl4-C3H8-H2 System (vapor
transport, film)
Matsunami, H.; Nishino, S.; Tanaka, T.
J. Cryst. Growth 45, 138-43 (1978)
1978
430

<617>
Nitrogen Doping into GaAs1-xPx Using Ionized Beam in Molecular Beam Epitaxy
Matsushima, Y.; Gonda, S.; Makita, Y.; Mukai, S.
J. Cryst. Growth 43, 281-86 (1978)
1978
417

<618>
Surface Photovoltage Experiments on SrTiO3 Electrodes (theory)
Mavroides, J.G.; Kolesar, D.F.
J. Vac. Sci. Tech. 15, 538-41 (1978)
1978
421

<619>
An Investigation of Thin Silver Films on Cleaved Silicon Surfaces (Ag, theory, evaporation)
McKinley, A.; Williams, R.H.; Parke, A.W.
J. Phys. C: Solid State Phys. 12, 2447-63 (1979)
1979
450

<620>

<620>
Single Crystal Growth by the Horizontal Levitation Zone Melting Method (DyFe2, TbFe2, Nd, Pr, Ce, cold crucible)
McMasters, O.D.; Holland, G.E.; Gschneidner, K.A.,Jr.
J. Cryst. Growth 43, 577-83 (1978)
1978
419

<621>
X-Ray Diffuse Scattering from Alkali, Silver, and Europium beta-Alumina (NaAl11O17, KAl11O17, RbAl11O17, AgAl11O17, EuAl11O17, flux, theory)
McWhan, D.B.; Dernier, P.D.; Vettier, C.; Cooper, A.S.; Remeika, J.P.
Phys. Rev. B 17, 4043-59 (1978)
1978
420

<622>
Polytypism in Vapor Grown Crystals of Cadmium Bromide (CdBr2)
Mehrotra, K.
J. Cryst. Growth 44, 45-49 (1978)
1978
432

<623>
Crystallization of Salts from Supersaturated Solutions: Diffusion Kinetics
Melikhov, I.V.; Berliner, L.B.
J. Cryst. Growth 46, 79-84 (1979)
1979
430

<624>
The Electrical, Optical and Photoconducting Properties of Fe2-xCrxO3 (vapor transport, electrolytic deposition)
Merchant, P.; Collins, R.; Kershaw, R.; Dwight, K.; Wold, A.
J. Solid State Chem. 27, 307-15 (1979)
1979
448

<625>
Preparation of Single Crystals of the Magnetic Semiconductor CdCr2Se4 (vapor transport, doped)
Merkulova, A.I.; Radautsan, S.I.; Tezlevan, V.E.
Inorg. Mater. 14, 1202-03 (1978)
1978
439

<626>
Preparation, Crystal Growth and Characterization of Mixed Bromides of Rare Earths, Sodium, and Cesium (Ca2NaNdBr6, Cs2NaCeBr6, Cs2NaGdBr6, Cs2NaYBr6, Bridgman-Stockbarger)
Mermant, G.; Primont, J.
Mat. Res. Bull. 14, 45-50 (1979)
1979
433

<627>
Linear Growth Rate of Calcite in Aqueous Solution (CaCO3, kinetics, theory)
Meyer, H.J.
J. Cryst. Growth 47, 21-28 (1979) (in German)
1979
443

<628>

<628>
Molecular (beam) Processes of Condensation and Evaporation of Alkali Halides (KCl, LiF)
Meyer, H.J.; Dabringhaus, H.
pp. 47-78 in Current Topics in Materials Science, Volume 1, E. Kaldis (ed.), North-Holland.
Publishing Company (1978)
1978
452

<629>
Growth from Skull-Melting of Zirconia-Rare Earth Oxide Crystals (ZrO2-RE2O3, eutectics,
directional solidification)
Michel, D.; Perez Y Jorba, M.; Collongues, R.
J. Cryst. Growth 43, 546-48 (1978)
1978
419

<630>
Ceramic Eutectics in the Systems ZrO2-Ln2O3 (Ln = Lanthanide): Unidirectional Solidification,
Microstructural and Crystallographic Characterization (Ln = Nd, Sm, Dy, skull-melting)
Michel, D.; Rouaux, Y.; Perez Y Jorba, M.
J. Mat. Sci. 15, 61-66 (1980)
1980
453

<631>
Magnetization, Magnetocrystalline Anisotropy, Magnetostriction and Elastic Constants of a
Ruthenium-Nickel Single Crystal (RuNi, Bridgman)
Michelutti, B.; Perrier de la Bathie, R.; du Tremolet de Lacheisserie, E.; Waintal, A.
Solid State Commun. 28, 879-82 (1978)
1978
434

<632>
The Melt Growth and Thermal Expansion of Na2CO3 Crystals (Bridgman)
Midorikawa, M.; Hashimoto, T.; Ishibashi, Y.; Takagi, Y.
J. Cryst. Growth 44, 505-6 (1978)
1978
429

<633>
Optical and Dilatometric Studies of KCaCl3 and RbCaCl3 Crystals (hygroscopic, Bridgman)
Midorikawa, M.; Ishibashi, Y.; Takagi, Y.
J. Phys. Soc. Japan 46, 1240-44 (1979)
1979
453

<634>
Bridgman-Stockbarger Crystal Growth of Li2Ti3O7
Mikkelsen, J.C.,Jr.
J. Cryst. Growth 47, 659-65 (1979)
1979
454

<635>
Phase Studies, Crystal Growth, and Optical Properties of CdGe(As1-xPx)2 and AgGa(Se1-xSx)2 Solid
Solutions (Bridgman)
Mikkelsen, J.C.,Jr.; Kildal, H.
J. Appl. Phys. 49, 426-31 (1978)
1978
415

<636>

<636>
Creation of Defects During the Growth of Semiconductor Single Crystals and Films (Si, Ge, GaAs, InSb, epitaxy, theory, Czochralski)
Mil'vidskii, M.G.; Bochkarev, E.P.
J. Cryst. Growth 44, 61-74 (1978)
1978
432

<637>
The Strain Configuration Within the Facet of Czochralski Grown Indium Antimonide (InSb)
Miller, D.C.
J. Cryst. Growth 46, 31-34 (1979)
1979
430

<638>
The Effect of Melt Flow Phenomena on the Perfection of Czochralski Grown Gadolinium Gallium Garnet (Gd3Ga5O12)
Miller, D.C.; Valentino, A.J.; Shick, L.K.
J. Cryst. Growth 44, 121-34 (1978)
1978
424

<639>
Alkali Ion Exchange Reactions with RbAlSiO4: A New Metastable Polymorph of KAlSiO4 (flux)
Minor, D.B.; Roth, R.S.; Brower, W.S.; McDaniel, C.L.
Mat. Res. Bull. 13, 575-81 (1978)
1978
436

<640>
Observations on Step Bunching for the Layer Perovskite (C3H7NH3)2MnCl4 Using a Compound Holographic Interference and Conventional Microscope (solution)
Mischgofsky, F.H.
J. Cryst. Growth 43, 549-66 (1978)
1978
419

<641>
Face Stability and Growth Rate Variations of the Layer Perovskite (C3H7NH3)2CuCl4 (solution, theory)
Mischgofsky, F.H.
J. Cryst. Growth 44, 223-34 (1978)
1978
424

<642>
Polytypism and Amorphousness in Silicon Whiskers (Si, vapor transport)
Miyamoto, Y.; Hirata, M.
J. Phys. Soc. Japan 44, 181-90 (1978)
1978
417

<643>
Epitaxial Growth of KNdF4O12 Laser Waveguides (KLaP4O12, solution)
Miyazawa, S.; Koizumi, H.; Kubodera, K.; Iwasaki, H.
J. Cryst. Growth 47, 351-56 (1979)
1979
454

<644>
Interface Shape Transitions in the Czochralski Growth of Dy3Al5O12
Miyazawa, Y.; Mori, M.; Honma, S.
J. Cryst. Growth 43, 541-42 (1978)
1978
419

<645>

<645>
Interface Shape Transitions in Czochralski Grown YAG Crystals (Y3Fe5012, theory)
Miyazawa, Y.; Mori, Y.; Homma, S.; Kitamura, K.
Mat. Res. Bull. 13, 675-80 (1978)
1978
430

<646>
The Growth of Dy3Al5012 (Czochralski)
Miyazawa, Y.; Mori, Y.; Homma, S.
J. Mat. Sci. 13, 2272-73 (1978)
1978
429

<647>
Vapor Phase Transport and Stoichiometry Control of Cadmium Sulfo-Selenide (Cd1+y(SxSe1-x),
sublimation)
Mochizuki, K.; Igaki, K.
J. Cryst. Growth 45, 218-23 (1978)
1978
430

<648>
Vapor Phase Transport and Stoichiometry Control of Cadmium Sulfide (CdS, sublimation)
Mochizuki, K.; Igaki, K.
Japan. J. Appl. Phys. 18, 1447-54 (1979)
1979
451

<649>
Properties of VPE-Grown GaN Doped with Al and Some Iron-Group Metals
Monemar, B.; Lagerstedt, O.
J. Appl. Phys. 50, 6480-91 (1979)
1979
454

<650>
Influence of the CaCO3/GeO2 Ratio on Y Sm Lu Ca Fe Ge Garnet Films Properties (LPE)
Moriceau, H.; Ferrand, E.; Daval, J.; Challeton, D.
Mat. Res. Bull. 15, 107-11 (1980)
1980
454

<651>
Ionized-Cluster Beam Epitaxial Growth of GaP Films on GaP and Si Substrates
Morimoto, K.; Watanabe, H.; Itoh, S.
J. Cryst. Growth 45, 334-39 (1978)
1978
430

<652>
Observation of Growth Defects in Synthetic Quartz Crystals by Light-Scattering Tomography (SiO2,
hydrothermal)
Moriya, K.; Ogawa, T.
J. Cryst. Growth 44, 53-60 (1978)
1978
432

<653>
A Thermodynamic Factor Influencing the Growth Rate and Purity of Epitaxial Layers in the
Ga_AsCl3-H2 System (GaAs, vapor transport, kinetics, theory)
Morizane, K.; Mori, Y.
J. Cryst. Growth 45, 164-70 (1978)
1978
430

<654>

<654>
Preparation of Fine Sodium Chloride Crystals (NaCl)
Motegi, H.; Emoto, T.; Czaki, T.; Nakamura, E.
Japan. J. Appl. Phys. 17, 717-18 (1978)
1978
427

<655>
A Quick Survey Method for the Study of CVD Conditions (TiN, vapor transport)
Motojima, S.; Ozaki, K.; Takahashi, Y.; Sugiyama, K.
J. Cryst. Growth 43, 264-66 (1978)
1978
416

<656>
Chemical Vapor Growth of LaB6 Whiskers and Crystals Having a Sharp Tip
Motojima, S.; Takahashi, Y.; Sugiyama, K.
J. Cryst. Growth 44, 106-9 (1978)
1978
432

<657>
A New Preparation Technique in Solid State Chemistry. I. High Temperature Reactions with
CO2-Laser
Muller-Buschbaum, Hk.; Fausch, H.
Z. Naturforsch. 34b, 371-74 (1979) (in German)
1979
444

<658>
A New Preparative Technique in Solid State Chemistry. II. High Pressure - High Temperature
Reactions with CO2-Lasers
Muller-Buschbaum, Hk.; Fausch, H.
Z. Naturforsch. 34b, 375-77 (1979) (in German)
1979
444

<659>
Growth Spirals in a Chemical Potential Varying in Space or Time (theory, kinetics, solution)
Muller-Krumbhaar, H.
J. Cryst. Growth 44, 135-38 (1978)
1978
424

<660>
Kinetics of Crystal Growth. Microscopic and Phenomenological Theories (review)
Muller-Krumbhaar, H.
Chapter 1, pp. 1-47 in Current Topics in Materials Science, Vol. 1, E. Kaldis (e d.),
North-Holland Publishing Company (1978)
1978
452

<661>
Synthesis and Crystal Structure of a New Mixed Oxide "FeV308" (FexV1-xO2 - x approx. = 0.25)
(melt, slow cooling)
Muller, J.; Joubert, J.C.; Marezio, M.
J. Solid State Chem. 27, 191-99 (1979) (in French)
1979
448

<662>

<662>
Synthesis and Crystal Structure of a New Mixed Oxide FeV2O6OH0.5: Relation to the Diaspore
Structure Type (hydrothermal)
Muller, J.; Joubert, J.C.; Marezio, M.
J. Solid State Chem. 27, 367-82 (1979) (in French)
1979
448

<663>
Aging Studies on Hydrous Lutetium Oxide (cubic Lu(OH)3)
Mullica, D.F.; Milligan, W.O.; Dillin, D.R.
J. Cryst.Growth 47, 635-38 (1979)
1979
454

<664>
Cerium Trihydroxide (Ce(OH)3, gel)
Mullica, D.F.; Oliver, J.D.; Milligan, W.O.
Acta Cryst. B35, 2668-70 (1979)
1979
453

<665>
Flux Growth of Bi2WO6 Single Crystal Below the Transformation Temperature (slow cooling)
Muramatsu, K.; Watanabe, A.; Goto, M.
J. Cryst. Growth 44, 50-55 (1978)
1978
432

<666>
Growth Temperature Dependence in Molecular Beam Epitaxy of Gallium Arsenide (GaAs)
Murotani, T.; Shimanoe, T.; Mitsui, S.
J. Cryst. Growth 45, 302-8 (1978)
1978
430

<667>
The Thermal Expansion of 2H-MoS2 and 2H-WSe2 Between 10 and 320 K
Murray, R.; Evans, B.L.
J. Appl. Cryst. 12, 312-15 (1979)
1979
441

<668>
Investigation of Liquidus Surface in Pb-Sn-Te System by Simplex Lattice Method (flux, theory,
Pb1-xSnxTe)
Maszynski, Z.; Davarashvili, O.I.; Riabtsev, N.G.; Shotov, A.P.
J. Cryst. Growth 46, 487-90 (1979)
1979
443

<669>
Epitaxial Growth of ZnSe by Vapor Flow Method
Mutsukura, N.; Machi, Y.
Japan. J. Appl. Phys. 17, 1123-24 (1978)
1978
428

<670>
Preparation and Optical Properties of ZnSe Epitaxial Layers by a Close-Spaced Technique
Mutsukura, N.; Machi, Y.
Japan. J. Appl. Phys. 18, 233-38 (1979)
1979
445

<671>

<671>
Control of Structure and Properties. Controlling Microstructures and Properties of Superalloys
via Use of Precipitated Phases (eutectics, carbides, review)
Muzyka, D.R.
pp. 325-33 in Source Book on Materials for Elevated-Temperature Applications, El ihu F. Bradley
(ed.), American Society for Metals (1979)
1979
452-TC

<672>
Dispersion of Acoustic Phonons in Layered SrGa2 (melt, slow cooling)
Myer, H.; Weiss, A.; Dorner, B.
Solid State Commun. 25, 1093-95 (1978)
1978
419

<673>
A Simple Analysis of Vapor Phase Growth - Citing an Instance of GaxIn1-xAs (theory)
Nagai, H.
J. Electrochem. Soc. 126, 1400-03 (1979)
1979
451

<674>
Synthesis of Nonstoichiometric Zirconium Carbide Whiskers by Chemical Vapor Deposition (ZrC)
Naito, K.; Kamegashira, N.; Fujiwara, N.
J. Cryst. Growth 45, 506-10 (1978)
1978
430

<675>
Growth of Single Crystals of Na3Al2Li3F12 Garnet Under Hydrothermal Conditions
Naka, S.; Takeda, Y.; Kawada, K.; Inagaki, M.
J. Cryst. Growth 46, 461-62 (1979)
1979
443

<676>
Unit Cell Step Structures on Vapour-Grown Cr2S3 Single Crystals
Nakada, I.; Kubota, M.
J. Cryst. Growth 43, 711-18 (1978)
1978
432

<677>
Flux Growth of LiNdP4O12 Single Crystals
Nakano, J.; Miyazawa, S.; Yamada, T.
Mat. Res. Bull. 14, 21-26 (1979)
1979
433

<678>
Top-Seeded Flux Growth cf LiNdP4O12 Single Crystals
Nakano, J.; Yamada, T.; Miyazawa, S.
J. Cryst. Growth 47, 693-98 (1979)
1979
454

<679>
The Hardness of NbB2 Single Crystals (flux, slow cooling)
Nakano, K.; Higashi, I
J. Less-Common Metals 67, 485-492 (1979)
1979
458

<680>
Single-Crystal Growth of Tantalum Diboride (TaB2, zone melting, solution)
Nakano, K.; Kumashiro, Y.; Sakuma, E.
J. Less-Common Metals 65, 27-34 (1979)
1979
445

<681>
Vapor-Phase Growth of Epitaxial Ga1-xInxSb Using Open-Tube Flow System (vapor transport)
Nakatani, I.; Masumoto, K.
J. Cryst. Growth 46, 205-8 (1979)
1979
442

<682>
Enantiomorphism of alpha-LiIO3 Single Crystals (solution)
Nalbandyan, A.G.; Nalbardyan, H.G.; Sharkhatunyan, R.O.
J. Cryst. Growth 47, 427-28 (1979)
1979
454

<683>
The Growth of Crystals in Solution (review, theory, alkaline earth sulfates, calcium phosphates)
Nancollas, G.H.
Adv. Colloid Interface Sci. 10, 215-52 (1979)
1979
433

<684>
Growth of Niobium Single Crystals by Pulling from a Melt on a Pedestal. III. Characteristic
Arrangement of Dislocations (Nb)
Naramoto, H.
J. Cryst. Growth 44, 475-82 (1978)
1978
429

<685>
The History and Present Status of Synthetic Diamond (C, review, doped)
Nassau, K.; Nassau, J.
J. Cryst. Growth 46, 157-71 (1979)
1979
442

<686>
Temperature Range for Growth of Autoepitaxial GaAs Films by MBE (molecular beam epitaxy, theory)
Neave, J.H.; Joyce, B.A.
J. Cryst. Growth 43, 204-8 (1978)
1978
416

<687>
Influence of Growth Conditions on Mirodefect Distribution in Disloation-Free Single Crystals of Si
Neimark, K.N.; Sheikhet, E.G.; Litvinova, I.Yu.; Fal'kevih, E.S.
Inorg. Mater. 15, 143-47 (1979)
1979
457

<688>
Ag5GaS4 -- A New Phase in the System Ag-Ga-S (AgGaS2)
Nenasheva, S.N.; Sinyakcva, E.F.; Sinyakov, I.V.; Bogdanova, V.I.
Inorg. Mater. 14, 661-63 (1978)
1978
432

<689>

<689>
Appearance of Non-Singular Surfaces on Vapour-Grown Ice Crystals (H2O, theory)
Nenow, D.; Stoyanova, V.
J. Cryst. Growth 46, 779-82 (1979)
1979
445

<690>
Growth and Electrical Properties of Epitaxial CuInS2 Thin Films on GaAs Substrates
Neumann, H.; Schumann, E.; Peters, D.; Tempel, A.; Kuhn, G.
Kristall Tech. 14, 379-88 (1979)
1979
441

<691>
Thin Films of LiNbO3, Doped with Na+ and Co2+ plus Zr4+, Grown by Liquid-Phase Epitaxy (flux)
Neurgaonkar, R.R.; Kalisher, M.H.; Staples, E.J.; Lim, T.C.
Appl. Phys. Lett. 35, 606-8 (1979)
1979
452

<692>
Crystallographic Studies and Structural Systematics of the C2/c Alkali Metal Metavanadates
(NaxLi1-xVO3)
Ng, H.N.; Calvo, C.; Idler, K.L.
J. Solid State Chem. 27, 357-66 (1979)
1979
448

<693>
Solution Growth of Sparingly Soluble Single Crystals from Soluble Complexes. I. General
Introduction (review, theory, kinetics)
Nicolau, I.F.
J. Cryst. Growth 48, 45-50 (1980)
1980
454

<694>
Solution Growth of Sparingly Soluble Single Crystals from Soluble Complexes. II. Growth of
alpha-HgI2 Single Crystals from Iodomecurate complexes (kinetics)
Nicolau, I.F.
J. Cryst. Growth 48, 51-60 (1980)
1980
454

<695>
Solution Growth of Sparingly Soluble Single Crystals from Soluble Complexes. III. Growth of
alpha-HgI2 Single Crystals from Dimethylsulfoxide Complexes (theory)
Nicolau, I.F.; Joly, J.F.
J.Cryst.Growth 48, 61-73 (1980)
1980
454

<696>
Growth, Morphology and Slip System of alpha-Si3N4 Single Crystal (vapor transport)
Niihara, K.; Hirai, T.
J. Mat. Sci. 14, 1952-60 (1979)
1979
445

<697>
Some Optical Properties of GeS2 Single Crystals (Bridgman)
Nikolic, P.M.; Popovic, Z.V.
J. Phys. C: Solid State Phys. 12, 1151 (1979)
1979
450

<698>
Studies of LPE Ripple Based on Morphological Stability Theory
Nishinaga, T.; Pak, K.; Uchiyama, S.
J. Cryst. Growth 43, 85-92 (1978)
1978
417

<699>
Growth and Morphology of 6H-SiC Epitaxial Layers by CVD (film)
Nishinc, S.; Matsunami, H.; Tanaka, T.
J. Cryst. Growth 45, 144-49 (1978)
1978
430

<700>
Vapor Growth of Zinc Telluride by Closed Tube Method (ZnTe, epitaxy)
Nishio, M.; Ogawa, H.; Yamada, E.
Japan. J. Appl. Phys. 17, 571-72 (1978)
1978
428

<701>
Aspects of Silicon Epitaxy (review, Si)
Nishizawa, J.
pp. 57-108 in Crystal Growth, Vol. 2, C.H.L. Goodman, Plenum Press, New York (1978)
1978
429-T

<702>
Mechanisms of Chemical Vapor Deposition of Silicon (Si, epitaxy)
Nishizawa, J.; Nihira, E.
J. Cryst. Growth 45, 82-89 (1978)
1978
430

<703>
Observations of Defects in LPE GaAs Revealed by New Chemical Etchant (flux, film)
Nishizawa, J.; Oyama, Y.; Tadano, H.; Inokuchi, K.; Okuno, Y.
J. Cryst. Growth 47, 434-36 (1979)
1979
454

<704>
An X-Ray Double-Crystal Method Utilizing Non-Parallel Setting for Measuring Local Lattice
Mismatches Between Epitaxial Films and Substrates (RE garnets)
Nittono, O.; Shimizu, S.
J. Cryst. Growth 45, 476-81 (1978)
1978
430

<705>
Surface Morphology of GaAs Grown by Vapor Phase Epitaxy
Nonomura, Y.; Okuno, Y.; Nishizawa, J.
J. Cryst. Growth 46, 795-800 (1979)
1979
445

<706>
Structural Studies of the Solid Electrolyte High-LiTa3O8 (electrolytic deposition)
Nord, A.G.; Thomas, J.O.
Acta Chem. Scand. A32, 539-44 (1978)
1978
433

<707>

<707>
Heteroepitaxial Homogeneous CdxHg1-xTe Films (vapor transport)
Nowak, Z.; Pictrowski, J.; Piotrowksi, T.; Sadowski, J.
Thin Solid Films 52, 405-13 (1978)
1978
437

<708>
Crystal Growth and Characterization of the Transition-Metal Phosphides CuP2, NiP2, and RhP3
(flux, vapor transport)
Odile, J.P.; Soled, S.; Castro, C.A.; Wold, A.
Inorg. Chem. 17, 283-86 (1978)
1978
412

<709>
Growth and Properties of Pb(WO4)1-x(MoO4)x Mixed Crystals (Czochralski)
Oeder, R.; Scharmann, A.; Schwabe, D.; Vitt, B.
J. Cryst. Growth 43, 537-40 (1978)
1978
419

<710>
Edge Emission in Melt-Grown ZnSexS1-x (Bridgman)
Ohmori, K.; Ohishi, M.; Okuda, T.; Hiramatsu, M.
J. Appl. Phys. 49, 4506-8 (1978)
1978
436

<711>
Chemical Vapor Deposition of Single-Crystalline ZnO Film with Smooth Surface on Intermediately
Sputtered ZnO Thin Film on Sapphire (epitaxy)
Ohnishi, S.; Hirokawa, Y.; Shiosaki, T.; Kawabata, A.
Japan. J. Appl. Phys. 17, 773-78 (1978)
1978
429

<712>
Horizontal Zone-Melting of Indium Antimonide in Vacuum. I. Basic Technique (InSb)
Ohno, I.; Iwasaki, M.; Yoneyama, T.
Japan. J. Appl. Phys. 18, 285-93 (1979)
1979
445

<713>
Si Contamination in Epitaxial Boron Monophosphide (BP)
Ohsawa, J.; Nishinaga, T.; Uchiyama, S.
Japan. J. Appl. Phys. 17, 1579-86 (1978)
1978
426

<714>
Observations of Etch Pits on As-Grown Faces of Brushite Crystals (CaHPO4.2H2O, gel)
Ohta, M.; Tsutsumi, M.; Ueno, S.
J. Cryst. Growth 47, 135-36 (1979)
1979
443

<715>
The Effects of M(M=Mg, Ca and Ba) Ions on the KCl Flux Growth of MWO4 Needle Crystals
Oishi, S.; Endo, Y.; Kobayashi, T.; Tate, I.
Nippon Kagaku Kaishi, No. 9, 1191-97 (1979)
1979
457

<716>

<716>
Growth of NaNd(MoO4)2 Single Crystals from the Ternary System Nd2O3-MoO3-Na2CO3 (flux, slow cooling)
Oishi, S.; Kanai, W.; Tate, I.
Nippon Kagaku Kaishi No. 9, 1181-84 (1979)
1979
457

<717>
Growth of Single Crystals of Lead Monoxide (PbO, flux)
Oka, K.; Unoki, H.; Sakudo, T.
J. Cryst. Growth 47, 568-72 (1979)
1979
453

<718>
Growth of alpha-Ag2S Whiskers in a VLS System (vapor transport)
Okabe, T.; Nakagawa, M.
J. Cryst. Growth 46, 504-10 (1979)
1979
443

<719>
The Crystal Structure of Cs6W11O36
Okada, K.; Marumo, F.; Iwai, S.
Acta Cryst. B34, 50-54 (1978)
1978
415

<720>
The Crystal Structure of K2W4O13
Okada, K.; Marumo, F.; Iwai, S.
Acta Cryst. B34, 3193-95 (1978)
1978
433

<721>
New Ferroelectric Na3Sc2(PO4)3 (melt)
Okonenko, S.A.; Stefanovich, S.Yu.; Kalinin, V.B.; Venevtsev, Yu.N.
Sov. Phys. Solid State 20, 1647-48 (1978)
1978
439

<722>
Formation of Mixed Crystals in the System CuxZn1-xCr2Se4
Okonska-Kozlowska, I.; Krok, J.
Z. Anorg. Allg. Chem. 447, 235-43 (1978)
1978
438

<723>
Relation Between Growth Conditions and Dislocation Structure in Gallium Phosphide Crystals (GaP, Czochralski, flux)
Okunev, Yu.A.; Akhin'ko, A.T.; Buzynin, A.N.; Selin, V.V.; Bletskan, N.I.
Inorg. Mater. 15, 905-7 (1979)
1979
457

<724>
Recarburization of Monocrystalline Ditungsten Carbide as Revealed by Field Electron Microscopy (W microcrystal, crystal growth study)
Okuyama, F.
Applications of Surface Science 3, 1-12 (1979)
1979
449

<725>

<725>
Phase Diagram of Pseudobinary System CdTe-ZnSe (Bridgman)
Oleinik, G.S.; Tomashik, V.N.; Mizetskaya, I.B.; Novitskaya, G.N.; Chalyi, V.P.
Inorg. Mater. 13, 1583-85 (1978)
1978
419

<726>
Phase Equilibria and Properties of Single Crystals of Solid Solutions in the Systems
Zn3As2-ZnSe(ZnTe) (Zn3As2, ZnSe, ZnTe, Zn3As2-ZnSe, Zn3As2-ZnTe)
Olekseyuk, I.D.; Stoika, I.M.; Gerasimenko, V.S.
Inorg. Mater. 14, 1078-82 (1978)
1978
439

<727>
Single-Crystal Growth of Mixed (La, Eu, Y, Ce, Ba, Cs) Hexaborides for Thermionic Emission (LaB6,
EuB6, CeB6, BaB6, (La,Eu)B6, La1-xEuB6, (La,Y)B6, La,Ba)B6, (La,Cs)B6, (Eu,Y)B6, (Eu,Ba)B6, flux)
Olsen, G.H.; Cafiero, A.V.
J. Cryst. Growth 44, 287-90 (1978)
1978
426

<728>
Growth Effects in the Heteroepitaxy of III-V Compounds (review)
Olsen, G.H.; Ettenberg, M.
pp. 1-56 in Crystal Growth, Vol. 2, C.H.L. Goodman (ed.), Plenum Press, New York (1978)
1978
429-T

<729>
Crystal Growth and Properties of Binary, Ternary and Quaternary (In,Ga)(As,P) Alloys Grown by the
Hydride Vapor Phase Epitaxy Technique (film)
Olsen, G.H.; Zamerowski, T.J.
Prog. Cryst. Growth Charact. 2, 309-75 (1979)
1979
000

<730>
Anisotropy of Upper Critical Field in Superconducting 2H-NbS2 (theory)
Onabe, K.; Naito, M.; Tanaka, S.
J. Phys. Soc. Japan 45, 50-58 (1978)
1978
436

<731>
Raman Scattering in 3R-NbS2 (vapor transport, theory)
Onari, S.; Arai, T.; Aoki, R.; Nakamura, S.
Solid State Commun. 31, 577-79 (1979)
1979
446

<732>
Synthesis of Potassium Hexatitanate Fibers by the Hydrothermal Dehydration Method (K2Ti6O13)
Oota, T.; Saito, H.; Yamai, I.
J. Cryst. Growth 46, 331-38 (1979)
1979
443

<733>
Synthesis of Single Crystal LnPO4 (vapor transport)
Orlovskii, V.P.; Khalikov, B.; Kurbanov, Kh.M.; Bugakov, V.I.; Kargareteli, L.N.
Zh. Neorg. Khim. SSSR, 23, 316-18 (1978) (in Russian)
1978
431-A

<734>

<734>
n-Type Doping Techniques in Silicon Molecular Beam Epitaxy by Simultaneous Arsenic Ion
Implantation and by Antimony Evaporation (Si, film)
Ota, Y.
J. Electrochem. Soc. 126, 1761-65 (1979)
1979
452

<735>
A New Technique for the Preparation of Copper Single Crystals (Cu)
Otooni, M.A.
J. Cryst. Growth 43, 391-93 (1978)
1978
417

<736>
A New Technique for Preparation of Thin Single Crystal Films of Copper (Cu, evaporation)
Otooni, M.A.
J. Cryst. Growth 46, 205-12 (1979)
1979
442

<737>
A New Technique for Preparation of Copper Single Crystals from the Bulk Using Four Points (Cu,
melt)
Otooni, M.A.; McKelvy, E.J.
J. Cryst. Growth 44, 505-11 (1978)
1978
429

<738>
Crystal Data and Crystal Growth of PbGe3O7 (melt)
Otto, H.H.
J. Appl. Cryst. 11, 157-58 (1978)
1978
425

<739>
Crystal Data and Crystal Growth of Pb5GeO7 (melt)
Otto, H.H.
J. Appl. Cryst. 12, 251-52 (1979)
1979
441

<740>
Physicochemical Properties and Crystal Growth of A(II)B(VI)-MnB(VI) Systems (review)
Pajaczkowska, A.
Prog. Cryst. Growth Charact. 1, 289-326 (1978)
1978
426

<741>
Physics of Magnetic Garnets (review)
Paoletti, A.(ed.)
Proceedings of the International School of Physics "Enrico Fermi", Course LXX, 27 June - 9 July,
1977, North Holland Publishing Company, Amsterdam-New York, 1978
1978
426

<742>
Interrelation Between Interface and Vapour Transport Kinetics in Chemical Vapour Deposition
(theory)
Paorici, C.
J. Cryst. Growth 46, 523-26 (1979)
1979
443

<743>

<743>
Crystal Growth and Properties of CuGaxIn1-xSe2 Chalcopyrite Compound (vapor transport)
Paorici, C.; Zanotti, L.; Romeo, N.; Sberveglieri, G.; Tarricone, L.
Solar Energy Materials 1, 3-9 (1979)
1979
446

<744>
A Temperature Variation Method for the Growth of Chalcopyrite Crystals by Iodine Vapour Transport
(CuInS2, CuGaS2, AgIn5Se)
Paorici, C.; Zanotti, L.; Zuccalli, G.
J. Cryst. Growth 43, 705-10 (1978)
1978
432

<745>
Properties of Homogeneously Doped ZnSe Single Crystals Obtained by a New Growth Method (vapor transport)
Papadopoulo, A.C.; Jean-Louis, A.M.; Charil, J.
J. Cryst. Growth 44, 587-92 (1978)
1978
427

<746>
Dependence of Degree of Perfection of Single Crystals of FeGe2 on Their Growth Rate and Direction
Papushina, T.I.; Frolov, A.A.; Bagaev, V.N.
Inorg. Mater. 15, 582-85 (1979)
1979
457

<747>
Growth and Characterization of Mixed Sr:Ca Tartrates (Sr1-xCaxC4H4O6.4H2O, gel)
Patel, A.R.; Arora, S.K.
Kristall Tech. 13, 899-904 (1978)
1978
432

<748>
Overgrowths on Diamond at Atmospheric Pressure (C, flux)
Patel, A.R.; Cherian, K.A.
J. Cryst. Growth 46, 706-8 (1979)
1979
443

<749>
Modified Gel Technique to Grow Single Crystals of KClO4
Patel, A.R.; Rao, A.V.
Ind. J. Pure Appl. Phys. 16, 544-45 (1978)
1978
441

<750>
Crystal Growth of Potassium Perchlorate in Gelatin Gels (KClO4)
Patel, A.R.; Rao, A.V.
Kristall Tech. 14, 151-57 (1979)
1979
444

<751>
Nucleation and Growth of KClO4 Single Crystals in Silica Gels
Patel, A.R.; Venkateswara Rao, A.
J. Cryst. Growth 43, 351-56 (1978)
1978
417

<752>
Gel Growth and Perfection of Orthorhombic Potassium Perchlorate Single Crystals (KClO4)
Patel, A.R.; Venkateswara Rao, A.
J. Cryst. Growth 47, 213-18 (1979)
1979
443

<753>
Interface Roughness of Vapor-Grown Adamantine Crystals (C10H16)
Pavlovska, A.
J. Cryst. Growth 46, 551-56 (1979) (in German)
1979
443

<754>
The Growth of GaxIn1-xAs on (100) InP by Liquid-Phase Epitaxy (flux)
Pearsall, T.P.; Bisaro, R.; Ansel, R.; Merenda, P.
Appl. Phys. Lett. 32, 497-99 (1978)
1978
418

<755>
A New Technique for Determining the Kinetics of Crystal Growth from the Melt (organics, benzophenone)
Pech, S.; Vignes-Adler, M.
J. Cryst. Growth 43, 123-25 (1978)
1978
417

<756>
Optical Properties of Cs2NaBiCl6
Pelle, F.; Jacquier, B.; Denis, J.P.; Blanzat, B.
J. Luminescence 17, 61-72 (1978)
1978
422

<757>
New Thiochlorides of Molybdenum II: Mo6Cl10Y (Y = S, Se, Te): Structure and Magnetic and Electrical Properties (Mo6Cl10Se, melt)
Perrin, C.; Sergent, M.; Le Traon, F.; LeTraon, A.
J. Solid State Chem. 25, 197-204 (1978) (in French)
1978
422

<758>
The Optical Absorption Edge of Single-Crystal AlN Prepared by a Close-Spaced Vapor Process
Perry, P.B.; Rutz, R.F.
Appl. Phys. Lett. 33, 319-21 (1978)
1978
427

<759>
Synthesis and Crystal Growing of La3NbO7
Peshev, P.; Khrusanova, M.
Mat. Res. Bull. 15, 195-200 (1980)
1980
456

<760>
On the Preparation of Nickel Ferrite Single Crystals by Chemical Transport (NiFe2O4, theory)
Peshev, P.; Tcshev, A.
J. Mat. Sci. 13, 143-48 (1978)
1978
409

<761>
Molecular Beam Epitaxy of Ge and Ga1-xAlxAs Ultra Thin Film Superlattices
Petroff, P.M.; Gossard, A.C.; Savage, A.; Wiegmann, W.
J. Cryst. Growth 46, 172-78 (1979)
1979
442

<762>
Crystal Growth Kinetics in (GaAs)n-(AlAs)m Superlattices Deposited by Molecular Beam Epitaxy. I.
Growth on Singular (100) GaAs Substrates (film, kinetics)
Petroff, P.M.; Gossard, A.C.; Wiegmann, W.; Savage, A.
J. Cryst. Growth 44, 5-13 (1978)
1978
432

<763>
Basic Consideration for an Analytic Theory of Two-Component Crystal Growth
Pfeiffer, H.
Phys. Stat. Sol. (b) 93, K149 (1979)
1979
449

<764>
On the Analytic Treatment of the Growth of Rough Interfaces (theory)
Pfeiffer, H.
Phys. Stat. Sol. (b) 94, K11-14 (1979)
1979
449

<765>
Analytical Theory of Crystal Growth of Two-Component Systems I. Basic Features and Equilibrium
Properties (melt)
Pfeiffer, H.; Haubenreisser, W.
Phys. Stat. Sol. (b) 96, 287-299 (1979)
1979
456

<766>
Crystal Chemistry of a Tellurate IV of Indium III: In2Te3O9
Philippot, E.; Astier, E.; Loeksmanto, W.; Maurin, M.; Moret, J.
Rev. Chem. Miner. 15, 283-91 (1978) (in French)
1978
437

<767>
Magnetic Properties of a Single Crystal of NdP3 Between 2 and 4.2K (Stockbarger)
Picard, J.; Guillot, M.; Le Gall, H.; Leycuras, C.; Feldmann, P.
C. R. Acad. Sci. 288B, 175-77 (1979)
1979
444

<768>
Growth of High Perfection Ag-Sn Crystals (in Ag solvent)
Pichaud, B.; Minari, F.; Bernardini, J.
J. Cryst. Growth 43, 273-76 (1978)
1978
416

<769>
Dissociation Processes and Crystal Growth of Gadolinium Gallium Garnet (Gd3Ga5O12, Czochralski,
kinetics, theory)
Piekarczyk, W.; Pajaczkowska, A.
J. Cryst. Growth 46, 483-86 (1979)
1979
443

<770>

<770>
On the Preparation of Pseudobrookite Single Crystals by Chemical Transport (Fe2TiO5,
electrolytic deposition)
Piekarczyk, W.; Peshev, P.; Toshev, A.; Pajaczkowska, A.
Mat. Res. Bull. 13, 587-93 (1978)
1978
436

<771>
The Czochralski Growth of Bismuth-Germanium Oxide Single Crystals (Bi12GeO20)
Piekarczyk, W.; Swirkowics, M.; Gazda, S.
Mat. Res. Bull. 13, 889-94 (1978)
1978
429

<772>
Hydrothermal Growth of the Systems MF-AlF3 and NH4F-FeF3 in Hydrofluoric Acid Solution (TlAlF4,
KAlF4, NH4AlF4, RbAlF4, CsAlF4, NH4Fe2F6)
Plet, F.; Fourquet, J.L.; Courbion, G.; Leblanc, M.; De Pape, R.
J. Cryst. Growth 47, 699-702 (1979)
1979
454

<773>
Investigation of Structural Formation and Growth Kinetics of Organic Crystals (solution, melt)
Podolinski, V.V.
J. Cryst. Growth 46, 511-22 (1979)
1979
443

<774>
Solubility of AlPO4 in Hydrothermal Solutions of H3PO4
Poignant, H.; Le Marechal, L.; Toudic, Y.
Mat. Res. Bull. 14, 603-12 (1979) (in French)
1979
442

<775>
Preparation of Photoluminescent and Conductive ZnSe:I Crystals by Iodine Vapor Transport
Poindessault, R.
J. Electron. Mater. 8, 619-33 (1979)
1979
442

<776>
Preparation and Properties of Monocrystalline Films of CdSb (melt)
Ponomarev, V.F.; Padalko, A.G.; Sanygin, V.M.
Inorg. Mater. 14, 157-59 (1978)
1978
436

<777>
Synthesis and Hydrothermal Growth of the Ferroelectric SbNbO4, SbTaO4 and Pb5S2I6 Single Crystals
Popolitov, V.I.; Lobachev, A.N.; Peskin, V.F.; Mininzon, Yu.M.
Ferroelectrics 21, 421-22 (1978)
1978
439

<778>
Anisotropy of Ion Transport in Stoichiometric Rutile Single Crystals (TiO2)
Popov, V.P.; Shvaiko-Shvaikovskii, V.E.; Andreev, A.A.
Sov. Phys. Solid State 21, 228-31 (1979)
1979
457

<779>

<779>
Effect of S-35 Tracer, Particle Size and Imperfection on the Solubility of Barium Sulphate
(BaSO4, solution)
Powell, J.L.; Coller, B.A.W.; Jones, A.L.
J. Cryst. Growth 43, 185-96 (1978)
1978
416

<780>
Preparation and Certain Electrical Properties of CdCr2S4 Single Crystals (vapor transport)
Radautsan, S.I.; Nikiforov, K.G.; Tezlevan, V.E.
Inorg. Mater. 14, 128-29 (1978)
1978
425

<781>
The Preparation and Crystal Structures of BiReO4 and BiRe2O6 (vapor transport)
Rae Smith, A.R.; Cheetham, A.K.
J. Solid State Chem. 30, 345-352 (1979)
1979
455

<782>
The Growth and Structure of Epitaxial Films of the Rare-Earth Dihydrides (GdH2, TbH2, DyH2, ErH2,
HoH2)
Rahman Khan, M.S.; Miller, R.F.
J. Phys. D: Appl. Phys. 12, 271-75 (1979)
1979
450

<783>
Liquid Phase Epitaxial Growth of Magnetic Garnets (review)
Randles, M.H.
in Crystals For Magnetic Applications, Vol. 1, C.J.M. Rooijmans (ed), Springer-V erlag, New York
(1978)
1978
431-A

<784>
Crystal Growth and Selective Etching Studies on Sodium Chlorate Crystals (NaClO3, solution)
Rao, K.K.; Sirdeshmukh, D.B.
J. Cryst. Growth 44, 533-36 (1978)
1978
427

<785>
Preparation, Optical and Dielectric Properties of Na5W3O9F5 (melt)
Ravez, J.; Elaatmani, M.; Chaminade, J.P.
Solid State Communications 32, 749-754 (1979) (in French)
1979
455

<786>
The Stage II Transient in Off-Eutectic Solidification (theory)
Ravishankar, P.S.; Wilcox, W.R.
J. Cryst. Growth 43, 480-87 (1978)
1978
419

<787>
New Observations in the Chemical Transport of GeO2. I. Transporting Agent Chlorine
Redlich, W.; Gruehn, R.
Z. Anorg. Allg. Chem. 438, 25-36 (1978)
1978
421

<788>

<788>
The Epitaxial Growth of Reactively Sputtered CuI Films on Rocksalt Substrates
Reichelt, K.; Florian, B.
J. Cryst. Growth 44, 507-8 (1978)
1978
429

<789>
Preparation of Single V3O5 Crystals by Chemical Transport Reactions (vapor transport)
Reichelt, W.; Oppermann, H.; Terukov, E.I.
Inorg. Mater. 14, 219-21 (1978)
1978
436

<790>
Single Crystal Preparation and Characterization of the Group VA Metal Hydrides and Deuterides
(V-H, V-D, Nb-H, Nb-D, Ta-H, Ta-D)
Reidinger, F.; Reilly, J.J.; Stoenner, R.W.
J. Less Common Metals 66, 23-32 (1979)
1979
450

<791>
Superconductivity of Single-Crystal NbGe2
Remeika, J.P.; Cooper, A.S.; Fisk, Z.; Johnston, D.C.
J. Less-Common Metals 62, 211-13 (1978)
1978
431

<792>
Preparation, Analysis and Properties of High Purity Zn(1-x)MgxTe Alloys (distillation)
Revel, G.; Pastol, J.L.; Rouchaud, J.C.; Fedoroff, M.; Guillaume, J.C.; Chevallier, J.;
Rommeluere, J.F.
J. Electron. Mater. 9, 165-184 (1979)
1979
458

<793>
Evaluation of Verneuil Sapphire as a Substrate for Silicon on an Insulator (Czochralski, epitaxy,
Al2O3)
Ricard, J.
DGRST-7670669 (1978), 44 p. (in French)
1978
431-A

<794>
Crystal Growth of Bi3B5O12 From its Stoichiometric Melt (kinetics)
Richards, E.A.; Bergeron, C.G.
J. Cryst. Growth 44, 112-13 (1978)
1978
432

<795>
Preparation, Structures, and Properties of Niobium Chalcogenide Halides, NbX2Y2 (X=S, Se - Y=Cl,
Br, I) (vapor transport)
Rijnsdorp, J.; De Lange, G.J.; Wiegers, G.A.
J. Solid State Chem. 30, 365-373 (1979)
1979
455

<796>
Effects of Rotation on Czochralski Crystal Growth (theory)
Riley, N.; Sweet, D.A.
J. Cryst. Growth 47, 557-67 (1979)
1979
453

<797>

<797>
On the Chemical Transport of Some Niobium Oxides with TeCl4. I. Calculation of the System
Nb/O/Te/Cl and the Transport of NbO2/NbO2.417 (NbO2, NbO2.417, theory)
Ritschel, M.; Oppermann, H.
Kristall Tech. 13, 1035-43 (1978)
1978
432

<798>
On the Chemical Transport of Some Niobium Oxides with TeCl4 (II) The Transport Behaviour of the
Equilibrium NbO2.417 to NbO2.5
Ritschel, M.; Oppermann, H.; Mattern, N.
Kristall Tech. 13, 1421-29 (1979)
1979
444

<799>
The Preparation and Properties of Single Crystal Copper Phosphide (Cu3P, Czochralski)
Robertson, D.S.; Snowball, G.; Webber, H.
J. Mat. Sci. 15, 256-58 (1980)
1980
453

<800>
The Effect of Impurities on the Piezoelectric Properties of Lithium Germanate (Li2GeO3,
Czochralski)
Robertson, D.S.; Young, I.M.; Ainger, F.W.; O'Hara, C.; Glazer, A.M.
J. Phys. D: Appl. Phys. 12, 611-17 (1979)
1979
450

<801>
The Cadmium Oxide-Tungsten Oxide Phase System and Growth of Cadmium Tungstate Single Crystals
(CdWO4, Czochralski)
Robertson, D.S.; Young, I.M.; Telfer, J.R.
J. Mat. Sci. 14, 2967-74 (1979)
1979
453

<802>
Liquid Phase Epitaxy of Garnets (review)
Robertson, J.M.
J. Cryst. Growth 45, 233-42 (1978)
1978
430

<803>
Liquid Phase Epitaxial Growth of Ga1-xInxSb on GaSb by Stepwise Grading (melt)
Rode, J.R.; Gertner, E.E.; Andrews, A.M.; Cheung, D.T.; Tennant, W.E.
J. Electron. Mater. 7, 337-45 (1978)
1978
434

<804>
Unstable Growth of ADP Crystals (NH4H2PO4, theory, solution, kinetics, slow cooling)
Rodriguez, R.; Aguilo, M.; Tejada, J.
J. Cryst. Growth 47, 518-26 (1979)
1979
453

<805>
Some Thermodynamic Considerations for the Growth of Layered Dichalcogenides by Vapour Transport
Processes (kinetics, theory)
Rogers, T.; Balchin, A.A.
J. Cryst. Growth 44, 398-402 (1978)
1978
429

<806>
The Crystal Structure of ThBC
Rogl, P.
J. Nucl. Mater. 73, 198-203 (1978)
1978
451-A

<807>
Crystals for Magnetic Applications, Vol. 1 (garnets, review, hydrothermal, epitaxy, flux,
Bridgman, film)
Rooijmans, C.J.M.(ed.)
Springer-Verlag, New York (1978)
1978
431-TC

<808>
A Gas Handling System for Exacting Crystal Growth Preparations (TiO2, vapor transport)
Rosenberger, F.; Olson, J.M.; Delong, M.C.
J. Cryst. Growth 47, 321-25 (1979)
1979
454

<809>
Low-Stress Physical Vapor Growth (PVT) (KCl, KBr, KCl:KCN, KCN:KOH, KCl:KOH, theory)
Rosenberger, F.; Westphal, G.H.
J. Cryst. Growth 43, 148-52 (1978)
1978
416

<810>
Kinetics of Crystal Growth of Calcium Tungstate from Solutions in Sodium Tungstate Melts by
Continuous Cooling (CaWC4, flux)
Roy, B.N.; Appalasami, S.
J. Amer. Ceram. Soc. 61, 38-41 (1978)
1978
416

<811>
Kinetics of Crystal Growth of Strontium Tungstate from Solutions in Sodium Tungstate Melts by
Continuous Cooling (SrWC4, flux)
Roy, B.N.; Appalasami, S.
Kristall Tech. 14, 19-27 (1979)
1979
452

<812>
Kinetics of Crystallization of Barium Tungstate from Solutions in Sodium Tungstate Melts by
Continuous Cooling (BaWC4, flux, kinetics, slow cooling)
Roy, B.N.; Banda, V.K.
J. Cryst. Growth 47, 577-82 (1979)
1979
453

<813>

<813>
Crystal-Chemistry, Crystal-Growth, and Phase-Equilibria of Apatites (review)
Roy, D.M.; Drafall, L.E.; Roy, R.
pp. 185-240 in Phase Diagrams, Pt. 5, Materials Science and Technology, (Refractory Materials: A Series of Monographs, Vol. 6, Pt. 5), edited by A. M. Alper, Academic Press, New York (1978)
1978
440-T

<814>
The Crystal Structure of Niobium Trisulfide, NbS3
Runsdorp, J.; Jellinek, F.
J. Solid State Chem. 25, 325-28 (1978)
1978
428

<815>
The Growth of CdS in Sealed Silica Capsules (vapor transport)
Russell, G.J.; Woods, J.
J. Cryst. Growth 46, 323-30 (1979)
1979
443

<816>
Vapour Growth and Defect Characterisation of Large Single Crystals of ZnS and Zn(S,Se)
Russell, G.J.; Woods, J.
J. Cryst. Growth 47, 647-53 (1979)
1979
454

<817>
Synthesis, Preparation of Single Crystals, and Physicochemical Investigation of CrPr2S4 and CrPr2Se4 (vapor transport)
Rustamov, P.G.; Guseinov, G.G.; Kurbanov, T.Kh.; Einullaev, A.V.; Aliev, O.M.
Inorg. Mater. 14, 504-6 (1978)
1978
430

<818>
In-Situ Carbon-Silicon Ribbon Thermocouple (Si, equipment)
Sachs, E.M.
J. Cryst. Growth 47, 477-85 (1979)
1979
453

<819>
Effect of Addition of Vanadium on the Chemical Transport of TiSx, x = 1.40-1.70 (TiyV1-ySx)
Saeki, M.; Ishii, M.; Kawada, I.; Nakahira, M.
J. Cryst. Growth 45, 25-28 (1978)
1978
430

<820>
Growing Single Crystals of Silicon-Germanium Solid Solutions by Electron-Beam Zone Melting Without a Crucible (Si-Ge)
Saidov, M.S.; Umerov, R.S.
Sov. Phys. Cryst. 23, 130-33 (1978)
1978
440

<821>
Crystal Structure and Habit of Fine Metal Particles Formed by Gas-Evaporation Technique: bcc Metals (V, Fe, Cr, Mo and W)
Saito, Y.; Yatsuya, S.; Mihama, K.; Uyeda, R.
J. Cryst. Growth 45, 501-505 (1978)
1978
430

<822>

<822>
Structure and Stability of Alkali Germanate Crystals and Glasses (Na2Ge4O9, Na4Ge9O20, Na2Ge2O5)
Sakka, S.; Kamiya, K.
Revue de Chimie Minerale 16, 293-307 (1979)
1979
456

<823>
Heteroepitaxial Structures of Si on Sapphire, Obtained by Vacuum Deposition (film, doped)
Sakvarelidze, L.G.; Bakhiya, T.N.; Chkheidze, I.M.; Mirtskhulava, A.A.
Inorg. Mater. 14, 1384-87 (1978)
1978
439

<824>
Structural Phase Transition in Mixed Crystals WxMo1-xO3 (melt)
Salje, E.; Gehlig, R.; Viswanathan, K.
J. Solid State Chem. 25, 239-50 (1978)
1978
437

<825>
The Crystal Structure of a New High -Nd- Concentration Laser Material: Na3Nd(PO4)2
Salmon, R.; Parent, C.; Vlasse, M.; Le Flem, G.
Mat. Res. Bull. 13, 439-44 (1978)
1978
424

<826>
Dynamics of Non-Equilibrium Crystallization of an Alloy (theory, melt, slow cooling)
Samojlovich, Yu.A.
Fiz. Khim. Obrabot. Mater. SSSR No. 3, 85-92 (1978) (in Russian)
1978
451-A

<827>
Selective Etching of Cesium Iodide Crystals and Comparison of Growth and Dissolution Processes
(CsI, solution)
Sangwal, K.; Urusovskaja, A.A.; Smirnov, A.E.
Ind. J. Pure Appl. Phys. 16, 501-7 (1978)
1978
441

<828>
Liquid Phase Epitaxial Growth of CdTe in the CdTe-CdCl2 System (flux)
Saraie, J.; Kitagawa, M.; Ishida, M.; Tanaka, T.
J. Cryst. Growth 43, 13-16 (1978)
1978
417

<829>
Growth of Single Crystals of High Melting Metal Alloys and Compounds by Plasma Heating (W, W-C)
Savitsky, E.M.; Burkhanov, G.S.
J. Cryst. Growth 43, 457-62 (1978)
1978
419

<830>
Ferroelectricity in (N(CH3)4)2ZnCl4 (evaporation)
Sawada, S.; Shiroishi, Y.; Yamamoto, A.; Takashige, M.; Matsuo, M.
J. Phys. Soc. Japan 44, 687-88 (1978)
1978
417

<831>

<831>
Ferroelectricity in (N(CH3)4)2CoCl4
Sawada, S.; Shiroishi, Y.; Yamamoto, A.; Takashige, M.; Matsuo, M.
Phys. Lett. 67A, 56-58 (1978)
1978
425

<832>
The Validity of Steady-State Dendrite Growth Models (theory)
Schaefer, R.J.
J. Cryst. Growth 43, 17-20 (1978)
1978
417

<833>
Reply to Comment of R. Trivedi on "The Validity of Steady-State Dendrite Growth Models" (theory)
Schaefer, R.J.
J. Cryst. Growth 44, 365 (1978)
1978
426

<834>
State-of-The-Art of Crystal Growth and Nuclear Spectroscopic Evaluation of Mercuric Iodide
Radiation Detectors. Part II. HgI2 (vapor transport, review)
Schieber, M.; Beinglass, I.; Dishon, G.; Holzer, A.; Yaron, G.
Nucl. Instr. Meth. 150, 71-77 (1978)
1978
420

<835>
Metal-Insulator Transitions and Phase Diagram of (Ti1-xVx)4O7: Electrical, Calorimetric,
Magnetic and EPR Studies (vapor transport)
Schlenker, C.; Ahmed, S.; Buder, R.; Gourmala, M.
J. Phys. C: Solid State Phys. 12, 3503-21 (1979)
1979
450

<836>
Pyroelectricity and Related Properties in the Fresnoite Pseudobinary System Ba2TiGe2O8-Ba2TiSi2O8
(Czochralski)
Schmid, H.; Genequand, P.; Tippmann, H.; Pouilly, G.; Guedu, H.
J. Mat. Sci. 13, 2257-65 (1978)
1978
429

<837>
Gas Phase Synthesis of Epitaxial Layers of Nickel-Chlorine Boracite on Chromium-Chlorine Boracite
(Ni3B7O13Cl, vapor transport)
Schmid, H.; Tippman, H.
J. Cryst. Growth 46, 723-42 (1979)
1979
445

<838>
Dislocation Propagation in Czochralski Grown Gadolinium Gallium Garnet (GGG, theory, Gd3Ga5O12)
Schmidt, W.; Weiss, R.
J. Cryst. Growth 43, 515-25 (1978)
1978
419

<839>
LPE Growth of Hg0.60Cd0.40Te from Te-Rich Solution (film, vapor transport)
Schmit, J.L.; Bowers, J.E.
Appl. Phys. Lett. 35, 457-58 (1979)
1979
452

<840>

<840>
On the Magnetic Structure and Magnetic Phase Transitions of Tb5Ge4. A Neutron Diffraction Study
(melt)
Schobinger-Papamantellos, P.
J. Phys. Chem. Solids 39, 197-205 (1978)
1978
418

<841>
Measurements of Growth Parameters During the Crystal Growth from the Vapor in Closed Ampules
(GeS, sublimation)
Schonherr, E.
J. Cryst. Growth 44, 604-8 (1978)
1978
427

<842>
Czochralski Growth of Li3N Crystals
Schonherr, E.; Muller, G.; Winckler, E.
J. Cryst. Growth 43, 469-72 (1978)
1978
419

<843>
Epitaxial Layers of CuIrSe2 on GaAs
Schumann, B.; Georgi, C.; Tempel, A.; Kuhn, G.; van Nam, N.; Neumann, H.; Horig, W.
Thin Solid Films 52, 45-52 (1978)
1978
437

<844>
Structural and Electrical Properties of CuGaSe2 Thin Films on GaAs Substrates (epitaxy)
Schumann, B.; Tempel, A.; Kuhn, G.; Neumann, H.; van Nam, N.; Hansel, T.
Kristall Tech 13, 1285-95 (1978)
1978
433

<845>
Magnetic Properties of beta-Phase Iron-Germanium (FexGe, slow cooling)
Schurer, P.J.; Hall, N.J.G.; Morrish, A.H.
Phys. Rev. B 18, 4860-74 (1978)
1978
434

<846>
Progress of Temperature and Concentration in the Region of Solidification of Pb-Bi Alloys Rich in
Single-Phase Pb in the Course of Solidification (Czochralski)
Schurmann, E.; Wedeking, G.
Giessereiforschung 30, 31-36 (1978) (in German)
1978
451-A

<847>
Some Evidence for the Existence and Magnitude of a Critical Marangoni Number for the Onset of
Oscillatory Flow in Crystal Growth Melts (NaNO3, theory)
Schwabe, D.; Scharmann, A.
J. Cryst. Growth 46, 125-31 (1979)
1979
430

<848>
Experiments on Surface Tension Driven Flow in Floating Zone Melting (NaNO3, theory)
Schwabe, D.; Scharmann, A.; Preisser, F.; Oeder, R.
J. Cryst. Growth 43, 305-12 (1978)
1978
417

<849>

<849>
Solution Growth of Indium-Doped Silicon (Si)
Scott, W.; Hager, R.J.
J. Electron. Mater. 8, 581-602 (1979)
1979
442

<850>
Polytypism and the Vibrational Properties of PbI2
Sears, W.M.; Klein, M.L.; Morrison, J.A.
Phys. Rev. B 19, 2305-13 (1979)
1979
447

<851>
Low Temperature Thermal Properties of PbI2 (Bridgman)
Sears, W.M.; Morrison, J.A.
J. Phys. Chem. Solids 40, 503-8 (1979)
1979
445

<852>
GaxIn1-xSb Crystals Grown by Liquid Phase Epitaxy (melt)
Segawa, K.; Otsubo, M.; Miki, H.; Fujibayashi, K.
Japan. J. Appl. Phys. 17, 165-70 (1978)
1978
428

<853>
Spacelab and Material Processing Facilities and Experiments (review)
Seibert, G.
Proc. Royal Scc. London A 361, 131-42 (1978)
1978
423

<854>
Spacelab and Past and Future Crystal Growth Experiments in Space (review)
Seibert, G.
Prog. Cryst. Growth. Charact. 1, 95-116 (1978)
1978
424

<855>
Solute Partitioning During Silicon Dendritic Web Growth (Si, theory)
Seidensticker, R.G.; Stewart, A.M.; Hopkins, R.H.
J. Cryst. Growth 46, 51-54 (1979)
1979
430

<856>
Impurity Effect on Grown-In Dislocation Density of InP and GaAs Crystals (Czochralski)
Seki, Y.; Watanabe, H.; Matsui, J.
J. Appl. Phys. 49, 822-28 (1978)
1978
416

<857>
Scanning Electron Microscopy of Growth Surfaces of Synthetic Quartz (SiO2, hydrothermal)
Sekiguchi, Y.; Funakubo, H.; Champier, G.
J. Cryst. Growth 47, 751-54 (1979)
1979
454

<858>

<858>
Vapor Phase Growth of Thin Plates of Cr2O3 Single Crystals
Sekiya, T.; Nishiyama, G.; Hayashi, H.; Okuda, H.
J. Cryst. Growth 43, 645-47 (1978)
1978
419

<859>
Growth of Cr2O3 Crystals by the Decomposition of a K2Cr2O7 Melt (flux)
Sekiya, T.; Okuda, H.
J. Cryst. Growth 46, 410-14 (1979)
1979
443

<860>
Flux Growth of Cr2O3 Crystals from a K2Cr2O7-K2CrO4-K2B4O7 Mixture
Sekiya, T.; Okuda, H.
J. Cryst. Growth 47, 551-56 (1979)
1979
453

<861>
Growth Characteristics of Crystals of Certain Metallides (alpha-FeSi2.33, CoSi2, FeGe2, Mn5Si3,
Fe1.76Ge, melt, theory, kinetics)
Semenov, V.N.; Frolov, B.A.; Povzner, A.A.
Sov. Phys. Cryst. 23, 209-11 (1978)
1978
434

<862>
Temperature Field Dynamics in the Synthesis of Diamonds (kinetics, theory, C)
Semenova-Tyan-Shanskaya, A.S.; Fedoseev, D.V.; Bokii, G.B.
Dokl. Akad. Nauk SSSR 240, 582-84 (1978) (in Russian)
1978
451-A

<863>
Influence of Magnetic Field on Vertical Bridgman-Stockbarger Growth of InxGa1-xSb
Sen, S.; Lefever, R.A.; Wilcox, W.R.
J. Cryst. Growth 43, 526-30 (1978)
1978
419

<864>
Non-Constant Distribution Coefficients for Directionally Solidified InSb-GaSb (magnetic field
effects, eutectics, Bridgman)
Sen, S.; Wilcox, W.R.
Mat. Res. Bull. 13, 293-303 (1978)
1978
424

<865>
Preparation and Characterization of Ruthenium Dioxide Crystals (RuO2, vapor transport)
Shafer, M.W.; Figat, R.A.; Olson, B.; LaPlaca, S.J.; Angilello, J.
J. Electrochem. Soc. 126, 1625-28 (1979)
1979
451

<866>
Electron Diffraction Study of Phase Transformations of the Compound CuSe
Shafizade, R.B.; Ivanova, I.V.; Kazinets, M.M.
Thin Solid Films 55, 211-20 (1978)
1978
447

<867>

<867>
Polymorphism in Bi2Sn207 (flux)
Shannon, R.D.; Bierlein, J.D.; Gillson, J.L.; Jones, G.A.; Sleight, A.W.
J. Phys. Chem. Solids 41, 117-22 (1980)
1979
456

<868>
Electroless Deposition of Semiconductor Films (PbS, PbSe, Pb1-xHgxS, solution)
Sharma, N.C.; Kainthla, R.C.; Pandya, D.K.; Chopra, K.L.
Thin Solid Films 60, 55-59 (1979)
1979
449

<869>
Electroless Deposition of Epitaxial Pb1-xHgxS Films
Sharma, N.C.; Pandya, D.K.; Sehgal, H.K.; Chopra, K.L.
Thin Solid Films 59, 157-64 (1979)
1979
445

<870>
Growth and Polytypism in Vapor-Grown Mixed Cadmium Halide Crystals (CdI2-xBrx, CdI2-xClx)
Sharma, S.D.; Mehrotra, K.; Agrawal, V.K.
J. Electrochem. Soc. 126, 325-28 (1979)
1979
432

<871>
An Inexpensive Zone Melting Furnace for Purification and Crystal Growth (NaNO3, PbCl2, organics)
Sharon, M.; Pradhananga, R.; Prasad, B.; Bholagir, A.
J. Cryst. Growth 46, 715-17 (1979)
1979
443

<872>
The Structure of K2Er(NC3)5 (flux)
Sherry, E.G.
J. Inorg. Nucl. Chem. 40, 257-68 (1978)
1978
421

<873>
Capture of Migrating Adatoms by Point Capture Centres for Different Two-Dimensional Substrate Lattices (theory)
Shevelev, V.V.; Temkin, D.E.
J. Cryst. Growth 43, 173-84 (1978)
1978
416

<874>
Morphology and Structure of ZnS Grown from the Vapour Phase
Shichiri, T.; Aikami, T.; Kakinoki, J.
J. Cryst. Growth 43, 320-28 (1978)
1978
417

<875>
Epitaxial Thin Iron Films on Antimony (Fe, vapor transport)
Shigematsu, T.; Hine, S.; Takada, T.
J. Cryst. Growth 43, 531-32 (1978)
1978
419

<876>

<876>
Czochralski Growth of Tellurium Single Crystals (Te)
Shih, I.; Champness, C.H.
J. Cryst. Growth 44, 492-98 (1978)
1978
429

<877>
Etch Pit Orientation in Czochralski-Grown Tellurium Crystals (Te)
Shih, I.; Champness, C.H.
J. Cryst. Growth 46, 456-60 (1979)
1979
443

<878>
Epitaxial Growth of Some II-VI Compound Films Evaporated Onto Electron-Bombarded NaCl Substrates
(CdS, CdSe, CdTe, ZnSe)
Shimaoka, G.
J. Cryst. Growth 45, 313-17 (1978)
1978
430

<879>
Single Crystal Growth of Substituted Yttrium Iron Garnets Y3Fe5-x(Ga,Al)xO12 by the Floating Zone
Method
Shindo, I.; Ii, N.; Kitamura, K.; Kimura, S.
J. Cryst. Growth 46, 307-13 (1979)
1979
443

<880>
Growth of Mg2TiO4 Single Crystals by the Floating Zone Method
Shindo, I.; Kimura, S.; Kitamura, K.
J. Mat. Sci. 14, 1901-06 (1979)
1979
445

<881>
Growth of Thallium Iodide Crystals in Silica Hydrogels (TlI, Tl2I3, gel)
Shiojiri, M.; Kaito, C.; Saito, Y.; Murakami, M.; Kawamoto, J.
J. Cryst. Growth 43, 61-70 (1978)
1978
417

<882>
High Rate Epitaxial Growth of ZnO Films on Sapphire by Planar Magnetron rf Sputtering System
Shiosaki, T.; Ohnishi, S.; Murakami, Y.; Kawabata, A.
J. Cryst. Growth 45, 346-49 (1978)
1978
430

<883>
Molecular Beam and Solid-Phase Epitaxies of Silicon Under Ultra-High Vacuum (film)
Shiraki, Y.; Katayama, Y.; Kobayashi, K.L.I.; Komatsubara, K.F.
J. Cryst. Growth 45, 287-91 (1978)
1978
430

<884>
Crystal Growth of Boron Monophosphide Using a B2H6-PH3-H2 System (BP, vapor transport)
Shohno, K.; Ohtake, H.; Bloem, J.
J. Cryst. Growth 45, 187-91 (1978)
1978
430

<885>

<885>
Growth of Pb0.8Sn0.2Te Films in a Quasi-Closed Space (vacuum evaporation)
Sidorov, Yu.G.; Sabinina, I.V.; Gavrilova, T.A.
Inorg. Mater. 14, 47-50 (1978)
1978
425

<886>
New Methods of Making Silicon Solar Cells (review)
Siffert, P.
Rev. Phys. Appl. 14, 165-92 (1979) (in French)
1979
444

<887>
The Growth of Thin Bismuth Single Crystals with High Resistivity Ratios (Bi, Czochralski)
Sima, V.; Vasek, P.; Lejcek, P.
Kristall Tech. 13, K36-37 (1978)
1978
451-A

<888>
Kinetics of Growth and Dissolution of Sodium Chlorate in Diffusion and Convection Regimes
(solution, NaClO3)
Simon, B.
J. Cryst. Growth 43, 640-42 (1978)
1978
419

<889>
The Occurrence of High Period Polytype of SiC at the Interface of Two Interacting Spirals
Singh, S.R.; Singh, G.
J. Cryst. Growth 47, 755-57 (1979)
1979
454

<890>
Crystal Growth of the Solid Electrolyte (C5H5NH)2Cu5Br7 From the Melt (Bridgman)
Skarstad, P.M.; Parker, H.S.
J. Cryst. Growth 43, 613-17 (1978)
1978
419

<891>
Crystal Growth and Structure of BiVO4 (Czochralski)
Sleight, A.W.; Chen, H.-Y.; Ferretti, A.; Cox, D.E.
Mat. Res. Bull. 14, 1571-81 (1979)
1979
454

<892>
An Automated System for the Growth of Multilayered Structures in the (GaAl)As System by LPE
(Ga1-xAlxAs, flux)
Small, M.B.; Blackwell, J.C.; Potemski, R.M.
J. Cryst. Growth 46, 253-61 (1979)
1979
443

<893>
Conditions for Constant Growth Rate by LPE from a Cooling, Static Solution (theory, melt)
Small, M.B.; Ghez, R.
J. Cryst. Growth 43, 512-14 (1978)
1978
419

<894>

<894>
Growth and Dissolution Kinetics of III-V Heterostructures Formed by LPE (GaxAl1-xAs)
Small, M.B.; Ghez, R.
J. Appl. Phys. 50, 5322-30 (1979)
1979
447

<895>
Some Aspects of the Design and Construction of the Cold Crucible
Smith, E.A.
Mat. Res. Bull. 14, 431-39 (1979)
1979
442

<896>
The Crystal Growth and Atomic Structure of As2(Se,S)3 Compounds (As2S3, As2Se3, vapor transport)
Smith, E.A.; Cowlam, N.; Shamah, A.M.
Phil. Mag. 39B, 111-32 (1979)
1979
444

<897>
Nonstoichiometry and Carrier Concentration Control in MBE of Compound Semiconductors (II-VI,
IV-VI, III-V, review)
Smith, D.L.
Prog. Cryst. Growth Charact. 2, 33-47 (1979)
1979
453

<898>
Impurity Dopant Incorporation and Diffusion During Molecular Beam Epitaxial Growth of IV-VI
Semiconductors (PbTe, Pb1-xSnxSe, Pb1-xSnxTe)
Smith, D.L.; Pickhardt, V.Y.
J. Electrochem. Soc. 125, 2042-50 (1978)
1978
432

<899>
Oriented Crystal Growth on Amorphous Substrates Using Artificial Surface-Relief Gratings (KCl,
films, solution)
Smith, H.I.; Flanders, D.C.
Appl. Phys. Lett. 32, 349-50 (1978)
1978
421

<900>
Flux Growth and Characterization by X-ray Topography of Rare Earth Arsenates (REAsO4, slow
cooling)
Smith, S.H.; Garton, G.; Tanner, B.K.; Midgley, D.
J. Mat. Sci. 13, 620-26 (1978)
1978
410

<901>
The Growth of LaAlO3 Crystals Enriched in Isotope 17O (flux)
Smith, S.H.; Walker, P.J.
J. Cryst. Growth 47, 315-16 (1979)
1979
443

<902>
Crystal Structure of Scandium Metaphosphate Sc(PO3)3 (solution)
Smolin, Yu.I.; Shepelev, Yu.F.; Domanskii, A.I.; Belov, N.V.
Sov. Phys. Cryst. 23, 100-101 (1978)
1978
440

<903>

<903>
Dependence of the Optical Homogeneity of Single Germanosillenite Crystals on the Growth Conditions (Bi12GeO20, Czochralski)
Sobolev, A.T.; Kopylov, Yu.L.; Kravchenko, V.B.; Kucha, V.V.
Sov. Phys. Cryst. 23, 93-96 (1978)
1978
440

<904>
The Thermodynamic Driving Force for Crystallization from Solution (theory, kinetics)
Sohnel, O.; Garside, J.
J. Cryst. Growth 46, 238-40 (1979)
1979
442

<905>
A Method for the Determination of Precipitation Induction Periods
Sohnel, O.; Mullin, J.W.
J. Cryst. Growth 44, 377-82 (1978)
1978
429

<906>
Reactive Pulse Plasma Crystallization of Diamond and Diamond-Like Carbon (C, film)
Sokolowski, M.; Sokolowska, A.; Gokieli, B.; Michalski, A.; Rusek, A.; Romanowski, Z.
J. Cryst. Growth 47, 421-26 (1979)
1979
454

<907>
Bridgman and Czochralski Growth of beta-Phase NaxV2O5
Sol, N.; Merenda, P.
J. Cryst. Growth 46, 557-62 (1979)
1979
443

<908>
Growth and Dissolution of BaFBr Crystals (flux)
Somaiah, K.; Hari Babu, V.
J. Cryst. Growth 46, 711-14 (1979)
1979
443

<909>
Metastable Pb1-xCdxS Epitaxial Films. I. Growth and Physical Properties
Sood, A.K.; Wu, K.; Zemel, J.N.
Thin Solid Films 48, 73-86 (1978)
1978
411

<910>
Matastable Pb1-xCdxS Epitaxial Films. II. Electrical Properties
Sood, A.K.; Wu, K.; Zemel, J.N.
Thin Solid Films 48, 87-94 (1978)
1978
411

<911>
Preparation and Properties of Two Indium Antimony Selenides (In2Se3:Sb, In2-xSbxSe3, InSbSe3, Bridgman)
Spiesser, M.; Gruska, R.P.; Subbarao, S.N.; Castro, C.A.; Wold, A.
J. Solid State Chem. 26, 111-14 (1978)
1978
436

<912>

<912>
Crystal Growth of Actinide Dioxides by Chemical Transport
Spirlet, J.C.; Bednarczyk, E.; Ray, I.; Muller, W.
J. Phys. (Paris) 40, Suppl. 4, 108-10 (1979)
1979
448-T

<913>
Dodecatungstophosphoric Acid-21-Water by Neutron Diffraction (evaporation, H3PW12O40.21H2O)
Spirlet, M.R.; Busing, W.R.
Acta Cryst. B34, 907-10 (1978)
1978
421

<914>
Optical and Magneto-Optical Properties of Sm0.55Tb0.45FeO3 (flux)
Stepniak, A.; Skwarcz, J.
Phys. Stat. Sol. (a) 52, K77-79 (1979)
1979
448

<915>
Study of Melting and Freezing in the Gaussian Core Model by Molecular Dynamics Simulation (melt, slow cooling, theory)
Stillinger, F.H.; Weber, T.A.
J. Chem. Phys. 68, 3837-44 (1978)
1978
450-A

<916>
Nd:Y2O3 Single-Crystal Fiber Laser: Room-Temperature cw Operation at 1.07- and 1.35-micrometer Wavelength (laser melting, Czochralski)
Stone, J.; Burrus, C.A.
J. Appl. Phys. 49, 2281-87 (1978)
1978
419

<917>
Crystalline Vanadium (II) Fluoride, VF2. Preparation, Structure, Heat Capacity from 5 to 300 K and Magnetic Ordering
Stout, J.W.; Boo, W.O.J.
J. Chem. Phys. 71, 1-8 (1979)
1979
444

<918>
VPE Growth of AlxGa1-xAs (theory)
Stringfellow, G.B.; Hall, H.T.,Jr.
J. Cryst. Growth 43, 47-60 (1978)
1978
417

<919>
Organometallic VPE Growth of AlxGa1-xAs (vapor transport)
Stringfellow, G.B.; Hall, H.T.,Jr.
J. Electron. Mater. 8, 201-26 (1979)
1979
446

<920>
Electrical and Optical Properties of High-Purity p-type Single Crystals of GeFe2O4 (vapor transport)
Strobel, P.; Koffyberg, F.P.; Wold, A.
J. Solid State Chem. 31, 209-216 (1980)
1980
456

<921>

<921>
Organometallic Vapor Deposition of Epitaxial ZnSe Films on GaAs Substrates
Stutius, W.
Appl. Phys. Lett. 33, 656-58 (1978)
1978
427

<922>
Direct Measurement of Temperature Dependence of Lattice Mismatches Between LPE-Grown Li(Nb,Ta)O3
Film and LiTaO3 Substrate (flux)
Sugii, K.; Kondo, S.
J. Cryst. Growth 46, 607-14 (1979)
1979
443

<923>
Large-Sized InP Single Crystals by the Synthesis, Solute Diffusion Technique (vapor transport)
Sugii, K.; Kubota, E.; Iwasaki, H.
J. Cryst. Growth 46, 289-92 (1979)
1979
443

<924>
Magnetic Spinel Single Crystals by Bridgman Technique (review)
Sugimoto, M.
in Crystals for Magnetic Applications, C.J.M. Rooijmans (ed.), Springer-Verlag, New York (1978)
1978
431-A

<925>
Low Temperature Growth of GaP LPE Layers from Indium Solvent (flux)
Sugiura, T.; Tanaka, A.; Sukegawa, T.
J. Cryst. Growth 46, 595-600 (1979)
1979
443

<926>
Chemical Vapor Growth of Titanium Diboride by a Modified Hot Wire Method (TiB2)
Sugiyama, K.; Iwakoshi, S.; Motojima, S.; Takahashi, Y.
J. Cryst. Growth 43, 533-36 (1978)
1978
419

<927>
Single Crystal Growth of Zirconium Carbide by a Modified Hot-Filament Method (de Boer, Van Arkel,
vapor transport, ZrC)
Sugiyama, K.; Mizuno, H.; Motojima, S.; Takahashi, Y.
J. Cryst. Growth 44, 617-20 (1978)
1978
427

<928>
Single Crystal Growth of Titanium Carbide from the Vapor by a Modified Hot Wire Method (TiC)
Sugiyama, K.; Mizuno, H.; Motojima, S.; Takahashi, Y.
J. Cryst. Growth 46, 788-94 (1979)
1979
445

<929>
On the Relations Between the Number of Gold Drops and TiP Whiskers in VLS Growth
Sugiyama, K.; Takigawa, M.; Motojima, S.; Takahashi, Y.
J. Cryst. Growth 44, 499-501 (1978)
1978
429

<930>

<930>
Gel Growth of Single Crystals of Some Rubidium and Cesium Tin Halides (CsSnCl3, RbSnCl3,
Cs2SnCl6, Rb2SnCl6, CsSnBr3, RbSnBr3, Rb2SnBr6, Cs2SnBr6, CsSnI3, RbSnI3, CsSn2I5, RbSn2I5)
Suib, S.L.; Weller, P.F.
J. Cryst. Growth 48, 155-60 (1980)
1980
454

<931>
Crystal Growth of Co Ferrite on Fine Acicular gamma-Fe2O3 Particles (solution)
Sumiya, K.; Matsumoto, T.; Watatani, S.; Hayama, F.
J. Phys. Chem. Solids 40, 1097-1102 (1979)
1979
455

<932>
Vapor Growth and Epitaxy of Minerals and Synthetic Crystals (review, theory)
Sunagawa, I.
J. Cryst. Growth 45, 3-12 (1978)
1978
430

<933>
Modes of Vibrations in Step Trains: Rhythmical Bunching (theory)
Sunagawa, I.; Bennema, P.
J. Cryst. Growth 46, 451-57 (1979)
1979
443

<934>
Electron Microscopic Characterization of Cubic BN Prepared by Water Solvent (solution)
Susa, K.; Kobayashi, T.; Nagata, F.; Taniguchi, S.
J. Cryst. Growth 43, 345-50 (1978)
1978
417

<935>
Photoluminescence Properties of CuGaSe2 Grown by Iodine Vapour Transport
Susaki, M.; Miyauchi, T.; Horinaka, H.; Yamamoto, N.
Japan. J. Appl. Phys. 17, 1555-59 (1978)
1978
426

<936>
Asymmetric 180 degree Bloch Walls in Fe Single Crystal Films (theory)
Suzuki, T.
Japan. J. Appl. Phys. 17, 141-48 (1978)
1978
428

<937>
In-Process Monitoring and Control of Doping Gas Concentration During Vapor Phase Silicon
Epitaxial Growth by Using a Flame Photometric Detector (Si)
Suzuki, T.; Inoue, Y.; Ura, M.; Ogawa, T.; Sugita, Y.
J. Cryst. Growth 45, 108-17 (1978)
1978
430

<938>
Growth of High-Purity ZnTe Single Crystals by the Sublimation Travelling Heater Method (vapor
transport)
Taguchi, T.; Fujita, S.; Inuishi, Y.
J. Cryst. Growth 45, 204-13 (1978)
1978
430

<939>

<939>
Growth by Travelling Heater Method and Characteristic of Undoped High-Resistivity CdTe
Taguchi, T.; Shirafuji, J.; Inuishi, Y.
Japan. J. Appl. Phys. 17, 1331-42 (1978)
1978
426

<940>
Growth Kinetics of Epitaxial Layers of Silicon Carbide Obtained by Sublimation in Vacuum (SiC)
Tairov, Yu.M.; Taranets, V.A.; Tsvetkov, V.F.
Inorg. Mater. 14, 1122-25 (1978)
1978
439

<941>
Investigation of Growth Processes of Ingots of Silicon Carbide Single Crystals (SiC, melt)
Tairov, Yu.M.; Tsvetkov, V.F.
J. Cryst. Growth 43, 209-12 (1978)
1978
416

<942>
Investigations of Kinetic and Thermal Conditions of Silicon Carbide Epitaxial Layer Growth from
the Vapour Phase (SiC, film, vapor transport, kinetics)
Tairov, Yu.M.; Tsvetkov, V.F.
J. Cryst. Growth 46, 403-409 (1979)
1979
443

<943>
Analysis of Dislocations in Gd3Ga5O12 by a Repeated Etching Technique
Takagi, K.; Fukazawa, T.; Ishii, M.
J. Cryst. Growth 48, 19-24 (1980)
1980
454

<944>
New Developments in Ionized-Cluster Beam and Reactive Ionized-Cluster Beam Deposition Techniques
(vapor transport)
Takagi, T.; Matsubara, K.; Takaoka, H.; Yamada, I.
Thin Solid Films 63, 41-51 (1979)
1979
458

<945>
Ionized-Cluster Beam Deposition and Epitaxy (review)
Takagi, T.; Yamada, I.; Matsubara, K.
Thin Solid Films 58, 9-19 (1979)
1979
445

<946>
Vapour Growth and Epitaxy (review)
Takahashi, K.(ed.); Ariumi, T.(ed.); Blom, G.M.(ed.); Kaldis, E.(ed.)
Proceedings of the Fourth International Conference on Vapour Growth and Epitaxy, Nagoya, Japan,
9-13 July 1978, J. Cryst. Growth, Vol. 45 (1978)
1978
430

<947>
Growth Parameters and Crystal Morphology of Vapor-Deposited Niobium Nitride (NbN)
Takahashi, T.; Itoh, H.; Yamaguchi, T.
J. Cryst. Growth 46, 69-74 (1979)
1979
430

<948>

<948>
Liquid Phase Epitaxy of AlxGa1-xP on GaP and Its Application to Double Heterostructure Light
Modulators
Takakura, H.; Yamamoto, M.; Hamakawa, Y.; Kariya, T.
J. Cryst. Growth 45, 267-71 (1978)
1978
430

<949>
Roles of Lattice Fitting in Epitaxy (theory, Ni, Cu, Pd, Pt, Fe, Al, Au, Ag, In, Pb, PbS, PbSe,
PbTe, SnTe, LiF, NaF, KF, LiBr, NaCl, NaBr, LiI, KCl, NaI, KBr, KI)
Takayanagi, K.; Yagi, K.; Honjo, G.
Thin Solid Films 48, 137-52 (1978)
1978
416

<950>
Techniques for Routine UHV in situ Electron Microscopy of Growth Processes of Epitaxial Thin Films
Takayanagi, K.; Yagi, K.; Kobayashi, K.; Honjo, G.
J. Phys. E: Sci. Instrum. 11, 441-48 (1978)
1978
427

<951>
Quasi-Grown Layers in Liquid Phase Epitaxial Growth (In1-xGaxAs)
Takeda, Y.; Imamura, Y.; Sasaki, A.
J. Cryst. Growth 46, 75-78 (1979)
1979
430

<952>
Composition Latching Phenomenon and Lattice Mismatch Effects in LPE-Grown In1-xGaxAs on InP
Substrate
Takeda, Y.; Sasaki, A.
J. Cryst. Growth 45, 257-61 (1978)
1978
430

<953>
Growth of Fayalite (Fe2SiO4) Single Crystals by the Floating Zone Method (arc image)
Takei, H.
J. Cryst. Growth 43, 463-68 (1978)
1978
419

<954>
Growth of FeTiO3 (Ilmenite) Crystals by the Floating-Zone Method
Takei, H.; Kitamura, K.
J. Cryst. Growth 44, 625-31 (1978)
1978
427

<955>
Growth and Properties of Strontium Cobaltate Single Crystals (SrCoO2.70, float zone melting)
Takei, H.; Oda, H.; Watanabe, H.; Shindo, I.
J. Mat. Sci. 13, 519-22 (1978)
1978
410

<956>
Magnetic Structure of GdCu1-xZnx System (melt)
Takei, K.; Ishikawa, Y.; Watanabe, N.; Tajima, K.
J. Phys. Soc. Japan 47, 88-94 (1979)
1979
453

<957>

<957>
Growth of GaAs1-xSbx Crystals by Steady-State Liquid Phase Epitaxy
Takenaka, N.; Inoue, M.; Shirafuji, J.; Inuishi, Y.
J. Phys. D: Appl. Phys. 11, L91-95 (1978)
1978
417

<958>
Catalytic Effect of Nickel on the Growth of Zirconium Carbide Whiskers by Chemical Vapor
Deposition (ZrC)
Tamari, N.; Kato, A.
J. Less-Common Metals 58, 147-60 (1978)
1978
418

<959>
Catalytic effects of Various Metals and Refractory Oxides on the Growth of TiC Whiskers by
Chemical Vapor Deposition
Tamari, N.; Kato, A.
J. Cryst. Growth 46, 221-37 (1979)
1979
442

<960>
Non-Seeded Growth of Large Single Pb1-xSnxTe Crystals on a Quartz Surface (sublimation)
Tamari, N.; Shtrikman, E.
J. Cryst. Growth 43, 378-80 (1978)
1978
417

<961>
Growth Study of Large Non-Seeded Pb1-xSnxTe Single Crystals (vapor transport, sublimation)
Tamari, N.; Shtrikman, E.
J. Electron. Mater. 8, 269-88 (1979)
1979
446

<962>
Improved Nucleation and a Planar Interface in LPE Grown Pb1-xSnxTe Heterostructures
Tamari, N.; Shtrikman, E.
J. Cryst. Growth 47, 463-66 (1979)
1979
454

<963>
Isotope Effects on the Central Peak Phenomena in KD3(SeO3)2 (solution)
Tanaka, H.; Yagi, T.; Tatsuzaki, I.
J. Phys. Soc. Japan 44, 1257-60 (1978)
1978
420

<964>
Fermi Surface Measurements of ZrB2 by the de Haas van Alphen Effect (float zone melting)
Tanaka, T.; Ishizawa, Y.; Bannai, E.; Kawai, S.
Solid State Commun. 26, 879-82 (1978)
1978
436

<965>
Geometrical Analysis of Crystallization of the Soft-Core Model in an FCC Crystal Formation
(theory)
Tanemura, M.; Hiwatari, Y.; Matsuda, H.; Ogawa, T.; Ogita, N.; Ueda, A.
Prog. Theor. Phys. 59, 323-24 (1978)
1978
451-A

<966>
Preparation of Silicon Plates by Pulling from a Molten Bath According to the Stepanov Procedure
(Si)
Tatarchenko, V.A.; Brantov, S.K.; Abrasimov, N.V.
Fiz. Khim. Obrabot. Mater. SSSR No. 1, 79-84 (1978) (in Russian)
1978
451-A

<967>
Some Peculiarities in the Growth and Properties of Beta-Boron Crystals (B, float zone melting)
Tavadze, F.N.; Gabunia, D.L.; Khvedelidze, A.G.
J. Less-Common Metals 67, 101-106 (1979)
1979
458

<968>
Shape of the Crystallization Front During Electron-Beam Zone Refining of beta-B Without a Crucible
Tavadze, F.N.; Gabuniya, D.L.; Tsagareishvili, G.V.; Khvedelidze, A.G.
Inorg. Mater. 14, 518-20 (1978)
1978
430

<969>
Region of Metastability During Normal Growth in a Binary Alloy (kinetics, theory)
Temkin, D.E.
Dokl. Akad. Nauk SSSR, 240, 833-35 (1978) (in Russian)
1978
451-A

<970>
Variation of Liquid and Solid Compositions During LPE Growth from a Ternary Component Solution
(Ga1-xAlxAs, flux, theory)
Teramoto, I.; Kazumura, M.; Yamanaka, H.
Japan. J. Appl. Phys. 10, 1509-16 (1979)
1979
451

<971>
Low-Defect InSb Crystal Growth by InN Doping (Czochralski)
Terashima, K.
J. Cryst. Growth 47, 746-48 (1979)
1979
454

<972>
Growth and X-Ray Study of Cu2SrGeS4 and Cu2BaGeS4 (flux)
Teske, C.L.
Z. Naturforsch. 34b, 386-89 (1979) (in German)
1979
444

<973>
The Crystal Structure of the Solid Electrolyte Ag44I53(C11H30N3)3 (solution)
Thackeray, M.M.; Coetzer, J.
Acta Cryst. B34, 71-75 (1978)
1978
415

<974>
Ellipsometric Assessment of (Ga,Al)As/GaAs Epitaxial Layers During Their Growth in an
Organometallic VPE System (Ga1-xAlxAs, GaAs, vapor transport)
Theeten, J.B.; Hottier, F.; Hallais, J.
J. Cryst. Growth 46, 245-52 (1979)
1979
443

<975>

<975>
Some Electrical Properties of GeBi2Te4 Single Crystals (Bridgman)
Tichy, L.; Frumar, M.; Klikorka, J.
Phys. Stat. Sol. (a) 56, 323-326 (1979)
1979
455

<976>
Rost Kristallov iz Rastvorov-Rasplavov (Crystal Growth from High-Temperature Solutions) (review, flux)
Timofeeva, V.A.; Belyaev, L.M.(ed.)
Nauka, Moscow (1978), 267 pages (in Russian)
1978
454-A

<977>
Growth of Bulk Yttrium Iron Garnet Crystals from High Temperature Solutions (Y3Fe5O12)
Timofeeva, V.A.; Bykov, A.B.
Prog. Cryst. Growth. Charact. 2, 377-404 (1979)
1979
000

<978>
Kinetic Characteristics of Growth of Y3Al5O12 Single Crystals from Melt-Solutions on Oriented Seeds (flux)
Timofeeva, V.S.; Bykov, A.B.; Rozin, K.M.; Heifets, A.Ya.
Sov. Phys. Cryst. 23, 248-49 (1978)
1978
434

<979>
X-Ray and Dielectric Investigations of Pb2CrO5 (melt)
Titov, A.V.; Bush, A.A.; Venevtsev, Yu.N.
Sov. Phys. Cryst. 23, 354-55 (1978)
1978
434

<980>
Correlation of Diffusion Behaviour with the Entropy of Fusion (theory)
Tiwari, G.P.
Trans. Jap. Inst. Metals 19, 125-28 (1978)
1978
450-A

<981>
The Compounds Ag7PS6 and Ag7PSe6
Toffoli, P.; Khodadad, P.
C.R. Acad. Sci. 286C, 349-51 (1978) (in French)
1978
420

<982>
Crystal Growth of Magnetic Garnets from High-Temperature Solutions (review)
Tolksdorf, W.; Welz, F.
in Crystals for Magnetic Applications, Vol. 1, edited by C. J. M. Rooijmans, Springer-Verlag, New York (1978)
1978
431-A

<983>
Crystal Structure of the Spinel CdErS4 (slow cooling)
Tomas, A.; Shilo, I.; Guittard, M.
Mat. Res. Bull. 13, 857-59 (1978)
1978
430

<984>

<984>
Epitaxial Growth of BaTiO3 Vacuum Condensates on SrTiO3 Crystals
Tomashpol'skii, Yu.Ya.; Sevast'yanov, M.A.
Sov. Phys. Cryst. 23, 246-48 (1978)
1978
434

<985>
The Bending of Silicon Wafers by Thin Polycrystalline Silicon Film Deposition and by Film Doping
Using Boron Diffusion (Si)
Toncheva, L.T.; Vassilev, I.S.
Thin Solid Films 60, 353-59 (1979)
1979
449

<986>
Crystal Structure of Morophosphate AgCoPO4
Tordjman, I.; Guitel, J.C.; Durif, A.; Averbuch, M.T.; Masse, R.
Mat. Res. Bull. 13, 983-88 (1978)
1978
430

<987>
Phase Diagram and Crystal Growth of Lead Phosphovanadates (Pb3(PxV1-xO4)2, Pb3(PO4)2)
Torres, J.; Morin, D.; Fateau, L.; Ossart, P.; Aubree, J.; Primot, J.
Mat. Res. Bull. 13, 811-18 (1978)
1978
430

<988>
Impurity Effect on the Formation of Terraces in GaAs LPE Growth (flux)
Toyoda, N.; Mihara, M.; Hara, T.
Phys. Stat. Sol. (a) 54, 225-30 (1979)
1979
448

<989>
Crystallization Kinetics of Lithium and Strontium Nitrate Crystal Hydrates (LiNO3.3H2O,
Sr(NO3)2.4H2O)
Treivus, E.B.; Franke, V.D.; Donakova, A.V.
Sov. Phys. Cryst. 23, 127-28 (1978)
1978
440

<990>
Morphological Stability of a Solid Particle Growing from a Binary Alloy Melt (dendrites, theory)
Trivedi, R.
J. Cryst. Growth 48, 93-99 (1980)
1980
454

<991>
Effects of Interface Kinetics on the Growth Rate of Dendrites (Sn, theory)
Trivedi, R.; Franke, H.; Lacmann, R.
J. Cryst. Growth 47, 389-96 (1979)
1979
454

<992>
Interface Morphology During Crystallization - I. Single Filament, Unconstrained Growth from a
Pure Melt (theory)
Trivedi, R.; Tiller, W.A.
Acta Met. 26, 671-78 (1978)
1978
416

<993>

<993>
Interface Morphology During Crystallization - II. Single Filament, Unconstrained Growth from a
Binary Alloy Melt (Sn-Pt, theory)
Trivedi, R.; Tiller, W.A.
Acta Met. 26, 679-87 (1978)
1978
416

<994>
Fractionation of Stable Isotopes and Impurities During Zone Recrystallization of Metals and
Preparation of Single Crystals Rich in Individual Isotopes
Troitskii, O.A.; Spitsyr, V.I.; Moiseenko, M.M.; Tsiganov, A.D.
Phys. Stat. Scl. 54(a), 651-54 (1979)
1979
457

<995>
Simulation of the Growth of a Metallic Face-Centered Cubic Lattice in a Crystal from a Molten
Mass in the (100) Orientation (theory)
Trunov, N.N.
Fiz. Metallov Metalloved., SSSR 45, 803-9 (1978) (in Russian)
1978
451-A

<996>
Grain Size Dependence in a Self-Implanted Silicon Layer on Laser Irradiation Energy Density (Si)
Tseng, W.F.; Mayer, J.W.; Campisano, S.U.; Foti, G.; Rimini, E.
Appl. Phys. Lett. 32, 824-26 (1978)
1978
434

<997>
Variations of Impurity Distribution in the Modified Czochralski Method of Crystal Growth (Si)
Turovskii, B.M.; Popov, A.I.
Inorg. Mater. 15, 939-40 (1979)
1979
457

<998>
Growth and Properties of Silicon Crystals With Automatic Diameter Control (Si)
Turovskii, B.M.; Shenderovich, I.L.; Popov, A.I.
Cvetn. Metally 6, 48-49 (1978) (in Russian)
1978
450-A

<999>
The Molten State of Matter. Melting and Crystal Structure (melt, review, theory)
Ubbelohde,A.R.
Wiley-Interscience, New York (1979)
1979
440-A

<1000>
Optical Phonon Modes and Localized Effective Charges of Transition-Metal Dichalcogenides (MoS2,
MoSe2, WS2, WSe2, HfSe2, ZrSe2, TiSe2, vapor transport)
Uchida, S.; Tanaka, S.
J. Phys. Soc. Japan 45, 153-62 (1978)
1978
436

<1001>

<1001>
Crystal Growth, and Magnetic and Mossbauer Studies of Sr(Fe0.766W0.234)O3 and its Related
Compounds (SrFe1-xWxO3, flux)
Uchino, K.; Nomura, S.
J. Phys. Soc. Japan 46, 432-39 (1979)
1979
453

<1002>
A Model for Macroscopic Czochralski Growth: Theoretical and Experimental Investigations (theory)
Uelhoff, W.; van der Hart, A.W.A.; Fattah, A.; Hanke, G.; Jedamzik, D.; Knook, B.; van der Berg,
G.J.; Wenzl, H.
Report JUL-1554 (November 1978)
1978
451

<1003>
Heterogeneous Equilibria and Mass Transfer Kinetics in the System TiP-I2 (TiP, vapor transport)
Ugai, Ya.A.; Gukov, O.Ya.; Illarionov, A.A.
Inorg. Mater. 14, 794-96 (1978)
1978
436

<1004>
Raman and Infrared Spectra of CdIn2S4 and ZnIn2S4 (vapor transport)
Unger, W.K.; Farnworth, B.; Irwin, J.C.; Pink, H.
Solid State Commun. 25, 913-15 (1978)
1978
419

<1005>
Single Crystal Growth of Alexandrite in a V2O5 Flux (BeAl2O4)
Ushio, M.
Nippon Kagaku Kaishi 9, 1186-90 (1979) (in Japanese)
1979
451

<1006>
Growth of Polyhedral Metal Crystallites in Inactive Gas (Mg, review)
Uyeda, R.
J. Cryst. Growth 45, 485-89 (1978)
1978
430

<1007>
Theory of Melting, Vacancy Model
Vaidya, S.N.
Pramana 12, 23-32 (1979)
1979
444

<1008>
Scandium Hydrogenselenite (Sc(HSeO3)3, evaporation, solution)
Valkonen, J.; Leskela, M.
Acta Cryst. B34, 1323-26 (1978)
1978
421

<1009>
Morphological Stability Analysis of Growth from the Vapour (theory)
van den Brekel, C.H.J.
Philips J. Res. 33, 20-30 (1978)
1978
432

<1010>

<1010>
Morphological Stability Analysis in Chemical Vapour Deposition Processes. I (theory)
van den Brekel, C.H.J.; Jansen, A.K.
J. Cryst. Growth 43, 364-70 (1978)
1978
417

<1011>
Interface Morphology in Chemical Vapour Deposition on Profiled Substrates (theory)
van den Brekel, C.H.J.; Jansen, A.K.
J. Cryst. Growth 43, 488-96 (1978)
1978
419

<1012>
Luminescence and Ionic Conductivity of Bariumfluorochloride (BaFCl, slow cooling)
van den Heuvel, G.P.M.; Blasse, G.; Schoonman, J.
Solid State Communications 28, 689-91 (1978)
1978
435

<1013>
Survey of Monte Carlo Simulations of Crystal Surfaces and Crystal Growth (review, theory, kinetics)
van der Eerden, J.P.; Bennema, P.; Cherepanova, T.A.
Prog. Cryst. Growth Charact. 1, 219-54 (1978)
1978
426

<1014>
The Role of Lattice Misfit in Epitaxy (review, theory)
van der Merwe, J.H.
CRC Crit. Rev. Solid State Mat. Sci. 7, 209-31 (1978)
1978
428

<1015>
Surface Morphology of HCl Etched Silicon Wafers (Si)
van der Putte, P.; Van Enckevort, W.J.P.; Giling, L.J.; Bloem, J.
J. Cryst. Growth 43, 659-75 (1978)
1978
432

<1016>
Liquid Phase Epitaxial Growth of Copper Ferrite Films (CuFe2O4)
van der Straten, P.J.M.; Metselaar, R.
IEEE Trans. Magnetics MAG-14, 421-23 (1978)
1978
426

<1017>
Liquid Phase Epitaxial Growth of Lithium Ferrite-Aluminate Films (flux, Li0.5Fe2.5-xAlxO4)
Van der Straten, P.J.M.; Metselaar, R.
J. Cryst. Growth 48, 114-20 (1980)
1980
454

<1018>
Flux Growth of ZnGa2O4 Single Crystals
van der Straten, P.J.M.; Metselaar, R.; Jonker, H.D.
J. Cryst. Growth 43, 270-72 (1978)
1978
416

<1019>

<1019>
The Influence of Adsorption and Step Reconstruction on the Growth and Etching Vectors of Silicon
(111) (theory, vapor transport, epitaxy)
Van Enckevort, W.J.P.; Giling, L.J.
J. Cryst. Growth 45, 90-96 (1978)
1978
430

<1020>
Monte Carlo Simulation of a (111) Diamond Face Around the Roughening Transition (C, theory)
Van Enckevort, W.J.P.; van der Eerden, J.P.
J. Cryst. Growth 47, 501-8 (1979)
1979
453

<1021>
The Growth Kinetics of Garnet Liquid Phase Epitaxy Using Horizontal Dipping (flux)
van Erk, W.
J. Cryst. Growth 43, 446-56 (1978)
1978
419

<1022>
A Solubility Model for Rare-Earth Iron Garnets in a PbO/B2O3 Solution (Y3Fe5O12,
RExY3-xGayFe5-yO12, flux)
van Erk, W.
J. Cryst. Growth 46, 539-50 (1979)
1979
443

<1023>
Optical Properties of V6O13 Single Crystals in the Metallic and the Semiconducting Phase
Van Hove, W.; Clauws, P.; Vennik, J.
Solid State Communications 33, 11-16 (1980)
1980
456

<1024>
On the Driving Force for Crystallization: The Growth Affinity (theory, kinetics, melt, solution)
van Leeuwen, C.
J. Cryst. Growth 46, 91-95 (1979)
1979
430

<1025>
On the Presentation of Growth Curves for Growth from Solution (theory)
van Leeuwen, C.; Blomen, L.J.M.J.
J. Cryst. Growth 46, 96-104 (1979)
1979
430

<1026>
Computation of Striated Impurity Distributions in Melt-Grown Crystals, Taking Account of Periodic
Remelt (Si, theory)
Van Run, A.M.J.G.
J. Cryst. Growth 47, 680-92 (1979)
1979
454

<1027>
Coarsening of Eutectic Structures During and After Unidirectional Growth (theory)
van Suchtelen, J.
J. Cryst. Growth 43, 28-46 (1978)
1978
417

<1028>

<1028>
Growth Striations in Fe-3wt%Si Single Crystals Grown by Floating Zone Melting at Various Growth
Rates
Vanek, P.; Kadeckova, S.
J. Cryst. Growth 47, 456-62 (1979)
1979
454

<1029>
Crystal Structure of Li2CaUF8 (flux)
Vedrine, A.; Trottier, D.; Cousseins, J.C.; Chevalier, R.
Mat. Res. Bull. 14, 583-87 (1979)
1979
442

<1030>
Kinetic Studies of Nucleation and Growth at Surfaces (review)
Venables, J.A.
Thin Solid Films 50, 357-69 (1978)
1978
423

<1031>
Vanadium (III) Oxychloride: Magnetic, Optical and Electrical Properties -Lithium and Molecular
Intercalations (VOCl)
Venien, J.P.; Palvadeau, P.; Schleich, D.; Rouxel, J.
Mat. Res. Bull. 14, 891-97 (1979)
1979
445

<1032>
Synthesis of Single Crystals of Cubic Boron Nitride in Systems Containing Hydrogen (BN)
Vereshchagin, L.F.; Gladkaya, I.S.; Dubitskii, G.A.; Slesarev, V.N.
Inorg. Mater. 15, 201-203 (1979)
1979
457

<1033>
X-Ray and Neutron Diffraction of Na2MnFeF7: A Variety of Trigonal Weberite (flux)
Verscharen, W.; Babel, D.
J. Solid State Chem. 24, 405-21 (1978) (in German)
1978
418

<1034>
Preparation, Stability, and Preliminary Crystallographic Study of Bi2Sn2O7 (flux)
Vetter, G.; Queyroux, F.; Gilles, J.-C.
Mat. Res. Bull. 13, 211-16 (1978)
1978
424

<1035>
Crystallization of Bi-Sb Solid Solutions in Capillaries (zone melting)
Vigdorovich, V.N.; Ukhlinov, G.A.; Dolinskaya, N.Yu.; Marychev, V.V.
Izv. Akad. Nauk SSSR, Metally, No. 1, 96-99 (1978) (in Russian)
1978
441-A

<1036>
Crystal Structure of CuTa2O6 (flux)
Vincent, H.; Bochu, B.; Aubert, J.J.; Joubert, J.C.; Marezio, M.
J. Solid State Chem. 24, 245-53 (1978) (in French)
1978
418

<1037>
Photoferromagnetic Effect and Photoconductivity of Cd1-xHgxCr2Se4 Single Crystals (vapor transport, melt)
Vinogradova, G.I.; Veselago, V.G.; Makhotkin, V.E.; Kovaleva, I.S.; Levshin, V.A.; Shabunina, G.G.; Kalinnikov, V.T.
Sov. Phys. Solid State 20, 827-28 (1978)
1978
430

<1038>
A New Technique for Simulating Two-Dimensional Nucleation Models of Crystal Growth (theory)
Viola, M.S.; Van Wormer, K.A.; Botsaris, G.D.
J. Cryst. Growth 47, 127-29 (1979)
1979
443

<1039>
Effect of Mineralizers on the Growth Morphology of Cubic Thulium Oxide (Tm2O3, hydrothermal)
Viswanathiah, M.N.; Narayanan Kutty, T.R.; Rao, N.; Tareen, J.A.K.
J. Cryst. Growth 44, 366-67 (1978)
1978
426

<1040>
Synthesis and Crystal Growth of CdGeP2 (melt)
Vohl, P.
ESD-TR-78-27, pp. 43-47, MIT Lincoln Laboratory, Lexington, Massachusetts
1978
421

<1041>
Synthesis and Crystal Growth of CdGeP2 (vapor transport)
Vohl, P.
J. Electron. Mater. 8, 517-28 (1979)
1979
446

<1042>
Dynamics of Laser-Induced Formation of Palladium Silicide (Pd2Si)
von Allmen, M.; Wittmer, M.
Appl. Phys. Lett. 34, 66-70 (1979)
1979
441

<1043>
Preparation of Pure and Doped Silicon Carbide by Pyrolysis of Silane Compounds (SiC, vapor transport, De Boer-Van Arkel, doped)
von Muench, W.; Pettenpaul, E.
J. Electrochem. Soc. 125, 294-99 (1978)
1978
411

<1044>
AlSiP3, A Compound with a Novel Wurtzite-Pyrite Intergrowth Structure (flux)
von Schnering, H.G.; Menge, G.
J. Solid State Chem. 28, 13-19 (1979)
1979
448

<1045>
Recent Developments in the Microstructural Characterization of Corrosion Processes (film, epitaxy, Cu2O)
Vook, R.W.; Ho, J.H.
Society for the Advancement of Material and Process Engineering (SAMPE) Quarterly 9, 7 (1978)
1978
429

<1046>

<1046>
Mass Transfer at the Surface of a Crystal Near to its Boundary with the Melt, and its Influence
on the Shape of the Growing Crystal (Czochralski, kinetics, theory)
Voronkov, V.V.
Sov. Phys. Cryst. 23, 137-41 (1978)
1978
434

<1047>
Crystallization of LiNbO3 from Solution in Borate, Vanadate, and Tungstate Melts (flux)
Voronkova, V.I.; Evianova, N.F.; Yanovskii, V.K.
Sov. Phys. Cryst. 23, 129-30 (1978)
1978
440

<1048>
Mixed Oxysulfides of Chromium III and Rare Earths (LaCrOS2, CeCrOS2, PrCrOS2, NdCrOS2, SmCrOS2)
Vovan, T.; Dugue, J.; Grittard, M.
Mat. Res. Bull. 13, 1163-66 (1978)
1978
429

<1049>
Crystallography of Aligned Cu-In Eutectoid (Cu-Cu7In3)
Vrolijk, J.W.G.A.; Wolff, L.R.
J. Cryst. Growth 48, 85-92 (1980)
1980
454

<1050>
Interface Growth Feature and Voids in Sapphire Ribbon Crystals (alpha-Al2O3, Czochralski)
Wada, K.; Hoshikawa, K.
Japan. J. Appl. Phys. 17, 449 (1978)
1978
428

<1051>
Dislocations in Sapphire Ribbon Crystals Grown by the Edge-Defined, Film-Fed Growth Technique
(alpha-Al2O3)
Wada, K.; Hoshikawa, K.
J. Cryst. Growth 44, 502-4 (1978)
1978
429

<1052>
Growth of Stacking Faults by Bardeen-Herring Mechanism in Czochralski Silicon (Si)
Wada, K.; Takaoka, H.; Inoue, N.; Kohra, K.
Japan. J. Appl. Phys. 18, 1629-30 (1979)
1979
451

<1053>
Crystal Growth of CdTe for X-Ray Detectors (review)
Wald, F.V.; Entine, G.
Nucl. Instr. Meth. 150, 13-23 (1978)
1978
420

<1054>
Crystal Growth of SrCl2 and Solid Solutions of SrCl2-PrCl3 and SrCl2-GdCl3 (Bridgman-Stockbarger,
Czochralski)
Walker, P.J.
J. Cryst. Growth 44, 187-89 (1978)
1978
424

<1055>
Crystal Growth of K2CuF4 and Some Solid Solutions of K2CuF4-K2ZnF4 (Bridgman)
Walker, P.J.
J. Cryst. Growth 46, 709-10 (1979)
1979
443

<1056>
Crystal Growth of Sodium Sulphide, Na2S (hygroscopic, Bridgman)
Walker, P.J.
J. Cryst. Growth 47, 598-600 (1979)
1979
453

<1057>
GaAs Vapour Phase Epitaxy Using Computer Control
Walline, R.E.; Heaton, J.L.; Holtz, J.E.; Kinzel, D.L.; Moysenko, A.E.
Ind. J. Pure Appl. Phys. 16, 875-80 (1978)
1978
441

<1058>
Indium Polytelluride In2Te5(II) (melt)
Walton, P.D.; Sutherland, H.H.; Hogg, J.H.C.
Acta Cryst. B34, 41-45 (1978)
1978
415

<1059>
Liquid Phase Growth of HgCdTe Epitaxial Layers
Wang, C.C.; Shin, S.H.; Chu, M.; Lanir, M.; Vanderwyck, A.H.B.
J. Electrochem. Soc. 127, 175-79 (1980)
1980
454

<1060>
Effects of Modifying Starting Compositions for Flux Growth (DyPO4, PrBO3, LaBO3, DyKMo2O8,
ErKMo2O8, NdKMo2O8, DyVC4, TbVO4, TbPO4, DyPO4, Ho2Ge2O7, Dy2GeMoO8, ZrSiO4, ThGeO4, Er2Si2O7)
Wanklyn, B.M.
J. Cryst. Growth 43, 336-44 (1978)
1978
417

<1061>
The Flux Growth of Some Rare-Earth and Iron Group Complex Oxides (NdBO3, LaBO3/Pr, LaBO3,
(La,Pb)MnO3, (Nd,Pb)MnO3, (Eu,Pb)MnO3, Pb3Mn7O15, Tm2Ti3O7, Tb2Ti2O7, Pr2Ti2O7, Tm2Ge2O7,
Er2SiO5, Dy2SiO5, Ho2SiC5, Pr2MoO6, LiCoPO4)
Wanklyn, B.M.; Maqsood, A.
J. Mat. Sci. 14, 1975-81 (1979)
1979
445

<1062>
Flux Growth of Crystals of RKMo2O8, R2MoO6 and R6MoO12 in the Systems R2O3-K2O-MoO3 (DyKMo2O8,
ErKMo2O8, NdKMo2O8, Dy2MoO6, Sm2MoO6, Pr2MoO6, Ho6MoO12)
Wanklyn, B.M.; Wondre, F.R.
J. Cryst. Growth 43, 93-100 (1978)
1978
417

<1063>

<1063>
The Flux Growth of Some New Rare Earth and Iron Group Complex Oxides (GdVO3, Gd4Ba6(BO3)9,
(Ba0.93Gd0.07)TiO3, Nd4Ea6(BO3)9, Pr4Ba6(BO3)9, Ni8Nb3O15, Ni2V2PbO8, Ni3Nb2O8, Ni3V2O8,
Ta2Co4O9, PbCr2.3Ti2.3O9, Pb2CrO5)
Wanklyn, B.M.; Wondre, F.R.; Garrard, B.J.; Smith, S.H.; Davison, W.
J. Mat. Sci. 13, 89-96 (1978)
1978
409

<1064>
Flux Growth of Crystals of Some Transition Metal Fluorides. Part 2 (KAlF4, KMnF3, RbFeF4,
Rb2FeF5, Rb3FeF6, RbxFeF3, CsFeF4, CsFe2F7, Cs3Fe2F9, CsxFeF3, Na2CoFeF7, Na2NiFeF7, Na2NiAlF7,
Na2ZnCrF7, NaCrF4, Rb2Cr5F17)
Wanklyn, B.M.; Wondre, F.R.; Maqsood, A.; Yanagisawa, K.; Davison, W.
J. Mat. Sci. 14, 1447-56 (1979)
1979
445

<1065>
Solution Growth of ZnS, ZnSe, CdS and Their Mixed Compounds Using Tellurium as a Solvent
Washiyama, M.; Sato, K.; Aoki, M.
Japan. J. Appl. Phys. 18, 869-72 (1979)
1979
445

<1066>
Preparation of Some Chalcogenide Spinel Single Crystals and Their Electronic Properties (FeCr2S4,
CoCr2S4, vapor transport)
Watanabe, T.; Nakada, I.
Japan. J. Appl. Phys. 17, 1745-54 (1978)
1978
429

<1067>
Cold Working Nb3Al in the bcc Structure and then Converting to the A-15 Structure
Webb, G.W.
Appl. Phys. Lett. 32, 773-75 (1978)
1978
419

<1068>
Magneto-Optical Properties of KTb3F10 and LiTbF4 Crystals (Czochralski)
Weber, M.J.; Morgret, R.; Leung, S.Y.; Griffin, J.A.; Gabbe, D.; Linz, A.
J. Appl. Phys. 49, 3464-69 (1978)
1978
420

<1069>
Thermodynamic Properties of Surface Steps (theory)
Weeks, J.D.; Gilmer, G.E.
J. Cryst. Growth 43, 385-87 (1978)
1978
417

<1070>
Materials Behaviour in Low Gravity Environment (melt, solution, vapor transport, zone melting,
doped, review)
Weiss, H.
Proc. Royal Soc. London A 361, 157-64 (1978)
1978
423

<1071>

<1071>
Measurements of the Contact Angle Between Melt and Crystal During Czochralski Growth of Gallium
and Germanium (Ga, Ge, theory)
Wenzl, H.; Pattah, A.; Gustin, D.; Mihelcic, M.; Uelhoff, W.
J. Cryst. Growth 43, 607-12 (1978)
1978
419

<1072>
Properties and Synthesis of Nb-H Interstitial Alloys (electron beam melting, float zone melting,
review)
Wenzl, H.; Welter, J.-M.
pp. 603-56 in Current Topics in Materials Science, Volume 1, E. Kaldis (ed.), North Holland
Publishing Company (1978)
1978
423

<1073>
Czochralski Growth of A15 Structure Intermetallic Compounds (V3Si, V3Ge)
Wernick, J.H.; Hull, G.W.; Geballe, T.H.; Bernardini, J.E.; Buehler, E.
J. Cryst. Growth 47, 73-76 (1979)
1979
443

<1074>
On Diffusive-Advective Interfacial Mass Transfer (theory, solution, vapor transport, flux)
Westphal, G.H.; Rosenberger, F.
J. Cryst. Growth 43, 687-93 (1978)
1978
432

<1075>
Some Notes on the Growth Kinetics and Morphology of VLS Silicon Crystals Grown with Platinum and
Gold as Liquid-Forming Agents (vapor transport, Si)
Weyher, J.
J. Cryst. Growth 43, 235-44 (1978)
1978
416

<1076>
Ca12Al14O33: A New Piezoelectric Material (Czochralski)
Whatmore, R.W.; O'Hara, C.; Cockayne, B.; Jones, G.R.; Lent, B.
Mat. Res. Bull. 14, 967-72 (1979)
1979
443

<1077>
Structural Studies of Compounds in the Series GaSxSe(1-x) (0 less than or equal to x less than or
equal to 1) Grown by Iodine Vapour Transport
Whitehouse, C.R.; Balchin, A.A.
J. Mat. Sci. 13, 2394-2402 (1978)
1978
429

<1078>
Optimization of Conditions for the Growth of Gallium Selenide and Gallium Sulphide by Iodine
Vapour Transport (GaS, GaSe)
Whitehouse, C.R.; Balchin, A.A.
J. Cryst. Growth 43, 727-33 (1978)
1978
432

124

<1079>

<1079>
Transport Properties of the Systems SnS2-SnI4 and SnS2-I2 (SnS2, vapor transport, kinetics, theory)
Wiedemeier, H.; Csillag, F.J.
J. Cryst. Growth 46, 189-97 (1979)
1979
442

<1080>
Non-Constant Distribution Coefficients from Experimental Data: Theoretical (directional crystallization, zone melting, eutectics)
Wilcox, W.R.
Mat. Res. Bull. 13, 287-91 (1978)
1978
424

<1081>
Preparation and Properties of Solid State Materials, Vol. 4, Morphological Stability, Convection, Graphite, and Integrated Optics (review, theory, C, GaAs)
Wilcox, W.R.(ed.)
Marcel Dekker, Inc., 270 Madison Avenue, New York, N.Y. 10016 (1979)
1979
430-TC

<1082>
A Novel Ruthenium Bronze: KRu4O8 (solution)
Wilhelm, M.; Hoppe, R.
Z. Anorg. Allg. Chem. 438, 90-96 (1978)
1978
421

<1083>
On Interpreting a Quantity in the Burton, Prim and Slichter Equation as a Diffusion Boundary Layer Thickness (effective distribution coefficient, melt, theory)
Wilson, L.O.
J. Cryst. Growth 44, 247-50 (1978)
1978
424

<1084>
A New Look at the Burton, Prim, and Slichter Model of Segregation During Crystal Growth from the Melt (Czochralski, theory)
Wilson, L.O.
J. Cryst. Growth 44, 371-76 (1978)
1978
429

<1085>
Growing and Polishing Large Thallium Single Crystals (Tl, melt)
Windgassen, N.R.
Mat. Res. Bull. 14, 717-20 (1979)
1979
442

<1086>
Partition of Principal Components in the Course of Preparation of II-IV-V2 Compounds by Gas Phase Transport, with ZnSiP2 as an Example
Winkler, K.; Schulz, U.; Hein, K.
Kristall Tech. 13, 137-44 (1978) (in German)
1978
451-A

<1087>

<1087>
Quantum-Mechanical Relations in the Calculation of Heats of Fusion for Ionic Crystals (review, theory)
Winzer, A.
Kristall Tech. 14, 51-61 (1979)
1979
452

<1088>
Crystal Growth and Segregation Under Zero Gravity: Ge (Bridgman)
Witt, A.F.; Gatos, H.C.; Lichtensteiger, M.; Herman, C.J.
J. Electrochem. Soc. 125, 1832-40 (1978)
1978
432

<1089>
Selected Papers on Polymer Crystallization (Review)
Wunderlich, B. (ed.)
J. Cryst. Growth, 48, No. 2, (1978)
1978
456

<1090>
Intercalation-Induced Shift of the Absorption Edge in ZrS2 and HfS2 (CuxZrS2, CuxHfS2, FexZrS2, FexHfS2, vapor transport, theory)
Yacobi, B.G.; Boswell, F.W.; Corbett, J.M.
J. Phys. C: Solid State Phys. 12, 2189-96 (1979)
1979
450

<1091>
Crystal Growth and Magnetic Properties of NdFeTiO5 (flux)
Yaeger, I.
Mat. Res. Bull. 13, 819-25 (1978)
1978
430

<1092>
Preparation of TiCx Single Crystal with Homogeneous Compositions (float zone melting)
Yajima, F.; Tanaka, T.; Bannai, E.; Kawai, S.
J. Cryst. Growth 47, 493-500 (1979)
1979
453

<1093>
Morphology of Iron Pyrite Crystals (FeS2, vapor transport)
Yamada, S.; Nanjo, J.; Komua, S.; Hara, S.
J. Cryst. Growth 46, 10-14 (1979)
1979
430

<1094>
Vapor Phase Growth of Alumina Whiskers by Hydrolysis of Aluminum Fluoride (Al2O3)
Yamai, I.; Saito, H.
J. Cryst. Growth 45, 511-16 (1978)
1978
430

<1095>
InP Single Crystal Growth by the Synthesis, Solute Diffusion Method
Yamamoto, A.; Uemura, C.
Japan. J. Appl. Phys. 17, 1869-70 (1978)
1978
429

<1096>

<1096>
Bismuth Tellurides: BiTe and Bi4Te3 (melt)
Yamana, K.; Kihara, K.; Matsumoto, T.
Acta Cryst. B35, 147-49 (1979)
1979
433

<1097>
Flux Growth of Double Oxides of Niobium and Rare-Earth Elements (Ln3NbO7) (Ho3NbO7, Y3NbO7,
Er3NbO7, Dy3NbO7, La3NbC7, Y3TaO7, Er3TaO7, Gd3TaO7)
Yamasaki, Y.; Sugitani, Y.
Bull. Chem. Scc. Japan 51, 3077-78 (1978)
1978
432

<1098>
Characterization of the Optical Properties of LPE InxGa1-xAsyP1-y Thin Layers Grown on InP (film)
Yamazoe, Y.; Takakura, H.; Nishino, T.; Hamakawa, Y.; Kariya, T.
J. Cryst. Growth 45, 454-58 (1978)
1978
430

<1099>
Analysis of Vapor-Deposited Silicon Carbide Films on Silicon Ribbon Surfaces (SiC)
Yang, K.H.; Schwuttke, G.H.
Phys. Stat. Sol. 48 (a) 335-43 (1978)
1978
426

<1100>
Molecular Beam Epitaxy cf GaSb and GaSbxAs1-x (films)
Yano, M.; Suzuki, Y.; Ishii, T.; Matsushima, Y.; Kimata, M.
Japan. J. Appl. Phys. 17, 2091-96 (1978)
1978
436

<1101>
Molecular Beam Epitaxy cf InxGa1-xSb (x = 0 to 1) (GaSb)
Yano, M.; Takase, T.; Kimata, M.
Japan. J. Appl. Phys. 18, 387-88 (1979)
1979
445

<1102>
Molecular Beam Epitaxy cf ZnSexTe1-x (film)
Yao, T.; Makita, Y.; Maekawa, S.
J. Cryst. Growth 45, 309-12 (1978)
1978
430

<1103>
Formation of Amorphous Mn-Bi Films and Their Crystallization Process (vapor transport)
Yoshida, K.; Dejima, K.; Nishijima, M.; Yamada, T.
J. Cryst. Growth 45, 376-82 (1978)
1978
430

<1104>
Cathodoluminescence of Impurity-Doped Aluminum Nitride Films Produced by Reactive Evaporation
(AlN)
Yoshida, S.; Misawa, S.; Gonda, S.
Thin Solid Films 58, 55-59 (1979)
1979
445

<1105>
The Central Region of the Calcium Oxide-Gallium Oxide System (CaGa2O4, Czochralski)
Young, I.H.
J. Mat. Sci. 14, 3008-1C (1979)
1979
453

<1106>
Electrical and Structural Characteristics of Laser-Induced Epitaxial Layers in Silicon (Si)
Young, R.T.; Narayan, J.; Wood, R.F.
Appl. Phys. Lett. 35, 447-49 (1979)
1979
452

<1107>
Magneto-Optical Properties of Some Rare Earth Iron Garnets (flux, (GdPrBi)3(FeGaIn)5O12,
(YLaBi)3(FeGa)5O12)
Yu-chuan, H.; Shih-chun, C.; Tie-lian, K.
IEEE Trans. Magnetics MAG-14, 457-60 (1978)
1978
426

<1108>
The Crystal Structure of Lithium Fluorosulphate LiSO3F (solution)
Zak, Z.; Kosicka, M.
Acta Cryst. B34, 38-40 (1978)
1978
415

<1109>
Crystal Growth of Gallium Arsenide by Chemical Vapor Deposition from the Complex
Monochlorodiethylgallium-Triethylarsine (GaAs)
Zaouk, A.; Lebugle, A.; Constant, G.
J. Cryst. Growth 46, 415-20 (1979) (in French)
1979
443

<1110>
On the Theory of Normal Growth of Crystals in Binary Systems (kinetics)
Zelenev, Yu.V.; Baikov, Yu.A.; Molotkov, A.P.
Kristall Tech. 14, 389-98 (1979)
1979
441

<1111>
Proceedings of the 8th All Union Meeting on the Preparation by Stepanov Method of Profiled
Crystals and Products and Their Application in National Economy, Leningrad, March 14-17, 1979
(review)
Bull. Acad. Sci. USSR, Phys. Ser. Vol. 43, No. 9 (1979)
1979
451-TC

<1112>
Boron and Borides (review)
J. Less Common Metals, Vol. 67, Nos. 1 and 2 (1979) (Proceedings of the 6th International
Symposium on Boron and Borides, Varna, Bulgaria, 1978)
1979
452

PERMUTED TITLE INDEX

Growth (theory)*   #Fast Growth of Ordered   AB Crystals: A Monte Carlo Simulation for Ionic Crystal   000182
FexZrS2, FexHfS2, #Intercalation-Induced Shift of the   Absorption Edge in ZrS2 and HfS2 (CuxZrS2, CuxHfS2,   001090
Close-Spaced Vapor Process*   #The Optical   Absorption Edge of Single-Crystal AlN Prepared by a   000758
Impurity Distributions in Melt-Grown Crystals, Taking   Account of Periodic Remelt (Si, theory)*   of Striated   001026
and the Thermal Expansion Coefficients of VO2 near the #   Accurate X-ray Determination of the Lattice Parameters   000513
                      #Crystal Growth of Co Ferrite on Fine   Acicular gamma-Fe2O3 Particles (solution)*   000931
of the Systems MF-AlF3 and NH4F-FeF3 in Hydrofluoric   Acid Solution (TlAlF4, KAlF4, NH4AlF4, RbAlF4, CsAlF4,   000772
Influence on Crystal   #Nitrilotri(Methylenephosphonic   Acid) Adsorption on Barium Sulfate Crystals and its   000553
H3PW12O40.21H2O)*   #Dodecatungstophosphoric   Acid-21-Water by Neutron Diffraction (evaporation,   000913
of NdP5O14 Single Crystals Grown from Polyphosphoric   Acids (solution)*   #Morphology Control   000466
                                     #Dispersion of   Acoustic Phonons in Layered SrGa2 (melt, slow cooling)*   000672
Czochralski)*   #   Acoustooptic Characteristics of LiBi(MoO4)2 (000016
                            #Crystal Growth of   Actinide Dioxides by Chemical Transport*   000912
*   #Single Crystal and Film Growth of   Actinide Pnictides by Chemical Vapour Transport (review)   000160
Conditions (#The Variability of the Rate of Growth of   Adamantane Crystals from the Vapour Under Constant   000572
            #Interface Roughness of Vapor-Grown   Adamantine Crystals (C10H16)*   000753
Dimensional Substrate Lattices (#Capture of Migrating   Adatoms by Point Capture Centres for Different Two-   000873
slow cooling)*   #Unstable Growth of   ADP Crystals (NH4H2PO4, theory, solution, kinetics,   000804
NH4H2PO4, solution)*   #Mechanism of Habit Change of   ADP Crystals by Fe3+, Based on Mossbauer Studies (000296
Dihydrogen Phosphate Crystals from Aqueous Solution (ADP, NH4H2PO4)* Dislocations in the Growth of Ammonium 000219
Etching Vectors of Silicon (111) (#The Influence of   Adsorption and Step Reconstruction on the Growth and   001019
(theory)*   #Localized   Adsorption in One Layer on a Crystal-Solution Interface   000575
                                     #The   Adsorption of Zr onto W(100) Surfaces*   000221
on Crystal Growth #Nitrilotri(Methylenephosphonic Acid)   Adsorption on Barium Sulfate Crystals and its Influence   000553
GaAs, theory,   #Theoretical Analysis of Equilibrium   Adsorption Layers in CVD Systems (Si-H-Cl, Ga-As-H-Cl) (000184
Fluorides (review, hydrothermal, theory)*   #   Advances in Preparative Chemistry of Oxides and   000398
vapor transport, flux)*   #On Diffusive-   Advective Interfacial Mass Transfer (theory, solution,   001074
#On the Driving Force for Crystallization: The Growth   Affinity (theory, kinetics, melt, solution)*   001024
Solution with Special Reference to Calcium   #The   Affinity of Crystal Growth and Dissolution in Aqueous   000193
Deposition From Beams of Ionized Atoms and Clusters (Ag on Si, epitaxy)*   #Thin Film   000516
#Growth of High Perfection Ag-Sn Crystals (in   Ag solvent)*   000768
            #Ag5GaS4 -- A New Phase in the System   Ag-Ga-S (AgGaS2)*   000688
of Irregular Silver-Germanium Eutectic Alloys (Ag-Ge)*   #Unidirectional Solidification 000549
          #Growth of High Perfection   Ag-Sn Crystals (in Ag solvent)*   000768
Temperature High Purity Source for Metal Beam Epitaxy (Ag, film)*   #A High   000287
of Thin Silver Films on Cleaved Silicon Surfaces (Ag, theory, evaporation)*   #An Investigation   000619
Growth on Bicrystalline Substrates of NaCl (Au,   Ag, vapor transport)* Boundaries Produced by Epitaxial 000207
Fitting in Epitaxy (theory, Ni, Cu, Pd, Pt, Fe, Al, Au,   Ag, In, Pb, PbS, PbSe, PbTe, SnTe, LiF, NaF, KF, LiBr,   000949
Europium beta-Alumina (NaAl11O17, KAl11O17, RbAl11O17,   AgAl11O17, EuAl11O17, flux, theory)*Alkali, Silver, and   000621
                   #Crystal Structure of Monophosphate   AgCoPO4*   000986
Growth, and Optical Properties of CdGe(As1-xPx)2 and   AgGa(Se1-xSx)2 Solid Solutions (Bridgman)*   Crystal   000635
#Ag5GaS4 -- A New Phase in the System Ag-Ga-S (AgGaS2)*   000688
#Crystal Structure of AgH2PO4. Crystallography of   AgH2AsO4 (solution)*   000608
                         #Crystal Structure of   AgH2PO4. Crystallography of AgH2AsO4 (solution)*   000608
                                                       #   Aging Studies on Hydrous Lutetium Oxide (cubic Lu(OH)3)* 000663
Crystals by Iodine Vapour Transport (CuInS2, CuGaS2,   AgIn5S8)*   #Method for the Growth of Chalcopyrite   000744
of Nonmetallic Vanadium Bronze (NaxV2O5, KxV2O5,   AgxV2O5, theory)*   Heat and Magnetic Susceptibility   000171
                 #Structure of Silver Dichromate: Ag2Cr2O7 (solution)*   000260
                               #Growth of alpha-   Ag2S Whiskers in a VLS System (vapor transport)*   000718
             #Growth of Single Crystals of   Ag2SeO4 in Silica Gels*   000122
#The Crystal Structure of the Solid Electrolyte   Ag44I53(C11H30N3)3 (solution)*   000973
                                                            Ag5GaS4 -- A New Phase in the System Ag-Ga-S (AgGaS2)*   000688
             #The Compounds Ag7PS6 and   Ag7PSe6*   000981
                       #The Compounds   Ag7PS6 and Ag7PSe6*   000981
        #Epitaxial Growth of CrO2 on Sapphire in   Air (vapor transport)*   001043
   #Epitaxial Growth of Ferromagnetic CrO2 Films in   Air*   000414
the Floating Zone Method*   #Single Crystal Growth of   Akermanite (Ca2MgSi2O7) and Gehlenite (Ca2Al2SiO7) by   000404
Grown by #Interface Shape and Horizontal Variations of   Al and Ga Contents in Substituted YIG Single Crystals   000483
           #Properties of VPE-Grown GaN Doped with   Al and Some Iron-Group Metals*   000649
of the fcc Phase in Uni-Directionally Solidified Zn-   Al Eutectic*   #Crystallography   000421
            #Crystallography of Aligned Fe-   Al Eutectoid (FeAl, FeAl2)*   000078
     #Solid Phase Epitaxial Growth of Si Through   Al Film*   000584
              #Growth Kinetics of (111) Si Through an   Al Layer by Solid Phase Epitaxy*   000585
Growth of Substituted Yttrium Iron Garnets Y3Fe5-x(Ga,   Al)xO12 by the Floating Zone Method*   #Single Crystal   000879
Organometallic VPE   #Ellipsometric Assessment of (Ga,   Al)As/GaAs Epitaxial Layers During Their Growth in an   000974
Directional Solidification Process (theory, eutectics,   Al-Al3Ni)*   #Effect of Some Thermal Parameters on the 000268
#Fivefold Twinned Silicon Crystals Grown in an   Al-16 wt.% Si Melt (Si, flux)*   000489
Prepared by Molecular-Beam Epitaxy (MBE) on GaAs (Al, films)*   Crystal-Aluminum Schottky-Barrier Diodes   000190
Aluminum Single Crystals with Low Dislocation Density (Al, zone melting, theory)*   #Method of High Purity   000228
Lattice Fitting in Epitaxy (theory, Ni, Cu, Pd, Pt, Fe,   Al, Au, Ag, In, Pb, PbS, PbSe, PbTe, SnTe, LiF, NaF,   000949
           #Orientation Dependence of 1-   Alanine Incorporation in TGS Crystals (solution)*   000559
                                          #GaAs/   AlAs Layered Films (epitaxy)*   000353
Epitary. I. Growth #Crystal Growth Kinetics in (GaAs)n-(AlAs)m Superlattices Deposited by Molecular Beam   000762
                       #Single Crystal Growth of   Alexandrite in a V2O5 Flux (BeAl2O4)*   001005
TlAlF4, KAlF4, #Hydrothermal Growth of the Systems MF-   AlF3 and NH4F-FeF3 in Hydrofluoric Acid Solution (000772
#Preparation by Vapor Transport of Monocrystals of   AlF3, FeF3, and TiF3*   000124
                              #Crystallography of   Aligned Cu-In Eutectoid (Cu-Cu7In3)*   001049
                              #Crystallography of   Aligned Fe-Al Eutectoid (FeAl, FeAl2)*   000078
Na4Ge9O20, Na2Ge2O5)*   #Structure and Stability of   Alkali Germanate Crystals and Glasses (Na2Ge4O9,   000822
(beam) Processes of Condensation and Evaporation of   Alkali Halides (KCl, LiF)*   #Molecular   000628
Metastable Polymorph of KAlSiO4 (flux)*   #   Alkali Ion Exchange Reactions with RbAlSiO4: A New   000639
Preparation and Principal Physicochemical Properties of   Alkali Metal Indates and Thioindates (LiInS2, Bridgman-   000509
Studies and Structural Systematics of the C2/c   Alkali Metal Metavanadates (MaxLi1-xVO3)*   000692
Potassium, Rubidium and Cesium (review, KxVO3, #On the   Alkali Metal Tungsten Bronzes, in Particular those of   000402
#Crystal Structure, Growth, and Optical Properties of   Alkali-Trihalogenoplumbates M(PbX3) (CsPbCl3, CsPbBr3,   000377
KAl11O17, RbAl11O17, #X-Ray Diffuse Scattering from   Alkali, Silver, and Europium beta-Alumina (NaAl11O17,   000621
#The Growth of Crystals in Solution (review, theory,   alkaline earth sulfates, calcium phosphates)*   000683
Sr2LaAlO5, Sr2PrAlO5,   #Crystal Growth of Some   Alkaline-Earth Rare-Earth Pentaoxometallates (flux,   000253

PERMUTED TITLE INDEX

of Metastability During Normal Growth in a Binary Alloy (kinetics, theory)*   #Region 000969
#Dynamics of Non-Equilibrium Crystallization of an Alloy (theory, melt, slow cooling)*   000826
During Unidirectional Solidification of a Binary Alloy with a Curved Solid-Liquid Interface (Ge:Ga, 000206
#Growth and Cleavage of Se-Te Alloy Crystals (Bridgman)*   000111
Mixture (Cu-Ni, #Growth and Morphology of Copper-Nickel Alloy Crystals by Hydrogen Reduction of a CuI-NiBr2 000364
Controlling the Growth of GaxIn1-xAs and GaxIn1-xP Alloy Films by MBE (kinetics)*   #Surface Processes 000301
Stability of a Solid Particle Growing from a Binary Alloy Melt (dendrites, theory)*   #Morphological 000990
II. Single Filament, Unconstrained Growth from a Binary Alloy Melt (Sn-Pb, theory)*   During Crystallization - 000993
for X-Ray Work Obtained by Plasma Remelting of Alloy Powder (Cu9Al4)*   #Spherical Monocrystals 000043
of the Cellular Growth of a Dilute Binary Alloy Single Crystal (Fe-Ni, eutectics, theory, 000113
Analysis and Properties of High Purity Zn(1-x)HgxTe Alloys (distillation)*   #Preparation, 000792
review)* #Properties and Synthesis of Nb-H Interstitial Alloys (electron beam melting, float zone melting, 001072
Growth of Bulk Single Crystals of Quaternary III-V Alloys (review, kinetics)*   #Melt and Solution 000054
Solidification of Irregular Silver-Germanium Eutectic Alloys (Ag-Ge)*   #Unidirectional 000549
#Elastic Constants and Stability of bcc In-Tl Alloys (Bridgman)*   000581
and Bandgap Studies of High Purity HgxZn1-xTe Ternary Alloys (Bridgman)*   #Luminescence 000600
Structures of Monocrystals of Dilute Binary Metallic Alloys (Cu - 0.5%Si, Cu1-xSbx - x = 0.0015-0.01)* 000210
Growth of Semiconductors: Elements, III-V Compounds and Alloys (Ge, Si, GaAs, BP)*   of the Epitaxial Vapour 000041
Constants of Niobium-Zirconium, Hafnium, and Tungsten Alloys (Nb-Zr, Nb-Hf, electron beam melting, zone 000307
#Growth of Single Crystals of High Melting Metal Alloys and Compounds by Plasma Heating (W, W-C)* 000829
#Preparation of Ge-Si Epitaxial Alloys by Sputtering*   000420
#Kinetics of Crystallization of Alloys of Multiple Constituents (theory)*   000181
of Binary, Ternary and Quaternary (In,Ga)(As,P) Alloys Grown by the Hydride Vapor Phase Epitaxy 000729
Concentration in the Region of Solidification of Pb-Bi Alloys Rich in Single-Phase Pb in the Course of 000846
Scattering #Growth of Large Single Crystals of Heusler Alloys, Ni2MnSn and Ni2Mn1-xVxSn, for Neutron Inelastic 000321
#Local Modes in Rb1-cKc Alloys: A Neutron Scattering Study (Bridgman, theory)* 000459
#The Optical Absorption Edge of Single-Crystal AlN Prepared by a Close-Spaced Vapor Process* 000758
Nitride Films Produced by Reactive Evaporation (AlN)*   #Cathodoluminescence of Impurity-Doped Aluminum 001104
and Crystallographic Relationships Between and alpha- and beta-phases (Cu-Zn, theory)*   I. Morphology 000372
Reduction Growth (Cu, #Morphological Change from Copper/ alpha-copper-Zinc Whiskers to Dendrites in Zinc 000481
Diffusion Method. II. Diffusion-Induced Twins in the alpha-phase (theory)*   Produced by a Solid-Solid 000371
#Growth of alpha-Ag2S Whiskers in a VLS System (vapor transport)* 000718
#Melting and Crystallization of Corundum (alpha-Al2O3)*   000563
Diameter Grown by the Verneuil Technique (MgAl2O4, alpha-Al2O3)*   #Powder Feeder for Crystals of Large 000284
Grown by the Edge-Defined, Film-Fed Growth Technique (alpha-Al2O3)* #Dislocations in Sapphire Ribbon Crystals 001051
#Profile-Sapphire and Its Structural Perfection (alpha-Al2O3, melt)*   000246
Materials by RF Melting in a Cold Container (review, alpha-Al2O3, Al2O3:Cr2O3)* Crystal Growth of Refractory 000020
Growth Feature and Voids in Sapphire Ribbon Crystals (alpha-Al2O3, Czochralski)*   #Interface 001050
#Structure and Luminescence Properties of alpha-CsNd(PO3)4 Crystals*   000587
Whiskers to Dendrites in Zinc Reduction Growth (Cu, alpha-Cu-Zn)*   Change from Copper/alpha-copper-Zinc 000481
ferrous halides, float zone #Controlled Growth of alpha-Fe Single Crystal Whiskers (hydrogen reduction of 000318
Characteristics of Crystals of Certain Metallides alpha-FeSi2.33, CoSi2, FeGe2, Mn5Si3, Fe1.76Ge, melt, 000861
Single Crystals from Soluble Complexes. III. Growth of alpha-HgI2 Single Crystals from Dimethylsulfoxide 000695
Single Crystals from Soluble Complexes. II. Growth of alpha-HgI2 Single Crystals from Iodomecurate complexes (000694
#Growth of Single Crystals of Cinnabar (alpha-HgS) at Rather Low Temperature (solution, doped)* 000157
#Far Infrared Optical Study of alpha-In2Se3 Compound (melt, slow cooling)*   000453
alpha-In2Te3 (melt)*   000593
#Enantiomorphism of alpha-LiIO3 Single Crystals (solution)*   000682
#Growth, Morphology and Slip System of alpha-Si3N4 Single Crystal (vapor transport)*   000696
#About SnF2 Stannous Flouride. I. Crystallochemistry of alpha-SnF2*   000235
Diffusion Method. II. Diffusion-Induced   #Two-Phase (alpha/beta)-Brass Bicrystals Produced by a Solid-Solid 000371
Diffusion Method. I. Morphology and   #Two-Phase (alpha/beta)-Brass Bicrystals Produced by a Solid-Solid 000372
#Fermi Surface Measurements of ZrB2 by the de Haas van Alphen Effect (float zone melting)*   000964
#Solubility of AlPO4 in Hydrothermal Solutions of H3PO4*   000774
#Hydrothermal Synthesis of Aluminum Orthophosphate (AlPO4)*   000496
Intergrowth Structure (flux)* # AlSiP3, A Compound with a Novel Wurtzite-Pyrite 001044
Quality of Solution Grown Crystals -- Case of Potash Alum (KAl(SO4)2.12H2O)*   of Stirring on Crystalline 000337
#Ionic Conductivity in N30+ Beta Alumina (solution)*   000286
Scattering from Alkali, Silver, and Europium beta- Alumina (NaAl11O17, KAl11O17, RbAl11O17, AgAl11O17, 000621
Al2O3)*   #Vapor Phase Growth of Alumina Whiskers by Hydrolysis of Aluminum Fluoride (001094
5Al2O3, theory)*   #Hydronium Beta" Alumina: A Fast Proton Conductor (0.84Na2O.0.84MgO. 000285
#Liquid Phase Epitaxial Growth of Lithium Ferrite- Aluminate Films (flux, LiO.5Fe2.5-xAlxO4)*   001017
Epitaxy Technique (#Preparation of Barium Lead Hexa- Aluminate Single Crystal Layers by the Liquid Phase 000071
5AlSi6O14.xH2O)*   #Growth of Fibrous Sodium Aluminosilicate Crystals in Silica Gel (zeolite, NaO. 000444
Transitions and Phase Composition of Thin Films of Aluminum with Silicon*   #Structural 000473
Vapor Phase Growth of Alumina Whiskers by Hydrolysis of Aluminum Fluoride (Al2O3)*   # 001094
and Stimulated Emission of Crystals of Yttrium Holmium Aluminum Garnets ((Y1-xHox)3Al5O12, Czochralski)* 000406
(AlN)*   #Cathodoluminescence of Impurity-Doped Aluminum Nitride Films Produced by Reactive Evaporation 001104
#Hydrothermal Synthesis of Aluminum Orthophosphate (AlPO4)*   000496
Aluminum Redistribution in EFG of Silicon Ribbon (Si)* 000452
Beam Epitaxy (MBE) on GaAs (Al, films)* #Single-Crystal Aluminum Schottky-Barrier Diodes Prepared by Molecular- 000190
Al, zone melting,   #Preparation Method of High Purity Aluminum Single Crystals with Low Dislocation Density (000228
#Growing Optical films of Aluminum Yttrium Garnet (RE substituted Y3Al5O12, flux)* 000044
#VPE Growth of AlxGa1-xAs (theory)*   000918
#Organometallic VPE Growth of AlxGa1-xAs (vapor transport)*   000919
Heterostructure Light   #Liquid Phase Epitaxy of AlxGa1-xP on GaP and Its Application to Double 000948
by the Liquid Phase Epitaxy Technique (flux, (Ba,Pb) Al12O19)*   Lead Hexa-Aluminate Single Crystal Layers 000071
#Single-Crystal Elastic Constants of Al2Cu (Bridgman, arc melting)*   000274
#Melting and Crystallization of Corundum (alpha- Al2O3)*   000563
of Alumina Whiskers by Hydrolysis of Aluminum Fluoride (Al2O3)*   #Vapor Phase Growth 001094
Grown by the Verneuil Technique (MgAl2O4, alpha- Al2O3)*   #Powder Feeder for Crystals of Large Diameter 000284
by the Edge-Defined, Film-Fed Growth Technique (alpha- Al2O3)* #Dislocations in Sapphire Ribbon Crystals Grown 001051
for Silicon on an Insulator (Czochralski, epitaxy, Al2O3)* #Evaluation of Verneuil Sapphire as a Substrate 000793
Zone Melting. I. Pure Rotation Driven Flows (Si, Al2O3, kinetics, float zone melting, theory)* Floating 000173
#Profile-Sapphire and Its Structural Perfection (alpha- Al2O3, melt)*   000246
in Crystal Growth (Si, Ge, GaAs, InAs, InSb, Al2O3, melt, theory)*   of Stress and Dislocations 000406
Zone Geometry and Prandtl Number at Zero Gravity (Si, Al2O3, theory, float zone melting, kinetics, Si)* of 000175
Simulation of Czochralski Bulk Flow in Microgravity (Al2O3, theory, kinetics)*   #Digital 000536

PERMUTED TITLE INDEX

by RF Melting in a Cold Container (review, alpha-	Al2O3, Al2O3:Cr2O3)* Growth of Refractory Materials	000020
Feature and Voids in Sapphire Ribbon Crystals (alpha-	Al2O3, Czochralski)* #Interface Growth	001050
Growth of Small-Diameter Sapphire and Silicon Crystals (	Al2O3, Si)* #Maximum Stable Zone Length in Float-Zone	000479
by RF Melting in a Cold Container (review, alpha-Al2O3:	Cr2O3)*and Crystal Growth of Refractory Materials	000020
from Solutions in Molten Potassium Trimolybdate ((Y,Nd)	Al3(BO3), flux)*into YAl-Borate Crystals in Preparation	000551
Solidification Process (theory, eutectics, Al-	Al3Ni)* of Some Thermal Parameters on the Directional	000268
#	Americium Ditelluride (AmTe2-x*	000153
of the Para- and Antiferroelectric Phases of Deuterium	Ammonium Dihydrogen Arsenate (DADA) (NH4(1-x)D4xH2(1-x)	000100
Solution (#The Role of Dislocations in the Growth of	Ammonium Dihydrogen Phosphate Crystals from Aqueous	000219
Laser Annealing (Ge film)**Heteroepitaxy of a Deposited	Amorphous Germanium Layer on a Silicon Substrate by	000343
(vapor transport)* #Formation of	Amorphous Mn-Bi Films and Their Crystallization Process	001103
#Solar Furnace Annealing of	Amorphous Si Layers (films)*	000537
Energy Band Structure of Eulytine on Transition to the	Amorphous State (Czochralski, Bi4Si3O12)**Change in the	000294
Gratings (KCl, films, #Oriented Crystal Growth on	Amorphous Substrates Using Artificial Surface-Relief	000899
#Polytypism and	Amorphousness in Silicon Whiskers (Si, vapor transport)*	000642
for Growing Mercuric Iodide Crystals in Small Opened	Ampoules (HgI2)* #Dynamic Sublimation Technique	000376
During the Crystal Growth from the Vapor in Closed	Ampules (GeS, sublimation)* of Growth Parameters	000841
#Americium Ditelluride	AmTe2-x*	000153
Alloys (distillation)* #Preparation,	Analysis and Properties of High Purity Zn(1-x)HgxTe	000792
theory)* #Morphological Stability	Analysis in Chemical Vapour Deposition Processes. I (	001010
theory)* #Morphological Stability	Analysis in Chemical Vapour Deposition Processes. II (	000435
Sulphide Crystals with Large Temperature Gradients (#	Analysis of the Static Method of the Growth of Cadmium	000289
an FCC Crystal Formation (theory)* #Geometrical	Analysis of Crystallization of the Soft-Core Model in	000965
Etching Technique* #	Analysis of Dislocations in Gd3Ga5O12 by a Repeated	000943
Systems (Si-H-Cl, Ga-As-H-Cl) (GaAs, #Theoretical	Analysis of Equilibrium Adsorption Layers in CVD	000184
#Morphological Stability	Analysis of Growth from the Vapour (theory)*	001009
vapor transport)* #Growth and X-Ray	Analysis of Single Crystals of EuSb4S7 and EuSb2Se4 (	000022
Growth (Si, kinetics, theory)* #Thermal	Analysis of Solidification in Web-Dendritic Ribbon	000369
GaxIn1-xAs (theory)* #A Simple	Analysis of Vapor Phase Growth - Citing an Instance of	000673
Silicon Ribbon Surfaces (SiC)* #	Analysis of Vapor-Deposited Silicon Carbide Films on	001099
#Basic Consideration for an	Analytic Theory of Two-Component Crystal Growth*	000763
theory)* #On the	Analytic Treatment of the Growth of Rough Interfaces (	000764
Installation for Preparation of Single Crystals by the #	Analytical Study of the Thermal Transfer Process in an	000057
Systems I. Basic Features and Equilibrium Properties (#	Analytical Theory of Crystal Growth of Two-Component	000765
with HBr and HCl* #Growth of	Anatase (TiO2) Crystals by Chemical Transport Reactions	000431
Growth of Gallium and #Measurements of the Contact	Angle Between Melt and Crystal During Czochralski	001071
Si, Czochralski)* #Impurity Effects on Low-	Angle Boundary Formation in Silicon Single Crystals (	000218
of the Hydrothermal Growth of Synthetic 11	Angstroms Tobermorite (Ca6Si6O18.4H2O) Crystals*	000366
vapor transport)* #Growth of Phthalic	Anhydride in a Closed Crystal-Vapour System (organics,	000327
Crystals, Grown from Aqueous TGS Solutions Containing	Aniline* #The Physical Properties of TGS Single	000266
#Structure of New Rhenium Apatites Containing Large	Anions: Ba10(ReO5)6Br2 and Ba10(ReO5)6I2 (vapor	000080
#Linear Expansion	Anisotropy of CdGa2Se4 (melt)*	000476
Single Crystals (TiO2)* #	Anisotropy of Ion Transport in Stoichiometric Rutile	000778
Surfaces (NaCl, theory)* #	Anisotropy of Round Evaporation Spirals on Rocksalt	000530
2H-NbS2 (theory)* #	Anisotropy of Upper Critical Field in Superconducting	000730
Ruthenium-Nickel #Magnetization, Magnetocrystalline	Anisotropy, Magnetostriction and Elastic Constants of a	000631
of Large Spherical Iron Single Crystals (Fe, strain-	anneal)* #The Preparation	000573
Germanium Layer on a Silicon Substrate by Laser	Annealing (Ge film)* of a Deposited Amorphous	000343
Properties of Single #Effect of High-Temperature	Annealing in Cadmium on the Electrical Transport	000612
#Solar Furnace	Annealing of Amorphous Si Layers (films)*	000537
#Positron	Annihilation in ReO3 (vapor transport)*	000015
Dihydrogen Arsenate (#Crystal Growth of the Para- and	Antiferroelectric Phases of Deuterium Ammonium	000100
#Specific Heat of a Quasi-Two-Dimensional	Antiferromagnet Ni(OH)2, hydrothermal)*	000273
#Flux Growth of Single Magnetic Sublattice	Antiferromagnetic Garnets Ca3Fe2Ge3O12 and Ca3Mn2Ge3O12*	000238
* #Magnetic and Neutron Scattering Experiments on the	Antiferromagnetic Layer- Type Compounds K2Mn1-xMxF4 (M	000106
#Crystal Structure and	Antiferromagnetism of EuSb2 (melt)*	000401
Within the Facet of Czochralski Grown Indium	Antimonide (InSb)* #The Strain Configuration	000637
#Horizontal Zone-Melting of Indium	Antimonide in Vacuum. I. Basic Technique (InSb)*	000712
#Epitaxial Thin Iron Films on	Antimony (Fe, vapor transport)*	000875
#Growth of	Antimony on Bismuth (Sb film, epitaxy)*	000501
Surfaces of Gold and Silver (Sb)* #Epitaxial Growth of	Antimony Electrodeposits on Oriented Single Crystal	000540
Epitaxy by Simultaneous Arsenic Ion Implantation and by	Antimony Evaporation (Si, film)* Silicon Molecular Beam	000734
Bridgman)* #Preparation and Properties of Two Indium	Antimony Selenides (In2Se3:Sb, In2-xSbxSe3, InSbSe3,	000911
#Crystal Growth of	Antimony Sulphur Iodide (SbSi, flux)*	000109
Chemistry, Crystal-Growth, and Phase-Equilibria of	Apatites (review)* #Crystal-	000813
Ba10(ReO5)6I2 (vapor #Structure of New Rhenium	Apatites Containing Large Anions: Ba10(ReO5)6Br2 and	000080
Crystals (H2O, theory)* #	Appearance of Non-Singular Surfaces on Vapour-Grown Ice	000689
Structure of a New Mixed Oxide "FeV3O8" (FexV1-xO2 - x	approx. = 0.25) (melt, slow cooling)* and Crystal	000661
#Solubility of KLa(MoO4)2 in	Aqueous K2MoO4 Solutions in Hydrothermal Conditions*	000475
Growth of Ammonium Dihydrogen Phosphate Crystals from	Aqueous Solution (ADP, NH4H2PO4)*of Dislocations in the	000219
#Linear Growth Rate of Calcite in	Aqueous Solution (CaCO3, kinetics, theory)*	000627
#The Growth Kinetics of Urea Monocrystals from	Aqueous Solution (CO(NH2)2, theory)*	000392
Crystallization of Potassium Bromate from Its Quiescent	Aqueous Solution by Differential Scanning Calorimeter.	000367
#The Affinity of Crystal Growth and Dissolution in	Aqueous Solution with Special Reference to Calcium	000193
KMg3AlSi3O10(OH)2)* #Solubility of Phlogopite in Basic	Aqueous Solutions Under Hydrothermal Conditions (mica,	000070
in the #Crystallization of Sodium Perborate from	Aqueous Solutions. I. Nucleation Rates in Pure Solution	000257
Faces in Pure #Crystallization of Sodium Perborate from	Aqueous Solutions. II. Growth Kinetics of Different	000258
Physical Properties of TGS Single Crystals, Grown from	Aqueous TGS Solutions Containing Aniline* #The	000266
(Fe2SiO4) Single Crystals by the Floating Zone Method (	arc image)* #Growth of Fayalite	000953
#Growth of Oriented Bicrystals of Nickel Oxide (NiO,	arc image, float zone, Verneuil)*	000239
#Single-Crystal Elastic Constants of Al2Cu (Bridgman,	arc melting)*	000274
Dioxide Crystals Using Tellurium Tetrachloride and	Argon (UO2, vapor transport)* #Growth of Uranium	000280
in Zinc Selenide Crystals Grown from the Melt Under	Argon Pressure (ZnSe, Bridgman)* #Formation of Defects	000477
of Silane Compounds (SiC, vapor transport, De Boer-Van	Arkel, doped)* and Doped Silicon Carbide by Pyrolysis	001043
Carbide by a Modified Hot-Filament Method (de Boer, Van	Arkel, vapor transport, ZrC)* Growth of Zirconium	000927
Pulling from a Melt on a Pedestal. III. Characteristic	Arrangement of Dislocations (Nb)* Single Crystals by	000684
Phases of Deuterium Ammonium Dihydrogen	Arsenate (DADA) (NH4(1-x)D4xH2(1-x)D2xAsO4, solution)*	000100
and Characterization by X-ray Topography of Rare Earth	Arsenates (REAsO4, slow cooling)* #Flux Growth	000900

PERMUTED TITLE INDEX

in Sodium Tungstate Melts by Continuous Cooling (	BaWO4, flux, kinetics, slow cooling)* from Solutions	000812
and Iron Group Complex Oxides (GdVO3, Gd4Ba6(BO3)9, (	Ba0.93Gd0.07)TiO3, Nd4Ba6(BO3)9, Pr4Ba6(BO3)9,	001063
of New Rhenium Apatites Containing Large Anions:	Ba10(ReO5)6Br2 and Ba10(ReO5)6I2 (vapor transport)*	000080
Apatites Containing Large Anions: Ba10(ReO5)6Br2 and	Ba10(ReO5)6I2 (vapor transport)* of New Rhenium	000080
#Crystal Structure of	Ba2Fe6O11 (flux)*	000120
Related Properties in the Fresnoite Pseudobinary System	Ba2TiGe2O8-Ba2TiSi2O8 (Czochralski)*Pyroelectricity and	000836
in the Fresnoite Pseudobinary System Ba2TiGe2O8-	Ba2TiSi2O8 (Czochralski)* and Related Properties	000836
#The Crystal Structure of	Ba2V2O7 (melt)*	000379
in the System BaO-GeO2-Na2O-H2O (barium germanates,	Ba3Ge9O21.H2O, BaGeO3.H2O, Na4Ge9O20)* Crystallization	000592
#Elastic Constants and Stability of	bcc In-Tl Alloys (Bridgman)*	000581
Metal Particles Formed by Gas-Evaporation Technique:	bcc Metals (V, Fe, Cr, Mo and W)* and Habit of Fine	000821
#Cold Working Nb3Al in the	bcc Structure and then Converting to the A-15 Structure*	001067
#Single Crystal Growth of Alexandrite in a V2O5 Flux (	BeAl2O4)*	001005
Doping Structure in GaAs (GaAs, Ga1-xAlxAs, molecular	beam epitaxy, review)* #Periodic	000248
Growth of Autoepitaxial GaAs Films by MBE (molecular	beam epitaxy, theory)* #Temperature Range for	000686
Method for Growing V3Si Single Crystals (electron	beam melting)* #A Simple	000162
and Rotation Driven Flows (kinetics, theory, electron	beam melting* Floating Zone Melting. II. Combined Free	000174
by High-Dose Si+-Ion Implantation of Pd (electron	beam melting, evaporation)* #Silicide Formation	000177
and Synthesis of Nb-H Interstitial Alloys (electron	beam melting, float zone melting, review)* #Properties	001072
in Vacuum onto Silicon (vapor transport, electron	beam melting, kinetics, theory)*Films of CdS Evaporated	000073
Hafnium, and Tungsten Alloys (Nb-Zr, Nb-Hf, electron	beam melting, zone melting)* of Niobium-Zirconium,	000307
Alkali Halides (KCl, LiF)* #Molecular (	beam) Processes of Condensation and Evaporation of	000628
Techniques (vapor #New Developments in Ionized-Cluster	Beam and Reactive Ionized-Cluster Beam Deposition	000944
High Vacuum (film)* #Molecular	Beam and Solid-Phase Epitaxies of Silicon Under Ultra-	000883
#Nitrogen Doping into GaAs1-xPx Using Ionized	Beam in Molecular Beam Epitaxy*	000617
#Ionized-Cluster	Beam Deposition and Epitaxy (review)*	000945
in Ionized-Cluster Beam and Reactive Ionized-Cluster	Beam Deposition Techniques (vapor transport)*	000944
Substrates* #Ionized-Cluster	Beam Epitaxial Growth of GaP Films on GaP and Si	000651
Dopant Incorporation and Diffusion During Molecular	Beam Epitaxial Growth of IV-VI Semiconductors (PbTe,	000898
#A High Temperature High Purity Source for Metal	Beam Epitaxy (Ag, film)*	000287
Aluminum Schottky-Barrier Diodes Prepared by Molecular-	Beam Epitaxy (MBE) on GaAs (Al, films)* #Single-Crystal-	000190
and by #n-Type Doping Techniques in Silicon Molecular	Beam Epitaxy by Simultaneous Arsenic Ion Implantation	000734
#Growth Temperature Dependence in Molecular	Beam Epitaxy of Gallium Arsenide (GaAs)*	000666
#Molecular	Beam Epitaxy of GaSb and GaSbAs1-x (films)*	001100
Superlattices* #Molecular	Beam Epitaxy of Ge and Ga1-xAlxAs Ultra Thin Film	000761
#Molecular	Beam Epitaxy of InxGa1-xSb (x = 0 to 1) (GaSb)*	001101
#Molecular	Beam Epitaxy of ZnSexTe1-x(film)*	001102
#Design Considerations for Molecular	Beam Epitaxy Systems (review)*	000576
in (GaAs)n-(AlAs)m Superlattices Deposited by Molecular	Beam Epitaxy. I. Growth on Singular (100) GaAs	000762
Doping into GaAs1-xPx Using Ionized Beam in Molecular	Beam Epitaxy* #Nitrogen	000617
#Growth of Cobalt Single Crystals by the Electron	Beam Floating-Zone Method (Co)*	000381
of Silicon-Germanium Solid Solutions by Electron-	Beam Zone Melting Without a Crucible (Si-Ge)* Crystals	000820
#Shape of the Crystallization Front During Electron-	Beam Zone Refining of beta-B Without a Crucible*	000968
of Silicon Epitaxial Layer Growth from Ion-Molecular	Beams (Si)* #The Mechanism	000019
#Thin Film Deposition From	Beams of Ionized Atoms and Clusters (Ag on Si, epitaxy)*	000516
Ge by the Czochralski Method* #	Behavior of Be as an Impurity During Crystallization of	000346
Czochralski Method* #	Behavior of Boron in Ge Single Crystals Grown by the	000345
#The Crystal Structure and Twinning	Behavior of Ferric Molybdate, Fe2(MoO4)3 (hydrothermal)*	000178
to Czochralski Growth. Part I. Growth and Segregation	Behavior of Ga-Doped Ge (theory)* of a Heat Pipe	000604
vapor transport, zone melting, doped, review)*Materials	Behaviour in Low Gravity Environment (melt, solution,	001070
of Some Niobium Oxides with TeCl4 (II) The Transport	Behaviour of the Equilibrium NbO2.417 to NbO2.5*	000798
#A Complexity in the Solidification	Behaviour of Molten Y3Al5O12 (Czochralski)*	000202
#Correlation of Diffusion	Behaviour with the Entropy of Fusion (theory)*	000980
Silicon Film Deposition and by Film Doping Using #The	Bending of Silicon Wafers by Thin Polycrystalline	000985
the Kinetics of Crystal Growth from the Melt (organics,	benzophenone)* #A New Technique for Determining	000755
Diffusion Method. II. Diffusion- #Two-Phase (alpha/	beta)-Brass Bicrystals Produced by a Solid-Solid	000371
Diffusion Method. I. Morphology and #Two-Phase (alpha/	beta)-Brass Bicrystals Produced by a Solid-Solid	000372
Crystallographic Relationships Between and alpha- and	beta-phases (Cu-Zn, theory)* Method. I. Morphology and	000372
Diffuse Scattering from Alkali, Silver, and Europium	beta-alumina (NaAl11O17, KAl11O17, RbAl11O17,	000621
Front During Electron-Beam Zone Refining of	beta-B Without a Crucible**#Shape of the Crystallization	000968
#Crystal Structure of	beta-Eu2(II)SiO4 (melt)*	000597
#Crystal Structure of	beta-KEr3F10 (melt, slow cooling)*	000021
#Magnetic Properties of	beta-Phase Iron-Germanium (FexGe, slow cooling)*	000845
#Bridgman and Czochralski Growth of	beta-Phase NaxV2O5*	000907
System (vapor transport, #Heteroepitaxial Growth of	beta-SiC on Silicon Substrate Using SiCl4-C3H8-H2	000616
#The Growth of	beta-Si3N4 Single Crystals*	000356
#Ionic Conductivity in N3O+	Beta Alumina (solution)*	000286
B)* #Growth of Single Crystals of	Beta Rhombohedral Boron by a Modified Verneuil Method (	000213
#Some Peculiarities in the Growth and Properties of	Beta-Boron Crystals (B, float zone melting)*	000967
84MgO.5Al2O3, theory)* #Hydronium	Beta" Alumina: A Fast Proton Conductor (0.84Na2O.0.	000285
and Concentration in the Region of Solidification of Pb-	Bi Alloys Rich in Single-Phase Pb in the Course of	000846
transport)* #Formation of Amorphous Mn-	Bi Films and Their Crystallization Process (vapor	001103
Mechanism of Vapor Grown Cd Crystals as Affected by	Bi Impurity* #Morphology and Growth	000375
#Magneto-Seebeck Effect of Pb-Doped	Bi-Sb (Bridgman)*	000034
#Crystallization of	Bi-Sb Solid Solutions in Capillaries (zone melting)*	001035
Bismuth Single Crystals with High Resistivity Ratios (	Bi, Czochralski)* #The Growth of Thin	000887
of the Mismatch of Crystal Lattice Parameters((Y,	Bi, Sm, Ca)3(Fe, Ge)5 O12, flux)* Layers as Functions	000251
(001) Tilt Boundaries Produced by Epitaxial Growth on	Bicrystalline Substrates of NaCl (Au, Ag, vapor	000207
Verneuil)* #Growth of Oriented	Bicrystals of Nickel Oxide (NiO, arc image, float zone,	000239
II. Diffusion-Induced #Two-Phase (alpha/beta)-Brass	Bicrystals Produced by a Solid-Solid Diffusion Method.	000371
I. Morphology and #Two-Phase (alpha/beta)-Brass	Bicrystals Produced by a Solid-Solid Diffusion Method.	000372
of the Quaternary Chalcogenides Cu2(I)B(II)C(IV)X4 (	BII=Zn, Mn, Fe, Co - CIV = Si, Ge, Sn - X = S,Se) (	000357
#Region of Metastability During Normal Growth in a	Binary Alloy (kinetics, theory)*	000969
Segregation During Unidirectional Solidification of a	Binary Alloy with a Curved Solid-Liquid Interface (Ge:	000206
Stability of a Solid Particle Growing from a	Binary Alloy Melt (dendrites, theory)* #Morphological	000990
- II. Single Filament, Unconstrained Growth from a	Binary Alloy Melt (Sn-Pb, theory)* Crystallization	000993
Thermodynamics of the Cellular Growth of a Dilute	Binary Alloy Single Crystal (Fe-Ni, eutectics, theory,	000113
of Cellular Structures of Monocrystals of Dilute	Binary Metallic Alloys (Cu - 0.5%Si, Cu1-xSbx - x = 0.	000210

PERMUTED TITLE INDEX

Quantities and Vacancy and Diffusion Characteristics in
#On the Theory of Normal Growth of Crystals in
Compared (CoGa3, NiGa4, flux)*   #The
Grown by the Hydride #Crystal Growth and Properties of

#The Preparation and Crystal Structures of
#The Preparation and Crystal Structures of BiReO4 and
#Growth of Antimony on
Lithium Niobate Single-Crystal Films on Potassium
Lithium Niobate Single-Crystal Films on Potassium
Bi, Czochralski)*   #The Growth of Thin

#The Czochralski Growth of
#Bismuth Tellurides:
#Preparation and Basic Physical Properties of
#Crystal Growth and Structure of
Growth of Bismuth-Germanium Oxide Single Crystals (
Germanosillenite Crystals on the Growth Conditions (
#Growth of
#Polymorphism in
Stability, and Preliminary Crystallographic Study of
Temperature (slow cooling)*   #Flux Growth of
#Crystal Growth of
#Growth, Spectroscopy, and Stimulated Emission of Cubic
on Transition to the Amorphous State (Czochralski,
#Bismuth Tellurides: BiTe and
#Asymmetric 180 degree
#Electron Microscopic Characterization of Cubic
of Cubic Boron Nitride in Systems Containing Hydrogen (
Pyrolysis of Silane Compounds (SiC, vapor transport, De
Zirconium Carbide by a Modified Hot-Filament Method (de
of Some II-VI Compound Films Evaporated Onto Electron-
A New Cubic KSbO3 Derivative Oxide with No Metal-Metal
System: The Chemical Vapour Transport of Ni-Cl
Layers of Nickel-Chlorine Boracite on Chromium-Chlorine
Phase Synthesis of Epitaxial Layers of Nickel-Chlorine
#Preparation and Physical Properties of
Potassium   #Neodymium Incorporation into YAl-
#Flux Growth of Lanthanum
#Crystallization of LiNbO3 from Solution in
#Boron and
#The Preparation of Single Crystals of the Rare Earth
#
#Growth of Single Crystals of Beta Rhombohedral
Method*   #Behavior of
Some Peculiarities in the Growth and Properties of Beta-
Silicon Film Deposition and by Film Doping Using
#Si Contamination in Epitaxial
vapor transport)*   #Crystal Growth of
#Synthesis of Single Crystals of Cubic
Bicrystalline   #Characterization of (001) Tilt
of Polycrystalline Snow Crystals with a Specific Grain
#Mass Transfer at the Surface of a Crystal Near to its
Czochralski)*   #Impurity Effects on Low-Angle
the Bruton, Prim and Slichter Equation as a Diffusion
Solutions in Molten Potassium Trimolybdate ((Y,Nd)Al3(
#Epitaxial Layers of the Laser Material Nd(Ga,Cr)3(
Rare Earth and Iron Group Complex Oxides (GdVO3, Gd4Ba6(
Gd4Ba6 (BO3)9, (Ba0.93Gd0.07)TiO3, Nd4Ba6(BO3)9, Pr4Ba6(
Oxides (GdVO3, Gd4Ba6(BO3)9, (Ba0.93Gd0.07)TiO3, Nd4Ba6(
#Si Contamination in Epitaxial Boron Monophosphide (
Elements, III-V Compounds and Alloys (Ge, Si, GaAs,
of Boron Monophosphide Using a B2H6-PH3-H2 System (
Divalent Rare Earths (M = Sm, Eu, Tm, and Yb - X = Cl,
of Niobium Chalcogenide Halides, NbX2Y2(X=S, Se - Y=Cl,
Families with M = Cd, Cu, Fe, Mn or Pd and X = Cl or
Method. II. Diffusion-Induced   #Two-Phase (alpha/beta)-
Method. I. Morphology and   #Two-Phase (alpha/beta)-
#
Growth of (MnxZn1-x)FeO4 by Liquid Phase Epitaxy on
Volatile Components (#An Improved System for the
Thermic Screen Translation (TST) Technique: A Modified
Mixed Crystals with a Large Segregation Tendency by the
#Magnetic Spinel Single Crystals by
#Single Crystal Growth of Stannous Selenide (SnSe,
#Magneto-Seebeck Effect of Pb-Doped Bi-Sb
#Growth and Cleavage of Se-Te Alloy Crystals (
#Optical and Electrical Properties of CdGeAs2 (
#Crystal Growth of Mn15Si26 (vapor transport,
#Elastic Constants and Stability of bcc In-Tl Alloys (
#Some Optical Properties of GeS2 Single Crystals (
#Edge Emission in Melt-Grown ZnSexS1-x (
#Phase Diagram of Pseudobinary System CdTe-ZnSe (
#Low Temperature Thermal Properties of PbI2 (
#Crystal Growth of Sodium Sulphide, Na2S (hygroscopic,
#Crystal Growth and Segregation Under Zero Gravity: Ge (
Some Electrical Properties of GeBi2Te4 Single Crystals (

Binary Metallic Solid Solutions (theory)* Thermodynamic   000090
Binary Systems (kinetics)*   001110
Binary Systems Cobalt-Gallium and Nickel-Gallium   000290
Binary, Ternary and Quaternary (In,Ga)(As,P) Alloys   000729
Birefringence in CdSiP2 (solution)*   000425
BiReO4 and BiRe2O6 (vapor transport)*   000781
BiRe2O6 (vapor transport)*   000781
Bismuth (Sb film, epitaxy)*   000501
Bismuth Niobate Single Crystals by the EGM Technique (000011
Bismuth Niobate Single Crystals by RF Sputtering (KxLi1-   000009
Bismuth Single Crystals with High Resistivity Ratios (000887
Bismuth Tellurides: BiTe and Bi4Te3 (melt)*   001096
Bismuth-Germanium Oxide Single Crystals (Bi12GeO20)*   000771
BiTe and Bi4Te3 (melt)*   001096
BiTeI Single Crystals (Bridgman)*   000240
BiVO4 (Czochralski)*   000891
Bi12GeO20)*   #The Czochralski   000771
Bi12GeO20, Czochralski)* Optical Homogeneity of Single   000903
Bi2(SeO4)3 Single Crystals in Silica Gels*   000121
Bi2Sn2O7 (flux)*   000867
Bi2Sn2O7 (flux)*   #Preparation,   001034
Bi2WO6 Single Crystal Below the Transformation   000665
Bi3B5O12 From its Stoichiometric Melt (kinetics)*   000794
Bi4Ge3O12 Crystals Doped with Dy3+, Ho3+, Er3+, Tm3+,   000458
Bi4Si3O12)*   in the Energy Band Structure of Eulytine   000294
Bi4Te3 (melt)*   001096
Bloch Walls in Fe Single Crystal Films (theory)*   000936
BN Prepared by Water Solvent (solution)*   000934
BN)*   #Synthesis of Single Crystals   001032
Boer-Van Arkel, doped)*   and Doped Silicon Carbide by   001043
Boer, Van Arkel, vapor transport, ZrC)*   Growth of   000927
Bombarded NaCl Substrates (CdS, CdSe, CdTe, ZnSe)*   000878
Bonding (flux)*   #The Crystal Structure of La3Ru3O11:   000209
Boracite (no materials grown - Ni3B7O13Cl theory)*   000236
Boracite (Ni3B7O13Cl, vapor transport)*   of Epitaxial   000837
Boracite on Chromium-Chlorine Boracite (Ni3B7O13Cl,   000837
Boracite Crystals (Chemical transport, M3B7O13X)*   000151
Borate Crystals in Preparation from Solutions in Molten   000551
Borate, LaBO3*   000508
Borate, Vanadate, and Tungstate Melts (flux)*   001047
Borides (review)*   001112
Borides by the Solution Method and a Study of Their   000359
Boron and Borides (review)*   001112
Boron by a Modified Verneuil Method (B)*   000213
Boron in Ge Single Crystals Grown by the Czochralski   000345
Boron Crystals (B, float zone melting)*   # 000967
Boron Diffusion (Si)*   Wafers by Thin Polycrystalline   000985
Boron Monophosphide (BP)*   000713
Boron Monophosphide Using a B2H6-PH3-H2 System (BP,   000884
Boron Nitride in Systems Containing Hydrogen (BN)*   001032
Boundaries Produced by Epitaxial Growth on   000207
Boundary (H2O)*   #On the Growth Mechanism   000314
Boundary with the Melt, and its Influence on the Shape   001046
Boundary Formation in Silicon Single Crystals (Si,   000218
Boundary Layer Thickness (effective distribution   001083
BO3), flux)*   YAl-Borate Crystals in Preparation from   000551
BO3)4 (flux)*   000578
BO3)4 For Thin Film Miniature Lasers (flux)*   000577
BO3)9, (Ba0.93Gd0.07)TiO3, Nd4Ba6(BO3)9, Pr4Ba6(BO3)9,   001063
BO3)9, Ni8Nb3O15, Ni2V2PbO8, Ni3Nb2O8, Ni3V2O8,   001063
BO3)9, Pr4Ba6(BO3)9, Ni8Nb3O15, Ni2V2PbO8, Ni3Nb2O8,   001063
BP)*   000713
BP)* of the Epitaxial Vapour Growth of Semiconductors: 000041
BP, vapor transport)*   #Crystal Growth   000884
Br, and I) (melt, SmFCl, SmFBr, SmFI, EuFCl, EuFBr,   000088
Br, I) (vapor transport)*   Structures, and Properties   000795
Br: Importance, Solubilities and Simple Growth   000039
Brass Bicrystals Produced by a Solid-Solid Diffusion   000371
Brass Bicrystals Produced by a Solid-Solid Diffusion   000372
Bridgman and Czochralski Growth of beta-Phase NaxV2O5*   000907
Bridgman Grown Zn2TiO4 Substrates (flux)*   #Thin Film   000216
Bridgman Growth of Crystals with Toxic and/or Highly   000164
Bridgman Method (ZnF2, CoF2, BaF2, MgF2, KY3F10)*by the   000541
Bridgman Method - Pseudobinary CdxHg1-xTe*   Homogeneous 000244
Bridgman Technique (review)*   000924
Bridgman)*   000014
Bridgman)*   000034
Bridgman)*   000111
Bridgman)*   000412
Bridgman)*   000495
Bridgman)*   000551
Bridgman)*   000697
Bridgman)*   000710
Bridgman)*   000725
Bridgman)*   000851
Bridgman)*   001056
Bridgman)*   # 001088
Bridgman)*   # 000975

PERMUTED TITLE INDEX

#Influence of the CaCO3/GeO2 Ratio on Y Sm Lu Ca Fe Ge Garnet Films Properties (LPE)* 000650
#Growth and Characterization of Mixed Sr: Ca Tartrates (Sr1-xCazC4H4O6.4H2O, gel)* 000747
#The Crystal Structure of Synthetic Lautarite, Ca(IO3)2 (flux) 000331
the Mismatch of Crystal Lattice Parameters((Y, Bi, Sm, Ca)3(Fe, Ge)5 O12, flux)* Garnet Layers as Functions of 000251
Epitaxy #Electrical Properties and Charge Imbalance for Ca,Ge-Substituted Garnet Films grown by Liquid Phase 000386
Properties (LPE)* #Influence of the CaCO3/GeO2 Ratio on Y Sm Lu Ca Fe Ge Garnet Films 000650
#Linear Growth Rate of Calcite in Aqueous Solution (CaCO3, kinetics, theory)* 000627
#Growth Morphology of Weddellite (CaC2O4.2XH2O, gel)* 000302
Single #Effect of High-Temperature Annealing in Cadmium on the Electrical Transport Properties of 000612
#Polytypism in Vapor Grown Crystals of Cadmium Bromide (CdBr2)* 000622
#Growth and Polytypism in Vapor-Grown Mixed Cadmium Halide Crystals (CdI2-xBrx, CdI2-xClx)* 000870
#Effect of Growth Conditions of Polytypism in Cadmium Iodide Crystals (CdI2, theory, solution, 000432
#An X-Ray Topographic Assessment of Cadmium Mercury Telluride (CdxHg1-xTe, Bridgman)* 000158
Speed on the Compositional Variations in Crystals of Cadmium Mercury Telluride (Cd1-xHgxTe, slow cooling)* 000074
Cadmium Tungstate Single Crystals (CdWO4, #The Cadmium Oxide-Tungsten Oxide Phase System and Growth of 000801
on the Sublimation Properties of Cadmium, Selenium, and Cadmium Selenide (CdSe, epitaxy, kinetics, films, 000365
and Structure of Epitaxial Cadmium Sulphide Layers on Cadmium Single Crystals (CdS, evaporation)* Synthesis 000429
#Surface Charge and Ledge Dynamics in Cadmium Sulfide (CdS, evaporation)* 000529
#Vapor Phase Transport and Stoichiometry Control of Cadmium Sulfide (CdS, sublimation)* 000648
#Vapor Phase Transport and Stoichiometry Control of Cadmium Sulfo-Selenide (Cd1+y(SxSe1-x), sublimation)* 000647
Electrical Transport Properties of Single Crystals of Cadmium Sulphide (CdS, vapor transport, theory)* on the 000612
#Analysis of the Static Method of the Growth of Cadmium Sulphide Crystals with Large Temperature 000289
CdS, #Direct Synthesis and Structure of Epitaxial Cadmium Sulphide Layers on Cadmium Single Crystals (000429
A Possible Method for the Growth of Homogeneous Mercury Cadmium Telluride Single Crystals (Hg1-xCdxTe, Bridgman)* 000292
Cadmium Oxide-Tungsten Oxide Phase System and Growth of Cadmium Tungstate Single Crystals (CdWO4, Czochralski)* 000801
and the Source Shape on the Sublimation Properties of Cadmium, Selenium, and Cadmium Selenide (CdSe, epitaxy, 000365
#Flux Growth and Characterization of Fluorite Crystals (CaF2)* 000237
Region of the Calcium Oxide-Gallium Oxide System (CaGa2O4, Czochralski)* #The Central 001105
#Crystal Quality of Gel Grown CaHPO4.2H2O Crystals* 000545
of Etch Pits on As-Grown Faces of Brushite Crystals (CaHPO4.2H2O, gel)* #Observations 000714
#Linear Growth Rate of Calcite in Aqueous Solution (CaCO3, kinetics, theory)* 000627
in Solution (review, theory, alkaline earth sulfates, calcium phosphates)* #The Growth of Crystals 000683
#Simultaneous Diffusion of Calcium and Strontium in KCl Single Crystals* 000579
Temperature Ferroelectric (Czochralski)* #Calcium Orthovanadate, Ca3(VO4)2 - A New High- 000339
Czochralski)* #The Central Region of the Calcium Oxide-Gallium Oxide System (CaGa2O4, 001105
solution, CaSO4)* #Kinetics of Growth of Calcium Sulfate Crystals at Heated Metal Surfaces (000334
in Aqueous Solution with Special Reference to Calcium Sulphate Dihydrate (CaSO4.2H2O, kinetics, 000193
Melts by Continuous #Kinetics of Crystal Growth of Calcium Tungstate from Solutions in Sodium Tungstate 000810
Its Quiescent Aqueous Solution by Differential Scanning Calorimeter. Growth Rate of Crystallite (KBrO3)* from 000367
and Phase Diagram of (Ti1-xVx)4O7: Electrical, Calorimetric, Magnetic and EPR Studies (vapor transport) 000835
#Crystallization of Bi-Sb Solid Solutions in Capillaries (zone melting)* 001035
#The Growth of CdS in Sealed Silica Capsules (vapor transport)* 000815
for Different Two-Dimensional Substrate Lattices (Capture of Migrating Adatoms by Point Capture Centres 000873
Lattices (theory) #Capture of Migrating Adatoms by Point Capture Centres for Different Two-Dimensional Substrate 000873
#Recarburization of Monocrystalline Ditungsten Carbide as Revealed by Field Electron Microscopy (W 000724
Arkel, vapor #Single Crystal Growth of Zirconium Carbide by a Modified Hot-Filament Method (de Boer, Van 000927
transport, De #Preparation of Pure and Doped Silicon Carbide by Pyrolysis of Silane Compounds (SiC, vapor 001043
TiC)* #Single Crystal Growth of Titanium Carbide from the Vapor by a Modified Hot Wire Method (000928
SiC, theory)* #Study of Silicon Carbide Epitaxial Growth Kinetics in the SiC-C System (000562
of Kinetic and Thermal Conditions of Silicon Carbide Epitaxial Layer Growth from the Vapour Phase (000942
#Analysis of Vapor-Deposited Silicon Carbide Films on Silicon Ribbon Surfaces (SiC)* 001099
Distribution in High-Temperature Furnaces for Silicon Carbide Monocrystal and Epitaxial Layer Growing (SiC, 000561
#Crystal Growth of Niobium Nitride and Niobium Carbide Nitride (NbN, NbN-NbC, NbxNyCz, zone melting)* 000192
#Growth Kinetics of Epitaxial Layers of Silicon Carbide Obtained by Sublimation in Vacuum (SiC)* 000940
#Investigation of Growth Processes of Ingots of Silicon Carbide Single Crystals (SiC, melt)* 000941
#Synthesis of Nonstoichiometric Zirconium Carbide Whiskers by Chemical Vapor Deposition (ZrC)* 000674
#Catalytic Effect of Nickel on the Growth of Zirconium Carbide Whiskers by Chemical Vapor Deposition (ZrC)* 000958
Superalloys via Use of Precipitated Phases (eutectics, carbides, review)* Microstructures and Properties of 000671
Plasma Crystallization of Diamond and Diamond-Like Carbon (C, film)* #Reactive Pulse 000906
vapor transport)* #Growth of Carbon Tetrabromide Crystals for Optical Studies (CBr4, 000263
#In-Situ Carbon-Silicon Ribbon Thermocouple (Si, equipment)* 000818
GdOHCO3, ErOHCO3, Y2O(OH)2CO3, #Synthesis of Rare-Earth Carbonates Under Hydrothermal Conditions (YOHCO3, 000169
#Fast Growth of Ordered AB Crystals: A Monte Carlo Simulation for Ionic Crystal Growth (theory)* 000182
Roughening Transition (C, theory)* #Monte Carlo Simulation of a (111) Diamond Face Around the 001020
Lennard-Jones Substance (theory)* #Monte Carlo Simulation of the Crystal-Melt Interface of a 000123
Growth (review, theory, kinetics)* #Survey of Monte Carlo Simulations of Crystal Surfaces and Crystal 001013
Semiconductors (II-VI, IV-VI, III-#Nonstoichiometry and Carrier Concentration Control in MBE of Compound 000897
Redistribution During a Modified Form of Zone Refining- Cascade Purification (metals)* #Theoretical Solute 000201
Crystal Growth for Metallic One-Component Systems. Two Cases of Modelling the Melt Growth Interface Motion* 000059
with Special Reference to Calcium Sulphate Dihydrate (CaSO4.2H2O, kinetics, theory)* in Aqueous Solution 000193
Effects and the Growth Kinetics of Ionic Crystals (CaSO4.2H2O, solution, theory)* #Solvent 000134
Sulfate Crystals at Heated Metal Surfaces, CaSO4)* #Kinetics of Growth of Calcium 000334
#A Study of Casting in the CdxHg1-xTe System* 000075
Oxides on the Growth of TiC Whiskers by Chemical Vapor # Catalytic effects of Various Metals and Refractory 000959
Carbide Whiskers by Chemical Vapor Deposition (ZrC)* # Catalytic Effect of Nickel on the Growth of Zirconium 000958
Films Produced by Reactive Evaporation (AlN)* # Cathodoluminescence of Impurity-Doped Aluminum Nitride 001104
New Structural Family M2"M'Ta5O15. Crystal Structure of CaTlTa5O15 (flux)* #A 000317
in Sodium Tungstate Melts by Continuous Cooling (CaWO4, flux) #Growth of Calcium Tungstate from Solutions 000810
and Structure of a New Pyrochlore Containing Rhenium: Ca1-xRe2O6(OH)2x (x = 0.3)* #High Pressure Synthesis 000104
Ca12Al14O33: A New Piezoelectric Material (Czochralski)* 001076
Growth of Akermanite (Ca2MgSi2O7) and Gehlenite (Ca2Al2SiO7) by the Floating Zone Method* #Single Crystal 000404
Zone Method* #Single Crystal Growth of Akermanite (Ca2MgSi2O7) and Gehlenite (Ca2Al2SiO7) by the Floating 000404
of Mixed Bromides of Rare Earths, Sodium, and Cesium (Ca2NaNdBr6, Cs2NaCeBr6, Cs2NaGdBr6, Cs2NaYBr6, Bridgman- 000626
Czochralski)* #Calcium Orthovanadate, Ca3(VO4)2 - A New High-Temperature Ferroelectric (000339
of Single Magnetic Sublattice Antiferromagnetic Garnets Ca3Fe2Ge3O12 and Ca3Mn2Ge3O12* #Flux Growth 000238
of V2O5 (Y3Fe5O12, Ho3Fe5Cl2, Dy3Fe5O12, Er3Fe5O12, Ca3Fe2Ge3O12, HoFeO3, DyFeO3, ErFeO3)* with Addition 000565
Sublattice Antiferromagnetic Garnets Ca3Fe2Ge3O12 and Ca3Mn2Ge3O12* #Flux Growth of Single Magnetic 000238
Single Crystals by Hydrothermal Recrystallization (Ca5(PO4)3OH)* #Preparation of Pure Hydroxyapatite 000040

PERMUTED TITLE INDEX

Growth of Synthetic 11 Angstroms Tobermorite (	Ca6Si6018.4H2O) Crystals* of the Hydrothermal	000366
of Carbon Tetrabromide Crystals for Optical Studies (	CBr4, vapor transport)* #Growth	000263
#Morphology and Growth Mechanism of Vapor Grown	Cd Crystals as Affected by Bi Impurity*	000375
Morphology of Spherical Crystals of omega-Phase in the	Cd-Hg System* #Growth and	000512
(CnH2n+1NH3)2MX4 and NH3(CH2)mNH3MX4 Families with M =	Cd, Cu, Fe, Mn or Pd and X = Cl or Br: Importance,	000039
#The Growth of	CdAs2 and ZnAs2 Single Crystals from the Vapor Phase*	000598
#Polytypism in Vapor Grown Crystals of Cadmium Bromide (	CdBr2)*	000622
#Liquid Phase Epitaxial Growth of CdTe in the CdTe-	CdCl2 System (flux)*	000828
of Single Crystals of the Magnetic Semiconductor	CdCr2Se4 (vapor transport, doped)* #Preparation	000625
#Magnetic Properties of	CdCr2Se4-ZnCr2Se4 Single Crystals (melt)*	000586
#Preparation and Certain Electrical Properties of	CdCr2S4 Single Crystals (vapor transport)*	000780
#Crystal Structure of the Spinel	CdErS4 (slow cooling)*	000983
#Linear Expansion Anisotropy of	CdGa2Se4 (melt)*	000476
Studies, Crystal Growth, and Optical Properties of	CdGe(As1-xPx)2 and AgGa(Se1-xSx)2 Solid Solutions (	000635
#Optical and Electrical Properties of	CdGeAs2 (Bridgman)*	000412
#Synthesis and Crystal Growth of	CdGeP2 (melt)*	001040
#Synthesis and Crystal Growth of	CdGeP2 (vapor transport)*	001041
#Raman and Infrared Spectra of	CdIn2S4 and ZnIn2S4 (vapor transport)*	001004
in Vapor-Grown Mixed Cadmium Halide Crystals (	CdI2-xBrx, CdI2-xClx)* #Growth and Polytypism	000870
Vapor-Grown Mixed Cadmium Halide Crystals (CdI2-xBrx,	CdI2-xClx)* #Growth and Polytypism in	000870
#Some Aspects of Polytypism in Crystals (SiC, ZnS,	CdI2, mica, theory, review)*	000069
Conditions of Polytypism in Cadmium Iodide Crystals (	CdI2, theory, solution, evaporation)* #Effect of Growth	000432
#Growth of	CdP4 Single Crystals (evaporation, vapor transport)*	000599
Solvent* #Solution Growth of ZnS, ZnSe,	CdS and Their Mixed Compounds Using Tellurium as a	001065
#Seeded Growth of Large Single Crystals of	CdS from the Vapor Phase*	000242
#The Growth of	CdS in Sealed Silica Capsules (vapor transport)*	000815
#Epitaxial Growth of	CdS on CuInSe2*	000556
electron beam melting, kinetics, #Growth of Films of	CdS Evaporated in Vacuum onto Silicon (vapor transport,	000073
#Improving the Electrical Characteristics of	CdS Single Crystals Grown from the Melt*	000588
#Surface Charge and Ledge Dynamics in Cadmium Sulfide (	CdS, evaporation)*	000529
Cadmium Sulphide Layers on Cadmium Single Crystals (	CdS, evaporation)* Synthesis and Structure of Epitaxial	000429
Sulphide Crystals with Large Temperature Gradients (	CdS, kinetics, theory, vapor transport)* of Cadmium	000289
Transport and Stoichiometry Control of Cadmium Sulfide (	CdS, sublimation)* #Vapor Phase	000648
Properties of Single Crystals of Cadmium Sulphide (	CdS, vapor transport, theory)* the Electrical Transport	000612
Evaporated Onto Electron-Bombarded NaCl Substrates (	CdS, CdSe, CdTe, ZnSe)* of Some II-VI Compound Films	000878
in II-VI Compounds Along their Polar Axis (ZnO, CdSe,	CdS, ZnS, kinetics)* #Crystal Growth and Sublimation	000428
#Preparation and Properties of Monocrystalline Films of	CdSb (melt)*	000776
Properties of Cadmium, Selenium, and Cadmium Selenide (	CdSe, epitaxy, kinetics, films, theory)*the Sublimation	000365
in II-VI Compounds Along their Polar Axis (ZnO,	CdSe, CdS, ZnS, kinetics)* Growth and Sublimation	000428
Onto Electron-Bombarded NaCl Substrates (CdS,	CdSe, CdTe, ZnSe)* Some II-VI Compound Films Evaporated	000878
#Birefringence in	CdSiP2 (solution)*	000425
#Crystal Growth of	CdTe for X-Ray Detectors (review)*	001053
#Liquid Phase Epitaxial Growth of	CdTe in the CdTe-CdCl2 System (flux)*	000828
Growth and X-Ray Topographic Characterization of	CdTe Single Crystals* #Sublimation	000148
#	CdTe Thin Films Grown by Hot Wall Epitaxy*	000569
Method and Characteristic of Undoped High-Resistivity	CdTe* #Growth by Travelling Heater	000939
#Liquid Phase Epitaxial Growth of CdTe in the	CdTe-CdCl2 System (flux)*	000828
#Phase Diagram of Pseudobinary System	CdTe-ZnSe (Bridgman)*	000725
for Zone Crystallization Experiments in Spacelab (	CdTe, zone melting, sublimation)* Mirror Furnace	000276
Onto Electron-Bombarded NaCl Substrates (CdS, CdSe,	CdTe, ZnSe)* of Some II-VI Compound Films Evaporated	000878
cooling)* #Solid State Reactions to	CdTeMoO6 and Its Structural Characterization (slow	000297
System and Growth of Cadmium Tungstate Single Crystals (	CdWO4, Czochralski)* Cadmium Oxide-Tungsten Oxide Phase	000801
#Heteroepitaxial Homogeneous	CdxHg1-xTe Films (vapor transport)*	000707
#A Study of Casting in the	CdxHg1-xTe System*	000075
Tendency by the Bridgman Method - Pseudobinary	CdxHg1-xTe Mixed Crystals with a Large Segregation	000244
Topographic Assessment of Cadmium Mercury Telluride (	CdxHg1-xTe, Bridgman)* #An X-Ray	000158
and Stoichiometry Control of Cadmium Sulfo-Selenide (	Cd1+y(SxSe1-x), sublimation)* #Vapor Phase Transport	000647
#Photoferromagnetic Effect and Photoconductivity of	Cd1-xHgxCr2Se4 Single Crystals (vapor transport, melt)*	001037
Variations in Crystals of Cadmium Mercury Telluride (	Cd1-xHgxTe, slow cooling)* Speed on the Compositional	000074
Close-Spaced #Epitaxial Growth of Single Crystal	Cd1-xZnxS Layers on (111) GaAs Substrates Using the	000305
#Cerium Trihydroxide	Ce(OH)3, gel)*	000664
Levitation Zone Melting Method (DyFe2, TbFe2, Nd, Pr,	Ce, cold crucible)* Crystal Growth by the Horizontal	000620
EuB6, CeB6, #Single-Crystal Growth of Mixed (La, Eu, Y,	Ce, Ba, Cs) Hexaborides for Thermionic Emission (LaB6,	000727
Temperature Thermal and Magnetic Properties of CeP and	CeAs (sublimation)* #Low	000400
Cs) Hexaborides for Thermionic Emission (LaB6, EuB6,	CeB6, BaB6, (La, Eu)B6, La1-xEuB6, (La, Y)B6, La, Ba)B6, (	000727
Oxysulfides of Chromium III and Rare Earths (LaCrOS2,	CeCrOS2, PrCrOS2, NdCrOS2, SmCrOS2)* #Mixed	001048
#Performance of a New High-Intensity Silicon Solar	Cell (Si, float zone melting)*	000303
Grown RTR Silicon Sheet and Their Effects on Solar	Cell Processing (zone melting, Si)* of Rapidly-	000546
Crystals* #Unit	Cell Step Structures on Vapour-Grown Cr2S3 Single	000676
Single Crystals. I* #Growth of Triple Unit	Cell Steps on the Surface of Vapour Grown FeCr2S4	000418
#New Methods of Making Silicon Solar	Cells (review)*	000886
Efficiency p+-n+ Back-Surface-Field Silicon Solar	Cells (Si)* #High-	000298
Growth of Silicon for Integrated Circuits and Solar	Cells (Si, review)* #Problems in the Melt- and Vapor	000117
#High Efficiency GaAs Thin Film Solar	Cells by Peeled Film Technology (epitaxy)*	000497
Czochralski, zone melting, vapor transport)* #Solar	Cells from Zone-Refined Metallurgical Silicon (Si,	000195
(Fe-Ni, eutectics, #Irreversible Thermodynamics of the	Cellular Growth of a Dilute Binary Alloy Single Crystal	000113
#Observation by Chemical Attack and X-Ray Topography of	Cellular Structures of Monocrystals of Dilute Binary	000210
in the (#Simulation of the Growth of a Metallic Face-	Centered Cubic Lattice in a Crystal from a Molten Mass	000995
#Isotope Effects on the	Central Peak Phenomena in KD3(SeO3)2 (solution)*	000963
System (CaGa2O4, Czochralski)* #The	Central Region of the Calcium Oxide-Gallium Oxide	001105
Lattices #Capture of Migrating Adatoms by Point Capture	Centres for Different Two-Dimensional Substrate	000873
of Crystals from the Plasmas of ZrO2, HfO2, ThO2 and	CeO2 (electrolytic deposition)* #Deposition	000472
#Low Temperature Thermal and Magnetic Properties of	CeP and CeAs (sublimation)*	000400
Lanthanide): Unidirectional Solidification,	# Ceramic Eutectics in the Systems ZrO2-Ln2O3 (Ln =	000630
9H2O, solution)* #Growth and Some Properties of	Cerium Sulphate Enneahydrate Single Crystals (Ce2(SO4)3.	000227
	# Cerium Trihydroxide (Ce(OH)3, gel)*	000664
Bronzes, in Particular those of Potassium, Rubidium and	Cesium (review, KxWO3, RbxWO3, CsxWO3)* Metal Tungsten	000402
of Mixed Bromides of Rare Earths, Sodium, and	Cesium (Ca2NaNdBr6, Cs2NaCeBr6, Cs2NaGdBr6, Cs2NaYBr6,	000626

PERMUTED TITLE INDEX

Dissolution Processes (CsI,   #Selective Etching of
#The Growth and Properties of Crystalline Rubidium and
#Gel Growth of Single Crystals of Some Rubidium and
of Cerium Sulphate Enneahydrate Single Crystals (
#Preparation, Structures, and Properties of Niobium
Electronic Properties (FeCr2S4,   #Preparation of Some
CIV = Si, Ge,   #Physical Properties of the Quarternary
#Crystal Growth and Properties of CuGaxIn1-xSe2
#Liquid Phase Epitaxy of II-IV-V2,
#A Temperature Variation Method for the Growth of
Growth (InP, CuInS2)*   #Simple Pressurized
S,Se)*   #Vapour Growth and Defect
#Solid State Reactions to CdTeMoO6 and Its Structural
#Growth of CuInS2 and Its
yAsy)*   #Semiconductor Superlattices by MBE and Their
Solidification, Microstructural and Crystallographic
To-Ribbon (RTR) Technique: Process Update and Material
Arsenates (REAsO4, slow cooling)*   #Flux Growth and
Epitaxial Growth on Bicrystalline Substrates of NaCl (#
Phase EuBr1.5Cl0.5 (melt, slow   #Preparation and
Deuterides (V-H, V-D, #Single Crystal Preparation and
Synthetic 11 Angstroms   #Electron Microscopic
InxGa1-xAsyP1-y Thin Layers Grown on InP (film)*
slow cooling, high pressure, Bridgman)* #Synthesis and
CuP2, NiP2, and RhP3 (flux, vapor   #Crystal Growth and
#Sublimation Growth and X-Ray Topographic
Cu2O)*   #Recent Developments in the Microstructural
solution)*   #Electron Microscopic
and Eu2B5O9Br (melt)*   #Synthesis and
#Flux Growth and
#Crystal Growth and
Czochralski)*   #Crystal Growth and
Czochralski)*   #Purification and
Sodium, and Cesium (#Preparation, Crystal Growth and
xCazC4H4O6.4H2O, gel)*   #Growth and
Vapor Phase (review,   #Physical Methods Used for the
vapor transport)*   #Preparation and
Wetting and Non-Wetting Dies (Si, EDFG)* #Growth and
chemical vapor transport, solution, cold #Growth and
Obtained by Organometallic Chemical Vapour #Growth and
evaporation)*   #Surface
grown by Liquid Phase   #Electrical Properties and
MoSe2, #Optical Phonon Modes and Localized Effective
and Characterization of Transition Metal Silicides (
and Physical Properties of Boracite Crystals (
Structures of Monocrystals of Dilute #Observation by
#Observations of Defects in LPE GaAs Revealed by New
#Kinetics of Gallium Phosphide Synthesis by
kinetics, solution)*   #Growth Spirals in a
V5S8 Phases (vapor transport)*   #Physico-
Praseodymium Pentaphosphate (PrP5O14)*   #
On the Preparation of Pseudobrookite Single Crystals by
On the Preparation of Nickel Ferrite Single Crystals by
Compositions in the Range TiB1.89 - TiB1.96 Prepared by
Chlorine*   #New Observations in the
II) The Transport Behaviour of the Equilibrium #On the
I. Calculation of the System Nb/O/Te/Cl and the #On the
#Effect of Addition of Vanadium on the
#Mass Spectrometric Study on
#Preparation of Single V3O5 Crystals by
#Growth of Anatase (TiO2) Crystals by
of Single Crystals of U1-xThxO2 Solid Solutions by
#Crystal Growth of Actinide Dioxides by
of HgCdTe Mixed Crystals from the Vapour Phase by
of Nonstoichiometric Zirconium Carbide Whiskers by
Nickel on the Growth of Zirconium Carbide Whiskers by
#Crystal Growth of Gallium Arsenide by
epitaxy, film)*   #Mechanisms of
Film with Smooth Surface on Intermediately Sputtered #
and Refractory Oxides on the Growth of TiC Whiskers by
Closed Systems*   #Modified
Having a Sharp Tip*   #
Modified Hot Wire Method (TiB2)*
Between Interface and Vapour Transport Kinetics in
#Mechanisms of the
theory)*   #Interface Morphology in
#Morphological Stability Analysis in
#Morphological Stability Analysis in
ZnSe Epitaxial Layers Obtained by Organometallic
Single Crystal and Film Growth of Actinide Pnictides by
and Single-Crystal Growth of Protactinium Arsenides by
Flux Function Method to a Five-Component System: The
#Crystal
Hydroxide #Preparation, Phase Equilibria, and Crystal
Czochralski)*   #Crystal Growth and Crystal
hydrothermal, theory)*   #Advances in Preparative
25NbO3-NaNbO3 (NaxTh0.25(1-x) NbO3, melt, slow #Crystal

Cesium Iodide Crystals and Comparison of Growth and   000827
Cesium Niobates (Czochralski, Cs8Nb22O59, Rb8Nb22O59)*   000440
Cesium Tin Halides (CsSnCl3, RbSnCl3, Cs2SnCl6,   000930
Ce2(SO4)3.9H2O, solution)* #Growth and Some Properties   000227
Chalcogenide Halides, NbX2Y2 (X=S, Se - Y=Cl, Br, I) (000795
Chalcogenide Spinel Single Crystals and Their   001066
Chalcogenides Cu2(I)B(II)C(IV)X4 (BII=Zn, Mn, Fe, Co -   000357
Chalcopyrite Compound (vapor transport)*   000743
Chalcopyrite Compounds (ZnSnAs2, solution)*   000089
Chalcopyrite Crystals by Iodine Vapor Transport (000744
Chambers for Liquid Encapsulated Czochralski Crystal   000150
Characterisation of Large Single Crystals of ZnS and Zn(000816
Characterization (slow cooling)*   000297
Characterization (vapor transport)*   000403
Characterization (GaAs, Ga1-xAlxAs, In1-xGaxAs, GaSb1-   000176
Characterization (Ln = Nd, Sm, Dy, skull-melting)*   000630
Characterization (Si, laser melting, review)*the Ribbon-   000360
Characterization by X-ray Topography of Rare Earth   000900
Characterization of (001) Tilt Boundaries Produced by   000207
Characterization of a Europium(II) Bromide-Chloride   000200
Characterization of the Group VA Metal Hydrides and   000790
Characterization of the Hydrothermal Growth of   000366
Characterization of the Optical Properties of LPE   001098
Characterization of the Ternary Compound ZnGeP2 (melt,   000336
Characterization of the Transition-Metal Phosphides   000708
Characterization of CdTe Single Crystals*   000148
Characterization of Corrosion Processes (film, epitaxy,   001045
Characterization of Cubic BN Prepared by Water Solvent (000934
Characterization of Europium(II)-Holoborates, Eu2B5O9Cl   000580
Characterization of Fluorite Crystals (CaF2)*   000237
Characterization of In1.9As0.1Se3 (vapor transport)*   000471
Characterization of Lithium Metasilicate (Li2SiO3,   000141
Characterization of Metallurgical Silicon (Si, melt,   000196
Characterization of Mixed Bromides of Rare Earths,   000626
Characterization of Mixed Sr:Ca Tartrates (Sr1-   000747
Characterization of Modes of Epitaxial Growth from the   000542
Characterization of Ruthenium Dioxide Crystals (RuO2,   000685
Characterization of Silicon Ribbon Crystals Grown with   000524
Characterization of Transition Metal Silicides (000607
Characterization of Undoped ZnSe Epitaxial Layers   000114
Charge and Ledge Dynamics in Cadmium Sulfide (CdS,   000529
Charge Imbalance for Ca,Ge-Substituted Garnet Films   000386
Charges of Transition-Metal Dichalcogenides (MoS2,   001000
chemical vapor transport, solution, cold crucible   000607
Chemical transport, M3B7O13X)*   #Preparation   000151
Chemical Attack and X-Ray Topography of Cellular   000210
Chemical Etchant (flux, film)*   000703
Chemical Gas Transport Reactions (GaP)*   000231
Chemical Potential Varying in Space or Time (theory,   000659
Chemical Properties of the Non-Stoichiometric VS2 and   000470
Chemical Synthesis and Crystal Growth of Laser Quality   000127
Chemical Transport (Fe2TiO5, electrolytic deposition)* #   000770
Chemical Transport (NiFe2O4, theory)*   #   000760
Chemical Transport (TiB2)*   #Single Crystals with   000291
Chemical Transport of GeO2. I. Transporting Agent   000787
Chemical Transport of Some Niobium Oxides with TeCl4 (000798
Chemical Transport of Some Niobium Oxides with TeCl4.   000797
Chemical Transport of TiSx, x = 1.40-1.70 (TiyV1-ySx)*   000819
Chemical Transport of VO2*   000063
Chemical Transport Reactions (vapor transport)*   000789
Chemical Transport Reactions with HBr and HCl*   000441
Chemical Transport Reactions*   #Growth   000455
Chemical Transport*   000912
Chemical Transport*   #Crystallization   000342
Chemical Vapor Deposition (ZrC)*   #Synthesis   000674
Chemical Vapor Deposition (ZrC)* #Catalytic Effect of   000958
Chemical Vapor Deposition from the Complex   001109
Chemical Vapor Deposition of Scandium Hydride (ScH2,   000494
Chemical Vapor Deposition of Silicon (Si, epitaxy)*   000702
Chemical Vapor Deposition of Single-Crystalline ZnO   000711
Chemical Vapor Deposition* effects of Various Metals   000959
Chemical Vapor Growth of Cinnabar (HgS) and GaP in   000279
Chemical Vapor Growth of LaB6 Whiskers and Crystals   000656
Chemical Vapor Growth of Titanium Diboride by a   000926
Chemical Vapour Deposition (theory)*   #Interrelation   000742
Chemical Vapour Deposition of Silicon (Si)*   000118
Chemical Vapour Deposition on Profiled Substrates (001011
Chemical Vapour Deposition Processes. I (theory)*   001010
Chemical Vapour Deposition Processes. II (theory)*   000435
Chemical Vapour Deposition* Characterization of Undoped   000114
Chemical Vapour Transport (review)*   #   000160
Chemical Vapour Transport (PaAs2, Pa3As4, PaAs)*   000161
Chemical Vapour Transport of Ni-Cl Boracite (no   000236
Chemistry of a Tellurate IV of Indium III: In2Te3O9*   000766
Chemistry of La, Pr, and Nd Hydroxide Bromides and   000531
Chemistry of NiAs-Type Compounds: MnSb, CoSb, and NiSb (000180
Chemistry of Oxides and Fluorides (review,   000398
Chemistry of Perovskite Phases of the Systems ThO.   000527

PERMUTED TITLE INDEX

#A New Preparation Technique in Solid State | Chemistry. I. High Temperature Reactions with CO2-Laser* | 000657
Reactions   #A New Preparative Technique in Solid State | Chemistry. II. High Pressure - High Temperature | 000658
Apatites (review)*   #Crystal- | Chemistry, Crystal-Growth, and Phase-Equilibria of | 000813
evaporation)*   #Flux Growth of Lead Neodymium | Chlorapatite (Na2Nd2Pb6(PO4)6Cl, slow cooling, | 000445
NaClO3)* #Kinetics of Growth and Dissolution of Sodium | Chlorate in Diffusion and Convection Regimes (solution, | 000888
#Crystal Growth and Selective Etching Studies on Sodium | Chlorate Crystals (NaClO3, solution)* | 000784
#Convection Phenomena During the Growth of Sodium | Chlorate Crystals from Solution (NaClO3)* | 000179
for Growing Large Needles and Single Crystals of Lead | Chloride (PbCl)*   #New Gel Method | 000003
#Preparation of Fine Sodium | Chloride Crystals (NaCl)* | 000654
and Characterization of a Europium(II) Bromide- | Chloride Phase EuBr1.5Cl0.5 (melt, slow cooling, | 000200
Layers of Nickel-Chlorine Boracite on Chromium- | Chlorine Boracite (Ni3B7O13Cl, vapor transport)* | 000837
#Gas Phase Synthesis of Epitaxial Layers of Nickel- | Chlorine Boracite on Chromium-Chlorine Boracite (| 000837
the Chemical Transport of GeO2. I. Transporting Agent | Chlorine*   #New Observations in | 000787
| Chromium Doped GaAs Vapor Phase Epitaxy* | 000469
PrCrOS2, NdCrOS2, SmCrOS2)*   #Mixed Oxysulfides of | Chromium III and Rare Earths (LaCrOS2, CeCrOS2, | 001048
#Growth of | Chromium Silicide, Cr3Si, Crystals (float zone melting)* | 000441
of Epitaxial Layers of Nickel-Chlorine Boracite on | Chromium-Chlorine Boracite (Ni3B7O13Cl, vapor transport) | 000837
X = #Layer Perovskites of the (CnH2n+1NH3)2MX4 and NH3(| CH2)mNH3MX4 Families with M = Cd, Cu, Fe, Mn or Pd and | 000039
#Ferroelectricity in (N(| CH3)4)2CoCl4* | 000831
#Ferroelectricity in (N(| CH3)4)2ZnCl4 (evaporation)* | 000830
solution, doped)*   #Growth of Single Crystals of | Cinnabar (alpha-HgS) at Rather Low Temperature (| 000157
#Modified Chemical Vapor Growth of | Cinnabar (HgS) and GaP in Closed Systems* | 000279
in the Melt- and Vapor Growth of Silicon for Integrated | Circuits and Solar Cells (Si, review)*   #Problems | 000117
#A Simple Analysis of Vapor Phase Growth - | Citing an Instance of GaxIn1-xAs (theory)* | 000673
Chalcogenides Cu2(I)B(II)C(IV)X4 (BII=Zn, Mn, Fe, Co - | CIV = Si, Ge, Sn - X = S,Se) (Cu2ZnGeS4, Cu2ZnSnS4, | 000357
theory)*   #Local Modes in Rb1- | cKc Alloys: A Neutron Scattering Study (Bridgman, | 000459
Oxides with TeCl4. I. Calculation of the System Nb/O/Te/ | Cl and the Transport of NbO2/NbO2.417 (NbO2, NbO2.417, | 000797
mNH3MX4 Families with M = Cd, Cu, Fe, Mn or Pd and X = | Cl or Br: Importance, Solubilities and Simple Growth | 000039
Component System: The Chemical Vapour Transport of Ni- | Cl Boracite (no materials grown - Ni3B7O13Cl theory)* | 000236
Adsorption Layers in CVD Systems (Si-H-Cl, Ga-As-H- | Cl) (GaAs, theory, epitaxy)*   Analysis of Equilibrium | 000184
of Divalent Rare Earths (M = Sm, Eu, Tm, and Yb - X = | Cl, Br, and I) (melt, SmFCl, SmFBr, SmFI, EuFCl, EuFBr, | 000088
of Niobium Chalcogenide Halides, NbX2Y2(X=S, Se - Y= | Cl, Br, I) (vapor transport) *Structures, and Properties | 000795
of Equilibrium Adsorption Layers in CVD Systems (Si-H- | Cl, Ga-As-H-Cl) (GaAs, theory, epitaxy)*   Analysis | 000184
#Growth and | Cleavage of Se-Te Alloy Crystals (Bridgman)* | 000111
#An Investigation of Thin Silver Films on | Cleaved Silicon Surfaces (Ag, theory, evaporation)* | 000619
Cd1-xZnxS Layers on (111) GaAs Substrates Using the | Close-Spaced Geometry (vapor transport)* Single Crystal | 000305
#Heteroepitaxial Growth of ZnS on GaP by the | Close-Spaced Technique (kinetics)* | 000482
and Optical Properties of ZnSe Epitaxial Layers by a | Close-Spaced Technique*   #Preparation | 000670
Absorption Edge of Single-Crystal AlN Prepared by a | Close-Spaced Vapor Process*   #The Optical | 000758
Parameters During the Crystal Growth from the Vapor in | Closed Ampules (GeS, sublimation)*   of Growth | 000841
*   #Growth of Phthalic Anhydride in a | Closed Crystal-Vapor System (organics, vapor transport) | 000327
LiCl Doped Rock Salt Crystals (NaCl,   #Semi-Periodic | Closed Growth Lamellae on the Crystal-Melt Interface of | 000532
#Growth of Pb0.8Sn0.2Te Films in a Quasi- | Closed Space (vacuum evaporation)* | 000885
Chemical Vapor Growth of Cinnabar (HgS) and GaP in | Closed Systems*   #Modified | 000279
#Vapor Growth of Zinc Telluride by | Closed Tube Method (ZnTe, epitaxy)* | 000700
Deposition Techniques (#New Developments in Ionized- | Cluster Beam and Reactive Ionized-Cluster Beam | 000944
#Ionized- | Cluster Beam Deposition and Epitaxy (review)* | 000945
in Ionized-Cluster Beam and Reactive Ionized- | Cluster Beam Deposition Techniques (vapor transport)* | 000944
Si Substrates*   #Ionized- | Cluster Beam Epitaxial Growth of GaP Films on GaP and | 000651
#Thin Film Deposition From Beams of Ionized Atoms and | Clusters (Ag on Si, epitaxy)* | 000516
#Ferroelasticity and Phase Transformation in Rb2Hg(| CN)4 Spinel (solution)* | 000378
Cd, Cu, Fe, Mn or Pd and X = #Layer Perovskites of the (| CnH2n+1NH3)2MX4 and NH3(CH2)mNH3MX4 Families with M = | 000039
Evaporation Technique (MoO3, WO3, theory)* | Coalescence Growth of Smoke Particles Prepared by a Gas- | 000448
Unidirectional Growth (theory)* | Coarsening of Eutectic Structures During and After | 001027
by the Hydrothermal Method on Single Crystal Seeds | Coated with Informative Interfacial Layers (| 000243
Zone Method (Co)*   #Growth of | Cobalt Single Crystals by the Electron Beam Floating- | 000381
NiGa4, flux)*   #The Binary Systems | Cobalt-Gallium and Nickel-Gallium Compared (CoGa3, | 000290
melting)*   #Growth and Properties of Strontium | Cobaltate Single Crystals (SrCoO2.70, float zone | 000955
Crystals and Their Electronic Properties (FeCr2S4, | CoCr2S4, vapor transport)*   Chalcogenide Spinel Single | 001066
Boundary Layer Thickness (effective distribution | coefficient, melt, theory)*   Equation as a Diffusion | 001083
the Gas Phase. I. Phosphorus Content and Segregation | Coefficient (Si, vapor transport, theory)*   Growth from | 000514
magnetic field effects,   #Non-Constant Distribution | Coefficients for Directionally Solidified InSb-GaSb (| 000864
directional crystallization, #Non-Constant Distribution | Coefficients from Experimental Data: Theoretical (| 001080
of the Lattice Parameters and the Thermal Expansion | Coefficients of VO2 near the Transition Temperature (| 000513
(TST) Technique: A Modified Bridgman Method (ZnF2, | CoF2, BaF2, MgF2, KY3F10)*   Thermic Screen Translation | 000541
Systems Cobalt-Gallium and Nickel-Gallium Compared (| CoGa3, NiGa4, flux)*   #The Binary | 000290
Metal Silicides (chemical vapor transport, solution, | cold crucible melting)*   Characterization of Transition | 000607
Zone Melting Method (DyFe2, TbFe2, Nd, Pr, Ce, | cold crucible)*   Growth by the Horizontal Levitation | 000620
Growth of Refractory Materials by RF Melting in a | Cold Container (review, alpha-Al2O3, Al2O3:Cr2O3)* | 000020
#Some Aspects of the Design and Construction of the | Cold Crucible* | 000895
Converting to the A-15 Structure* | # Cold Working Nb3Al in the bcc Structure and then | 001067
of Si2ON2 by Nitridization in the Si-SiO2 System in | Compacted Powders*   #Growth of a Crystal Needle | 000049
Point and the Vacancy Mechanism of Melting (theory, | compilation, 39 metals)*   of Metals Near the Melting | 000351
Gallium Arsenide by Chemical Vapor Deposition from the | Complex Monochlorodiethylgallium-Triethylarsine (GaAs)* | 001109
#The Flux Growth of Some New Rare Earth and Iron Group | Complex Oxides (GdVO3, Gd4Ba6(BO3)9, (Ba0.93Gd0.07) | 001063
Nd,   #The Flux Growth of Some Rare-Earth and Iron Group | Complex Oxides (NdBO3, LaBO3/Pr, LaBO3, (La,Pb)MnO3, (| 001061
Growth of alpha-HgI2 Single Crystals from Iodomecurate | complexes (kinetics)*   from Soluble Complexes. II. | 000694
of alpha-HgI2 Single Crystals from Dimethylsulfoxide | Complexes (theory)* from Soluble Complexes. III. Growth | 000695
of Sparingly Soluble Single Crystals from Soluble | Complexes. I. General Introduction (review, theory, | 000693
of Sparingly Soluble Single Crystals from Soluble | Complexes. II. Growth of alpha-HgI2 Single Crystals | 000694
of Sparingly Soluble Single Crystals from Soluble | Complexes. III. Growth of alpha-HgI2 Single Crystals | 000695
Y3Al5O12 (Czochralski)*   #A | Complexity in the Solidification Behaviour of Molten | 000202
#Basic Consideration for an Analytic Theory of Two- | Component Crystal Growth* | 000763
and Solid Compositions During LPE Growth from a Ternary | Component Solution (Ga1-xAlxAs, flux, theory)*of Liquid | 000970
#An Application of the Flux Function Method to a Five- | Component System: The Chemical Vapour Transport of Ni- | 000236
Properties (#Analytical Theory of Crystal Growth of Two- | Component Systems I. Basic Features and Equilibria | 000765
Theory of Normal Crystal Growth for Metallic One- | Component Systems. Two Cases of Modelling the Melt | 000059
Growth of Crystals with Toxic and/or Highly Volatile | Components (equipment)*Improved System for the Bridgman | 000164

PERMUTED TITLE INDEX

Compounds by Gas Phase   #Partition of Principal
         Conditions -- Microanalytical Evaluation of
    #Growth of Pb1-xSnXTe Single Crystals of Graded
Conditions of Constant Cooling #Improvement of Crystal
Fields (Czochralski)*   #Control of the
         #Structural Transitions and Phase
Solid Crystalline Lithium   #Influence of Isotopic
Effects in LPE-Grown In1-xGaxAs on InP Substrate*   #
Telluride (Cd1-   #The Effects of Growth Speed on the
    Ditelluride Crystals Grown from Near-Stoichiometric
    #Preparation of TiCr Single Crystal with Homogeneous
DyKMo2O8, ErKMo2O8,   #Effects of Modifying Starting
Chemical Transport (TiB2)*   #Single Crystals with
Solution (Ga1-xAlxAs,   #Variation of Liquid and Solid
Grown Crystals, Taking Account of Periodic Remelt (Si, #
Czochralski Growth (theory)*
         #GaAs Vapour Phase Epitaxy Using
theory)*
Melting. I. Pure Rotation Driven Flows (Si, Al2O3,
Melting. II. Combined Free and Rotation Driven Flows (#
Alloys Rich in Single-   #Progress of Temperature and
(II-VI, IV-VI, III-V,   #Nonstoichiometry and Carrier
Growth #In-Process Monitoring and Control of Doping Gas
         #The Crystal Structure of a New High -Nd-
         #Epitaxial Growth of BaTiO3 Vacuum
LiF)*   #Molecular (beam) Processes of
    #Preparation of Photoluminescent and
Single Crystal*   #Electrical
         #Ionic
cooling)*   #Luminescence and Ionic
powder) as a Function of Magnetic   #Low Temperature
Superionic Conductors (#Crystal Structure and Ionic
    #Telluride Halides of Copper with High Ionic Partial
         #Li3N: A Promising Li Ionic
    #Hydronium Beta" Alumina: A Fast Proton
    #A New Organic Low Temperature
of Li24Zn(GeO4)4 and other new Li+ Superionic
Indium Antimonide (InSb)*   #The Strain
kinetics)*
Vacancy and Diffusion Characteristics in Binary
*   #Determination of the Lattice
    #Correlation of the Interface Reaction
of Growth of Adamantane Crystals from the Vapour Under
in Ga1-xAlxAs LPE Layers Grown under Conditions of
Solidified InSb-GaSb (magnetic field effects,   #Non-
Data: Theoretical (directional crystallization,   #Non-
Solution (theory, melt)*   #Conditions for
         #Elastic
Anisotropy, Magnetostriction and Elastic
         #Single-Crystal Elastic
Alloys (Nb-Zr, Nb-Hf, electron beam melting,   #Elastic
#Kinetics of Crystallization of Alloys of Multiple
of InSb from the Melt #Morphological Instability Under
    #Some Aspects of the Design and
Growth of Refractory Materials by RF Melting in a Cold
Controlled Phosphorus Vapor Pressure and Reduced Oxygen
         #Si
#Interface Shape and Horizontal Variations of Al and Ga
 Properties of Silicon Crystals With Automatic Diameter
VI, III-V,   #Nonstoichiometry and Carrier Concentration
Electric and Magnetic Fields (Czochralski)*   #
         #Vapor Phase Transport and Stoichiometry
sublimation)*   #Vapor Phase Transport and Stoichiometry
Silicon Epitaxial Growth by   #In-Process Monitoring and
Polyphosphoric Acids (solution)*   #Morphology
Microstructures and Properties of Superalloys via Use #
PH3 Partial Pressure (flux)*
         #GaAs Vapour Phase Epitaxy Using Computer
hydrogen reduction of ferrous halides, float zone
         #Heat-Flow-
of Lattice-Matched GaP-Ga0.5Al0.5P-GaP Layer Under
Alloy Films by MBE (kinetics)*   #Surface Processes.
Superalloys via   #Control of Structure and Properties.
Driven Flows (Si, Al2O3,   #Computer Simulation of
and Rotation Driven Flows (#Computer Simulation of
Geometry and Prandtl Number at Zero   #Thermocapillary
of Liquid Phase in the Y3Fe5O12-BaO.B2O3 System on
Chlorate Crystals from Solution (NaClO3)*   #
    and Dissolution of Sodium Chlorate in Diffusion and
Solid State Materials, Vol. 4, Morphological Stability,
2MnCl4 Using a Compound Holographic Interference and
    #Cold Working Nb3Al in the bcc Structure and then
    #Crystal Structure of beta-KEr3F10 (melt, slow
         #Crystal Structure of Se2TiO6 (slow
    #The Hardness of NbB2 Single Crystals (flux, slow
    #Crystal Structure of the Spinel CdEr2S4 (slow
Properties of beta-Phase Iron-Germanium (FexGe, slow
Properties and Lithium Intercalates of CrPS4 (slow

Components in the Course of Preparation of II-IV-V2   001086
Composition (vapor transport)* Under Non-Stoichiometric   000082
Composition by the Travelling Solvent Method (theory)*   000564
Composition in Ga1-xAlxAs LPE Layers Grown under   000249
Composition of Crystals Grown in Electric and Magnetic   000511
Composition of Thin Films of Aluminum with Silicon*   000473
Composition on the Density and Molecular Volume of   000023
Composition Latching Phenomenon and Lattice Mismatch   000952
Compositional Variations in Crystals of Cadmium Mercury   000074
Compositions (doped, zone melting)*   of Copper Indium   000220
Compositions (float zone melting)*   001092
Compositions for Flux Growth (DyPO4, PrBO3, LaBO3,   001060
Compositions in the Range TiB1.89 - TiB1.96 Prepared by   000291
Compositions During LPE Growth from a Ternary Component   000970
Computation of Striated Impurity Distributions in Melt-   001026
Computational Simulation of the Melt Flow During   000491
Computer Control*   001057
Computer Model of Autoepitaxial Growth of Ge Films (000018
Computer Simulation of Convection in Floating Zone   000173
Computer Simulation of Convection in Floating Zone   000174
Concentration in the Region of Solidification of Pb-Bi   000846
Concentration Control in MBE of Compound Semiconductors   000897
Concentration During Vapor Phase Silicon Epitaxial   000937
Concentration Laser Material: Na3Nd(PO4)2*   000885
Condensates on SrTiO3 Crystals*   000984
Condensation and Evaporation of Alkali Halides (KCl,   000628
Conductive ZnSe:I Crystals by Iodine Vapor Transport*   000775
Conductivity and Galvanomagnetic Effects in (SN)x   000463
Conductivity in N3O+ Beta Alumina (solution)*   000286
Conductivity of Bariumfluorochloride (BaFCl, slow   001012
Conductivity of Gd2(SO4)38H2O (crystal) and Dy2Ti2O7 (000065
Conductivity of Li24Zn(GeO4)4 and other new Li+   000396
Conductivity. III. On the Phase Transition of CuTeBr   000025
Conductor (electrolytic deposition, review)*   000024
Conductor (0.84Na2O.0.84MgO.5Al2O3, theory)*   000285
Conductor: HNTSF-TNAP (solution, organics)*   000087
Conductors (kinetics, melt)*   and Ionic Conductivity   000396
Configuration Within the Facet of Czochralski Grown   000637
Congruent (Diffusionless) Vapor Transport (theory,   000007
Connections Between Thermodynamic Quantities and   000090
Constant of Epitaxial Layers of III-V Compounds (theory)   000399
Constant with Elementary Surface Processes (theory)*   000330
Constant Conditions (C10O16, theory, vapor transport)*   000572
Constant Cooling Rate (melt, theory)*   Composition   000249
Constant Distribution Coefficients for Directionally   000864
Constant Distribution Coefficients from Experimental   001080
Constant Growth Rate by LPE from a Cooling, Static   000893
Constants and Stability of bcc In-Tl Alloys (Bridgman)*   000581
Constants of a Ruthenium-Nickel Single Crystal (RuNi,   000631
Constants of Al2Cu (Bridgman, arc melting)*   000274
Constants of Niobium-Zirconium, Hafnium, and Tungsten   000307
Constituents (theory)*   000181
Constitutional Supercooling During the Crystal Growth   000478
Construction of the Cold Crucible*   000895
Container (review, alpha-Al2O3, Al2O3:Cr2O3)*   Crystal   000020
Contamination (GaP, Ga0.5Al0.5P, flux)* GaP Layer Under   000460
Contamination in Epitaxial Boron Monophosphide (BP)*   000713
Contents in Substituted YIG Single Crystals Grown by   000483
Control (Si)*   #Growth and   000998
Control in MBE of Compound Semiconductors (II-VI, IV-   000897
Control of the Composition of Crystals Grown in   000511
Control of Cadmium Sulfide (CdS, sublimation)*   000648
Control of Cadmium Sulfo-Selenide (Cd1+y(SxSe1-x),   000647
Control of Doping Gas Concentration During Vapor Phase   000937
Control of NdP5O14 Single Crystals Grown from   000466
Control of Structure and Properties.   000671
Control of Substrate Degradation in InP LPE Growth with   000199
Control*   001057
Controlled Growth of alpha-Fe Single Crystal Whiskers (000318
Controlled Growth During Li2B4O7 Crystallization (melt)*   000385
Controlled Phosphorus Vapor Pressure and Reduced Oxygen   000460
Controlling the Growth of GaxIn1-xAs and GaxIn1-xP   000301
Controlling Microstructures and Properties of   000671
Convection in Floating Zone Melting. I. Pure Rotation   000173
Convection in Floating Zone Melting. II. Combined Free   000174
Convection in Floating Zone Melting: Influence of Zone   000175
Convection of a Solution with a Temperature Difference*   000048
Convection Phenomena During the Growth of Sodium   000179
Convection Regimes (solution, NaClO3)*   of Growth   000888
Convection, Graphite, and Integrated Optics (review,   001081
Conventional Microscope (solution)* Perovskite (C3H7NH3)   000640
Converting to the A-15 Structure*   001067
cooling)*   000021
cooling)*   000547
cooling)*   000679
cooling)*   000983
cooling)*   #Magnetic   000845
cooling)*   #Physical   000571

PERMUTED TITLE INDEX

of Acoustic Phonons in Layered SrGa2 (melt, slow

Optical Study of alpha-In2Se3 Compound (melt, slow

Ionic Conductivity of Bariumfluorochloride (BaFCl,

of a New Variety of Nickel Pyrophosphate: Ni2P2O7 (slow

to CdTeMoO6 and Its Structural Characterization (slow

Crystals (NH4H2PO4, theory, solution, kinetics, slow

Crystallization of an Alloy (theory, melt, slow

Crystal Below the Transformation Temperature (slow

of Tl-Se-TlInSe2 (TlSe, zone melting, slow

X-ray Topography of Rare Earth Arsenates (REAsO4, slow

Compounds K2Mn1-xMxF4 (M + Fe, Co) (K2Mn1-xFexF4, slow

from the Ternary System Nd2O3-MoO3-Na2CO3 (flux, slow

by Continuous Cooling (BaWO4, flux, kinetics, slow

Crystals of Cadmium Mercury Telluride (Cd1-xHgxTe, slow

the Heat Exchange Method (HEM) (Y3Al5O12, melt, slow

"FeV3O8" (FexV1-xO2 - x approx. = 0.25) (melt, slow

Inelastic Scattering Experiments (Czochralski, slow

Th0.25NbO3-NaNbO3 (NaxTh0.25(1-x)NbO3, melt, slow

State of the Intermediate Valence Compound TmSe (slow

of Lead Neodymium Chlorapatite (Na2Nd2Pb6(PO4)6Cl, slow

Structure of the Double Molybdate K2Ni(MoO4)2 (slow

of the Ternary Compound ZnGeP2 (melt, slow

II) Bromide-Chloride Phase EuBr1.5Cl0.5 (melt, slow

of GaxIn1-xPyAs1-y Via the Gradient Freeze Method (slow

Core Model by Molecular Dynamics Simulation (melt, slow

from Solutions in Sodium Tungstate Melts by Continuous

from Solutions in Sodium Tungstate Melts by Continuous

from Solutions in Sodium Tungstate Melts by Continuous

xAlxAs LPE Layers Grown under Conditions of Constant

theory, kinetics, Czochralski) #On the

#Conditions for Constant Growth Rate by LPE from a

#Ultrasonic Velocities of K2PbCu(NO2)6 Near the

Growth (Cu, #Morphological Change from Copper/alpha-

for Preparation of Thin Single Crystal Films of

the Phase Transition of CuTeBr #Telluride Halides of

#Liquid Phase Epitaxial Growth of

Stoichiometric #Electrical and Optical Properties of

Earth (TbCu2, DyCu2, #Magnetic Properties of Rare Earth

Earth (PrCu2, NdCu2)##Magnetic Properties of Rare Earth

#Properties of Fe2P Crystals Prepared from a Liquid

of Impurities on the Crystallization of Supercooled

#The Preparation and Properties of Single Crystal

#A New Technique for the Preparation of

Cu, melt) #A New Technique for Preparation of

CuI-NiBr2 Mixture (Cu-Ni, #Growth and Morphology of

Reduction Growth (Cu, alpha- #Morphological Change from

cooling, #Study of Melting and Freezing in the Gaussian

#Geometrical Analysis of Crystallization of the Soft-

#Re-Investigation of the Re-Entrant

Elementary Surface Processes (theory)

Fusion (theory)

Developments in the Microstructural Characterization of

#Melting and Crystallization of

and Crystal Chemistry of NiAs-Type Compounds: MnSb,

of Crystals of Certain Metallides (alpha-FeSi2.33,

#Novel Reactor for High Volume Low-

#A New Method to Follow Crystal Growth by

#A New Method to Follow Crystal Growth by Coulter

#Thin Films of LiNbO3, Doped with Na+ and

State Chemistry. I. High Temperature Reactions with

II. High Pressure - High Temperature Reactions with

#Crystal Growth and EPR Study of

theory) #Magnetic Properties of (Mn,

#Epitaxial Layers of the Laser Material Nd(Ga,

Formed by Gas-Evaporation Technique: bcc Metals (V, Fe,

Single Crystals and Films (Si, Ge, GaAs, InSb,

#Anisotropy of Upper

#Upper and Lower

#Some Evidence for the Existence and Magnitude of a

#Epitaxial Growth of

#Epitaxial Growth of Ferromagnetic

and Physicochemical Investigation of CrPr2S4 and

Single Crystals, and Physicochemical Investigation of

#Physical Properties and Lithium Intercalates of

Silicides (chemical vapor transport, solution, cold

Zone Melting Method (DyFe2, TbFe2, Nd, Pr, Ce, cold

Solid Solutions by Electron-Beam Zone Melting (

Kinetics, theory) #Effect of

Some Aspects of the Design and Construction of the Cold

During Electron-Beam Zone Refining of beta-B Without a

Flux with Addition of V2O5 (Y3Fe5O12, Ho3Fe5O12,

by Differential Scanning Calorimeter. Growth Rate of

#Growth of Polyhedral Metal

#Protein

Ni3Pb2Se2 and Note on the Shandite Structure (

#About SnF2 Stannous Flouride. I.

and CsFeCl3.2D2O (solution)

: Unidirectional Solidification, Microstructural and

#Dispersion 000672

#Far Infrared 000453

#Luminescence and 001012

#Crystal Structure 000609

#Solid State Reactions 000297

#Unstable Growth of ADP 000804

#Dynamics of Non-Equilibrium 000826

#Flux Growth of Bi2WO6 Single 000665

#Structure and Growth Peculiarities 000361

#Flux Growth and Characterization by 000900

on the Antiferromagnetic Layer- Type 000106

#Growth of NaNd(MoO4)2 Single Crystals 000716

from Solutions in Sodium Tungstate Melts 000812

Speed on the Compositional Variations in 000074

#Melt Growth of Nd:Y3Al5O12(Nd:YAG) Using 000167

and Crystal Structure of a New Mixed Oxide 000661

Alloys, Ni2MnSn and Ni2Mn1-xVxSn, for Neutron 000321

Chemistry of Perovskite Phases of the Systems 000527

#Resistivity of the Magnetically Ordered Ground 000031

cooling, evaporation) #Flux Growth 000445

cooling, flux) #Crystal 000486

cooling, high pressure, Bridgman) and Characterization 000336

cooling, hygroscopic) Characterization of a Europium(000200

cooling, melt, doped) of In-Rich Bulk Single Crystals 000053

cooling, theory) Melting and Freezing in the Gaussian 000915

Cooling (BaWO4, flux, kinetics, slow cooling)*Tungstate 000812

Cooling (CaWO4, flux) Growth of Calcium Tungstate 000810

Cooling (SrWO4, flux) Growth of Strontium Tungstate 000811

Cooling Rate (melt, theory) Crystal Composition in Ga1- 000249

Cooling Rates of Large-Diameter Silicon Crystals (Si, 000165

Cooling, Static Solution (theory, melt) 000893

Cooperative Jahn-Teller Phase Transitions (solution) 000467

copper-Zinc Whiskers to Dendrites in Zinc Reduction 000481

Copper (Cu, evaporation) #A New Technique 000736

Copper with High Ionic Partial Conductivity. III. On 000025

Copper Ferrite Films (CuFe2O4) 001016

Copper Indium Ditelluride Crystals Grown from Near- 000220

Copper Intermetallic Compounds, RCu2. I. Heavy Rare 000373

Copper Intermetallic Compounds, RCu2. II. Light Rare 000374

Copper Medium 000030

Copper Melts (Cu) #The Influence 000485

Copper Phosphide (Cu3P, Czochralski) 000799

Copper Single Crystals (Cu) 000735

Copper Single Crystals from the Bulk Using Four Points (000737

Copper-Nickel Alloy Crystals by Hydrogen Reduction of a 000364

Copper/alpha-copper-Zinc Whiskers to Dendrites in Zinc 000481

Core Model by Molecular Dynamics Simulation (melt, slow 000915

Core Model in an FCC Crystal Formation (theory) 000965

Corner Effect in Twinned Crystals (theory) 000484

Correlation of the Interface Reaction Constant with 000330

Correlation of Diffusion Behaviour with the Entropy of 000980

Corrosion Processes (film, epitaxy, Cu2O) #Recent 001045

Corundum (alpha-Al2O3) 000563

CoSb, and NiSb (Czochralski) #Crystal Growth 000180

CoSi2, FeGe2, Mn5Si3, Fe1.76Ge, melt, theory, kinetics) 000861

Cost Silicon Epitaxy (Si, vapor transport, kinetics) 000062

Coulter Counter (solution, organics) 000601

Counter (solution, organics) 000601

Co2+ plus Zr4+, Grown by Liquid-Phase Epitaxy (flux) 000691

CO2-Laser #A New Preparation Technique in Solid 000657

CO2-Lasers Technique in Solid State Chemistry. 000658

Cr+-Doped MgTe Single Crystals (sublimation) 000133

Cr)P and (Mn,Fe)P Compounds (Mn1-xFexP, x = 0 to 0.15, 000430

Cr)3(BO3)4 (flux) 000578

Cr, Mo and W) and Habit of Fine Metal Particles 000821

Creation of Defects During the Growth of Semiconductor 000636

Critical Field in Superconducting 2H-NbS2 (theory) 000730

Critical Fields of TaS2(Pyridine)1/2 (vapor transport) 000468

Critical Marangoni Number for the Onset of Oscillatory 000847

CrO2 on Sapphire in Air (vapor transport) 000413

CrO2 Films in Air 000414

CrPr2Se4 (vapor transport) of Single Crystals, 000817

CrPr2S4 and CrPr2Se4 (vapor transport) Preparation of 000817

CrPS4 (slow cooling) 000571

crucible melting) Characterization of Transition Metal 000607

crucible) Crystal Growth by the Horizontal Levitation 000620

Crucible (Si-Ge) Single Crystals of Silicon-Germanium 000820

Crucible Shape on a Crystallization Front (Bridgman, 000004

Crucible* # 000895

Crucible #Shape of the Crystallization Front 000968

Crystallisation of Some Ferrite Materials from Lead 000565

Crystallite (KBrO3)*from Its Quiescent Aqueous Solution 000367

Crystallites in Inactive Gas (Mg, review) 001006

Crystallizations (solution, organics) 000382

Crystallochemical Study of the System Ni3Pb2S2- 000198

Crystallochemistry of alpha-SnF2 000235

Crystallographic and Magnetic Structure of RbFeCl3.2D2O 000077

Crystallographic Characterization (Ln = Nd, Sm, Dy, 000630

PERMUTED TITLE INDEX

PERMUTED TITLE INDEX

#Crystal Structure and Physical Properties of | Cu00.65VS2 (vapor transport)* | 000543
and Characterization of the Transition-Metal Phosphides | CuP2, NiP2, and RhP3 (flux, vapor transport)*   Growth | 000708
Current Topics in Materials Science, Volume 1 (review)* | 000450
Current Topics in Materials Science, Volume 2 (Review)* | 000449
Unidirectional Solidification of a Binary Alloy with a | Curved Solid-Liquid Interface (Ge:Ga, Bridgman, theory)* | 000206
#On the Presentation of Growth | Curves for Growth from Solution (theory)* | 001025
Single Crystals of Rare Earth Vanadates and #Solubility | Curves in High-Temperature Melts for the Growth of | 000265
Study of Phase Transformations of the Compound | CuSe*   #Electron Diffraction | 000866
#Crystal Structure of | CuTa2O6 (flux)* | 001036
Partial Conductivity. III. On the Phase Transition of | CuTeBr and the Structure of the High Temperature Phase (| 000025
#The Crystal Structure of | CuTeO4 (hydrothermal)* | 000282
magnetic susceptibility, and NMR) of the Spinel | CuV2S4 (vapor transport)*   Properties (transport, | 000544
Shift of the Absorption Edge in ZrS2 and HfS2 (CuxZrS2, | CuxHfS2, FexZrS2, FexHfS2, vapor transport, theory)* | 001090
#Formation of Mixed Crystals in the System | CuxZn1-xCr2Se4* | 000722
Induced Shift of the Absorption Edge in ZrS2 and HfS2 (| CuxZrS2, CuxHfS2, FexZrS2, FexHfS2, vapor transport, | 001090
of Dilute Binary Metallic Alloys (Cu - 0.5%Si, | Cu1-xSbx - x = 0.0015-0.01)* Structures of Monocrystals | 000210
#Electrical Resistivity in In-Heusler | Cu2.0Mn0.94In1.0 Single Crystal (Bridgman)* | 000499
#Physical Properties of the Quarternary Chalcogenides | Cu2(I)B(II)C(IV)X4 (BII=Zn, Mn, Fe, Co - CIV = Si, Ge, | 000357
#Growth and X-Ray Study of Cu2SrGeS4 and | Cu2BaGeS4 (flux)* | 000972
Cu2MnSnS4, Cu2MnGeSe4, Cu2MnSnSe4, Cu2FeGeS4, | Cu2CoGeS4, vapor transport)*   Cu2MnSiS4, Cu2MnGeS4, | 000357
Cu2MnGeS4, Cu2MnSnS4, Cu2MnGeSe4, Cu2MnSnSe4, | Cu2FeGeS4, Cu2CoGeS4, vapor transport)*   Cu2MnSiS4, | 000357
Cu2ZnSnSe4, Cu2MnSiS4, Cu2MnGeS4, Cu2MnSnS4, | Cu2MnGeSe4, Cu2MnSnSe4, Cu2FeGeS4, Cu2CoGeS4, vapor | 000357
= S,Se) (Cu2ZnGeS4, Cu2ZnSnS4, Cu2ZnGeSe4, Cu2ZnSnSe4, | Cu2MnGeS4, Cu2MnSnS4, Cu2MnGeSe4, Cu2MnSnSe4, | 000357
Cu2MnSiS4, Cu2MnGeS4, Cu2MnSnS4, Cu2MnGeSe4, | Cu2MnSiS4, Cu2MnGeS4, Cu2MnSnS4, Cu2MnGeSe4, | 000357
Cu2ZnGeSe4, Cu2ZnSnSe4, Cu2MnSiS4, Cu2MnGeS4, | Cu2MnSnSe4, Cu2FeGeS4, Cu2CoGeS4, vapor transport)* | 000357
#(111) | Cu2MnSnS4, Cu2MnGeSe4, Cu2MnSnSe4, Cu2FeGeS4, | 000357
Characterization of Corrosion Processes (film, epitaxy, | Cu2O Growth Modes on (111)Cu Surfaces (epitaxy)* | 000388
#Growth and X-Ray Study of | Cu2O)*   #Recent Developments in the Microstructural | 001045
- CIV = Si, Ge, Sn - X = S,Se) (Cu2ZnGeS4, Cu2ZnSnS4, | Cu2SrGeS4 and Cu2BaGeS4 (flux)* | 000972
X4 (BII=Zn, Mn, Fe, Co - CIV = Si, Ge, Sn - X = S,Se) (| Cu2ZnGeSe4, Cu2ZnSnSe4, Cu2MnSiS4, Cu2MnGeS4, | 000357
Ge, Sn - X = S,Se) (Cu2ZnGeS4, Cu2ZnSnS4, Cu2ZnGeSe4, | Cu2ZnSnSe4, Cu2MnSiS4, Cu2MnGeS4, Cu2MnSnS4, | 000357
Mn, Fe, Co - CIV = Si, Ge, Sn - X = S,Se) (Cu2ZnGeS4, | Cu2ZnSnS4, Cu2ZnGeSe4, Cu2ZnSnSe4, Cu2MnSiS4, | 000357
and Properties of Single Crystal Copper Phosphide (| Cu3P, Czochralski)*   #The Preparation | 000799
#The Crystal Structure of | Cu4(PO4)2O (flux)* | 000029
#Crystal Structure of | Cu4(PO4)2O (flux)* | 000142
#Crystallography of Aligned Cu-In Eutectoid (Cu- | Cu7In3)* | 001049
Ray Work Obtained by Plasma Remelting of Alloy Powder (| Cu9Al4)*   #Spherical Monocrystals for X- | 000043
#Growth and Morphology of 6H-SiC Epitaxial Layers by | CVD (film)* | 000699
#A Quick Survey Method for the Study of | CVD Conditions (TiN, vapor transport)* | 000655
3Fe5O12, vapor transport, epitaxy,   #Low Temperature | CVD Garnet Growth (Er3Fe5O12, (Eu,Y)3Fe5O12 and (Eu,Yb) | 000211
#Epitaxial Growth of ZnS by Zn-S-H2 | CVD Method* | 000615
Analysis of Equilibrium Adsorption Layers in | CVD Systems (Si-H-Cl, Ga-As-B-Cl) (GaAs, theory, | 000184
#Nd:Y2O3 Single-Crystal Fiber Laser: Room-Temperature | cw Operation at 1.07- and 1.35-micrometer Wavelength (| 000916
#Crystal Growth from Solution Using | Cylindrical Seeds (TGS, TBFB)* | 000566
kinetics)*   #Effect of the Buoyancy Parameter on | Czochralski Bulk Flow in Garnet Growth (theory, | 000535
kinetics)*   #Digital Simulation of | Czochralski Bulk Flow in Microgravity (Al2O3, theory, | 000536
#Effects of Rotation on | Czochralski Crystal Growth (theory)* | 000796
A Theoretical Study of Temperature Distributions During | Czochralski Crystal Growth (Cu, Si)*   # 000149
#Simple Pressurized Chambers for Liquid Encapsulated | Czochralski Crystal Growth (InP, CuInS2)* | 000150
YB4, TbB4, ErB4)* | # Czochralski Crystal Growth of Rare Earth Tetraborides (| 000137
#The Effect of Melt Flow Phenomena on the Perfection of | Czochralski Grown Gadolinium Gallium Garnet (Gd3Ga5O12)* | 000638
theory, Gd3Ga5O12)*   #Dislocation Propagation in | Czochralski Grown Gadolinium Gallium Garnet (GGG, | 000838
#The Strain Configuration Within the Facet of | Czochralski Grown Indium Antimonide (InSb)* | 000637
#Crystal Perfection in | Czochralski Grown Nickel Single Crystals (Ni, theory)* | 000521
#Interface Shape Transitions in | Czochralski Grown YAG Crystals (Y3Fe5O12, theory)* | 000645
#Computational Simulation of the Melt Flow During | Czochralski Growth (theory)* | 000491
#Dislocation-Free | Czochralski Growth of (110) Silicon Crystals (Si)* | 000262
#Bridgman and | Czochralski Growth of beta-Phase NaxV2O5* | 000907
Compounds (V3Si, V3Ge)* | # Czochralski Growth of A15 Structure Intermetallic | 001073
Crystals (BaAl12O19)* | # Czochralski Growth of Barium Hexaaluminate Single | 000611
Crystals (Bi12GeO20)*   #The | Czochralski Growth of Bismuth-Germanium Oxide Single | 000771
#Interface Shape Transitions in the | Czochralski Growth of Dy3Al5O12* | 000644
of the Contact Angle Between Melt and Crystal During | Czochralski Growth of Gallium and Germanium (Ga, Ge, | 001071
Czochralski Growth of Li3N Crystals* | 000842
Behavior of Ga-Doped Ge (#Application of a Heat Pipe to | # Czochralski Growth of Tellurium Single Crystals (Te)* | 000876
Investigations (theory)*   #A Model for Macroscopic | Czochralski Growth. Part I. Growth and Segregation | 000604
#Variations of Impurity Distribution in the Modified | Czochralski Growth: Theoretical and Experimental | 001002
#Growth of Crystalline Slabs of Layered InSe by the | Czochralski Method of Crystal Growth (Si)* | 000997
#Behavior of Boron in Ge Single Crystals Grown by the | Czochralski Method* | 000187
Be as an Impurity During Crystallization of Ge by the | Czochralski Method* | 000345
on the Defect Generation in KCl Crystals Grown by the | Czochralski Method*   #Behavior of | 000346
of Stacking Faults by Bardeen-Herring Mechanism in | Czochralski Method*   #Effects of Growth Rates | 000409
on Microdefect Formation in As-Grown Dislocation-Free | Czochralski Silicon (Si)*   #Growth | 001052
of Single Crystals of Some Organic Compounds by the | Czochralski Silicon Crystals*   #The Effect of Doping | 000223
#Acoustooptic Characteristics of LiBi(MoO4)2 (| Czochralski Technique and the Assessment of their | 000115
#Single Crystal Growth of 12CaO.7Al2O3 (| Czochralski)* | 000016
#Dielectric and Optical Properties of K5Nd(MoO4)4 (| Czochralski)* | 000203
#The Growth of Dy3Al5O12 (| Czochralski)* | 000582
#Crystal Growth and Structure of BiVO4 (| Czochralski)* | 000646
#Low-Defect InSb Crystal Growth by InN Doping (| Czochralski)* | 000891
#Ca12Al14O33: A New Piezoelectric Material (| Czochralski)* | 000991
of LiTaO3 Single Crystals for Saw Device Applications (| Czochralski)* | 001076
and Properties of Pb(WO4)1-x(MoO4)x Mixed Crystals (| Czochralski)*   #Growth | 000311
Optical Properties of KTb3F10 and LiTbF4 Crystals (| Czochralski)*   #Growth | 000709
in the Solidification Behaviour of Molten Y3Al5O12 (| Czochralski)*   #Magneto- | 001068
and Characterization of Lithium Metasilicate (Li2SiO3, | Czochralski)*   #A Complexity | 000202
| Czochralski)*   #Crystal Growth | 000141

PERMUTED TITLE INDEX

PERMUTED TITLE INDEX

```
           #Effects of Interface Kinetics on the Growth Rate of   Dendrites (Sn, theory)*                                       000991
                                     #On the Growth of NH4Cl       Dendrites at Very Low Supersaturation (solution)*            000172
              Change from Copper/alpha-copper-Zinc Whiskers to     Dendrites in Zinc Reduction Growth (Cu, alpha-Cu-Zn)*        000481
                                 #Evidence for a Universal Law of  Dendritic Growth Rates (theory)*                             000534
                     #Thermal Analysis of Solidification in Web-   Dendritic Ribbon Growth (Si, kinetics, theory)*             000369
                          #Solute Partitioning During Silicon      Dendritic Web Growth (Si, theory)*                           000855
         Purity Aluminum Single Crystals with Low Dislocation      Density (Al, zone melting, theory)*          Method of High  000228
       Implanted Silicon Layer on Laser Irradiation Energy         Density (Si)*              #Grain Size Dependence in a Self-  000996
Lithium          #Influence of Isotopic Composition on the        Density and Molecular Volume of Solid Crystalline            000023
                      #Impurity Effect on Grown-In Dislocation     Density of InP and GaAs Crystals (Czochralski)*              000856
on Convection of a Solution with a          #Influence of         Density of Liquid Phase in the Y3Fe5012-BaO.B203 System       000048
     Between Successive Monatomic Step Trains of Different         Density on Evaporating Crystal Surfaces (NaCl, theory)*      000105
           #Crystalline Defects in Solid Phase Epitaxy Si Films   Deposited at Elevated Temperatures (vapor transport)*        000194
         Growth Kinetics in (GaAs)n-(AlAs)m Superlattices          Deposited by Molecular Beam Epitaxy. I. Growth on            000762
     Substrate by Laser Annealing (Ge   #Heteroepitaxy of a       Deposited Amorphous Germanium Layer on a Silicon             000343
     #Growth Parameters and Crystal Morphology of Vapor-          Deposited Niobium Nitride (NbN)*                             000947
Surfaces (SiC)*                             #Analysis of Vapor-    Deposited Silicon Carbide Films on Silicon Ribbon            001099
                 #Growth and Electrical Properties of Sputter-    Deposited Single-Crystal GaSb Films on GaAs Substrates*      000269
           of the Solid Electrolyte High-LiTa308 (electrolytic    deposition)*                             #Structural Studies  000706
         the Plasmas of ZrO2, HfO2, ThO2 and CeO2 (electrolytic   deposition)*            #Deposition of Crystals from          000472
             (metals, Nb, NaxWO3, Czochralski, electrolytic       deposition)*            in Electrochemical Crystallization    000225
         Crystals by Chemical Transport (Fe2TiO5, electrolytic    deposition)*On the Preparation of Pseudobrookite Single       000770
       Properties of Fe2-xCrxO3 (vapor transport, electrolytic    deposition)*The Electrical, Optical and Photoconducting       000624
             Kinetics (Peltier effect, GaAs, theory, electrolytic deposition, kinetics)*           Phase Electroepitaxy: Growth 000437
           #Li3N: A Promising Li Ionic Conductor (electrolytic    deposition, review)*                                         000024
         Structures of Si on Sapphire, Obtained by Vacuum         Deposition (film, doped)*                   #Heteroepitaxial  000823
           and Vapour Transport Kinetics in Chemical Vapour       Deposition (theory)*   #Interrelation Between Interface       000742
           Zirconium Carbide Whiskers by Chemical Vapor           Deposition (ZrC)*    #Synthesis of Nonstoichiometric          000674
    Growth of Zirconium Carbide Whiskers by Chemical Vapor        Deposition (ZrC)*   #Catalytic Effect of Nickel on the        000958
  of Silicon Wafers by Thin Polycrystalline Silicon Film         Deposition and by Film Doping Using Boron Diffusion (Si)      000985
                              #Ionized-Cluster Beam               Deposition and Epitaxy (review)*                             000945
    #Crystal Growth of Gallium Arsenide by Chemical Vapor         Deposition from the Complex Monochlorodiethylgallium-        001109
  ThO2 and CeO2 (electrolytic deposition)*                      # Deposition of Crystals from the Plasmas of ZrO2, HfO2,      000472
                                     #Electroless                 Deposition of Epitaxial Pb1-xHgxS Films*                      000869
                          #Organometallic Vapor                  Deposition of Epitaxial ZnSe Films on GaAs Substrates*       000921
                          #Liquid Phase Epitaxial                 Deposition of GaP on GaAs*                                    000354
                          #Chemical Vapor                        Deposition of Scandium Hydride (ScH2, epitaxy, film)*        000494
  xHgxS, solution)*                  #Electroless                Deposition of Semiconductor Films (PbS, PbSe, Pb1-          000868
                      #Mechanisms of the Chemical Vapour          Deposition of Silicon (Si)*                                  000118
                      #Mechanisms of Chemical Vapor               Deposition of Silicon (Si, epitaxy)*                         000702
  Surface on Intermediately Sputtered ZnO #Chemical Vapor         Deposition of Single-Crystalline ZnO Film with Smooth        000711
                #Interface Morphology in Chemical Vapour          Deposition on Profiled Substrates (theory)*                 001011
  on Si, epitaxy)*                           #Thin Film           Deposition From Beams of Ionized Atoms and Clusters (Ag     000516
           #Morphological Stability Analysis in Chemical Vapor    Deposition Processes. I (theory)*                           001010
           #Morphological Stability Analysis in Chemical Vapour   Deposition Processes. II (theory)*                           000435
     Ionized-Cluster Beam and Reactive Ionized-Cluster Beam      Deposition Techniques (vapor transport)*Developments in      000944
  Oxides on the Growth of TiC Whiskers by Chemical Vapor          Deposition*    effects of Various Metals and Refractory      000959
     Layers Obtained by Organometallic Chemical Vapour           Deposition*    Characterization of Undoped ZnSe Epitaxial     000114
                                      #Development of             Depressions and Voids During LPE Growth of GaAs (film)*      000084
     #The Crystal Structure of La3Ru3011: A New Cubic KSbO3       Derivative Oxide with No Metal-Metal Bonding (flux)*         000209
                          #Some Aspects of the                   Design and Construction of the Cold Crucible*                000895
  Systems (review)*                                            # Design Considerations for Molecular Beam Epitaxy            000576
     Silicon Epitaxial Growth by Using a Flame Photometric        Detector (Si)*       Gas Concentration During Vapor Phase    000937
                #Crystal Growth of CdTe for X-Ray                Detectors (review)*                                         001053
     Spectroscopic Evaluation of Mercuric Iodide Radiation        Detectors. Part II. HgI2 (vapor transport, review)*         000834
                #Single Crystal Growth and Properties of         Deuterated Triglycine Fluoroberyllate (TGFB)*               000567
  solution)*          #Growth and Electrical Properties of       Deuterated TGS Produced by the Rotating Disc Technique (     000560
  and Characterization of the Group VA Metal Hydrides and         Deuterides (V-H, V-D, Nb-H, Nb-D, Ta-H, Ta-D)*              000790
     Growth of the Para- and Antiferroelectric Phases of         Deuterium Ammonium Dihydrogen Arsenate (DADA) (NH4(1-x)      000100
                #Growth of LiTaC3 Single Crystals for Saw         Device Applications (Czochralski)*                          000311
                #MBE Techniques for IV-VI Optoelectronic          Devices (review)*                                           000394
  film)*                              #Integrated Optical         Devices Fabricated by MBE (GaAs, Ga1-xAlxAs, review,         000152
           of LPE Growth for Lattice Matched GaInAsP/InP          DH Lasers with (100) Substrate in the Range of 1.2-1.5      000035
  Pb3(PxV1-xO4)2, Pb3(PO4)2*                  #Phase             Diagram and Crystal Growth of Lead Phosphovanadates (       000987
  Magnetic and EPR #Metal-Insulator Transitions and Phase        Diagram of (Ti1-xVx)407: Electrical, Calorimetric,          000835
                                             #Phase              Diagram of Pseudobinary System CdTe-ZnSe (Bridgman)*        000725
  and Properties of Silicon Crystals With Automatic              Diameter Control (Si)*                     #Growth            000998
  alpha-Al2O3)*       #Powder Feeder for Crystals of Large       Diameter Grown by the Verneuil Technique (MgAl2O4,           000284
     Stable Zone Length in Float-Zone Growth of Small-           Diameter Sapphire and Silicon Crystals (Al2O3, Si)*         000479
  Czochralski)*         #On the Cooling Rates of Large-           Diameter Silicon Crystals (Si, theory, kinetics,            000165
           #The History and Present Status of Synthetic          Diamond (C, review, doped)*                                 000685
           #Reactive Pulse Plasma Crystallization of             Diamond and Diamond-Like Carbon (C, film)*                  000906
                              #Overgrowths on                   Diamond at Atmospheric Pressure (C, flux)*                   000748
  theory)*                      #Monte Carlo Simulation of a (111) Diamond Face Around the Roughening Transition (C,            001020
     #Reactive Pulse Plasma Crystallization of Diamond and       Diamond-Like Carbon (C, film)*                              000906
           #Temperature Field Dynamics in the Synthesis of       Diamonds (kinetics, theory, C)*                             000862
     of a New Mixed Oxide FeV2O6OH0.5: Relation to the           Diaspore Structure Type (hydrothermal)*        Structure     000662
                #Single-Crystal Growth of Tantalum              Diboride (TaB2, zone melting, solution)*                    000680
                #Chemical Vapor Growth of Titanium               Diboride by a Modified Hot Wire Method (TiB2)*              000926
  and Localized Effective Charges of Transition-Metal            Dichalcogenides (MoS2, MoSe2, WS2, WSe2, HfSe2, ZrSe2,       001000
  Thermodynamic Considerations for the Growth of Layered        Dichalcogenides by Vapour Transport Processes (            000805
  Insulators (Si on sapphire, vapor transport)*               # Dichlorosilane as Silicon Source for Epitaxial Films on    000254
                              #Structure of Silver               Dichromate: Ag2Cr2O7 (solution)*                            000260
                                                              # Dielectric and Optical Properties of K5Nd(MoO4)4 (         000582
  Czochralski)*                      #X-Ray and                 Dielectric Investigations of Pb2CrO5 (melt)*                000979
                              #Thermal and                      Dielectric Properties of LiKSO4 and LiCsSO4*                000229
           #Preparation, Optical and                            Dielectric Properties of Na5W3O9F5 (melt)*                  000785
     Ribbon Crystals Grown with Wetting and Non-Wetting         Dies (Si, EDFG)**#Growth and Characterization of Silicon   000524
```

PERMUTED TITLE INDEX

System on Convection of a Solution with a Temperature | Difference*   of Liquid Phase in the Y3Fe5012-BaO.B203   000048
Silicon from the Gas Phase. II. Phosphorus Content and | Differential Molar Solution Enthalpies (Si, vapor   000515
Bromate from Its Quiescent Aqueous Solution by | Differential Scanning Calorimeter. Growth Rate of   000367
#Dodecatungstophosphoric Acid-21-Water by Neutron | Diffraction (evaporation, H3PW12040.21H20)*   000913
Weberite (flux)*   #X-Ray and Neutron | Diffraction of Na2MnFeF7: A Variety of Trigonal   001033
and Magnetic Phase Transitions of Tb5Ge4. A Neutron | Diffraction Study (melt)*   #On the Magnetic Structure   000840
Compound CuSe*   #Electron | Diffraction Study of Phase Transformations of the   000866
beta-Alumina (NaAl11017, KA11017, RbAl11017,   #X-Ray | Diffuse Scattering from Alkali, Silver, and Europium   000621
#Improvements in SSD (synthesis, solute | diffusion) Growth of GaP*   000461
Silicon Film Deposition and by Film Doping Using Boron | Diffusion (Si)* Silicon Wafers by Thin Polycrystalline   000985
of Growth and Dissolution of Sodium Chlorate in | Diffusion and Convection Regimes (solution, NaClO3)*   000888
Crystals*   #Simultaneous | Diffusion of Calcium and Strontium in KCl Single   000579
#Correlation of | Diffusion Behaviour with the Entropy of Fusion (theory)*   000090
Quantity in the Bruton, Prim and Slichter Equation as a | Diffusion Boundary Layer Thickness (effective   001083
Between Thermodynamic Quantities and Vacancy and | Diffusion Characteristics in Binary Metallic Solid   000090
VI Semiconductors (#Impurity Dopant Incorporation and | Diffusion During Molecular Beam Epitaxial Growth of IV-   000898
Crystallization of Salts from Supersaturated Solutions: | Diffusion Kinetics (#   000623
(alpha/beta)-Brass Bicrystals Produced by a Solid-Solid | Diffusion Method. I. Morphology and Crystallographic   000372
(alpha/beta)-Brass Bicrystals Produced by a Solid-Solid | Diffusion Method. II. Diffusion-Induced Twins in the   000371
#Growth of InP Crystals by the Synthesis Solute | Diffusion Method*   000603
#InP Single Crystal Growth by the Synthesis, Solute | Diffusion Method*   001095
Sized InP Single Crystals by the Synthesis, Solute | Diffusion Technique (vapor transport)*   #Large-   000923
x Layers on ZnS and ZnSe Substrates by a Solid-State- | Diffusion Technique*   #Growth of ZnSxSe1-   000045
Produced by a Solid-Solid Diffusion Method. II. | Diffusion-Induced Twins in the alpha-phase (theory)*   000371
#Congruent (| Diffusionless) Vapor Transport (theory, kinetics)*   000007
solution, vapor transport, flux)*   #On | Diffusive-Advective Interfacial Mass Transfer (theory,   001074
Microgravity (Al203, theory, kinetics)*   # | Digital Simulation of Czochralski Bulk Flow in   000536
Solution with Special Reference to Calcium Sulphate | Dihydrate (CaSO4.2H2O, kinetics, theory)*   in Aqueous   000193
and Structure of Epitaxial Films of the Rare-Earth | Dihydrides (GdH2, TbH2, DyH2, ErH2, HoH2)*   #The Growth   000782
and Antiferroelectric Phases of Deuterium Ammonium | Dihydrogen Arseanate (DADA) (NH4(1-x)D4xH2(1-x)D2xAsO4,   000100
#Single Crystal Growth of Ferroelectric Potassium | Dihydrogen Orthophosphate in Silica Gels (KDP)*   000442
#Nucleation in Supersaturated Potassium | Dihydrogen Orthophosphate Solutions (KDP)*   000443
#The Role of Dislocations in the Growth of Ammonium | Dihydrogen Phosphate Crystals from Aqueous Solution (000219
#Crystallization During Polymerization of Lithium | Dihydrogen Phosphate. II. Crystal Growth by Dimer   000097
hygroscopic, Bridgman)*   #Optical and | Dilatometric Studies of KCaCl3 and RbCaCl3 Crystals (000633
Irreversible Thermodynamics of the Cellular Growth of a | Dilute Binary Alloy Single Crystal (Fe-Ni, eutectics,   000113
Topography of Cellular Structures of Monocrystals of | Dilute Binary Metallic Alloys (Cu - 0.5%Si, Cu1-xSbx -   000210
#Specific Heat of a Quasi-Two- | Dimensional Antiferromagnet Ni(OH)2, hydrothermal)*   000273
#A New Technique for Simulating Two- | Dimensional Nucleation Models of Crystal Growth (theory)   001038
#Epitaxial Relationships During the Formation of Three- | Dimensional Snow Dendrites (H2O)*   000493
Adatoms by Point Capture Centres for Different Two- | Dimensional Substrate Lattices (theory)*   of Migrating   000873
#Structural and Magnetic Properties of "One- | Dimensional" Barium Vanadium Triselenide (BaVSe3)*   000474
of Lithium Dihydrogen Phosphate. II. Crystal Growth by | Dimer Addition (LiPO3)*   During Polymerization   000097
III. Growth of alpha-HgI2 Single Crystals from | Dimethylsulfoxide Complexes (theory)*Soluble Complexes.   000695
(Al, films)* #Single-Crystal-Aluminum Schottky-Barrier | Diodes Prepared by Molecular-Beam Epitaxy (MBE) on GaAs   000190
#Preparation and Characterization of Ruthenium | Dioxide Crystals (RuO2, vapor transport)*   000865
Argon (UO2, vapor transport)*   #Growth of Uranium | Dioxide Crystals Using Tellurium Tetrachloride and   000280
#Crystal Growth of Actinide | Dioxides by Chemical Transport*   000912
of Garnet Liquid Phase Epitaxy Using Horizontal | Dipping (flux)*   #The Growth Kinetics   001021
for Preparation of Single Crystals by the Method of | Directed Vertical Crystallization (theory, kinetics)*   000057
(vapor transport, film)*   #Crystal Growth in c | Direction and Crystallographic Polarity in ZnO Crystals   000426
Zone Method*   #Preferred Growth | Direction of Pure Cu Single Crystals by the Floating-   000380
of Single Crystals of FeGe2 on Their Growth Rate and | Direction*   #Dependence of Degree of Perfection   000746
Coefficients from Experimental Data: Theoretical (| directional crystallization, zone melting, eutectics)*   001080
#Single Crystal Growth of LiInS2 (| directional solidification)*   000456
Rare Earth Oxide Crystals (ZrO2-RE203, eutectics, | directional solidification)* Skull-Melting of Zirconia-   000629
Al-Al3Ni)*   #Effect of Some Thermal Parameters on the | Directional Solidification Process (theory, eutectics,   000268
effects,   #Non-Constant Distribution Coefficients for | Directionally Solidified InSb-GaSb (magnetic field   000864
#Crystallography of the fcc Phase in Uni- | Directionally Solidified Zn-Al Eutectic*   000421
Germanium (Ge, theory, kinetics)* #Preferred Growth | Directions and Perfection of Single Crystals of   000275
Properties of Deuterated TGS Produced by the Rotating | Disc Technique (solution)*   #Growth and Electrical   000560
Er2Si207,   #Flux Growth of Polymorphic Rare-Earth | Disilicates, R2Si2O7 (R = Tm, Er, Ho, Dy) (Tm2Si2O7,   000595
of Growth Conditions on Mirodefect Distribution in | Disloation-Free Single Crystals of Si*   #Influence   000687
Method of High Purity Aluminum Single Crystals with Low | Dislocation Density (Al, zone melting, theory)*   000228
Czochralski)*   #Impurity Effect on Grown-In | Dislocation Density of InP and GaAs Crystals (000856
Gallium Garnet (GGG, theory, Gd3Ga5012)*   # | Dislocation Propagation in Czochralski Grown Gadolinium   000838
Crystallization (SiO2)*   #Dislocation-Free and Low- | Dislocation Quartz Prepared by Hydrothermal   000068
GaP,   #Relation Between Growth Conditions and | Dislocation Structure in Gallium Phosphide Crystals (000723
#Rate Equations for Dislocation-Free and | Dislocation-Assisted Growth of Gallium (Ga, theory)*   000001
Gallium (Ga, theory)*   #Rate Equations for | Dislocation-Free and Dislocation-Assisted Growth of   000001
Hydrothermal Crystallization (SiO2)* | Dislocation-Free and Low-Dislocation Quartz Prepared by   000068
Crystals (Si)*   # | Dislocation-Free Czochralski Growth of (110) Silicon   000262
Effect of Doping on Microdefect Formation in As-Grown | Dislocation-Free Czochralski Silicon Crystals*   #The   000223
the Crystal Growth of Silicon Crystals, Free from | Dislocations (theory)*   and Crystal Quality During   000335
Melt on a Pedestal. III. Characteristic Arrangement of | Dislocations (Nb)*   Single Crystals by Pulling from a   000684
Phosphate Crystals from Aqueous Solution (#The Role of | Dislocations in the Growth of Ammonium Dihydrogen   000219
InSb, Al203, melt, theory)*   #Formation of Stress and | Dislocations in Crystal Growth (Si, Ge, GaAs, InAs,   000406
review)*   #Nature, Origin and Effect of | Dislocations in Epitaxial Semiconductor Layers (III-V,   000126
Technique*   #Analysis of | Dislocations in Gd3Ga5012 by a Repeated Etching   000943
Edge-Defined, Film-Fed Growth Technique (alpha-Al203)* # | Dislocations in Sapphire Ribbon Crystals Grown by the   001051
slow cooling)* | Dispersion of Acoustic Phonons in Layered SrGa2 (melt,   000672
Gallium Garnet (Gd3Ga5012, Czochralski, kinetics, | Dissociation Processes and Crystal Growth of Gadolinium   000769
#The Kinetics of | Dissolution (theory)*   000498
to Calcium Sulphate #The Affinity of Crystal Growth and | Dissolution in Aqueous Solution with Special Reference   000193
#Growth and | Dissolution of BaFBr Crystals (flux)*   000908
Autoclaves (SiO2, hydrothermal)*   #On the | Dissolution of Grown Synthetic Quartz Inside the   000507
Convection Regimes (solution,   #Kinetics of Growth and | Dissolution of Sodium Chlorate in Diffusion and   000888
flux, kinetics,   #The Liquidus Temperature and Growth- | Dissolution Kinetics of Garnet Liquid Phase Epitaxy (000333

PERMUTED TITLE INDEX

by LPE (GaxAl1-xAs)*   #Growth and
   of Cesium Iodide Crystals and Comparison of Growth and
      and Properties of High Purity Zn(1-x)MgxTe Alloys (
transport, film)*   #Tetragonal
   as a Diffusion Boundary Layer Thickness (effective
   of the Solidification Front Shape on the Temperature
Crystal Growth (Si)*   #Variations of Impurity
         #Influence of Growth Conditions on Mirodefect
Carbide Monocrystal and Epitaxial #Study of Temperature
                #Evaluation of Temperature
InSb-GaSb (magnetic field effects,   #Non-Constant
Theoretical (directional crystallization, #Non-Constant
Periodic Remelt (Si, #Computation of Striated Impurity
*   #A Theoretical Study of Temperature
                       #Americium
   Crystal Growth of Molybdenum Trioxide and Molybdenum
   #Electrical and Optical Properties of Copper Indium
Microscopy (W   #Recarburization of Monocrystalline
Br, and I) (melt,   #New Mixed Halide Compounds MPX of
                #Crystals of CTA-
Diffraction (evaporation, H3PW12O40.21H2O)*   #
Beam Epitaxial Growth of IV-VI Semiconductors #Impurity
      #Low-Defect InSb Crystal Growth by InN
         #Si-MBE: Growth and Sb
Phase. II.   #Temperature Dependence of Degree of
I. Phosphorus #Temperature Dependence of the Degree of
Beam Epitaxy*   #Nitrogen
Growth from the Gas Phase (Si)*   #
Free Czochralski Silicon Crystals*   #The Effect of
Epitaxial Growth #In-Process Monitoring and Control of
beam epitaxy, review)*   #Periodic
Simultaneous Arsenic Ion Implantation and by #n-Type
   Polycrystalline Silicon Film Deposition and by Film
evaporation)*   #Silicide Formation by High-
#Growth of Single Crystals of Cubic Zinc Sulphide by
   Epitaxy of AlxGa1-xP on GaP and Its Application to
            #Crystal Structure of
Ln3NbO7) (Ho3NbO7, Y3NbO7, Er3NbO7,   #Flux Growth of
Oxide La6Re4O18 (melt)*   #Crystal Structure of
for Measuring Local Lattice Mismatches   #An X-Ray
Crystallization Experiments in Spacelab (CdTe, zone #A
            #Experiments on Surface Tension
   Floating Zone Melting. II. Combined Free and Rotation
Convection in Floating Zone Melting. I. Pure Rotation
theory, kinetics)*   #The Thermodynamic
theory, kinetics, melt, solution)*   #On the
Morphology of Ice Dioxtals Grown from Supercooled Water
         #On the Relations Between the Number of Gold
            #Morphology of Ice
   Rare-Earth Disilicates, R2Si2O7 (R = Tm, Er, Ho,
   and Crystallographic Characterization (Ln = Nd, Sm,
      Compounds, RCu2. I. Heavy Rare Earth (TbCu2,
Ho3Fe5O12, Dy3Fe5O12, Er3Fe5C12, Ca3Fe2Ge3O12, HoFeO3,
   by the Horizontal Levitation Zone Melting Method (
Growth of Some Rare Earth Trifluorides (GdF3, TbF3,
   Films of the Rare-Earth Dihydrides (GdH2, TbH2,
Supercooling in Fluxed Melt Systems (Ho2Si2O7, TbVO 4,
#Viscosity Measurements in the System Dy2O3-K2O-MoO3 (
   Compositions for Flux Growth (DyPO4, PrBO3, LaBO3,
   R2MoO6 and R6MoO12 in the Systems R2O3-K2O-MoO3 (
Iodide Crystals in Small Opened Ampoules (HgI2)*
theory, C)*   #Temperature Field
            #Surface Charge and Ledge
Silicide (Pd2Si)*
(theory, melt, slow cooling)*
   and Freezing in the Gaussian Core Model by Molecular
transport, kinetics)*   #A Molecular
   DyKMo2O8, ErKMo2O8, NdKMo2O8, DyVO4, TbVO4, TbPO4,
   of Modifying Starting Compositions for Flux Growth
   (DyPO4, PrBO3, LaBO3, DyKMo2O8, ErKMo2O8, NdKMo2O8,
      NdKMo2O8, DyVO4, TbVO4, TbPO4, DyPO4, Ho2Ge2O7,
   Systems R2O3-K2O-MoO3 (DyKMo2O8, ErKMo2O8, NdKMo2O8,
         #Viscosity Measurements in the System
   Ta2Ti3O7, Tb2Ti2O7, Pr2Ti2O7, Tm2Ge2O7, Er2SiO5,
   (R = Tm, Er, Ho, Dy) (Tm2Si2O7, Er2Si2O7, Ho2Si2O7,
Temperature Conductivity of Gd2(SO4)38H2O (crystal) and
   Emission of Cubic Bi4Ge3O12 Crystals Doped with
            #The Growth of
   Shape Transitions in the Czochralski Growth of
Lead Flux with Addition of V2O5 (Y3Fe5O12, Ho3Fe5O12,
   Earth Elements (Ln3NbO7) (Ho3NbO7, Y3NbO7, Er3NbO7,
Ammonium Dihydrogen Arsenate (DADA) (NH4(1-x)D4xH2(1-x)
Deuterium Ammonium Dihydrogen Arsenate (DADA) (NH4(1-x)
   of Crystals in Solution (review, theory, alkaline
Copper Intermetallic Compounds, RCu2. II. Light Rare
   Copper Intermetallic Compounds, RCu2. I. Heavy Rare
9, (Ba0.93Gd0.07)   #The Flux Growth of Some New Rare
LaBO3, (La,Pb)MnO3, (Nd, #The Flux Growth of Some Rare-

Dissolution Kinetics of III-V Heterostructures Formed   000894
Dissolution Processes (CsI, solution)*Selective Etching   000827
distillation)*   #Preparation, Analysis   000792
Distortion in Heteroepitaxial Layers: Ge On GaAs (vapor   000362
distribution coefficient, melt, theory)*   Equation   001083
Distribution in the Crystal (NaCl, theory, melt)   000128
Distribution in the Modified Czochralski Method of   000997
Distribution in Disloation-Free Single Crystals of Si*   000687
Distribution in High-Temperature Furnaces for Silicon   000561
Distribution of Melt in Silicon Ribbon Growth (EDFG)*   000523
Distribution Coefficients for Directionally Solidified   000864
Distribution Coefficients from Experimental Data: 001080
Distributions in Melt-Grown Crystals, Taking Account of   001026
Distributions During Czochralski Crystal Growth (Cu, Si) 000149
Ditelluride (AmTe2-x*   000153
Ditelluride (MoO3-x, MoTe2)* #Vapor Phase Transport and   000299
Ditelluride Crystals Grown from Near-Stoichiometric   000220
Ditungsten Carbide as Revealed by Field Electron   000724
Divalent Rare Earths (M = Sm, Eu, Tm, and Yb - X = Cl,   000088
DNA: A Growth and Morphological Study (solution)*   000488
Dodecatungstophosphoric Acid-21-Water by Neutron   000913
Dopant Incorporation and Diffusion During Molecular   000898
Doping (Czochralski)*   000971
Doping (Si, film)*   000502
Doping in Growth of Epitaxial Silicon from the Gas   000515
Doping in Silicon Epitaxial Growth from the Gas Phase.   000514
Doping into GaAs1-xPx Using Ionized Beam in Molecular   000617
Doping of Silicon Crystals with Phosphorus During   000027
Doping on Microdefect Formation in As-Grown Dislocation- 000223
Doping Gas Concentration During Vapor Phase Silicon   000937
Doping Structure in GaAs (GaAs, Ga1-xAlxAs, molecular   000248
Doping Techniques in Silicon Molecular Beam Epitaxy by   000734
Doping Using Boron Diffusion (Si)*   Wafers by Thin   000985
Dose Si+-Ion Implantation of Pd (electron beam melting,   000177
Double Decomposition Reaction in Melt (ZnS, flux)*   000328
Double Heterostructure Light Modulators* #Liquid Phase   000948
Double Molybdate K2Ni(MoO4)2 (slow cooling, flux)*   000486
Double Oxides of Niobium and Rare-Earth Elements (001097
Double Oxides of Rhenium. II. The Lanthanum-Rhenium   000103
Double-Crystal Method Utilizing Non-Parallel Setting   000704
Double-Ellipsoid Mirror Furnace for Zone   000276
Driven Flow in Floating Zone Melting (NaNO3, theory)*   000848
Driven Flows (kinetics, theory, electron beam melting)*   000174
Driven Flows (Si, Al2O3, kinetics, float zone melting,   000173
Driving Force for Crystallization from Solution (000904
Driving Force for Crystallization: The Growth Affinity (001024
Droplets (H2O)*   # 000347
Drops and TiP Whiskers in VLS Growth*   000929
Droxtals Grown from Supercooled Water Droplets (H2O)*   000347
Dy) (Tm2Si2O7, Er2Si2O7, Ho2Si2O7, Dy2Si2O7)*   000595
Dy, skull-melting)*   Solidification, Microstructural 000630
DyCu2, HoCu2, ErCu2)*of Rare Earth Copper Intermetallic   000373
DyFeO3, ErFeO3)* Flux with Addition of V2O5 (Y3Fe5O12,   000565
DyFe2, TbFe2, Nd, Pr, Ce, cold crucible)*Crystal Growth   000620
DyF3, HoF3, ErF3, Bridgman-Stockbarger, Czochralski)*   000323
DyH2, ErH2, HoH2)*The Growth and Structure of Epitaxial   000782
DyKMo2O8)*   #Relative Supersaturation and   000594
DyKMo2O8, melt)*   000320
DyKMo2O8, ErKMo2O8, NdKMo2O8, DyVO4, TbVO4, TbPO4,   001060
DyKMo2O8, ErKMo2O8, NdKMo2O8, Dy2MoO6, Sm2MoO6,   001062
Dynamic Sublimation Technique for Growing Mercuric   000376
Dynamics in the Synthesis of Diamonds (kinetics,   000862
Dynamics in Cadmium Sulfide (CdS, evaporation)*   000529
Dynamics of Laser-Induced Formation of Palladium   001042
Dynamics of Non-Equilibrium Crystallization of an Alloy   000826
Dynamics Simulation (melt, slow cooling, theory)*   000915
Dynamics Study of Surface Melting (theory, vapor   000140
DyPO4, Ho2Ge2O7, Dy2GeMoO8, ZrSiO4, ThGeO4, Er2Si2O7)*   001060
DyPO4, PrBO3, LaBO3, DyKMo2O8, ErKMo2O8, NdKMo2O8,   001060
DyVO4, TbVO4, TbPO4, DyPO4, Ho2Ge2O7, Dy2GeMoO8,   001060
Dy2GeMoO8, ZrSiO4, ThGeO4, Er2Si2O7)*   ErKMo2O8,   001060
Dy2MoO6, Sm2MoO6, Pr2MoO6, Ho6MoO12)*and R6MoO12 in the   001062
Dy2O3-K2O-MoO3 (DyKMo2O8, melt)*   000320
Dy2SiO5, Ho2SiO5, Pr2MoO6, LiCoPO4*Pb) MnO3, Pb3Mn7O15,   001061
Dy2Si2O7)* Polymorphic Rare-Earth Disilicates, R2Si2O7   000595
Dy2Ti2O7 (powder) as a Function of Magnetic Field (000065
Dy3+, Ho3+, Er3+, Tm3+, or Yb3+ Ions (Czochralski)*   000458
Dy3Al5O12 (Czochralski)*   000646
Dy3Al5O12*   #Interface   000644
Dy3Fe5O12, Er3Fe5O12, Ca3Fe2Ge3O12, HoFeO3, DyFeO3,   000565
Dy3NbO7, La3NbO7, Y3TaO7, Er3TaO7, Gd3TaO7)*   and Rare- 001097
D2xAsO4, solution)*   Phases of Deuterium   000100
D4xH2(1-x)D2xAsO4, solution)*   Phases of   000100
earth sulfates, calcium phosphates)*   #The Growth   000683
Earth (PrCu2, NdCu2)**Magnetic Properties of Rare Earth   000374
Earth (TbCu2, DyCu2, HoCu2, ErCu2)*   of Rare Earth   000373
Earth and Iron Group Complex Oxides (GdVO3, Gd4Ba6(BO3)   001063
Earth and Iron Group Complex Oxides (NdBO3, LaBO3/Pr,   001061

PERMUTED TITLE INDEX

Growth and Characterization by X-ray Topography of Rare — Earth Arsenates ($REAsO_4$, slow cooling)* #Flux — 000900

Their #The Preparation of Single Crystals of the Rare — Earth Borides by the Solution Method and a Study of — 000359

$GdOHCO_3$, $ErOHCO_3$, $Y_2O(OH)_2CO_3$, #Synthesis of Rare- — Earth Carbonates Under Hydrothermal Conditions ($YOHCO_3$, — 000169

Rare Earth ($TbCu_2$, $DyCu_2$, #Magnetic Properties of Rare — Earth Copper Intermetallic Compounds, RCu_2. I. Heavy — 000373

Rare Earth ($PrCu_2$, $NdCu_2$)* #Magnetic Properties of Rare — Earth Copper Intermetallic Compounds, RCu_2. II. Light — 000374

The Growth and Structure of Epitaxial Films of the Rare- — Earth Dihydrides (GdH_2, TbH_2, DyH_2, ErH_2, HoH_2)* # — 000782

$Tm_2Si_2O_7$, $Er_2Si_2O_7$, #Flux Growth of Polymorphic Rare- — Earth Disilicates, $R_2Si_2O_7$ (R = Tm, Er, Ho, Dy) (— 000595

#Flux Growth of Double Oxides of Niobium and Rare- — Earth Elements (Ln_3NbO_7) (Ho_3NbO_7, Y_3NbO_7, Er_3NbO_7, — 001097

$3(FeGa)_5O_{12}$)* #Magneto-Optical Properties of Some Rare — Earth Iron Garnets (flux, $(GdPrBi)_3(FeGaIn)_5O_{12}$, (YLaBi) — 001107

$RExY_{3-x}GayFe_{5-y}O_{12}$, flux)* #A Solubility Model for Rare- — Earth Iron Garnets in a PbO/B_2O_3 Solution ($Y_3Fe_5O_{12}$, — 001022

in Gadolinium Molybdate and Other Isotypic Rare- — Earth Molybdates (Czochralski, melt, $Gd_2(MoO_4)_3$, Sm_2(— 000446

Data of Some $Ln_2(MoO_4)_3$ Type Mixed Rare — Earth Molybdates ($RE(MoO_4)_3$, Czochralski)* — 000447

directional #Growth from Skull-Melting of Zirconia-Rare — Earth Oxide Crystals (ZrO_2-RE_2O_3, eutectics, — 000629

Sr_2NdAlO_5, #Crystal Growth of Some Alkaline-Earth Rare — Earth Pentaoxometallates (flux, Sr_2LaAlO_5, Sr_2PrAlO_5, — 000253

Sr_2PrAlO_5, Sr_2NdAlO_5, #Crystal Growth of Some Alkaline- — Earth Rare-Earth Pentaoxometallates (flux, Sr_2LaAlO_5, — 000253

#Czochralski Crystal Growth of Rare — Earth Tetraborides (YB_4, TbB_4, ErB_4)* — 000137

Bridgman-Stockbarger, #Crystal Growth of Some Rare — Earth Trifluorides (GdF_3, TbF_3, DyF_3, HoF_3, ErF_3, — 000323

Melts for the Growth of Single Crystals of Rare — Earth Vanadates and Phosphates (YPO_4, YVO_4, $REPO_4$, flux) — 000265

#Mixed Oxysulfides of Chromium III and Rare — Earths ($LaCrOS_2$, $CeCrOS_2$, $PrCrOS_2$, $NdCrOS_2$, $SmCrOS_2$)* — 001048

melt, #New Mixed Halide Compounds MFX of Divalent Rare — Earths (M = Sm, Eu, Tm, and Yb - X = Cl, Br, and I) (— 000088

Growth and Characterization of Mixed Bromides of Rare — Earths, Sodium, and Cesium ($Ca_2NaNdBr_6$, $Cs_2NaCeBr_6$, — 000626

Crystals and Products and Their Application in National — Economy, Leningrad, March 14-17, 1979 (review)*Profiled — 001111

#Evaluation of Temperature — 000523

Distribution of Melt in Silicon Ribbon Growth (— EDFG)*

Crystals Grown with Wetting and Non-Wetting Dies (Si, — EDFG, #Growth and Characterization of Silicon Ribbon — 000524

#Wide Silicon Ribbon Crystals (— EDFG, Si)* — 000522

$FexHfS_2$, #Intercalation-Induced Shift of the Absorption — Edge in ZrS_2 and HfS_2 ($CuxZrS_2$, $CuxHfS_2$, $FexZrS_2$, — 001090

Vapor Process* #The Optical Absorption — Edge of Single-Crystal AlN Prepared by a Close-Spaced — 000758

Edge Emission in Melt-Grown $ZnSexS_{1-x}$ (Bridgman)* — 000710

#Dislocations in Sapphire Ribbon Crystals Grown by the — Edge-Defined, Film-Fed Growth Technique (alpha-Al_2O_3)* — 001051

Equation as a Diffusion Boundary Layer Thickness (— effective distribution coefficient, melt, theory)* — 001083

MoS_2, $MoSe_2$, WS_2, #Optical Phonon Modes and Localized — Effective Charges of Transition-Metal Dichalcogenides (— 001000

#Impurity Redistribution in — EFG (Si, ribbon, theory, kinetics)* — 000451

#Aluminum Redistribution in — EFG of Silicon Ribbon (Si)* — 000452

on Potassium Bismuth Niobate Single Crystals by the — EGM Technique ($KxLi_{1-x}NbO_3$)* Single-Crystal Films — 000011

Lithium Niobate (KLN) Crystals ($K_{2.89}Li_{1.55}Nb_{5.11}O_{15}$, # — Elastic and Piezoelectric Properties of Potassium — 000010

Bridgman)* — Elastic Constants and Stability of bcc In-Tl Alloys (— 000581

Magnetocrystalline Anisotropy, Magnetostriction and — Elastic Constants of a Ruthenium-Nickel Single Crystal (— 000631

#Single-Crystal — Elastic Constants of Al_2Cu (Bridgman, arc melting)* — 000274

Tungsten Alloys (Nb-Zr, Nb-Hf, electron beam melting, # — Elastic Constants of Niobium-Zirconium, Hafnium, and — 000307

Crystal Growth and Study of Their Piezoelectric and — Elastic Properties (hydrothermal)* #Na_2CoGeO_4 Single — 000256

#Control of the Composition of Crystals Grown in — Electric and Magnetic Fields (Czochralski)* — 000511

Ice Crystals #Influence of Environmental Saturation and — Electric Field on Growth and Evaporation of Epitaxial — 000028

Ditelluride Crystals Grown from Near-Stoichiometric # — Electrical and Optical Properties of Copper Indium — 000220

Single Crystals of $GeFe_2O_4$ (vapor transport)* # — Electrical and Optical Properties of High-Purity p-type — 000920

Induced Epitaxial Layers in Silicon (Si)* # — Electrical and Structural Characteristics of Laser- — 001106

from the Melt* #Improving the — Electrical Characteristics of CdS Single Crystals Grown — 000588

SN)x Single Crystal* — Electrical Conductivity and Galvanomagnetic Effects in (— 000463

review)* #Subsidiary — Electrical Heating for Verneuil Furnaces in the USSR (— 000349

$Mo_6Cl_{10}Y$ (Y = S, Se, Te): Structure and Magnetic and # — Electrical Properties ($Mo_6Cl_{10}Se$, melt)* Molybdenum II: — 000757

#Vanadium (III) Oxychloride: Magnetic, Optical and — Electrical Properties -Lithium and Molecular — 001031

Substituted Garnet Films grown by Liquid Phase Epitaxy # — Electrical Properties and Charge Imbalance for Ca,Ge- — 000386

Crystals (melt)* — Electrical Properties of p- and n-Type $CuInSe_2$ Single — 000410

#Magnetic and — Electrical Properties of $BaVS_3$ and $BaVxTi_{1-x}S_3$ (flux)* — 000610

transport)* #Preparation and Certain — Electrical Properties of $CdCr_2S_4$ Single Crystals (vapor — 000780

#Optical and — Electrical Properties of $CdGeAs_2$ (Bridgman)* — 000412

Substrates (epitaxy)* #Structural and — Electrical Properties of $CuGaSe_2$ Thin Films on GaAs — 000844

Rotating Disc Technique (solution)* #Growth and — Electrical Properties of Deuterated TGS Produced by the — 000560

GaAs Substrates* #Growth and — Electrical Properties of Epitaxial $CuInS_2$ Thin Films on — 000690

Bridgman)* #Some — Electrical Properties of $GeBi_2Te_4$ Single Crystals (— 000975

Selendies $VxTi_{1-x}Se_2$ and $TaxTi_{1-x}Se_2$ (#Structural and — Electrical Properties of Layered Transition Metal — 000555

vapor transport)* — Electrical Properties of $Pb_{1-x}CdxS$ Epitaxial Films (— 000439

Crystal GaSb Films on GaAs Substrates* #Growth and — Electrical Properties of Sputter-Deposited Single- — 000269

theory)* — Electrical Properties of TlTe Single Crystals (melt, — 000405

#Metastable $Pb_{1-x}CdxS$ Epitaxial Films. II. — Electrical Properties* — 000910

Bridgman)* #Effect of Pressure on — Electrical Resistivity and Thermoelectric Power of FeS (— 000465

Single Crystal (Bridgman)* — Electrical Resistivity in In-Heusler $Cu_{2.0}Mn_{0.94}In_{1.0}$ — 000499

#Effect of High-Temperature Annealing in Cadmium on the — Electrical Transport Properties of Single Crystals of — 000612

Transitions and Phase Diagram of $(Ti_{1-x}Vx)_4O_7$: — Electrical, Calorimetric, Magnetic and EPR Studies (— 000835

$Fe_{2-x}CrxO_3$ (vapor transport, electrolytic #The — Electrical, Optical and Photoconducting Properties of — 000624

Czochralski, electrolytic #Growth Rate Limitations in — Electrochemical Crystallization (metals, Nb, $NaxWO_3$, — 000225

from Molten Salts (flux)* — Electrocrystallization of NbO and Hollow NbO_2 Needles — 000204

theory)* — Electrocrystallization. Fundamental Aspects (review, — 000226

Stable Growth of Epitaxial GaP Layers by Molten Salt — Electrodeposition (flux)* #Conditions for — 000224

Gold and Silver (Sb)* #Epitaxial Growth of Antimony — Electrodeposits on Oriented Single Crystal Surfaces of — 000540

#Surface Photovoltage Experiments on $SrTiO_3$ — Electrodes (theory)* — 000618

theory, electrolytic deposition, kinetics)#Liquid-Phase — Electroepitaxy: Growth Kinetics (Peltier effect, GaAs, — 000437

Electroless Deposition of Epitaxial $Pb_{1-x}HgxS$ Films* — 000869

PbSe, $Pb_{1-x}HgxS$, solution)* — Electroless Deposition of Semiconductor Films (PbS, — 000868

#Crystal Growth of the Solid — Electrolyte $(C_5H_5NH)_2Cu_5Br_7$ From the Melt (Bridgman)* — 000890

#The Crystal Structure of the Solid — Electrolyte $Ag_{44}I_{53}(C_{11}H_{30}N_3)_3$ (solution)* — 000973

#Structural Studies of the Solid — Electrolyte High-$LiTa_3O_8$ (electrolytic deposition)* — 000706

solution)* #Raman Scattering Studies in the Solid — Electrolytes of the $RbAg_4I_5$ Family (KAg_4I_5, $NH_4Ag_4I_5$, — 000315

Studies of the Solid Electrolyte High-$LiTa_3O_8$ (— electrolytic deposition)* #Structural — 000706

Crystals from the Plasmas of ZrO_2, HfO_2, ThO_2 and CeO_2 (— electrolytic deposition)* #Deposition of — 000472

Properties of $Fe_{2-x}CrxO_3$ (vapor transport, — electrolytic deposition)* Optical and Photoconducting — 000624

Single Crystals by Chemical Transport (Fe_2TiO_5, — electrolytic deposition)* Preparation of Pseudobrookite — 000770

Crystallization (metals, Nb, $NaxWO_3$, Czochralski, — electrolytic deposition)*Limitations in Electrochemical — 000225

Growth Kinetics (Peltier effect, GaAs, theory, — electrolytic deposition, kinetics)* Electroepitaxy: — 000437

PERMUTED TITLE INDEX

#Li3N: A Promising Li Ionic Conductor (electrolytic deposition, review)* 000024
#A Simple Method for Growing V3Si Single Crystals (electron beam melting)* 000162
Free and Rotation Driven Epitaxy Flows (kinetics, theory, electron beam melting)* Zone Melting. II. Combined 000174
Formation by High-Dose Si+-Ion Implantation of Pd (electron beam melting, evaporation)* #Silicide 000177
#Properties and Synthesis of Nb-H Interstitial Alloys (electron beam melting, float zone melting, review)* 001072
CdS Evaporated in Vacuum onto Silicon (vapor transport, electron beam melting, kinetics, theory)* of Films of 000073
Zirconium, Hafnium, and Tungsten Alloys (Nb-Zr, Nb-Hf, electron beam melting, zone melting)* of Niobium- 000307
#Growth of Cobalt Single Crystals by the Electron Beam Floating-Zone Method (Co)* 000381
the Compound CuSe* # Electron Diffraction Study of Phase Transformations of 000866
transport)* # Electron Energy-Loss Spectroscopy of TiS2 (vapor 000288
of Solidification Interfaces with a Very High Voltage Electron Microscope* #In-Situ Observation 000163
Hydrothermal Growth of Synthetic 11 Angstroms # Electron Microscopic Characterization of the 000366
Prepared by Water Solvent (solution)* # Electron Microscopic Characterization of Cubic BN 000934
Monocrystalline Ditungsten Carbide as Revealed by Field Electron Microscopy (W microcrystal, crystal growth 000724
Thin Films* #Techniques for Routine UHV in situ Electron Microscopy of Growth Processes of Epitaxial 000950
Quartz (SiO2, hydrothermal)* #Scanning Electron Microscopy of Growth Surfaces of Synthetic 000857
ZnTe: Shape of the Tellurium Solidus Line #Transmission Electron Microscopy Observations of Precipitates in 000098
Crystals (MoO3, Mo4O11, microcrystals)* # Electron Microscopy Study of Some Molybdenum Oxide 000539
Single Crystals of Silicon-Germanium Solid Solutions by Electron-Beam Zone Melting Without a Crucible (Si-Ge)* 000820
* #Shape of the Crystallization Front During Electron-Beam Zone Refining of beta-B Without a Crucible 000968
Growth of Some II-VI Compound Films Evaporated Onto Electron-Bombarded NaCl Substrates (CdS, CdSe, CdTe, 000878
vapor transport)* # Electronic Bandgap Measurements of SnS2 Polytypes (000008
of Some Chalcogenide Spinel Single Crystals and Their Electronic Properties (FeCr2S4, CoCr2S4, vapor 001066
Vapour Grown FeCr2S4* #Step Growth by Elementary Steps and Multi-Steps on the Surface of 000419
#Correlation of the Interface Reaction Constant with Elementary Surface Processes (theory)* 000330
Defects in Solid Phase Epitaxy Si Films Deposited at Elevated Temperatures (vapor transport)* #Crystalline 000194
Experiments in Spacelab (CdTe, zone melting, #A Double- Ellipsoid Mirror Furnace for Zone Crystallization 000276
Layers During Their Growth in an Organometallic VPE # Ellipsometric Assessment of (Ga,Al)As/GaAs Epitaxial 000974
(La, Eu, Y, Ce, Ba, Cs) Hexaborides for Thermionic Emission (LaB6, EuB6, CeB6, BaB6, (La,Eu)B6, La1-xEuB6, 000727
#Edge Emission in Melt-Grown ZnSexS1-x (Bridgman)* 000710
Garnets ((Y1- #Structure, Spectroscopy, and Stimulated Emission of Crystals of Yttrium Holmium Aluminum 000046
Ho3+, Er3+, Tm3+, #Growth, Spectroscopy, and Stimulated Emission of Cubic Bi4Ge3O12 Crystals Doped with Dy3+, 000458
Spectroscopic Investigations, and Some New Stimulated Emission Data of Gd3Ga5O12:Nd3+ Single Crystals (000457
solution)* # Enantiomorphism of alpha-LiIO3 Single Crystals (000682
#Simple Pressurized Chambers for Liquid Encapsulated Czochralski Crystal Growth (InP, CuInS2)* 000150
transport, sublimation)* #Liquid Encapsulation Zone-Refining of PbS (zone melting, vapor 000583
Amorphous State (Czochralski, Bi4Si3O12)**Change in the Energy Band Structure of Eulytine on Transition to the 000294
in a Self-Implanted Silicon Layer on Laser Irradiation Energy Density (Si)* #Grain Size Dependence 000996
#Electron Energy-Loss Spectroscopy of TiS2 (vapor transport)* 000288
#Growth and Some Properties of Cerium Sulphate Enneahydrate Single Crystals (Ce2(SO4)3.9H2O, solution)* 000227
#The Growth of LaAlO3 Crystals Enriched in Isotope 17O (flux)* 000901
II. Phosphorus Content and Differential Molar Solution Enthalpies (Si, vapor transport, theory)*the Gas Phase. 000515
Heterogeneous Reactions (GaAs, vapor #The Modified Entrainment Method and Its Application to the Study of 000281
#Re-Investigation of the Re- Entrant Corner Effect in Twinned Crystals (theory)* 000484
#Percolation Theory and Crystallization (entropy, specific heat, melting, theory)* 000168
#Correlation of Diffusion Behaviour with the Entropy of Fusion (theory)* 000980
melting, doped, #Materials Behaviour in Low Gravity Environment (melt, solution, vapor transport, zone 001070
#Vapour Growth in a Microgravity Environment (review)* 000091
and Evaporation of Epitaxial Ice Crystals #Influence of Environmental Saturation and Electric Field on Growth 000028
#Preparation of Ge-Si Epitaxial Alloys by Sputtering* 000420
#Si Contamination in Epitaxial Boron Monophosphide (BP)* 000713
Crystals (CdS, #Direct Synthesis and Structure of Epitaxial Cadmium Sulphide Layers on Cadmium Single 000429
#Growth and Electrical Properties of Epitaxial CuInS2 Thin Films on GaAs Substrates* 000690
#Liquid Phase Epitaxial Deposition of GaP on GaAs* 000354
System* #Influence of the VPE Parameters on the Epitaxial Film Properties for the (Pb1-xSnx)1+yTe1-y 000433
#Electrical Properties of Pb1-xCdxS Epitaxial Films (vapor transport)* 000439
Setting for Measuring Local Lattice Mismatches Between Epitaxial Films and Substrates (RE garnets)* Parallel 000704
TbH2, DyH2, ErH2, HoH2)* #The Growth and Structure of Epitaxial Films of the Rare-Earth Dihydrides (GdH2, 000782
transport)* #Dichlorosilane as Silicon Source for Epitaxial Films on Insulators (Si on sapphire, vapor 000254
#Metastable Pb1-xCdxS Epitaxial Films. I. Growth and Physical Properties* 000909
#Metastable Pb1-xCdxS Epitaxial Films. II. Electrical Properties* 000910
Crystal Lattice #Growth Kinetics and Structure of Epitaxial Garnet Layers as Functions of the Mismatch of 000251
flux)* #Conditions for Stable Growth of Epitaxial GaP Layers by Molten Salt Electrodeposition (000224
transport)* #Vapor-Phase Growth of Epitaxial Ga1-xInxSb Using Open-Tube Flow System (vapor 000681
#In Situ X-Ray Topography of Epitaxial Ge Layers During Growth* 000363
#Quasi-Grown Layers in Liquid Phase Epitaxial Growth (In1-xGaxAs)* 000951
of Doping Gas Concentration During Vapor Phase Silicon Epitaxial Growth by Using a Flame Photometric Detector (000937
#Theory of Crystallization During Epitaxial Growth from a Solution in a Melt (GaP, flux)* 000951
Dependence of the Degree of Doping in Silicon Epitaxial Growth from the Gas Phase. I. Phosphorus 000514
Methods Used for the Characterization of Modes of Epitaxial Growth from the Vapor Phase (review, theory)* 000542
Graph (vapor transport, theory, Si)* #Model of Silicon Epitaxial Growth in SiCl4-HCl-H2 System Based on Flow 000503
Vapour Phase (theory)* #Flow-Graph Model of Epitaxial Growth of A(III)B(V) Compounds from the 000504
Vapour Phase (review, theory)* # Epitaxial Growth of A(III)B(V) Semiconductors from 000352
Oriented Single Crystal Surfaces of Gold and Silver (# Epitaxial Growth of Antimony Electrodeposits on 000540
Crystals* # Epitaxial Growth of BaTiO3 Vacuum Condensates on SrTiO3 000984
Epitaxial Growth of CdS on CuInSe2* 000556
* #Liquid Phase Epitaxial Growth of CdTe in the CdTe-CdCl2 System (flux) 000288
#Liquid Phase Epitaxial Growth of Copper Ferrite Films (CuFe2O4)* 001016
transport)* # Epitaxial Growth of CrO2 on Sapphire in Air (vapor 000413
O3) Thin Films* # Epitaxial Growth of Ferroelectric PLZT ((Pb,La)(Zr,Ti) 000415
Epitaxial Growth of Ferromagnetic CrO2 Films in Air* 000414
#Ionized-Cluster Beam Epitaxial Growth of GaP Films on GaP and Si Substrates* 000651
Grading (melt)* #Liquid Phase Epitaxial Growth of Ga1-xInxSb on GaSb by Stepwise 000803
Incorporation and Diffusion During Molecular Beam Epitaxial Growth of IV-VI Semiconductors (PbTe, Pb1- 000898
KLaP4O12, solution)* # Epitaxial Growth of KNdP4O12 Laser Waveguides (000643
Layer Under Controlled Phosphorus Vapor #Liquid Phase Epitaxial Growth of Lattice-Matched GaP-Ga0.5Al0.5P-GaP 000460
flux, Li0.5Fe2.5-xAlxO4)* #Liquid Phase Epitaxial Growth of Lithium Ferrite-Aluminate Films (001017
#Liquid Phase Epitaxial Growth of Magnetic Garnets (review)* 000783

PERMUTED TITLE INDEX

Crystal Films on Potassium Bismuth Niobate Single # Epitaxial Growth of Potassium Lithium Niobate Single- 000009
Crystal Films on Potassium Bismuth Niobate Single # Epitaxial Growth of Potassium Lithium Niobate Single- 000011
Rocksalt Substrates* #The Epitaxial Growth of Reactively Sputtered CuI Films on 000788
#Solid Phase Epitaxial Growth of Si Through Al Film* 000584
111) GaAs Substrates Using the Close-Spaced Geometry (# Epitaxial Growth of Single Crystal Cd1-xZnxS Layers on (000305
Evaporated Onto Electron-Bombarded NaCl Substrates (# Epitaxial Growth of Some II-VI Compound Films 000878
#Liquid Phase Epitaxial Growth of Zn and S Doped GaAs* 000416
Magnetron rf Sputtering System* #High Rate Epitaxial Growth of ZnO Films on Sapphire by Planar 000882
Epitaxial Growth of ZnS by Zn-S-H2 CVD Method* 000615
Flow (#Mass Transport Processes Associated with the Epitaxial Growth of ZnS in H2: The Effects of a Bypass 000558
#Vapour-Phase Epitaxial Growth of ZnS on GaP* 000613
Epitaxial Growth of ZnSe by Vapor Flow Method* 000669
#Characterization of (001) Tilt Boundaries Produced by Epitaxial Growth on Bicrystalline Substrates of NaCl (000207
theory)* #Study of Silicon Carbide Epitaxial Growth Kinetics in the SiC-C System (SiC, 000562
#Temperature Gradient Measurement in Liquid Epitaxial Growth Systems (flux)* 000574
and Electric Field on Growth and Evaporation of Epitaxial Ice Crystals (H2O, vapor transport, 000028
Furnaces for Silicon Carbide Monocrystal and Epitaxial Layer Growing (SiC, melt)*in High-Temperature 000561
of Kinetic and Thermal Conditions of Silicon Carbide Epitaxial Layer Growth from the Vapour Phase (SiC, 000942
#The Mechanism of Silicon Epitaxial Layer Growth from Ion-Molecular Beams (Si)* 000019
#Preparation and Optical Properties of ZnSe Epitaxial Layers by a Close-Spaced Technique* 000670
#Growth and Morphology of 6H-SiC Epitaxial Layers by CVD (film)* 000699
Factor Influencing the Growth Rate and Purity of Epitaxial Layers in the Ga_AsCl3-H2 System (GaAs, vapor 000653
and Structural Characteristics of Laser-Induced Epitaxial Layers in Silicon (Si)* #Electrical 001106
(flux)* # Epitaxial Layers of the Laser Material Nd(Ga,Cr)3(BO3)4 000578
Epitaxial Layers of CuInSe2 on GaAs* 000843
#Determination of the Lattice Constant of Epitaxial Layers of III-V Compounds (theory)* 000399
Chromium-Chlorine Boracite (#Gas Phase Synthesis of Epitaxial Layers of Nickel-Chlorine Boracite on 000837
Sublimation in Vacuum (SiC)* #Growth Kinetics of Epitaxial Layers of Silicon Carbide Obtained by 000940
#Ellipsometric Assessment of (Ga,Al)As/GaAs Epitaxial Layers During Their Growth in an 000974
Vapour #Growth and Characterization of Undoped ZnSe Epitaxial Layers Obtained by Organometallic Chemical 000114
#Liquid Phase Growth of HgCdTe Epitaxial Layers* 001059
#Electroless Deposition of Epitaxial Pb1-xHgxS Films* 000869
Triglycine Sulphate Thin Films Grown by a New Epitaxial Process (TGS)* Trichroism of Ferroelectric 000329
Dimensional Snow Dendrites (H2O)* # Epitaxial Relationships During the Formation of Three- 000493
#Nature, Origin and Effect of Dislocations in Epitaxial Semiconductor Layers (III-V, review)* 000126
Temperature Dependence of Degree of Doping in Growth of Epitaxial Silicon from the Gas Phase. II. Phosphorus 000515
Epitaxial Stratified Growth (kinetics, theory)* 000017
UHV in situ Electron Microscopy of Growth Processes of Epitaxial Thin Films* #Techniques for Routine 000950
transport)* # Epitaxial Thin Iron Films on Antimony (Fe, vapor 000875
III-V Compounds and Alloys (Ge, #Some Aspects of the Epitaxial Vapour Growth of Semiconductors: Elements, 000041
#Organometallic Vapor Deposition of Epitaxial ZnSe Films on GaAs Substrates* 000921
#Molecular Beam and Solid-Phase Epitaxies of Silicon Under Ultra-High Vacuum (film)* 000883
#Properties of GaN Grown on Sapphire Substrates (epitaxy)* 000212
#GaAs/AlAs Layered Films (epitaxy)* 000353
#(111) Cu2O Growth Modes on (111)Cu Surfaces (epitaxy)* 000388
#Growth of Antimony on Bismuth (Sb film, epitaxy)* 000501
Mechanisms of Chemical Vapor Deposition of Silicon (Si, epitaxy)* # 000702
Growth of Zinc Telluride by Closed Tube Method (ZnTe, epitaxy)* #Vapor 000700
Harmonic Generation in a Sputtered LiNbO3 Film on MgO (epitaxy)* #Second 000518
GaAs Thin Film Solar Cells by Peeled Film Technology (epitaxy)* #High Efficiency 000497
Mechanism and Kinetics of Growth of LPE GaAs (theory, epitaxy)* #Investigation of the 000144
From Beams of Ionized Atoms and Clusters (Ag on Si, epitaxy)* #Thin Film Deposition 000516
Properties of CuGaSe2 Thin Films on GaAs Substrates (epitaxy)* #Structural and Electrical 000844
on Intermediately Sputtered ZnO Thin Film on Sapphire (epitaxy)* Crystalline ZnO Film with Smooth Surface 000711
in CVD Systems (Si-H-Cl, Ga-As-H-Cl, epitaxy)* Analysis of Equilibrium Adsorption Layers 000184
Vectors of Silicon (111) (theory, vapor transport, epitaxy)* Step Reconstruction on the Growth and Etching 001019
#Chemical Vapor Deposition of Scandium Hydride (ScH2, epitaxy, film)* 000494
(Eu,Y)3Fe5O12, and (Eu,Yb)3Fe5O12, vapor transport, epitaxy, films)* CVD Garnet Growth (Er3Fe5O12, 000211
Applications, Vol. 1 (garnets, review, hydrothermal, epitaxy, flux, Bridgman, film)* #Crystals for Magnetic 000807
of Cadmium, Selenium, and Cadmium Selenide (CdSe, epitaxy, kinetics, films, theory)* Properties 000365
Structure in GaAs (GaAs, Ga1-xAlxAs, molecular beam epitaxy, review)* #Periodic Doping 000248
#Vapor Growth of HgCr2Se4 (epitaxy, theory)* 000332
of Autoepitaxial GaAs Films by MBE (molecular beam epitaxy, theory)* #Temperature Range for Growth 000686
Single Crystals and Films (Si, Ge, GaAs, InSb, epitaxy, theory, Czochralski)* Growth of Semiconductor 000636
Germanium (111) Single Crystals with GeBr4 Vapor (Ge, epitaxy, vapor transport, kinetics, theory)*Reaction of 000060
a Substrate for Silicon on an Insulator (Czochralski, epitaxy, Al2O3)* #Evaluation of Verneuil Sapphire as 000793
Characterization of Corrosion Processes (film, epitaxy, Cu2O) Developments in the Microstructural 001045
Growth - Theory and Techniques, Volume 2 (review, epitaxy, Verneuil, III-V compounds, Si)* #Crystal 000348
#The Growth of GaxIn1-xAs on (100) InP by Liquid-Phase Epitaxy (flux)* 000754
with Na+ and Co2+ plus Zr4+, Grown by Liquid-Phase Epitaxy (flux)* #Thin Films of LiNbO3, Doped 000691
and Growth-Dissolution Kinetics of Garnet Liquid-Phase Epitaxy (flux, kinetics, theory)* Liquidus Temperature 000333
Surface Morphology of GaAs Layers Grown by Liquid-Phase Epitaxy (flux, theory, kinetics)* # 000605
#GaxIn1-xSb Crystals Grown by Liquid Phase Epitaxy (melt)* 000852
#Ionized-Cluster Beam Deposition and Epitaxy (review)* 000945
#Vapour Growth and Epitaxy (review)* 000946
#The Role of Lattice Misfit in Epitaxy (review, theory)* 001014
#Aspects of Silicon Epitaxy (review, Si)* 000701
Pb, PbS, PbSe, PbTe, SnTe, #Roles of Lattice Fitting in Epitaxy (theory, Ni, Cu, Pd, Pt, Fe, Al, Au, Ag, In, 000949
#A High Temperature High Purity Source for Metal Beam Epitaxy (Ag, film)* 000287
compounds, evaporation, review, theory)* #Hot Wall Epitaxy (II-VI compounds, IV-VI compounds, III-V 000568
Schottky-Barrier Diodes Prepared by Molecular-Beam Epitaxy (MBE) on GaAs (Al, films)* Crystal-Aluminum 000190
#Novel Reactor for High Volume Low-Cost Silicon Epitaxy (Si, vapor transport, kinetics)* 000062
#n-Type Doping Techniques in Silicon Molecular Beam Epitaxy by Simultaneous Arsenic Ion Implantation and by 000734
Ca,Ge-Substituted Garnet Films grown by Liquid Phase Epitaxy from PbO-B2O3 Fluxed Melts*Charge Imbalance for 000386
Double Heterostructure Light Modulators* #Liquid Phase Epitaxy of AlxGa1-xP on GaP and Its Application to 000948
#Growth Temperature Dependence in Molecular Beam Epitaxy of Gallium Arsenide (GaAs)* 000666
#Liquid Phase Epitaxy of Garnets (review)* 000802
#Molecular Beam Epitaxy of GaSb and GaSbxAs1-x (films)* 001100

PERMUTED TITLE INDEX

Superlattices*   #Molecular Beam Epitaxy of Ge and Ga1-xAlxAs Ultra Thin Film   000761
                                                  #Molecular Beam Epitaxy of InxGa1-xSb (x = 0 to 1) (GaSb)*   001101
solution)*   #Liquid Phase Epitaxy of II-IV-V2, Chalcopyrite Compounds (ZnSnAs2,   000089
theory)*   #Vapor Growth and Epitaxy of Minerals and Synthetic Crystals (review, 000932
flux)*   # Epitaxy of NdAl3(BO3)4 For Thin Film Miniature Lasers (000577
                                                  #Liquid-Phase Epitaxy of ZnSe from Zn-Ga Solution (film)*   000308
                                                  #Molecular Beam Epitaxy of ZnSexTe1-x(film)*   001102
     #Thin Film Growth of (MnxZn1-x)Fe04 by Liquid Phase Epitaxy on Bridgman Grown Zn2TiO4 Substrates (flux)*   000216
vapor transport)* #Crystalline Defects in Solid Phase Epitaxy Si Films Deposited at Elevated Temperatures (000194
                  #Design Considerations for Molecular Beam Epitaxy Systems (review)*   000576
(In,Ga)(As,P) Alloys Grown by the Hydride Vapor Phase Epitaxy Technique (film)*Binary, Ternary and Quaternary   000729
Aluminate Single Crystal Layers by the Liquid Phase Epitaxy Technique (flux, (Ba,Pb)Al12019)*   Lead Hexa-   000071
                               #GaAs Vapour Phase Epitaxy Using Computer Control*   001057
      #The Growth Kinetics of Garnet Liquid Phase Epitaxy Using Horizontal Dipping (flux)*   001021
n-(AlAs)m Superlattices Deposited by Molecular Beam Epitaxy. I. Growth on Singular (100) GaAs Substrates (000762
          #A New Substrate Holder for Liquid Phase Epitaxy*   000072
                           #Chromium Doped GaAs Vapor Phase Epitaxy*   000469
                      #CdTe Thin Films Grown by Hot Wall Epitaxy*   000569
                                 #Reproductive Epitaxy*   000606
     #Surface Morphology of GaAs Grown by Vapor Phase Epitaxy*   #Growth   000585
Kinetics of (111) Si Through an Al Layer by Solid Phase Epitaxy*   #Growth   000957
of GaAs1-xSbx Crystals by Steady-State Liquid Phase Epitaxy*   #Nitrogen Doping   000617
into GaAs1-xPx Using Ionized Beam in Molecular Beam Epitaxy*   #Spectroscopic Properties   000245
of Er3Al5-xGax012 Films Obtained by Liquid-Phase Epitaxy*   and Phase Diagram   000835
of (Ti1-xVx)407: Electrical, Calorimetric, Magnetic and EPR Studies (vapor transport)*   000133
sublimation)*   #Crystal Growth and EPR Study of Cr+-Doped MgTe Single Crystals (001017
of Compounds in the Series GaSxSe(1-x) (0 less than or equal to x less than or equal to 1) Grown by Iodine Vapo   001077
GaSxSe(1-x) (0 less than or equal to 1) Grown by Iodine Vapour Transport*the Series   001083
a Quantity in the Bruton, Prim and Slichter Equation as a Diffusion Boundary Layer Thickness (000001
Growth of Gallium (Ga, theory)*   #Rate Equations for Dislocation-Free and Dislocation-Assisted   001003
I2 (TiP, vapor transport)*   #Heterogeneous Equilibria and Mass Transfer Kinetics in the System TiP-   000726
Solutions in the Systems Zn3As2-ZnSe(ZnTe) (#Phase Equilibria and Properties of Single Crystals of Solid   000159
Y3Fe5012)*   #Phase Equilibria in Solvent-Garnet Systems (flux, Y3Al5012,   000813
     #Crystal-Chemistry, Crystal-Growth, and Phase Equilibria of Apatites (review)*   000531
Hydroxide Bromides and Hydroxide #Preparation, Phase Equilibria, and Crystal Chemistry of La, Pr, and Nd   000191
theory, thermodynamics)*   #Germanium-Iodine Equilibrium from 700 to 1000 K (Ge, vapor transport,   000184
Ga-As-H-Cl) (GaAs, theory,   #Theoretical Analysis of Equilibrium Adsorption Layers in CVD Systems (Si-H-Cl,   000826
slow cooling)*   #Dynamics of Non- Equilibrium Crystallization of an Alloy (theory, melt,   000798
Oxides with TeCl4 (II) The Transport Behaviour of the Equilibrium Nb02.417 to Nb02.5*   of Some Niobium   000765
Growth of Two-Component Systems I. Basic Features and Equilibrium Properties (melt)*   Theory of Crystal   000818
     #In-Situ Carbon-Silicon Ribbon Thermocouple (Si, equipment)*   000164
Crystals with Toxic and/or Highly Volatile Components (equipment)* Improved System for the Bridgman Growth of   000595
of Polymorphic Rare-Earth Disilicates, R2Si207 (R = Tm, Er, Ho, Dy) (Tm2Si207, Er2Si207, Ho2Si207, Dy2Si207)*   000137
Crystal Growth of Rare Earth Tetraborides (YB4, TbB4, ErB4)*   #Czochralski   000373
RCu2. I. Heavy Rare Earth (TbCu2, DyCu2, HoCu2, ErCu2)*   of Rare Earth Copper Intermetallic Compounds,   000565
Dy3Fe5012, Er3Fe5012, Ca3Fe2Ge3012, HoFeO3, DyFeO3, ErFeO3)*   with Addition of V205 (Y3Fe5012, Ho3Fe5012,   000245
Some Rare Earth Trifluorides (GdF3, TbF3, DyF3, HoF3, ErF3, Bridgman-Stockbarger, Czochralski)*   Growth of   000137
Films of the Rare-Earth Dihydrides (GdH2, TbH2, DyH2, ErH2, HoH2)*   #The Growth and Structure of Epitaxial   000782
for Flux Growth (DyPO4, PrBO3, LaBO3, DyKMo208, ErKMo208, NdKMo208, DyVO4, TbVO4, TbPO4, DyPO4,   001060
and R6Mo012 in the Systems R203-K20-Mo03 (DyKMo208, ErKMo208, NdKMo208, Dy2Mo06, Sm2Mo06, Pr2Mo06, Ho6Mo012)   001062
Under Hydrothermal Conditions (YOHCO3, GdOHCO3, ErOHCO3, Y20(OH)2CO3, Er202CO3)* Rare-Earth Carbonates   000169
Conditions (YOHCO3, GdOHCO3, ErOHCO3, Y20(OH)2CO3, Er202CO3)* of Rare-Earth Carbonates Under Hydrothermal   000169
Pb3Mn7015, Tm2Ti307, Tb2Ti207, Pr2Ti207, Tm2Ge207, Er2Si05, Dy2Si05, Ho2Si05, Pr2Mo06, LiCoPO4)* Pb)Mn03,   001061
TbP04, DyP04, Ho2Ge207, Dy2GeMo08, ZrSi04, ThGe04, Er2Si207)* DyKMo208, ErKMo208, NdKMo208, DyVO4, TbVO4,   001060
Disilicates, R2Si207 (R = Tm, Er, Ho, Dy) (Tm2Si207, Er2Si207, Ho2Si207, Dy2Si207)*of Polymorphic Rare-Earth   000595
of Cubic Bi4Ge3012 Crystals Doped with Dy3+, Ho3+, Er3+, Tm3+, or Yb3+ Ions (Czochralski)*   Emission   000458
     #Spectroscopic Properties of Er3Al5-xGax012 Films Obtained by Liquid-Phase Epitaxy*   000245
transport, epitaxy, #Low Temperature CVD Garnet Growth (Er3Fe5012, (Eu,Y)3Fe5012 and (Eu,Yb)3Fe5012, vapor   000211
with Addition of V205 (Y3Fe5012, Ho3Fe5012, Dy3Fe5012, Er3Fe5012, Ca3Fe2Ge3012, HoFeO3, DyFeO3, ErFeO3)* Flux   000565
and Rare-Earth Elements (Ln3Nb07) (Ho3Nb07, Y3Nb07, Er3Nb07, Dy3Nb07, La3Nb07, Y3Ta07, Er3Ta07, Gd3Ta07)*   001097
(Ho3Nb07, Y3Nb07, Er3Nb07, Dy3Nb07, La3Nb07, Y3Ta07, Er3Ta07, Gd3Ta07)*   and Rare-Earth Elements (Ln3Nb07)   001097
Crystals (Te)*   # Etch Pit Orientation in Czochralski-Grown Tellurium   000877
CaHPO4.2H20, gel)*   #Observations of Etch Pits on As-Grown Faces of Brushite Crystals (000714
of Defects in LPE GaAs Revealed by New Chemical Etchant (flux, film)*   #Observations   000703
     #Surface Morphology of HCl Etched Silicon Wafers (Si)*   001015
Growth and Dissolution Processes (CsI,   #Selective Etching of Cesium Iodide Crystals and Comparison of   000827
                                    #Growth & Etching of InBi Single Crystals (melt, zone melting)*   000110
solution)*   #Crystal Growth and Selective Etching Studies on Sodium Chlorate Crystals (NaClO3,   000784
#Analysis of Dislocations in Gd3Ga5012 by a Repeated Etching Technique*   000943
of Adsorption and Step Reconstruction on the Growth and Etching Vectors of Silicon (111) (theory, vapor   001019
for Thermionic Emission (LaB6, EuB6, CeB6, BaB6, (La, Eu)B6, La1-xEuB6, (La,Y)B6, (La,Ba)B6, (La,Cs)B6, (Eu,Y)   000727
Halide Compounds MFX of Divalent Rare Earths (M = Sm, Eu, Tm, and Yb - X = Cl, Br, and I) (melt, SmFCl,   000088
LaB6, EuB6, CeB6,   #Single-Crystal Growth of Mixed (La, Eu, Y, Ce, Ba, Cs) Hexaborides for Thermionic Emission (000727
La1-xEuB6, (La,Y)B6, La,Ba)B6, (La,Cs)B6, (Eu,Y)B6, (Eu,Ba)B6, flux)*   (LaB6, EuB6, CeB6, BaB6, (La,Eu)B6,   000727
(NdBO3, LaBO3/Pr, LaBO3, (La,Pb)MnO3, (Nd,Pb)MnO3, (Eu,Pb)MnO3, Pb3Mn7015, Tm2Ti307, Tb2Ti207, Pr2Ti207,   001061
(La,Eu)B6, La1-xEuB6, (La,Y)B6, La,Ba)B6, (La,Cs)B6, (Eu,Y)B6, (Eu,Ba)B6, flux)* (LaB6, EuB6, CeB6, BaB6,   000727
     #Low Temperature CVD Garnet Growth (Er3Fe5012, (Eu,Y)3Fe5012 and (Eu,Yb)3Fe5012, vapor transport,   000211
CVD Garnet Growth (Er3Fe5012, (Eu,Y)3Fe5012 and (Eu,Yb)3Fe5012, vapor transport, epitaxy, films)*   000211
Alumina (NaAl11017, KAl11017, RbAl11017, AgAl11017, EuAl11017, flux, theory)*   Silver, and Europium beta-   000621
of a Europium(II) Bromide-Chloride Phase EuBr1.5Cl0.5 (melt, slow cooling, hygroscopic)*   000200
Transport, and Thermal Properties of Ferromagnetic EuB6 (review)*   #Magnetic,   000293
Ce, Ba, Cs) Hexaborides for Thermionic Emission (LaB6, EuB6, CeB6, BaB6, (La,Eu)B6, La1-xEuB6, (La,Y)B6, La,Ba)   000727
a Study of Their Properties (flux, LaB6, SmB6, SmB4, EuB6, YbB6)* Earth Borides by the Solution Method and   000359
- X = Cl, Br, and I) (melt, SmFCl, SmFBr, SmFI, EuFCl, EuFBr, EuFI, TmFCl, TmFBr, TmFI, YbFCl, YbFBr, YbFI)*Yb   000088
and Yb - X = Cl, Br, and I) (melt, SmFCl, SmFBr, SmFI, EuFCl, EuFBr, EuFI, TmFCl, TmFBr, TmFI, YbFCl, YbFBr,   000088
Cl, Br, and I) (melt, SmFCl, SmFBr, SmFI, EuFCl, EuFBr, EuFI, TmFCl, TmFBr, TmFI, YbFCl, YbFBr, YbFI)* Yb - X = 000088
CsLnF4 (melt, RbEu2F7)*   #The Systems RbF- EuF3 and CsF-EuF3. Study of the Phases RbLnF4 and   000037
RbEu2F7)*   #The Systems RbF-EuF3 and CsF- EuF3. Study of the Phases RbLnF4 and CsLnF4 (melt,   000037

PERMUTED TITLE INDEX

Czochralski,   #Change in the Energy Band Structure of   Eulytine on Transition to the Amorphous State (000294
    #X-Ray Diffuse Scattering from Alkali, Silver, and   Europium beta-Alumina (NaAl11017, KAl11017, RbAl11017,   000621
slow cooling,   #Preparation and Characterization of a   Europium(II) Bromide-Chloride Phase EuBr1.5C10.5 (melt,   000200
*   #Synthesis and Characterization of   Europium(II)-Holoborates, Eu2B509Cl and Eu2B509Br (melt)   000580
transport, theory)*   #Heat Capacity of   EuS near the Ferromagnetic Transition (Bridgman, vapor   000505
    #Crystal Structure and Antiferromagnetism of   EuSb2 (melt)   000401
and X-Ray Analysis of Single Crystals of EuSb4S7 and   EuSb2Se4 (vapor transport)   000022
    #Growth and X-Ray Analysis of Single Crystals of   EuSb4S7 and EuSb2Se4 (vapor transport)*   #Growth   000022
Solution-Grown Mixed Crystals of NaCl and KCl (NaCl-KCl   eutectic)*   Growth, X-Ray Studies and Microstructure of   000407
Solidification of Irregular Silver-Germanium   Eutectic Alloys (Ag-Ge)*   #Unidirectional   000549
    #The Stage II Transient in Off-   Eutectic Solidification (theory)*   000786
Growth (theory)*   #Coarsening of   Eutectic Structures During and After Unidirectional   001027
of the fcc Phase in Uni-Directionally Solidified Zn-Al   Eutectic*   #Crystallography   000421
Theoretical (directional crystallization, zone melting,   eutectics)*   Coefficients from Experimental Data: 001080
of Superalloys via Use of Precipitated Phases (eutectics, carbides, review)*   and Properties   000671
of Zirconia-Rare Earth Oxide Crystals (ZrO2-RE2O3,   eutectics, directional solidification)*   Skull-Melting   000629
Growth of a Dilute Binary Alloy Single Crystal (Fe-Ni,   eutectics, theory, dendrites)*   of the Cellular   000113
on the Directional Solidification Process (theory,   eutectics, Al-Al3Ni)**Effect of Some Thermal Parameters   000268
Solidified InSb-GaSb (magnetic field effects,   eutectics, Bridgman)*   Coefficients for Directionally   000864
Unidirectional Solidification, Microstructural #Ceramic   Eutectics in the Systems ZrO2-Ln2O3 (Ln = Lanthanide):   000630
    #Crystallography of Aligned Cu-In   Eutectoid (Cu-Cu7In3)*   001049
    #Crystallography of Aligned Fe-Al   Eutectoid (FeAl, FeAl2)*   000078
    #Crystal Structure of beta-   Eu2(II)SiO4 (melt)*   000597
of Europium(II)-Holoborates, Eu2B509Cl and   Eu2B509Br (melt)*   #Synthesis and Characterization   000580
and Characterization of Europium(II)-Holoborates,   Eu2B509Cl and Eu2B509Br (melt)*   #Synthesis   000580
electron beam melting,   #Growth of Films of CdS   Evaporated in Vacuum onto Silicon (vapor transport,   000073
CdS,   #Epitaxial Growth of Some II-VI Compound Films   Evaporated Onto Electron-Bombarded NaCl Substrates (000878
Monatomic Step Trains of Different Density on   Evaporating Crystal Surfaces (NaCl, theory)* Successive   000105
    #Optical Properties of Zn3P2 (evaporation)*   000278
    #Ferroelectricity in (N(CH3)4)2ZnCl4 (evaporation)*   000830
of Pb0.8Sn0.2Te Films in a Quasi-Closed Space (vacuum   evaporation)*   #Growth   000885
Charge and Ledge Dynamics in Cadmium Sulfide (CdS,   evaporation)*   #Surface   000529
Preparation of Thin Single Crystal Films of Copper (Cu,   evaporation)*   #A New Technique for   000736
Silver Films on Cleaved Silicon Surfaces (Ag, theory,   evaporation)*   #An Investigation of Thin   000619
Dose Si+-Ion Implantation of Pd (electron beam melting,   evaporation)*   #Silicide Formation by High-   000177
Chlorapatite (Na2Nd2Fb6(PO4)6Cl, slow cooling,   evaporation)*   #Flux Growth of Lead Neodymium   000445
Compounds in a Quasi-Sealed Volume (SnTe, PbTe, PbSe,   evaporation)*   #Preparation of Films of IV-VI   000306
Sulphide Layers on Cadmium Single Crystals (CdS,   evaporation)*   and Structure of Epitaxial Cadmium   000429
Ice Crystals (H2O, vapor transport, sublimation,   evaporation)*   on Growth and Evaporation of Epitaxial   000028
in Cadmium Iodide Crystals (CdI2, theory, solution,   evaporation)**Effect of Growth Conditions of Polytypism   000432
(II-VI compounds, IV-VI compounds, III-V compounds,   evaporation, review, theory)*   #Hot Wall Epitaxy   000568
    #Scandium Hydrogenselenite (Sc(HSeO3)3,   evaporation, solution)*   001008
    #Growth of CdP4 Single Crystals (evaporation, vapor transport)*   000599
Acid-21-Water by Neutron Diffraction (evaporation, H3PW12O40.21H2O)* #Dodecatungstophosphoric   000913
Doped Aluminum Nitride Films Produced by Reactive   Evaporation (AlN)*   #Cathodoluminescence of Impurity-   001104
Simultaneous Arsenic Ion Implantation and by Antimony   Evaporation (Si, film)*   Molecular Beam Epitaxy by   000734
    #Molecular (beam) Processes of Condensation and   Evaporation of Alkali Halides (KCl, LiF)*   000628
Saturation and Electric Field on Growth and   Evaporation of Epitaxial Ice Crystals (H2O, vapor   000028
    #Anisotropy of Round   Evaporation Spirals on Rocksalt Surfaces (NaCl, theory)*   000530
Coalescence Growth of Smoke Particles Prepared by a Gas-   Evaporation Technique (MoO3, WO3, theory)*   #   000448
and Habit of Fine Metal Particles Formed by Gas-   Evaporation Technique: bcc Metals (V, Fe, Cr, Mo and W)*   000821
transport)*   #A Gas Handling System for   Exacting Crystal Growth Preparations (TiO2, vapor   000808
for the Onset of Oscillatory   #Some Evidence for   Existence and Magnitude of a Critical Marangoni Number   000847
    #The Melt Growth and Thermal   Expansion of Na2CO3 Crystals (Bridgman)*   000632
    #Linear   Expansion of 2H-MoS2 and 2H-WSe2 Between 10 and 320 K*   000667
Determination of the Lattice Parameters and the Thermal   Expansion Anisotropy of CdGa2Se4 (melt)*   000476
    #Integrated Optical Devices   Expansion Coefficients of VO2 near the Transition   000513
    #Monte Carlo Simulation of a (111) Diamond   Fabricated by MBE (GaAs, Ga1-xAlxAs, review, film)*   000152
Perovskite (C3H7NH3)2CuCl4 (solution, theory)*   #   Face Around the Roughening Transition (C, theory)*   001020
Mass in the (#Simulation of the Growth of a Metallic   Face Stability and Growth Rate Variations of the Layer   000641
Aqueous Solutions. II. Growth Kinetics of Different   Face-Centered Cubic Lattice in a Crystal from a Molten   000995
    #Observations of Etch Pits on As-Grown   Faces in Pure Solution and in the Presence of a   000258
    #The Strain Configuration Within the   Faces of Brushite Crystals (CaHPO4.2H2O, gel)*   000714
organic)*   #On the Transition from   Facet of Czochralski Grown Indium Antimonide (InSb)*   000637
    #On the Transition from Faceted to Non-   Faceted to Non-Faceted Growth in Melt-Grown Crystals (000355
    #Spacelab and Material Processing   Faceted Growth in Melt-Grown Crystals (organic)*   000355
Perovskites of the (CnH2n+1NH3)2MX4 and NH3(CH2)mNH3MX4   Facilities and Experiments (review)*   000853
Studies in the Solid Electrolytes of the RbAg4I5   Families with M = Cd, Cu, Fe, Mn or Pd and X = Cl or   000039
flux)*   #A New Structural   Family (KAg4I5, NH4Ag4I5, solution)*   #Raman Scattering   000315
Silicon (Si)*   #Growth of Stacking   Family M2"M'Ta5O15. Crystal Structure of CaTlTa5O15 (000317
Method (arc image)*   #Growth of   Faults by Bardeen-Herring Mechanism in Czochralski   001052
*   #Crystallography of the   Fayalite (Fe2SiO4) Single Crystals by the Floating Zone   000953
of Crystallization of the Soft-Core Model in an   fcc Phase in Uni-Directionally Solidified Zn-Al Eutectic   000421
    #On the Ordering of   FCC Crystal Formation (theory)*   #Geometrical Analysis   000965
    #Influence of the CaCO3/GeO2 Ratio on Y Sm Lu Ca   Fe Atoms in FexNbS2 (vapor transport)*   000129
    #Asymmetric 180 degree Bloch Walls in   Fe Ge Garnet Films Properties (LPE)*   000650
ferrous halides, float zone #Controlled Growth of alpha-   Fe Single Crystal Films (theory)*   000936
    #Magnetic Properties of (Mn,Cr)P and (Mn,   Fe Single Crystal Whiskers (hydrogen reduction of   000318
    #Crystallography of Aligned   Fe)P Compounds (Mn1-xFexP, x = 0 to 0.15, theory)*   000430
Growth of a Dilute Binary Alloy Single Crystal (Fe-Al Eutectoid (FeAl, FeAl2)*   000078
Melting at Various Growth Rates*   #Growth Striations in   Fe-Ni, eutectics, theory, dendrites)*   of the Cellular   000113
Preparation of Large Spherical Iron Single Crystals (Fe-3wt%Si Single Crystals Grown by Floating Zone   001028
    #Epitaxial Thin Iron Films on Antimony (Fe, strain-anneal)*   #The   000573
    #Growth Rate of Iron Whiskers (Fe, vapor transport)*   000875
of Lattice Fitting in Epitaxy (theory, Ni, Cu, Pd, Pt,   Fe, vapor transport, theory)*   000462
Chalcogenides Cu2(I)B(II)C(IV)X4 (BII=Zn, Mn,   Fe, Al, Au, Ag, In, Pb, PbS, PbSe, PbTe, SnTe, LiF,   000949
Layer-Type Compounds K2Mn1-xMxF4 (M +   Fe, Co - CIV = Si, Ge, Sn - X = S,Se) (Cu2ZnGeS4,   000357
    Fe, Co) (K2Mn1-xFexF4, slow cooling)* Antiferromagnetic   000106

PERMUTED TITLE INDEX

Formed by Gas-Evaporation Technique: bcc Metals (V,	Fe, Cr, Mo and W)* and Habit of Fine Metal Particles	000821
Mismatch of Crystal Lattice Parameters((Y, Bi, Sm, Ca)3(	Fe, Ge)5 012, flux)* Garnet Layers as Functions of the	000251
1NH3)2MX4 and NH3(CH2)mNH3MX4 Families with M = Cd, Cu,	Fe, Mn or Pd and X = Cl or Br: Importance, Solubilities	000039
#An Improved Method of Stirring for LPE Growth (Sm,Y)3(	Fe,Ga)5012 (flux, theory, film)*	000222
#Crystallography of Aligned Fe-Al Eutectoid (	FeAl, FeAl2)*	000078
#Crystallography of Aligned Fe-Al Eutectoid (FeAl,	FeAl2)	000078
Triple Unit Cell Steps on the Surface of Vapour Grown	FeCr2S4 Single Crystals. I* #Growth of	000418
Mechanism of Multi-Steps on the Surface of Vapour Grown	FeCr2S4* #Growth	000417
Steps and Multi-Steps on the Surface of Vapour Grown	FeCr2S4* #Step Growth by Elementary	000419
Spinel Single Crystals and Their Electronic Properties (	FeCr2S4, CoCr2S4, vapor transport)*of Some Chalcogenide	001066
Ribbon Crystals Grown by the Edge-Defined, Film-	Fed Growth Technique (alpha-Al2O3)* in Sapphire	001051
Verneuil Technique (MgAl2O4, alpha-Al2O3)* #Powder	Feeder for Crystals of Large Diameter Grown by the	000284
#Hydrothermal Growth of the Systems MF-AlF3 and NH4F-	FeF3 in Hydrofluoric Acid Solution (TlAlF4, KAlF4,	000772
Preparation by Vapor Transport of Monocrystals of AlF3,	FeF3, and TiF3* #	000124
Iron Garnets (flux, (GdPrBi)3(FeGaIn)5012, (YLaBi)3(	FeGa)5012)* Optical Properties of Some Rare Earth	001107
of Some Rare Earth Iron Garnets (flux, (GdPrBi)3(	FeGaIn)5012, (YLaBi)3(FeGa)5012)* Optical Properties	001107
of Degree of Perfection of Single Crystals of	FeGe2 on Their Growth Rate and Direction* #Dependence	000746
Crystals of Certain Metallides (alpha-FeSi2.33, CoSi2,	FeGe2, Mn5Si3, Fe1.76Ge, melt, theory, kinetics)* of	000861
#Mossbauer Study of	FeMo2S4 (vapor transport)*	000006
#Magnetic and Mossbauer Studies of FeNi2BO5 and	FeNiBO4 (ferrites, vapor transport)*	000005
#Magnetic and Mossbauer Studies of	FeNi2BO5 and FeNiBO4 (ferrites, vapor transport)*	000005
Substrates (flux)* #Thin Film Growth of (MnxZn1-x)	FeO4 by Liquid Phase Epitaxy on Bridgman Grown Zn2TiO4	000216
#Synthesis and Crystal Structure of Iron Polyphosphide	FeP4*	000438
Alphen Effect (float zone melting)* #	Fermi Surface Measurements of ZrB2 by the de Haas van	000964
#The Crystal Structure and Twinning Behavior of	Ferric Molybdate, Fe2(MoO4)3 (hydrothermal)*	000178
solution)* #Crystal Growth of Co	Ferrite on Fine Acicular gamma-Fe2O3 Particles (	000931
5O4)* #LPE Growth of Lithium	Ferrite on Spinel Substrate Crystals (films, Li0.5Fe2.	000340
#Liquid Phase Epitaxial Growth of Copper	Ferrite Films (CuFe2O4)*	001016
Y3Fe5012, Ho3Fe5012, #Crystallisation of Some	Ferrite Materials from Lead Flux with Addition of V2O5 (	000565
theory)* #On the Preparation of Nickel	Ferrite Single Crystals by Chemical Transport (NiFe2O4,	000760
Melts with a Nonstoichiometric #Growth of Nickel-Zinc	Ferrite Single Crystals from BaO-B2O3-ZnO-NiO-Fe2O3	000107
#Liquid Phase Epitaxial Growth of Lithium	Ferrite-Aluminate Films (flux, Li0.5Fe2.5-xAlxO4)*	001017
ZnO-NiO-Fe2O3 Melts with a Nonstoichiometric Content of	Ferrite-Forming Oxides (NixZn1-xFe2O4, flux)* BaO-B2O3-	000107
Magnetic and Mossbauer Studies of FeNi2BO5 and FeNiBO4	(ferrites, vapor transport)* #	000005
Spinel (solution)*	Ferroelasticity and Phase Transformation in Rb2Hg(CN)4	000378
Orthovanadate, Ca3(VO4)2 - A New High-Temperature	Ferroelectric (Czochralski)* #Calcium	000339
#New	Ferroelectric (Pb,Ba)5Ge3O11 Crystals (flux)*	000155
Silica Gels (KDP)* #Single Crystal Growth of	Ferroelectric Na3Sc2(PO4)3 (melt)*	000721
#Epitaxial Growth of	Ferroelectric Potassium Dihydrogen Orthophosphate in	000442
* #Synthesis and Hydrothermal Growth of the	Ferroelectric PLZT ((Pb,La)(Zr,Ti)O3) Thin Films*	000415
New Epitaxial Process (TGS)* #Infrared Trichroism of	Ferroelectric SbNbO4, SbTaO4 and Pb5S2I6 Single Crystals	000777
	Ferroelectric Triglycine Sulphate Thin Films Grown by a	000329
#Epitaxial Growth of	Ferroelectricity in (N(CH3)4)2CoCl4*	000831
#Magnetic, Transport, and Thermal Properties of	Ferroelectricity in (N(CH3)4)2ZnCl4 (evaporation)*	000830
theory)* #Heat Capacity of EuS near the	Ferromagnetic CrO2 Films in Air*	000414
#Uniaxial	Ferromagnetic EuB6 (review)*	000293
alpha-Fe Single Crystal Whiskers (hydrogen reduction of	Ferromagnetic Transition (Bridgman, vapor transport,	000505
on Electrical Resistivity and Thermoelectric Power of	Ferromagnetism in NiZrF6.6H2O (solution)*	000464
of Crystals of Certain Metallides (alpha-	ferrous halides, float zone melting)* Growth of	000318
#Morphology of Iron Pyrite Crystals (	FeS (Bridgman)* #Effect of Pressure	000465
#Growth of	FeSi2.33, CoSi2, FeGe2, Mn5Si3, Fe1.76Ge, melt, theory,	000861
#Synthesis and Crystal Structure of a New Mixed Oxide	FeS2, vapor transport)*	001093
#Synthesis and Crystal Structure of a New Mixed Oxide "	FeTiO3 (Ilmenite) Crystals by the Floating-Zone Method*	000954
#Magnetic Properties of beta-Phase Iron-Germanium (	FeV2O6OH0.5: Relation to the Diaspore Structure Type (	000662
Edge in ZrS2 and HfS2 (CuxZrS2, CuxHfS2, FexZrS2,	FeV3O8" (FexV1-xO2 - x approx. = 0.25) (melt, slow	000661
#On the Ordering of Fe Atoms in	FexGe, slow cooling)*	000845
and Crystal Structure of a New Mixed Oxide "FeV3O8"	FexHfS2, vapor transport, theory)* of the Absorption	001090
the Absorption Edge in ZrS2 and HfS2 (CuxZrS2, CuxHfS2,	FexNbS2 (vapor transport)*	000129
Growth, and Magnetic and Mossbauer Studies of Sr	(FexV1-xO2 - x approx. = 0.25) (melt, slow cooling)*	000661
Metallides (alpha-FeSi2.33, CoSi2, FeGe2, Mn5Si3,	FexZrS2, FexHfS2, vapor transport, theory)* Shift of	001090
Structure and Twinning Behavior of Ferric Molybdate,	Fe0.766W0.234)O3 and its Related Compounds (SrFe1-	001001
Electrical, Optical and Photoconducting Properties of	Fe1.76Ge, melt, theory, kinetics)* Crystals of Certain	000861
Zinc Ferrite Single Crystals from BaO-B2O3-ZnO-NiO-	Fe2(MoO4)3 (hydrothermal)* #The Crystal	000178
#Crystal Growth of Co Ferrite on Fine Acicular gamma-	Fe2-xCrxO3 (vapor transport, electrolytic deposition)*	000624
#Properties of	Fe2O3 Melts with a Nonstoichiometric Content of Ferrite-	000107
arc image)* #Growth of Fayalite (	Fe2O3 Particles (solution)*	000931
Pseudobrookite Single Crystals by Chemical Transport (	Fe2P Crystals Prepared from a Liquid Copper Medium*	000030
#Mechanism of Habit Change of ADP Crystals by	Fe2SiO4) Single Crystals by the Floating Zone Method (	000953
#Skull Melter Growth of Magnetite (	Fe2TiO5, electrolytic deposition)*On the Preparation of	000770
1.35-micrometer Wavelength (#Nd:Y2O3 Single-Crystal	Fe3+, Based on Mossbauer Studies (NH4H2PO4, solution)*	000296
During Crystal Growth (solution)*	Fe3O4)*	000370
* #Synthesis of Potassium Hexatitanate	Fiber Laser: Room-Temperature cw Operation at 1.07- and	000916
zeolite, Na0.5AlSi6O14.xH2O)* #	Fiber Optic Probe for Thermal Profiling of Liquids	000395
for Directionally Solidified InSb-GaSb (magnetic	Fibers by the Hydrothermal Dehydration Method (K2Ti6O13)	000732
and Dy2Ti2O7 (powder) as a Function of Magnetic	Fibrous Sodium Aluminosilicate Crystals in Silica Gel (	000444
#Anisotropy of Upper Critical	field effects, eutectics, Bridgman)* Coefficients	000864
#Influence of Environmental Saturation and Electric	Field (solution)* of Gd2(SO4)38H2O (crystal)	000065
xSb* #Influence of Magnetic	Field in Superconducting 2H-NbS2 (theory)*	000730
theory, C)* #Temperature	Field on Growth and Evaporation of Epitaxial Ice	000028
of Monocrystalline Ditungsten Carbide as Revealed by	Field on Vertical Bridgman-Stockbarger Growth of InxGa1-	000863
#High-Efficiency p+-n-n+ Back-Surface-	Field Dynamics in the Synthesis of Diamonds (kinetics,	000862
Composition of Crystals Grown in Electric and Magnetic	Field Electron Microscopy (W microcrystal, crystal	000724
#Upper and Lower Critical	Field Silicon Solar Cells (Si)*	000298
#The Growth	Fields (Czochralski)* #Control of the	000511
Crystal Growth of Zirconium Carbide by a Modified Hot-	Fields of TaS2(Pyridine)1/2 (vapor transport)*	000468
#The Growth	Figures of MnAs (whiskers, theory)*	000067
Crystal Growth of Zirconium Carbide by a Modified Hot-	Filament Method (de Boer, Van Arkel, vapor transport,	000927
Morphology During Crystallization - II. Single	Filament, Unconstrained Growth from a Binary Alloy Melt	000993

PERMUTED TITLE INDEX

PERMUTED TITLE INDEX

System* #High Rate Epitaxial Growth of ZnO	Films on Sapphire by Planar Magnetron rf Sputtering	000882
#Analysis of Vapor-Deposited Silicon Carbide	Films on Silicon Ribbon Surfaces (SiC)*	001099
#Crystalline Defects in Solid Phase Epitaxy Si	Films Deposited at Elevated Temperatures (vapor	000194
Substrates (#Epitaxial Growth of Some II-VI Compound	Films Evaporated Onto Electron-Bombarded NaCl	000878
Trichroism of Ferroelectric Triglycine Sulphate Thin	Films Grown by a New Epitaxial Process (TGS)* #Infrared	000329
#CdTe Thin	Films Grown by Hot Wall Epitaxy*	000569
#Real Structure of Heteroepitaxial Zinc Oxide	Films Grown on Sapphire (vapor transport, ZnO)*	000526
#Spectroscopic Properties of Er3Al5-xGaxO12	Films Obtained by Liquid-Phase Epitaxy*	000245
#Cathodoluminescence of Impurity-Doped Aluminum Nitride	Films Produced by Reactive Evaporation (AlN)*	001104
of the CaCO3/GeO2 Ratio on Y Sm Lu Ca Fe Ge Garnet	Films Properties (LPE)* #Influence	000650
#Metastable Pb1-xCdxS Epitaxial	Films. I. Growth and Physical Properties*	000909
#Matastable Pb1-xCdxS Epitaxial	Films. II. Electrical Properties*	000910
#Electroless Deposition of Epitaxial Pb1-xHgxS	Films*	000869
Growth of Ferroelectric PLZT ((Pb,La)(Zr,Ti)O3) Thin	Films* #Epitaxial	000415
Microscopy of Growth Processes of Epitaxial Thin	Films* #Techniques for Routine UHV in situ Electron	000950
#Crystal Growth of Co Ferrite on	Fine Acicular gamma-Fe2O3 Particles (solution)*	000931
Technique: bcc Metals (#Crystal Structure and Habit of	Fine Metal Particles Formed by Gas-Evaporation	000821
#Preparation of	Fine Sodium Chloride Crystals (NaCl)*	000654
Ag, In, Pb, PbS, PbSe, PbTe, SnTe, #Roles of Lattice	Fitting in Epitaxy (theory, Ni, Cu, Pd, Pt, Fe, Al, Au,	000949
% Si Melt (Si, flux)* #	Fivefold Twinned Silicon Crystals Grown in an Al-16 wt.	000489
During Vapor Phase Silicon Epitaxial Growth by Using a	Flame Photometric Detector (Si)* Gas Concentration	000937
#Growth of Chromium Silicide, Cr3Si, Crystals (| float zone melting)*	000441
of a New High-Intensity Silicon Solar Cell (Si, | float zone melting)* #Performance	000303
of TiCx Single Crystal with Homogeneous Compositions (| float zone melting)* #Preparation	001092
Measurements of ZrB2 by the de Haas van Alphen Effect (| float zone melting)* #Fermi Surface	000964
in the Growth and Properties of Beta-Boron Crystals (B, | float zone melting)* #Some Peculiarities	000967
of Strontium Cobaltate Single Crystals (SrCoO2.70, | float zone melting)* #Growth and Properties	000955
Whiskers (hydrogen reduction of ferrous halides, | float zone melting)* Growth of alpha-Fe Single Crystal	000318
and Prandtl Number at Zero Gravity (Si, Al2O3, theory, | float zone melting, kinetics, Si)* of Zone Geometry	000175
of Nb-H Interstitial Alloys (electron beam melting, | float zone melting, review)* #Properties and Synthesis	001072
I. Pure Rotation Driven Flows (Si, Al2O3, kinetics, | float zone melting, theory) in Floating Zone Melting.	000173
(review, melt, flux, Czochralski, Kyropoulos, | float zone melting, Bridgman)* at High Temperatures	000139
of Oriented Bicrystals of Nickel Oxide (NiO, arc image, | float zone, Verneuil)* #Growth	000239
#Limitations in Using Kilohertz Radio Frequencies for	Float Zone Silicon Crystals (Si, theory)*	000358
Silicon Crystals (Al2O3, #Maximum Stable Zone Length in	Float-Zone Growth of Small-Diameter Sapphire and	000479
#Power Required to Form a	Floating Zone and the Zone Shape (theory)*	000492
#Experiments on Surface Tension Driven Flow in	Floating Zone Melting (NaNO3, theory)*	000848
Growth Striations in Fe-3wt%Si Single Crystals Grown by	Floating Zone Melting at Various Growth Rates* #	001028
Si, Al2O3, #Computer Simulation of Convection in	Floating Zone Melting. I. Pure Rotation Driven Flows (	000173
Driven Flows (#Computer Simulation of Convection in	Floating Zone Melting. II. Combined Free and Rotation	000174
Prandtl Number at Zero #Thermocapillary Convection in	Floating Zone Melting: Influence of Zone Geometry and	000175
#Growth of Fayalite (Fe2SiO4) Single Crystals by the	Floating Zone Method (arc image)*	000953
of Yttrium Iron Garnet Single Crystals Grown by the	Floating Zone Method (Y3Fe5O12)* #Evaluation	000480
in Substituted YIG Single Crystals Grown by the	Floating Zone Method (Y3Fe5O12, Y3Fe4Ga012, Y3Fe4Al012)*	000483
#Growth of Mg2TiO4 Single Crystals by the	Floating Zone Method*	000880
(Ca2MgSi2O7) and Gehlenite (Ca2Al2SiO7) by the	Floating Zone Method* Crystal Growth of Akermanite	000404
Yttrium Iron Garnets Y3Fe5-x(Ga,Al)xO12 by the	Floating Zone Method* Crystal Growth of Substituted	000379
#Growth of Cobalt Single Crystals by the Electron Beam	Floating-Zone Method (Co)*	000381
#Growth of FeTiO3 (Ilmenite) Crystals by the	Floating-Zone Method*	000954
Growth Direction of Pure Cu Single Crystals by the	Floating-Zone Method* #Preferred	000380
#About SnF2 Stannous	Flouride. I. Crystallochemistry of alpha-SnF2*	000235
Epitaxial Growth of ZnS in H2: The Effects of a Bypass	Flow (theory)* Transport Processes Associated with the	000558
Critical Marangoni Number for the Onset of Oscillatory	Flow in Crystal Growth Melts (NaNO3, theory)* of a	000847
#Experiments on Surface Tension Driven	Flow in Floating Zone Melting (NaNO3, theory)*	000848
#Effect of the Buoyancy Parameter on Czochralski Bulk	Flow in Garnet Growth (theory, kinetics)*	000535
#Digital Simulation of Czochralski Bulk	Flow in Microgravity (Al2O3, theory, kinetics)*	000536
#Computational Simulation of the Melt	Flow During Czochralski Growth (theory)*	000491
Epitaxial Growth in SiCl4-HCl-H2 System Based on	Flow Graph (vapor transport, theory, Si)* of Silicon	000503
#Epitaxial Growth of ZnSe by Vapor	Flow Method*	000669
Gadolinium Gallium Garnet (#The Effect of Melt	Flow Phenomena on the Perfection of Czochralski Grown	000638
Crystals #Orientation and Temperature Dependence of the	Flow Stress in the Intermetallic Compound Ni3Ge Single	000033
Phase Growth of Epitaxial Ga1-xInxSb Using Open-Tube	Flow System (vapor transport)* #Vapor-	000681
melt)* #Heat-	Flow-Controlled Growth During Li2B4O7 Crystallization (	000385
Compounds from the Vapour Phase (theory)* #	Flow-Graph Model of Epitaxial Growth of A(III)B(V)	000504
Zone Melting. II. Combined Free and Rotation Driven	Flows (kinetics, theory, electron beam melting)*	000174
in Floating Zone Melting. I. Pure Rotation Driven	Flows (Si, Al2O3, kinetics, float zone melting, theory)*	000173
Metallic One-Component Systems. Two Cases of #A	Fluctuation Theory of Normal Crystal Growth for	000059
#Lasing and	Fluorescence in K5NdLi2F10 (flux)*	000189
Growth of Alumina Whiskers by Hydrolysis of Aluminum	Fluoride (Al2O3)* #Vapor Phase	001094
ZrFeF6, ZrCrF6, ZrNiF6, ZrCoF6, #Flux Growth of Some	Fluoride Crystals Under Reducing Conditions. IV (	000319
from 5 to 300 K and Magnetic #Crystalline Vanadium (II)	Fluoride, VF2. Preparation, Structure, Heat Capacity	000917
#Advances in Preparative Chemistry of Oxides and	Fluorides (review, hydrothermal, theory)*	000398
#Flux Growth of Crystals of Some Transition Metal	Fluorides. Part 2 (KAlF4, KMnF3, RbFeF4, Rb2FeF5,	001064
#Flux Growth and Characterization of	Fluorite Crystals (CaF2)*	000237
Crystal Growth and Properties of Deuterated Triglycine	Fluoroberyllate (TGFB)* #Single	000567
#The Crystal Structure of Lithium	Fluorosulphate LiSO3F (solution)*	001108
#The Crystal Structure of Cu4(PO4)2O (| flux)*	000029
#Crystal Growth of Antimony Sulphur Iodide (SbSi, | flux)*	000109
#Crystal Structure of Ba2Fe6O11 (| flux)*	000120
#Crystal Structure of Cu4(PO4)2O (| flux)*	000142
#Ferroelectric (Pb,Ba)5Ge3O11 Crystals (| flux)*	000155
#Lasing and Fluorescence in K5NdLi2F10 (| flux)*	000189
#Twinning and Crystal Structure of Mn0.75Ga2.17S4 (| flux)*	000247
#Photoconduction of Lead Trititanate (PbTi3O7, | flux)*	000264
#The Crystal Structure of Lithium Uranate (Li2UO4, | flux)*	000324
#Optically Pumped LPE-Grown Hg1-xCdxTe Lasers (films, | flux)*	000368
#Crystal Structure of K5NdLi2F10 (| flux)*	000397
#Epitaxy of NdAl3(BO3)4 For Thin Film Miniature Lasers (| flux)*	000577

PERMUTED TITLE INDEX

#Growth of Single Crystals of Lead Monoxide (PbO,	flux)*	000717
#Overgrowths on Diamond at Atmospheric Pressure (C,	flux)*	000748
#Polymorphism in Bi2Sn2O7 (	flux)*	000867
#The Structure of K2Er(NO3)5 (	flux)*	000872
#The Growth of LaAlO3 Crystals Enriched in Isotope 17O (	flux)*	000901
#Growth and Dissolution of BaFBr Crystals (	flux)*	000908
#Growth and X-Ray Study of Cu2SrGeS4 and Cu2BaGeS4 (	flux)*	000972
#Crystal Structure of Li2CaUF8 (	flux)*	001029
#Crystal Structure of CuTa2O6 (	flux)*	001036
#Crystal Growth and Magnetic Properties of NdFeTiO5 (	flux)*	001091
The Crystal Structure of Synthetic Lautarite, Ca(IO3)2 (	flux)*	000331
and Magneto-Optical Properties of Sm0.55Tb0.45FeO3 (	flux)* #Optical	000914
Effect on the Formation of Terraces in GaAs LPE Growth (	flux)* #Impurity	000988
Measurements in the Pb-Sn-Te System (Pb1-xSnxTe, PbTe,	flux)* #Liquidus	000047
and Electrical Properties of BaVS3 and BaVxTi1-xS3 (	flux)* #Magnetic	000610
Layers of the Laser Material Nd(Ga,Cr)3(BO3)4 (	flux)* #Epitaxial	000578
Cr2O3 Crystals by the Decomposition of a K2Cr2O7 Melt (	flux)* #Growth of	000859
Structures and Crystal Growth (review, vapor transport,	flux)* #Nitrides--	000533
Phase Transitions in Single Crystals of (KxNa1-x)NbO3 (	flux)* #Growth and	000056
of GaxIn1-xAs on (100) InP by Liquid-Phase Epitaxy (	flux)* #The Growth	000754
Epitaxial Growth of CdTe in the CdTe-CdCl2 System (	flux)* #Liquid Phase	000828
Growth of GaP LPE Layers from Indium Solvent (	flux)* #Low Temperature	000925
Family M2"M'Ta5O15. Crystal Structure of CaTlTa5O15 (	flux)* #A New Structural	000317
Silicon Crystals Grown in an Al-16 wt.% Si Melt (Si,	flux)* #Fivefold Twinned	000489
of the Phosphate of Trivalent Manganese Mn(PO3)3 (	flux)* #Crystal Structure	000058
of the Double Molybdate K2Ni(MoO4)2 (slow cooling,	flux)* #Crystal Structure	000486
with a Novel Wurtzite-Pyrite Intergrowth Structure (	flux)* #AlSiP3, A Compound	001044
Measurement in Liquid Epitaxial Growth Systems (	flux)* #Temperature Gradient	000574
of Aluminum Yttrium Garnet (RE substituted Y3Al5O12,	flux)* #Growing Optical films	000044
of NbO and Hollow NbO2 Needles from Molten Salts (	flux)* #Electrocrystallization	000204
Garnet Liquid Phase Epitaxy Using Horizontal Dipping (	flux)* #The Growth Kinetics of	001021
and Preliminary Crystallographic Study of Bi2Sn2O7 (	flux)* #Preparation, Stability,	001034
from Solution in Borate, Vanadate, and Tungstate Melts (	flux)* #Crystallization of LiNbO3	001047
Gallium and Nickel-Gallium Compared (CoGa3, NiGa4,	flux)* #The Binary Systems Cobalt-	000290
During Epitaxial Growth from a Solution in a Melt (GaP,	flux)* #Theory of Crystallization	000591
with RbAlSiO4: A New Metastable Polymorph of KAlSiO4 (	flux)* #Alkali Ion Exchange Reactions	000639
of Na2MnFeF7: A Variety of Trigonal Weberite (	flux)* #X-Ray and Neutron Diffraction	001033
and Their Structural Perfection (vapor transport, melt,	flux)* #Growth of USb2 Single Crystals	000383
Epitaxial GaP Layers by Molten Salt Electrodeposition (	flux)* #Conditions for Stable Growth of	000224
in InP LPE Growth with PH3 Partial Pressure (	flux)* #Control of Substrate Degradation	000199
Na+ and Co2+ plus Zr4+, Grown by Liquid-Phase Epitaxy (	flux)* #Thin Films of LiNbO3, Doped with	000691
Ruthenium Oxide of a New Hexagonal Structure Type (	flux)* #La8Ru4O21: A Mixed-Valence Ternary	000208
Mass Transfer (theory, solution, vapor transport,	flux)* #On Diffusive-Advective Interfacial	001074
Transformation Phenomena to Its Low Form (Czochralski,	flux)* #High Temperature Form of Pb2WO5 and	000310
Pnictides with ThCr2Si2-Type Structure (phosphides,	flux)* #Ternary Lanthanoid-Transition Metal	000596
Sulphide by Double Decomposition Reaction in Melt (ZnS,	flux)* #Growth of Single Crystals of Cubic Zinc	000328
LPE-Grown Li(Nb,Ta)O3 Film and LiTaO3 Substrate (	flux)* Dependence of Lattice Mismatches Between	000922
in Sodium Tungstate Melts by Continuous Cooling (CaWO4,	flux)* Growth of Calcium Tungstate from Solutions	000810
Rare Earth Vanadates and Phosphates (YPO4, YVO4, REPO4,	flux)* Melts for the Growth of Single Crystals of	000265
in Gallium Phosphide Crystals (GaP, Czochralski,	flux)* Growth Conditions and Dislocation Structure	000723
766W0.234)O3 and its Related Compounds (SrFe1-xWxO3,	flux)* and Magnetic and Mossbauer Studies of Sr(Fe0.	001001
in Sodium Tungstate Melts by Continuous Cooling (SrWO4,	flux)* Growth of Strontium Tungstate from Solutions	000811
Phase Epitaxy on Bridgman Grown Zn2TiO4 Substrates (	flux)* #Thin Film Growth of (MnxZn1-x)FeO4 by Liquid	000216
(La,Y)B6, La,Ba)B6, (La,Cs)B6, (Eu,Y)B6, (Eu,Ba)B6,	flux)* (LaB6, EuB6, CeB6, BaB6, (La,Eu)B6, La1-xEuB6,	000727
Single Crystals from Melt-Solutions on Oriented Seeds (	flux)* #Kinetic Characteristics of Growth of Y3Al5O12	000978
Film Waveguides on LiTaO3 Substrates (LiNb1-xTax O3,	flux)* #LPE Growth of Li(Nb,Ta)O3 Solid-Solution Thin	000500
Lattice Parameters((Y, Bi, Sm, Ca)3(Fe, Ge)5 O12,	flux)* Layers as Functions of the Mismatch of Crystal	000251
in a PbO/B2O3 Solution (Y3Fe5O12, RExY3-xGayFe5-yO12,	flux)* #A Solubility Model for Rare-Earth Iron Garnets	001022
Structures in the (GaAl)As System by LPE (Ga1-xAlxAs,	flux)* Automated System for the Growth of Multilayered	000892
Growth from High-Temperature Solutions (review,	flux)* #Rost Kristallov iz Rastvorov-Rasplavov (Crystal	000976
KSbO3 Derivative Oxide with No Metal-Metal Bonding (	flux)* #The Crystal Structure of La3Ru3O11: A New Cubic	000209
Y3Al5O12 Crystals on Oriented Seeds in Melt-Solutions (	flux)* Crystallization Kinetics and Growth Mechanism of	000093
and Reduced Oxygen Contamination (GaP, Ga0.5Al0.5P,	flux)* Layer Under Controlled Phosphorus Vapor Pressure	000460
Content of Ferrite-Forming Oxides (NixZn1-xFe2O4,	flux)* B2O3-ZnO-NiO-Fe2O3 Melts with a Nonstoichiometric	000107
in Molten Potassium Trimolybdate ((Y,Nd)Al3(BO3),	flux)* YAl-Borate Crystals in Preparation from Solutions	000551
Crystal Layers by the Liquid Phase Epitaxy Technique (	flux, (Ba,Pb)Al12O19)*Barium Lead Hexa-Aluminate Single	000071
Optical Properties of Some Rare Earth Iron Garnets (	flux, (GdPrBi)3(FeGaIn)5O12, (YLaBi)3(FeGa)5O12)*	001107
Defects in LPE GaAs Revealed by New Chemical Etchant (	flux, film)* #Observations of	000703
in Sodium Tungstate Melts by Continuous Cooling (BaWO4,	flux, kinetics, slow cooling)* Tungstate from Solutions	000812
Dissolution Kinetics of Garnet Liquid Phase Epitaxy (	flux, kinetics, theory)*Liquidus Temperature and Growth-	000333
#The Hardness of NbB2 Single Crystals (	flux, slow cooling)*	000679
Crystals from the Ternary System Nd2O3-MoO3-Na2CO3 (	flux, slow cooling)* #Growth of NaNd(MoO4)2 Single	000716
Growth from a Ternary Component Solution (Ga1-xAlxAs,	flux, theory)* Liquid and Solid Compositions During LPE	000970
(NaAl11O17, KAl11O17, RbAl11O17, AgAl11O17, EuAl11O17,	flux, theory)*Alkali, Silver, and Europium beta-Alumina	000621
Method of Stirring for LPE Growth (Sm,Y)3(Fe,Ga)5O12 (	flux, theory, film)* #An Improved	000222
of GaAs Layers Grown by Liquid-Phase Epitaxy (	flux, theory, kinetics)* #Surface Morphology	000605
Surface in Pb-Sn-Te System by Simplex Lattice Method (	flux, theory, Pb1-xSnxTe)* #Investigation of Liquidus	000668
the Transition-Metal Phosphides CuP2, NiP2, and RhP3 (	flux, vapor transport)* Growth and Characterization of	000708
Vol. 1 (garnets, review, hydrothermal, epitaxy,	flux, Bridgman, film)* for Magnetic Applications,	000807
Growth from Liquids at High Temperatures (review, melt,	flux, Czochralski, Kyropoulos, float zone melting,	000139
by the Solution Method and a Study of Their Properties (	flux, LaB6, SmB6, SmB4, EuB6, YbB6)* Rare Earth Borides	000359
Epitaxial Growth of Lithium Ferrite-Aluminate Films (	flux, Li0.5Fe2.5-xAlxO4)* #Liquid Phase	001017
for Sodium-Titanium Bronze Crystals in NaBO2 (	flux, Na2Fe2Ti6O14, Na2NiTi7O16)* #Growth Conditions	000602
of Some Alkaline-Earth Rare-Earth Pentaoxometallates (	flux, Sr2LaAlO5, Sr2PrAlO5, Sr2NdAlO5, Sr2SmAlO5,	000253
#Phase Equilibria in Solvent-Garnet Systems (	flux, Y3Al5O12, Y3Fe5O12)*	000159
#Single Crystal Growth of Alexandrite in a V2O5	Flux (BeAl2O4)*	001005
#Crystallisation of Some Ferrite Materials from Lead	Flux with Addition of V2O5 (Y3Fe5O12, Ho3Fe5O12,	000565
Chemical Vapour Transport of Ni- #An Application of the	Flux Function Method to a Five-Component System: The	000236

PERMUTED TITLE INDEX

#Effects of Modifying Starting Compositions for Rare Earth Arsenates (REAsO4, slow cooling)*

CaF2)*

Transformation Temperature (slow cooling)*

in the Systems R2O3-K2O-MoO3 (DyKMo2O8, ErKMo2O8, Fluorides. Part 2 (KAlF4, KMnF3, RbFeF4, Rb2FeF5, K2B4O7 Mixture*

Elements (Ln3NbO7) (Ho3NbO7, Y3NbO7, Er3NbO7, Dy3NbO7,

PO4)6Cl, slow cooling, evaporation)*

#Top-Seeded

#The Effects of M(M=Mg, Ca and Ba) Ions on the KCl R2Si2O7 (R = Tm, Er, Ho, Dy) (Tm2Si2O7, Er2Si2O7, Antiferromagnetic Garnets Ca3Fe2Ge3O12 and Ca3Mn2Ge3O12# Conditions. IV (ZrFeF6, ZrCrF6, ZrNiF6, ZrCoF6, Complex Oxides (GdVO3, Gd4Ba6 (BO3)9, (Ba0.93Gd0.07)#The Oxides (NdBO3, LaBO3/Pr, LaBO3, (La,Pb)MnO3, (Nd, #The

#Relative Supersaturation and Supercooling in Films grown by Liquid Phase Epitaxy from PbO-B2O3 #The Role of B2O3 in PbO/PbF2/B2O3 organics)*   #A New Method to kinetics)*   #The Thermodynamic Driving kinetics, melt, solution)*   #On the Driving Form of Pb2WO5 and Transformation Phenomena to Its Low #Power Required to Form (Czochralski, flux)*   #High Temperature #Theoretical Solute Redistribution During a Modified of the Soft-Core Model in an FCC Crystal solution, melt)*   #Investigation of Structural electron beam melting, evaporation)*   #Silicide Silicon Crystals* #The Effect of Doping on Microdefect #Impurity Effects on Low-Angle Boundary Crystallization Process (vapor transport)* from the Melt Under Argon Pressure (ZnSe, Bridgman)* *

#Dynamics of Laser-Induced Si, Ge, GaAs, InAs, InSb, Al2O3, melt, theory)* #Impurity Effect on the #Epitaxial Relationships During the #Crystal Structure and Habit of Fine Metal Particles and Dissolution Kinetics of III-V Heterostructures Silicon Crystals Grown with Platinum and Gold as Liquid-Fe2O3 Melts with a Nonstoichiometric Content of Ferrite-Zone Recrystallization of Metals and Preparation of # Single Crystals of GaxIn1-xPyAs1-y Via the Gradient Dynamics Simulation (melt, slow #Study of Melting and *   #Limitations in Using Kilohertz Radio #Pyroelectricity and Related Properties in the #Effect of Crucible Shape on a Crystallization Without a Crucible*   #Shape of the Crystallization Crystal (NaCl,   #The Effect of the Solidification of Gd2(SO4)3 8H2O (crystal) and Dy2Ti2O7 (powder) as a Chemical Vapour Transport #An Application of the Flux Kinetics and Structure of Epitaxial Garnet Layers in #Electrocrystallization. PbCl2, organics)*   #An Inexpensive Zone Melting Spacelab (CdTe, zone   #A Double-Ellipsoid Mirror #Solar #Study of Temperature Distribution in High-Temperature #Subsidiary Electrical Heating for Verneuil #Correlation of Diffusion Behaviour with the Entropy of Mechanical Relations in the Calculation of Heats of #Spacelab and Past and #Interface Shape and Horizontal Variations of Al and #Liquid-Phase Epitaxy of ZnSe from Zn-and Properties of Binary, Ternary and Quaternary (In, Improved Method of Stirring for LPE Growth (Sm,Y)3 (Fe, Equilibrium Adsorption Layers in CVD Systems (Si-H-Cl, Growth. Part I. Growth and Segregation Behavior of #Ag5GaS4 -- A New Phase in the System Ag-Free and Dislocation-Assisted Growth of Gallium (a Binary Alloy with a Curved Solid-Liquid Interface (Ge: During Czochralski Growth of Gallium and Germanium (Growth of Substituted Yttrium Iron Garnets Y3Fe5-x(an Organometallic VPE   #Ellipsometric Assessment of #Epitaxial Layers of the Laser Material Nd (the Growth Rate and Purity of Epitaxial Layers in the for the Growth of Multilayered Structures in the (of Depressions and Voids During LPE Growth of of the Mechanism and Kinetics of Growth of LPE Gaseous Impurity and Its Effect on the Purity of LPE #Tetragonal Distortion in Heteroepitaxial Layers: Ge on Diodes Prepared by Molecular-Beam Epitaxy (MBE) on #Periodic Doping Structure in Effect on Grown-In Dislocation Density of InP and Organometallic #Ellipsometric Assessment of (Ga,Al)As/

Flux Growth (DyPO4, PrBO3, LaBO3, DyKMo2O8, ErKMo2O8,   001060
Flux Growth and Characterization by X-ray Topography of   000900
Flux Growth and Characterization of Fluorite Crystals (000237
Flux Growth of Bi2WO6 Single Crystal Below the   000665
Flux Growth of Crystals of RKMo2O8, R2MoO6 and R6MoO12   001062
Flux Growth of Crystals of Some Transition Metal   001064
Flux Growth of Cr2O3 Crystals from a K2Cr2O7-K2CrO4-   000860
Flux Growth of Double Oxides of Niobium and Rare-Earth   001097
Flux Growth of Lanthanum Borate, LaBO3*   000508
Flux Growth of Lead Neodymium Chloroapatite (Na2Nd2Pb6(000445
Flux Growth of LiNdP4O12 Single Crystals*   000677
Flux Growth of LiNdP4O12 Single Crystals*   000678
Flux Growth of MWO4 Needle Crystals*   000715
Flux Growth of Polymorphic Rare-Earth Disilicates,   000595
Flux Growth of Single Magnetic Sublattice   000238
Flux Growth of Some Fluoride Crystals Under Reducing   000319
Flux Growth of Some New Rare Earth and Iron Group   001063
Flux Growth of Some Rare-Earth and Iron Group Complex   001061
Flux Growth of ZnGa2O4 Single Crystals*   001018
Fluxed Melt Systems (Ho2Si2O7, TbVO 4, DyKMo2O8)*   000594
Fluxed Melts*   Imbalance for Ca,Ge-Substituted Garnet   000386
Fluxes (Y3Al15O12, theory)*   000272
Follow Crystal Growth by Coulter Counter (solution,   000601
Force for Crystallization from Solution (theory,   000904
Force for Crystallization: The Growth Affinity (theory,   001024
Form (Czochralski, flux)*   #High Temperature   000310
Form a Floating Zone and the Zone Shape (theory)*   000492
Form of Pb2WO5 and Transformation Phenomena to Its Low   000310
Form of Zone Refining-Cascade Purification (metals)*   000201
Formation (theory)*   Analysis of Crystallization   000965
Formation and Growth Kinetics of Organic Crystals (000773
Formation by High-Dose Si+-Ion Implantation of Pd (000777
Formation in As-Grown Dislocation-Free Czochralski   000223
Formation in Silicon Single Crystals (Si, Czochralski)*   000218
Formation of Amorphous Mn-Bi Films and Their   001103
Formation of Defects in Zinc Selenide Crystals Grown   000477
Formation of Mixed Crystals in the System CuxZn1-xCr2Se4   000722
Formation of Palladium Silicide (Pd2Si)*   001042
Formation of Stress and Dislocations in Crystal Growth (000406
Formation of Terraces in GaAs LPE Growth (flux)*   000988
Formation of Three-Dimensional Snow Dendrites (H2O)*   000493
Formed by Gas-Evaporation Technique: bcc Metals (V, Fe,   000821
Formed by LPE (GaxAl1-xAs)*   #Growth   000894
Forming Agents (vapor transport, Si)* Morphology of VLS   001075
Forming Oxides (NixZn1-xFe2O4, flux)* BaO-B2O3-ZnO-NiO-   000107
Fractionation of Stable Isotopes and Impurities During   000994
Freeze Method (slow cooling, melt, doped)* In-Rich Bulk   000053
Freezing in the Gaussian Core Model by Molecular   000915
Frequencies for Float Zone Silicon Crystals (Si, theory)   000358
Fresnoite Pseudobinary System Ba2TiGe2O8-Ba2TiSi2O8 (000836
Front (Bridgman, Kinetics, theory)*   000004
Front During Electron-Beam Zone Refining of beta-B   000968
Front Shape on the Temperature Distribution in the   000128
Function of Magnetic Field (solution)*   Conductivity   000065
Function Method to a Five-Component System: The   000236
Functions of the Mismatch of Crystal Lattice Parameters(000251
Fundamental Aspects (review, theory)*   000026
Furnace for Purification and Crystal Growth (NaNO3,   000871
Furnace for Zone Crystallization Experiments in   000276
Furnace Annealing of Amorphous Si Layers (films)*   000537
Furnaces for Silicon Carbide Monocrystal and Epitaxial   000561
Furnaces in the USSR (review)*   000349
Fusion (theory)*   000980
Fusion for Ionic Crystals (review, theory)* #Quantum-   001087
Future Crystal Growth Experiments in Space (review)*   000854
Ga Contents in Substituted YIG Single Crystals Grown by   000483
Ga Solution (film)*   000308
Ga) (As,P) Alloys Grown by the Hydride Vapor Phase   000729
Ga) 5O12 (flux, theory, film)*   #An   000222
Ga-As-H-Cl) (GaAs, theory, epitaxy)*   Analysis of   000184
Ga-Doped Ge (theory)*   of a Heat Pipe to Czochralski   000604
Ga-S (AgGaS2)*   000688
Ga, theory)*   #Rate Equations for Dislocation-   000001
Ga, Bridgman, theory)* Unidirectional Solidification of   000206
Ga, Ge, theory)* Contact Angle Between Melt and Crystal   001071
Ga,Al)xO12 by the Floating Zone Method* #Single Crystal   000879
Ga,Al)As/GaAs Epitaxial Layers During Their Growth in   000974
Ga,Cr)3 (BO3)4 (flux)*   000578
Ga_AsCl3-H2 System (GaAs, vapor transport, kinetics,   000653
GaAl)As System by LPE (Ga1-xAlxAs, flux)*   System   000892
GaAs (film)*   #Development   000004
GaAs (theory, epitaxy)*   #Investigation   000144
GaAs (vapor transport)*   #New   000313
GaAs (vapor transport, film)*   000362
GaAs (Al, films)*   Crystal-Aluminum Schottky-Barrier   000190
GaAs (GaAs, Ga1-xAlxAs, molecular beam epitaxy, review)*   000248
GaAs Crystals (Czochralski)*   #Impurity   000856
GaAs Epitaxial Layers During Their Growth in an   000974

PERMUTED TITLE INDEX

#Temperature Range for Growth of Autoepitaxial	GaAs Films by MBE (molecular beam epitaxy, theory)* 000686
#Surface Morphology of	GaAs Grown by Vapor Phase Epitaxy* 000705
theory, kinetics)* #Surface Morphology of	GaAs Layers Grown by Liquid-Phase Epitaxy (flux, 000605
#Impurity Effect on the Formation of Terraces in	GaAs LPE Growth (flux)* 000988
#Observations of Defects in LPE	GaAs Revealed by New Chemical Etchant (flux, film)* 000703
and Electrical Properties of CuGaSe2 Thin Films on	GaAs Substrates (epitaxy)* #Structural 000844
by Molecular Beam Epitaxy. I. Growth on Singular (100)	GaAs Substrates (film, kinetics)* Deposited 000762
Growth of Single Crystal Cd1-xZnxS Layers on (111)	GaAs Substrates Using the Close-Spaced Geometry (vapor 000305
Electrical Properties of Epitaxial CuInS2 Thin Films on	GaAs Substrates* #Growth and 000690
Vapor Deposition of Epitaxial ZnSe Films on	GaAs Substrates* #Organometallic 000921
of Sputter-Deposited Single-Crystal GaSb Films on	GaAs Substrates* #Growth and Electrical Properties 000269
epitaxy)* #High Efficiency	GaAs Thin Film Solar Cells by Peeled Film Technology (000497
#Chromium Doped	GaAs Vapor Phase Epitaxy* 000469
	# GaAs Vapour Phase Epitaxy Using Computer Control* 001057
vapor transport)* #Fast Growth in	GaAs VPE at Low Temperature and High Partial Pressures (000393
#The Thermal Oxidation of	GaAs* 000156
#Liquid Phase Epitaxial Deposition of GaP on	GaAs* 000354
#Liquid Phase Epitaxial Growth of Zn and S Doped	GaAs* 000416
#Epitaxial Layers of CuInSe2 on	GaAs* 000843
in Molecular Beam Epitaxy of Gallium Arsenide (	GaAs)* #Growth Temperature Dependence 000666
the Complex Monochlorodiethylgallium-Triethylarsine (	GaAs)* Arsenide by Chemical Vapor Deposition from 001109
Graphite, and Integrated Optics (review, theory, C,	GaAs)* Vol. 4, Morphological Stability, Convection, 001081
Beam Epitaxy. I. Growth on #Crystal Growth Kinetics in (	GaAs)n-(AlAs)m Superlattices Deposited by Molecular 000762
of Crystals of Solid Solutions of the System	GaAs-ZnSe (Bridgman)* #Preparation and Investigation 000185
	# GaAs/AlAs Layered Films (epitaxy)* 000353
Phase Electroepitaxy: Growth Kinetics (Peltier effect,	GaAs, theory, electrolytic deposition, kinetics)*Liquid- 000437
Adsorption Layers in CVD Systems (Si-H-Cl, Ga-As-H-Cl) (	GaAs, theory, epitaxy)* Analysis of Equilibrium 000184
Growth in an Organometallic VPE System (Ga1-xAlxAs,	GaAs, vapor transport)* Epitaxial Layers During Their 000974
Purity of Epitaxial Layers in the Ga_AsCl3-H2 System (	GaAs, vapor transport, kinetics, theory)* Rate and 000653
Application to the Study of Heterogeneous Reactions (	GaAs, vapor transport, theory)* Method and Its 000281
Elements, III-V Compounds and Alloys (Ge, Si,	GaAs, BP)* Epitaxial Vapour Growth of Semiconductors: 000041
#Periodic Doping Structure in GaAs (	GaAs, Ga1-xAlxAs, molecular beam epitaxy, review)* 000248
Surface and Crystal Physics Studied by MBE (	GaAs, Ga1-xAlxAs, review of MBE process)**Semiconductor 000055
#Integrated Optical Devices Fabricated by MBE (	GaAs, Ga1-xAlxAs, review, film)* 000152
Superlattices by MBE and Their Characterization (	GaAs, Ga1-xAlxAs, In1-xGaxAs, GaSb1-yAsy)*Semiconductor 000176
of Stress and Dislocations in Crystal Growth (Si, Ge,	GaAs, InAs, InSb, Al2O3, melt, theory)* #Formation 000406
of Semiconductor Single Crystals and Films (Si, Ge,	GaAs, InSb, epitaxy, theory, Czochralski)* the Growth 000636
#Nitrogen Doping into	GaAs1-xPx Using Ionized Beam in Molecular Beam Epitaxy* 000617
* #Growth of	GaAs1-xSbx Crystals by Steady-State Liquid Phase Epitaxy 000957
Flow Phenomena on the Perfection of Czochralski Grown	Gadolinium Gallium Garnet (Gd3Ga5O12)* Effect of Melt 000638
	# Gadolinium Gallium Garnet (Gd3Ga5O12, review)* 000143
kinetics, #Dissociation Processes and Crystal Growth of	Gadolinium Gallium Garnet (Gd3Ga5O12, Czochralski, 000769
#Dislocation Propagation in Czochralski Grown	Gadolinium Gallium Garnet (GGG, theory, Gd3Ga5O12)* 000838
Molybdates (#Crystal Growth and Imperfections in	Gadolinium Molybdate and Other Isotypic Rare-Earth 000446
of 1.2-1. #Conditions of LPE Growth for Lattice Matched	GaInAsP/InP DH Lasers with (100) Substrate in the Range 000035
#The Praseodymium-Gallium System from 0 to 50 At%	Gallium (melt, Pr2Ga, Pr5Ga3, PrGA)* 000197
for Dislocation-Free and Dislocation-Assisted Growth of	Gallium (Ga, theory)* #Rate Equations 000001
Between Melt and Crystal During Czochralski Growth of	Gallium and Germanium (Ga, Ge, theory)* Contact Angle 001071
* #The Binary Systems Cobalt-	Gallium and Nickel-Gallium Compared (CoGa3, NiGa4, flux) 000290
Temperature Dependence in Molecular Beam Epitaxy of	Gallium Arsenide (GaAs)* #Growth 000666
Complex Monochlorodiethylgallium- #Crystal Growth of	Gallium Arsenide by Chemical Vapor Deposition from the 001109
#The Binary Systems Cobalt-Gallium and Nickel-	Gallium Compared (CoGa3, NiGa4, flux)* 000290
on the Perfection of Czochralski Grown Gadolinium	Gallium Garnet (Gd3Ga5O12)* of Melt Flow Phenomena 000638
#Gadolinium	Gallium Garnet (Gd3Ga5O12, review)* 000143
Dissociation Processes and Crystal Growth of Gadolinium	Gallium Garnet (Gd3Ga5O12, Czochralski, kinetics, 000769
Dislocation Propagation in Czochralski Grown Gadolinium	Gallium Garnet (GGG, theory, Gd3Ga5O12)* # 000838
#The Central Region of the Calcium Oxide-	Gallium Oxide System (CaGa2O4, Czochralski)* 001105
Between Growth Conditions and Dislocation Structure in	Gallium Phosphide Crystals (GaP, Czochralski, flux)* 000723
Reactions (GaP)* #Kinetics of	Gallium Phosphide Synthesis by Chemical Gas Transport 000231
Transport #Optimization of Conditions for the Growth of	Gallium Selenide and Gallium Sulphide by Iodine Vapour 001078
of Conditions for the Growth of Gallium Selenide and	Gallium Sulphide by Iodine Vapour Transport (GaS, GaSe)* 001078
Pr5Ga3, PrGA)* #The Praseodymium-	Gallium System from 0 to 50 At% Gallium (melt, Pr2Ga, 000197
#Electrical Conductivity and	Galvanomagnetic Effects in (SN)x Single Crystal* 000463
#Crystal Growth of Co Ferrite on Fine Acicular	gamma-Fe2O3 Particles (solution)* 000931
#Properties of VPE-Grown	GaN Doped with Al and Some Iron-Group Metals* 000649
#Properties of	GaN Grown on Sapphire Substrates (epitaxy)* 000212
Synthesis by Chemical Gas Transport Reactions (	GaP)* #Kinetics of Gallium Phosphide 000231
Modulators* #Liquid Phase Epitaxy of AlxGa1-xP on	GaP and Its Application to Double Heterostructure Light 000948
#Ionized-Cluster Beam Epitaxial Growth of GaP Films on	GaP and Si Substrates* 000651
#Heteroepitaxial Growth of ZnS on	GaP by the Close-Spaced Technique (kinetics)* 000482
#Modified Chemical Vapor Growth of Cinnabar (HgS) and	GaP in Closed Systems* 000729
#Liquid Phase Epitaxial Deposition of	GaP on GaAs* 000354
#Solution Growth and Some Properties of	GaP Bulk Crystals* 000277
#Ionized-Cluster Beam Epitaxial Growth of	GaP Films on GaP and Si Substrates* 000651
Epitaxial Growth of Lattice-Matched GaP-Ga0.5Al0.5P-	GaP Layer Under Controlled Phosphorus Vapor Pressure 000460
#Conditions for Stable Growth of Epitaxial	GaP Layers by Molten Salt Electrodeposition (flux)* 000224
#Low Temperature Growth of	GaP LPE Layers from Indium Solvent (flux)* 000925
#Vapour-Phase Epitaxial Growth of ZnS on	GaP* 000613
in SSD (synthesis, solute diffusion) Growth of	GaP* #Improvements 000461
Vapor #Liquid Phase Epitaxial Growth of Lattice-Matched	GaP-Ga0.5Al0.5P-GaP Layer Under Controlled Phosphorus 000460
During Epitaxial Growth from a Solution in a Melt (	GaP, flux)* #Theory of Crystallization 000591
Dislocation Structure in Gallium Phosphide Crystals (	GaP, Czochralski, flux)* Between Growth Conditions and 000723
Vapor Pressure and Reduced Oxygen Contamination (	GaP, Ga0.5Al0.5P, flux)* Under Controlled Phosphorus 000460
of Single Crystals of Solid Solutions GaxIn1-xP (	GaP, InP)* #Preparation and Properties 000119
the Perfection of Czochralski Grown Gadolinium Gallium	Garnet (Gd3Ga5O12)*The Effect of Melt Flow Phenomena on 000638
#Gadolinium Gallium	Garnet (Gd3Ga5O12, review)* 000143
Processes and Crystal Growth of Gadolinium Gallium	Garnet (Gd3Ga5O12, Czochralski, kinetics, theory)* 000769
Propagation in Czochralski Grown Gadolinium Gallium	Garnet (GGG, theory, Gd3Ga5O12)* #Dislocation 000838

PERMUTED TITLE INDEX

```
                    #Growing Optical films of Aluminum Yttrium   Garnet (RE substituted Y3Al5012, flux)*                       000044
Y3Fe5012)*                        #Growth of Bulk Yttrium Iron    Garnet Crystals from High Temperature Solutions (          000977
      Properties and Charge Imbalance for Ca,Ge-Substituted       Garnet Films grown by Liquid Phase Epitaxy from PbO-        000386
     #Influence of the CaCO3/GeO2 Ratio on Y Sm Lu Ca Fe Ge       Garnet Films Properties (LPE)*                               000650
      of the Buoyancy Parameter on Czochralski Bulk Flow in       Garnet Growth (theory, kinetics)*                #Effect     000535
     3Fe5012, vapor transport, epitaxy, #Low Temperature CVD      Garnet Growth (Er3Fe5012, (Eu,Y)3Fe5012 and (Eu,Yb)         000211
     Lattice   #Growth Kinetics and Structure of Epitaxial        Garnet Layers as Functions of the Mismatch of Crystal       000251
      Liquidus Temperature and Growth-Dissolution Kinetics of     Garnet Liquid Phase Epitaxy (flux, kinetics, theory)*       000333
     flux)*                         #The Growth Kinetics of       Garnet Liquid Phase Epitaxy Using Horizontal Dipping (      001021
     Method (Y3Fe5012)*          #Evaluation of Yttrium Iron      Garnet Single Crystals Grown by the Floating Zone           000480
                      #Phase Equilibria in Solvent-               Garnet Systems (flux, Y3Al5012, Y3Fe5012)*                  000159
             #Growth of Single Crystals of Na3Al2Li3F12           Garnet Under Hydrothermal Conditions*                       000675
     Mismatches Between Epitaxial Films and Substrates (RE        garnets)*   Parallel Setting for Measuring Local Lattice    000704
     film)   #Crystals for Magnetic Applications, Vol. 1 (        garnets, review, hydrothermal, epitaxy, flux, Bridgman,     000807
       Emission of Crystals of Yttrium Holmium Aluminum           Garnets ((Y1-xHox)3Al5012, Czochralski)* and Stimulated    000046
                 #Magneto-Optical Properties of Some Rare Earth Iron  Garnets (flux, (GdPrBi)3(FeGaIn)5012, (YLaBi)3(FeGa)   001107
                          #Physics of Magnetic                    Garnets (review)*                                           000741
             #Liquid Phase Epitaxial Growth of Magnetic           Garnets (review)*                                           000783
                        #Liquid Phase Epitaxy of                  Garnets (review)*                                           000802
                     #Crystal Growth of Magnetic                  Garnets from High-Temperature Solutions (review)*           000982
     yO12, flux)*     #A Solubility Model for Rare-Earth Iron     Garnets in a PbO/B203 Solution (Y3Fe5012, RExY3-xGayFe5-     001022
     Growth of Single Magnetic Sublattice Antiferromagnetic       Garnets Ca3Fe2Ge3012 and Ca3Mn2Ge3012*          #Flux       000238
          #Single Crystal Growth of Substituted Yttrium Iron      Garnets Y3Fe5-x(Ga,Al)x012 by the Floating Zone Method*     000879
         #Growth of Polyhedral Metal Crystallites in Inactive     Gas (Mg, review)*                                           001006
     Growth by  #In-Process Monitoring and Control of Doping      Gas Concentration During Vapor Phase Silicon Epitaxial      000937
     Preparations (TiO2, vapor transport)*               #A       Gas Handling System for Exacting Crystal Growth             000808
                     #Crystallization of WO2 from the             Gas Phase (vapor transport)*                                000052
             #Growing Monocrystalline Bars of SiC from the        Gas Phase (vapor transport)*                                000554
     Silicon Crystals with Phosphorus During Growth from the      Gas Phase (Si)*                                #Doping of    000027
     Chlorine Boracite on Chromium-Chlorine Boracite (       #    Gas Phase Synthesis of Epitaxial Layers of Nickel-          000837
         in the Course of Preparation of II-IV-V2 Compounds by    Gas Phase Transport, with ZnSiP2 as an Example*            001086
     Microanalytical Evaluation of       #Growth of ZnSiP2 from   Gas Phase Under Non-Stoichiometric Conditions --            000082
     Degree of Doping in Silicon Epitaxial Growth from the        Gas Phase. I. Phosphorus Content and Segregation           000514
        of Doping in Growth of Epitaxial Silicon from the         Gas Phase. II. Phosphorus Content and Differential          000515
           #Kinetics of Gallium Phosphide Synthesis by Chemical   Gas Transport Reactions (GaP )*                             000231
         #Coalescence Growth of Smoke Particles Prepared by a     Gas-Evaporation Technique (MoO3, WO3, theory)*              000448
      Structure and Habit of Fine Metal Particles Formed by       Gas-Evaporation Technique: bcc Metals (V, Fe, Cr, Mo        000821
      and Gallium Sulphide by Iodine Vapour Transport (           GaS, GaSe)*         for the Growth of Gallium Selenide      001078
       Coefficients for Directionally Solidified InSb-            GaSb (magnetic field effects, eutectics, Bridgman)*        000864
                         #Molecular Beam Epitaxy of               GaSb and GaSbxAs1-x (films)*                                001100
     Horizontal Travelling Heater Method*                    #    GaSb and InSb Crystals Grown by a Vertical and             000099
          #Liquid Phase Epitaxial Growth of Ga1-xInxSb on         GaSb by Stepwise Grading (melt)*                            000803
         Near Irregularly Remelted Surfaces in Pulled             GaSb Crystals (Czochralski)*Observations of Microfacets     000517
     Properties of Sputter-Deposited Single-Crystal               GaSb Films on GaAs Substrates*   #Growth and Electrical     000269
          #Molecular Beam Epitaxy of InxGa1-xSb (x = 0 to 1) (    GaSb)*                                                      001101
                      #Molecular Beam Epitaxy of GaSb and         GaSbxAs1-x (films)*                                         001100
      Their Characterization (GaAs, Ga1-xAlxAs, In1-xGaxAs,       GaSb1-yAsy)*      #Semiconductor Superlattices by MBE and    000176
        and Gallium Sulphide by Iodine Vapour Transport (GaS,     GaSe)* of Conditions for the Growth of Gallium Selenide     001078
     GaAs (vapor transport)*                          #New        Gaseous Impurity and Its Effect on the Purity of LPE        000313
     equal to  #Structural Studies of Compounds in the Series     GaSxSe(1-x) (0 less than or equal to x less than or         001077
     melt, slow        #Study of Melting and Freezing in the      Gaussian Core Model by Molecular Dynamics Simulation (      000915
       Kinetics of III-V Heterostructures Formed by LPE (         GaxAl1-xAs)*                      #Growth and Dissolution    000894
     Analysis of Vapor Phase Growth - Citing an Instance of       GaxIn1-xAs (theory)*                      #A Simple         000673
        #Surface Processes Controlling the Growth of              GaxIn1-xAs and GaxIn1-xP Alloy Films by MBE (kinetics)*     000301
                           #The Growth of                        GaxIn1-xAs on (100) InP by Liquid-Phase Epitaxy (flux)*     000754
                #Preparation and Properties of Bulk               GaxIn1-xAs*                                                 000552
      and Properties of Single Crystals of Solid Solutions        GaxIn1-xP (GaP, InP)*                  #Preparation         000119
       Processes Controlling the Growth of GaxIn1-xAs and         GaxIn1-xP Alloy Films by MBE (kinetics)*     #Surface       000301
     cooling, #The Growth of In-Rich Bulk Single Crystals of      GaxIn1-xPyAs1-y Via the Gradient Freeze Method (slow        000053
     *                                                            GaxIn1-xSb Crystals Grown by Liquid Phase Epitaxy (melt)    000852
                            #LPE Growth of Quaternary             GayIn1-yAsxP1-x (film)*                                     000188
     #Liquid Phase Epitaxial Growth of Lattice-Matched GaP-       Ga0.5Al0.5P-GaP Layer Under Controlled Phosphorus Vapor     000460
     Vapor Pressure and Reduced Oxygen Contamination (GaP,        Ga0.5Al0.5P, flux)*   Layer Under Controlled Phosphorus     000460
     Constant Cooling #Improvement of Crystal Composition in      Ga1-xAlxAs LPE Layers Grown under Conditions of            000249
                   #Molecular Beam Epitaxy of Ge and              Ga1-xAlxAs Ultra Thin Film Superlattices*                   000761
     Multilayered Structures in the (GaAl)As System by LPE (      Ga1-xAlxAs, flux)*An Automated System for the Growth of     000892
     During LPE Growth from a Ternary Component Solution (        Ga1-xAlxAs, flux, theory)*Liquid and Solid Compositions     000970
                 #Periodic Doping Structure in GaAs (GaAs,        Ga1-xAlxAs, molecular beam epitaxy, review)*                000248
         Surface and Crystal Physics Studied by MBE (GaAs,        Ga1-xAlxAs, review of MBE process)*        #Semiconductor   000055
     #Integrated Optical Devices Fabricated by MBE (GaAs,         Ga1-xAlxAs, review, film)*                                  000152
     During Their Growth in an Organometallic VPE System (        Ga1-xAlxAs, GaAs, vapor transport)*   Epitaxial Layers      000974
       Superlattices by MBE and Their Characterization (GaAs,     Ga1-xAlxAs, In1-xGaxAs, GaSb1-yAsy)*    #Semiconductor      000176
                   #Liquid Phase Epitaxial Growth of             Ga1-xInxSb on GaSb by Stepwise Grading (melt)*              000803
     *                     #Vapor-Phase Growth of Epitaxial       Ga1-xInxSb Using Open-Tube Flow System (vapor transport)    000681
        of SrCl2 and Solid Solutions of SrCl2-PrCl3 and SrCl2-    GdCl3 (Bridgman-Stockbarger, Czochralski)*      Growth     001054
                             #Magnetic Structure of               GdCu1-xZnx System (melt)*                                   000956
       #Crystal Growth of Some Rare Earth Trifluorides (          GdF3, TbF3, DyF3, HoF3, ErF3, Bridgman-Stockbarger,         000323
        of Epitaxial Films of the Rare-Earth Dihydrides,          GdH2, TbH2, DyH2, ErH2, HoH2)*#The Growth and Structure    000782
     Earth Carbonates Under Hydrothermal Conditions (YOHCO3,      GdOHCO3, ErOHCO3, Y20(OH)2CO3, Er202CO3)*    of Rare-       000169
        Properties of Some Rare Earth Iron Garnets (flux,         GdPrBi)3(FeGaIn)5012, (YLaBi)3(FeGa)5012)*    Optical       001107
        of Some New Rare Earth and Iron Group Complex Oxides (    GdVO3, Gd4Ba6(BO3)9, (Ba0.93Gd0.07)TiO3, Nd4Ba6(BO3)9,     001063
        Isotypic Rare-Earth Molybdates (Czochralski, melt,        Gd2(MoO4)3, Sm2(MoO4)3, Tb2(MoO4)3)*Molybdate and Other   000446
     Function of Magnetic #Low Temperature Conductivity of        Gd2(SO4)38H2O (crystal) and Dy2Ti207 (powder) as a         000065
                     #Analysis of Dislocations in                 Gd3Ga5012 by a Repeated Etching Technique*                  000943
        Grown Gadolinium Gallium Garnet (GGG, theory,             Gd3Ga5012)*      #Dislocation Propagation in Czochralski    000838
        of Czochralski Grown Gadolinium Gallium Garnet (          Gd3Ga5012)*  of Melt Flow Phenomena on the Perfection      000638
                   #Gadolinium Gallium Garnet (                   Gd3Ga5012, review)*                                         000143
        and Crystal Growth of Gadolinium Gallium Garnet (         Gd3Ga5012, Czochralski, kinetics, theory)*   Processes     000769
```

PERMUTED TITLE INDEX

and Some New Stimulated Emission Data of
Y3NbO7, Er3NbO7, Dy3NbO7, La3NbO7, Y3TaO7, Er3TaO7,
New Rare Earth and Iron Group Complex Oxides (GdVO3,
Part I. Growth and Segregation Behavior of Ga-Doped
#Crystal Growth and Segregation Under Zero Gravity:
#Molecular Beam Epitaxy of
Behavior of Be as an Impurity During Crystallization of
Layer on a Silicon Substrate by Laser Annealing of
#Computer Model of Autoepitaxial Growth of
#Influence of the CaCO3/GeO2 Ratio on Y Sm Lu Ca Fe
#In Situ X-Ray Topography of Epitaxial
#Tetragonal Distortion in Heteroepitaxial Layers:
#Behavior of Boron in
of Irregular Silver-Germanium Eutectic Alloys (Ag-
by Electron-Beam Zone Melting Without a Crucible (Si-
in Germanium Under the Influence of the Peltier Effect (
of Crystal Lattice Parameters((Y, Bi, Sm, Ca)3(Fe,
#Preparation of
#Electrical Properties and Charge Imbalance for Ca,
of Germanium (111) Single Crystals with GeBr4 Vapor (
During Czochralski Growth of Gallium and Germanium (Ga,
and Perfection of Single Crystals of Germanium (
#Germanium-Iodine Equilibrium from 700 to 1000 K (
of Stress and Dislocations in Crystal Growth (Si,
Growth of Semiconductor Single Crystals and Films (Si,
Semiconductors: Elements, III-V Compounds and Alloys (
Cu2(I)B(II)C(IV)X4 (BII=Zn, Mn, Fe, Co - CIV = Si,
of a Binary Alloy with a Curved Solid-Liquid Interface (
#Some Electrical Properties of
#Reaction of Germanium (111) Single Crystals with
Properties of High-Purity p-type Single Crystals of
#Single Crystal Growth of Akermanite (Ca2MgSi2O7) and
#Growth Morphology of Weddellite (CaC2O4.2H2O,
#Cerium Trihydroxide (Ce(OH)3,
Iodide Crystals in Silica Hydrogels (TlI, Tl2I3,
on As-Grown Faces of Brushite Crystals (CaHPO4.2H2O,
of Mixed Sr:Ca Tartrates (Sr1-xCazC4H4O6.4H2O,
of Fibrous Sodium Aluminosilicate Crystals in Silica
of Barium Molybdate Single Crystals in Silica
#Crystal Quality of
Perchlorate Single Crystals (KClO4)*
Cesium Tin Halides (CsSnCl3, RbSnCl3, Cs2SnCl6,
Crystals of Lead Chloride (PbCl)*
#Modified
#Crystal Growth of Potassium Perchlorate in
#Crystal Growth of Potassium Perchlorate in Gelatin
Potassium Dihydrogen Orthophosphate in Silica
Tl2NiPb(NO2)6,   #Crystal Growth of Triple Nitrites in
#Growth of Bi2(SeO4)3 Single Crystals in Silica
#Growth of Single Crystals of Ag2SeO4 in Silica
and Growth of KClO4 Single Crystals in Silica
#Man-Made
#Second Harmonic
Method*   #Effects of Growth Rates on the Defect
Core Model in an FCC Crystal Formation (theory)*
Layers on (111) GaAs Substrates Using the Close-Spaced
Convection in Floating Zone Melting: Influence of Zone
#Vapor Growth of Hexagonal
LPE)*   #Influence of the CaCO3/
#New Observations in the Chemical Transport of
H2O,   #Hydrothermal Crystallization in the System BaO-
#Crystal Structure and Ionic Conductivity of Li24Zn(
Impurities on the Piezoelectric Properties of Lithium
Na2Ge2O5)*   #Structure and Stability of Alkali
Crystallization in the System BaO-GeO2-Na2O-H2O (barium
#Magnetic Properties of beta-Phase Iron-
and Crystal During Czochralski Growth of Gallium and
Growth Directions and Perfection of Single Crystals of
epitaxy, vapor transport, kinetics, theory)*#Reaction of
#Unidirectional Solidification of Irregular Silver-
Annealing (Ge #Heteroepitaxy of a Deposited Amorphous
#The Czochralski Growth of Bismuth-
Without a Crucible (#Growing Single Crystals of Silicon-
*   #Observations of the Movement of a Molten Zone in
vapor transport, theory, thermodynamics)*
#Dependence of the Optical Homogeneity of Single
the Crystal Growth from the Vapor in Closed Ampules (
#Some Optical Properties of Single
in Czochralski Grown Gadolinium Gallium Garnet (
and Stability of Alkali Germanate Crystals and
Electrodeposits on Oriented Single Crystal Surfaces of
of VLS Silicon Crystals Grown with Platinum and
#On the Relations Between the Number of
and the Crystal Structures of Two Modifications of
theory)*   #Growth of Pb1-xSnxTe Single Crystals of
Growth of InSb from the Melt Under Stabilizing Thermal
Tn-Rich Bulk Single Crystals of GaxIn1-xPyAs1-y Via the
(flux)*   #Temperature

Gd3Ga5O12:Nd3+ Single Crystals (Nd3Ga5O12, doped,   000457
Gd3TaO7)*   and Rare-Earth Elements (Ln3NbO7) (Ho3NbO7,   001097
Gd4Ba6(BO3)9, (Ba0.93Gd0.07)TiO3, Nd4Ba6(BO3)9, Pr4Ba6(001063
Ge (theory)*   of a Heat Pipe to Czochralski Growth.   000604
Ge (Bridgman)*   001088
Ge and Ga1-xAlxAs Ultra Thin Film Superlattices*   000761
Ge by the Czochralski Method*   # 000346
Ge film)*   of a Deposited Amorphous Germanium   000343
Ge Films (theory)*   000018
Ge Garnet Films Properties (LPE)*   000650
Ge Layers During Growth*   000363
Ge On GaAs (vapor transport, film)*   000362
Ge Single Crystals Grown by the Czochralski Method*   000345
Ge)*   #Unidirectional Solidification   000549
Ge)*   Crystals of Silicon-Germanium Solid Solutions   000820
Ge)*   #Observations of the Movement of a Molten Zone   000061
Ge)5 O12, flux)* Layers as Functions of the Mismatch   000251
Ge-Si Epitaxial Alloys by Sputtering*   000420
Ge-Substituted Garnet Films grown by Liquid Phase   000386
Ge, epitaxy, vapor transport, kinetics, theory)*   000060
Ge, theory)* the Contact Angle Between Melt and Crystal   001071
Ge, theory, kinetics)*   #Preferred Growth Directions   000275
Ge, vapor transport, theory, thermodynamics)*   000191
Ge, GaAs, InAs, InSb, Al2O3, melt, theory)* #Formation   000406
Ge, GaAs, InSb, epitaxy, theory, Czochralski)*   the   000636
Ge, Si, GaAs, BP)*   of the Epitaxial Vapour Growth of   000041
Ge, Sn - X = S,Se) (Cu2ZnGeS4, Cu2ZnSnS4, Cu2ZnGeSe4,   000357
Ge:Ga, Bridgman, theory)* Unidirectional Solidification   000206
GeBi2Te4 Single Crystals (Bridgman)*   000975
GeBr4 Vapor (Ge, epitaxy, vapor transport, kinetics,   000060
GeFe2O4 (vapor transport)*   #Electrical and Optical   000920
Gehlenite (Ca2Al2SiO7) by the Floating Zone Method*   000404
gel)*   000302
gel)*   000664
gel)*   #Growth of Thallium   000881
gel)*   #Observations of Etch Pits   000714
gel)*   #Growth and Characterization   000747
Gel (zeolite, Na0.5AlSi6O14.xH2O)*   #Growth   000444
Gel (BaMoO4)*   #Growth Kinetics and Habit Modification   000519
Gel Grown CaHPO4.2H2O Crystals*   000545
Gel Growth and Perfection of Orthorhombic Potassium   000752
Gel Growth of Single Crystals of Some Rubidium and   000930
Gel Method for Growing Large Needles and Single   000003
Gel Technique to Grow Single Crystals of KClO4*   000749
Gelatin Gels (KClO4)*   000750
Gels (KClO4)*   000750
Gels (KDP)*   #Single Crystal Growth of Ferroelectric   000442
Gels (K2CuPb(NO2)6, Tl2CuPb(NO2)6, K2NiPb(NO2)6,   000550
Gels*   000121
Gels*   000122
Gels*   #Nucleation   000751
Gemstones (review)*   000271
Generation in a Sputtered LiNbO3 Film on MgO (epitaxy)*   000518
Generation in KCl Crystals Grown by the Czochralski   000409
Geometrical Analysis of Crystallization of the Soft-   000965
Geometry (vapor transport)* of Single Crystal Cd1-xZnxS   000305
Geometry and Prandtl Number at Zero Gravity (Si, Al2O3,   000175
GeO2 Needle Crystals*   000423
GeO2 Ratio on Y Sm Lu Ca Fe Ge Garnet Films Properties (000650
GeO2. I. Transporting Agent Chlorine*   000787
GeO2-Na2O-H2O (barium germanates, Ba3Ge9O21.H2O, BaGeO3.   000592
GeO4)4 and other new Li+ Superionic Conductors (000396
Germanate (Li2GeO3, Czochralski)*   #The Effect of 000800
Germanate Crystals and Glasses (Na2Ge4O9, Na4Ge9O20,   000822
germanates, Ba3Ge9O21.H2O, BaGeO3.H2O, Na4Ge9O20)*   000592
Germanium (FexGe, slow cooling)*   000845
Germanium (Ga, Ge, theory)* Contact Angle Between Melt   001071
Germanium (Ge, theory, kinetics)*   #Preferred   000275
Germanium (111) Single Crystals with GeBr4 Vapor (Ge,   000060
Germanium Eutectic Alloys (Ag-Ge)*   000549
Germanium Layer on a Silicon Substrate by Laser   000343
Germanium Oxide Single Crystals (Bi12GeO20)*   000771
Germanium Solid Solutions by Electron-Beam Zone Melting   000820
Germanium Under the Influence of the Peltier Effect (Ge)   000061
Germanium-Iodine Equilibrium from 700 to 1000 K (Ge,   000191
Germanosillenite Crystals on the Growth Conditions (000903
GeS, sublimation)*   of Growth Parameters During   000841
GeS2 Single Crystals (Bridgman)*   000697
GGG, theory, Gd3Ga5O12)*   #Dislocation Propagation   000838
Glasses (Na2Ge4O9, Na4Ge9O20, Na2Ge2O5)*   #Structure   000822
Gold and Silver (Sb)*   #Epitaxial Growth of Antimony   000540
Gold as Liquid-Forming Agents (vapor transport, Si)*   001075
Gold Drops and TiP Whiskers in VLS Growth*   000929
Gold Monobromide, I-AuBr and P-AuBr (vapor transport)*   000436
Graded Composition by the Travelling Solvent Method (000564
Gradient (Bridgman, theory)*   During the Crystal   000478
Gradient Freeze Method (slow cooling, melt, doped)* of   000053
Gradient Measurement in Liquid Epitaxial Growth Systems   000574

PERMUTED TITLE INDEX

Growth of p-Type InP Single Crystals by the Temperature
of Cadmium Sulphide Crystals with Large Temperature
melt, Si)*   #Effects of Radial and Axial Temperature
Epitaxial Growth of Ga1-xInxSb on GaSb by Stepwise
of Polycrystalline Snow Crystals with a Specific
on Laser Irradiation Energy Density (Si)*
Epitaxial Growth in SiCl4-HCl-H2 System Based on Flow
from the Vapour Phase (theory)*   #Flow-
First Stage Graphite/ #The Synthesis of the First Stage
C8(+)AsF6(-) and Its Relationship to the First Stage
Materials, Vol. 4, Morphological Stability, Convection,
on Amorphous Substrates Using Artificial Surface-Relief
Influence of Zone Geometry and Prandtl Number at Zero
zone melting, doped, review)#Materials Behaviour in Low
#Crystal Growth and Segregation Under Zero
Growth and Phase Transitions in the NbxCr1-xO2 (x
slow   #On the Resistivity of the Magnetically Ordered
zone melting, review)* #Properties and Synthesis of Nb-
Adsorption Layers in CVD Systems (Si-H-Cl, Ga-As-
of Equilibrium Adsorption Layers in CVD Systems (Si-
Group VA Metal Hydrides and Deuterides (V-H, V-D, Nb-
Metal Hydrides and Deuterides (V-H, V-D, Nb-H, Nb-D, Ta-
of the Group VA Metal Hydrides and Deuterides (V-
#Fermi Surface Measurements of ZrB2 by the de
Technique: bcc Metals (V, Fe, #Crystal Structure and
theory)*   #Growth Morphology and Crystal
Mossbauer Studies (NH4H2PO4, solution)* #Mechanism of
in Silica Gel (BaMoO4)*   #Growth Kinetics and
beam melting, #Elastic Constants of Niobium-Zirconium,
Eu, Tm, and Yb - X = Cl, Br, and I) (melt, #New Mixed
#Growth and Polytypism in Vapor-Grown Mixed Cadmium
Single Crystal Whiskers (hydrogen reduction of ferrous
of Single Crystals of Some Rubidium and Cesium Tin
Processes of Condensation and Evaporation of Alkali
III. On the Phase Transition of CuTeBr and #Telluride
Structures, and Properties of Niobium Chalcogenide
#The
epitaxy)*   #Second
(TiO2) Crystals by Chemical Transport Reactions with
#Surface Morphology of
Crystals by Chemical Transport Reactions with HBr and
theory, Si)*#Model of Silicon Epitaxial Growth in SiCl4-
Theory and Crystallization (entropy, specific
Vanadium Bronze (NaxV2O5, #Low-Temperature Specific
2, hydrothermal)*   #Specific
theory)*   #Specific
Vanadium (II) Fluoride, VF2. Preparation, Structure,
Bridgman, vapor transport, theory)*   #
Vacancy Mechanism of Melting (theory, compilation, 39 #
cooling)* #Melt Growth of Nd:Y3Al5O12(Nd:YAG) Using the
Segregation Behavior of Ga-Doped Ge (#Application of a
Crystallization (melt)*
#Kinetics of Growth of Calcium Sulfate Crystals at
ZnTe Single Crystals by the Sublimation Travelling
Resistivity CdTe*   #Growth by Travelling
Crystals Grown by a Vertical and Horizontal Travelling
of High Melting Metal Alloys and Compounds by Plasma
#Subsidiary Electrical
#Quantum-Mechanical Relations in the Calculation of
of Rare Earth Copper Intermetallic Compounds, RCu2. I.
of Nd:Y3Al5O12(Nd:YAG) Using the Heat Exchange Method (
#Growth of Stacking Faults by Bardeen-
Using SiCl4-C3H8-H2 System (vapor transport, film)*
Spaced Technique (kinetics)*
transport)*   #Tetragonal Distortion in
film)*   #Tetragonal Distortion in
by Vacuum Deposition (film, doped)*
vapor transport, ZnO)*   #Real Structure of
on a Silicon Substrate by Laser Annealing (Ge film)*
#Growth Effects in the
the System TiP-I2 (TiP, vapor transport)*
Entrainment Method and Its Application to the Study of
of AlxGa1-xP on GaP and Its Application to Double
#Growth and Dissolution Kinetics of III-V
and a Planar Interface in LPE Grown Pb1-xSnxTe
Inelastic   #Growth of Large Single Crystals of
#Electrical Resistivity in In-
#Czochralski Growth of Barium
#Single-Crystal Growth of Mixed (La, Eu, Y, Ce, Ba, Cs)
#Vapor Growth of
A Mixed-Valence Ternary Ruthenium Oxide of a New
Tungstate Tl2W4O13: A Tunnel Structure Related to the
Method (K2Ti6O13)*   #Synthesis of Potassium
Zirconium, Hafnium, and Tungsten Alloys (Nb-Zr, Nb-
#The Crystal Structures of Two Modifications
#Deposition of Crystals from the Plasmas of ZrO2,
Metal Dichalcogenides (MoS2, MoSe2, WS2, WSe2,
Induced Shift of the Absorption Edge in ZrS2 and

Gradient Method*   # 000422
Gradients (CdS, kinetics, theory, vapor transport)*   000289
Gradients on Tensions in Crystals (theory, Czochralski,   000326
Grading (melt)*   #Liquid Phase   000803
Grain Boundary (H2O)*   #On the Growth Mechanism   000314
Grain Size Dependence in a Self-Implanted Silicon Layer   000996
Graph (vapor transport, theory, Si)* #Model of Silicon   000503
Graph Model of Epitaxial Growth of A(III)B(V) Compounds   000504
Graphite Salt C8(+)AsF6(-) and Its Relationship to the   000076
Graphite/AsF6 Intercalate (C8AsF6)* Stage Graphite Salt   000076
Graphite, and Integrated Optics (review, theory, C,   001081
Gratings (KCl, films, solution)*Oriented Crystal Growth   000899
Gravity (Si, Al2O3, float zone melting,   000175
Gravity Environment (melt, solution, vapor transport,   001070
Gravity: Ge (Bridgman)*   001088
greater than 0.5) System (vapor transport)*   #Crystal   000995
Ground State of the Intermediate Valence Compound TmSe (000031
H Interstitial Alloys (electron beam melting, float   001072
H-Cl) (GaAs, theory, epitaxy)* Analysis of Equilibrium   000184
H-Cl, Ga-As-H-Cl) (GaAs, theory, epitaxy)*   Analysis   000184
H, Nb-D, Ta-H, Ta-D)*   and Characterization of the   000790
H, Ta-D)*   and Characterization of the Group VA   000790
H, V-D, Nb-H, Nb-D, Ta-H, Ta-D)*   and Characterization   000790
Haas van Alphen Effect (float zone melting)*   000964
Habit of Fine Metal Particles Formed by Gas-Evaporation   000821
Habit of Struvite Crystals (MgNH4PO4.6H2O) (solution,   000002
Habit Change of ADP Crystals by Fe3+, Based on   000296
Habit Modification of Barium Molybdate Single Crystals   000519
Hafnium, and Tungsten Alloys (Nb-Zr, Nb-Hf, electron   000307
Halide Compounds MFX of Divalent Rare Earths (M = Sm,   000088
Halide Crystals (CdI2-xBrx, CdI2-xClx)*   000870
halides, float zone melting)*   Growth of alpha-Fe   000318
Halides (CsSnCl3, RbSnCl3, Cs2SnCl6, Rb2SnCl6, CsSnBr3,   000930
Halides (KCl, LiF)*   #Molecular (beam)   000628
Halides of Copper with High Ionic Partial Conductivity.   000025
Halides, NbX2Y2(X=S, Se - Y=Cl, Br, I) (vapor transport)   000795
Hardness of NbB2 Single Crystals (flux, slow cooling)*   000679
Harmonic Generation in a Sputtered LiNbO3 Film on MgO (000518
HBr and HCl*   #Growth of Anatase   000431
HCl Etched Silicon Wafers (Si)*   001015
HCl*   #Growth of Anatase (TiO2)   000431
HCl-H2 System Based on Flow Graph (vapor transport,   000503
heat, melting, theory)*   #Percolation   000168
Heat and Magnetic Susceptibility of Nonmetallic   000171
Heat of a Quasi-Two-Dimensional Antiferromagnet Ni(OH)   000273
Heat of V6O11 Between 0.4 and 50 K (vapor transport,   000215
Heat Capacity from 5 to 300 K and Magnetic Ordering*   000917
Heat Capacity of EuS near the Ferromagnetic Transition (000505
Heat Capacity of Metals Near the Melting Point and the   000351
Heat Exchange Method (HEM) (Y3Al5O12, melt, slow   000167
Heat Pipe to Czochralski Growth. Part I. Growth and   000604
Heat-Flow-Controlled Growth During Li2B4O7   000385
Heated Metal Surfaces (solution, CaSO4)*   000334
Heater Method (vapor transport)* #Growth of High-Purity   000938
Heater Method and Characteristic of Undoped High-   000939
Heater Method*   #GaSb and InSb   000099
Heating (W, W-C)*   #Growth of Single Crystals   000829
Heating for Verneuil Furnaces in the USSR (review)*   000349
Heats of Fusion for Ionic Crystals (review, theory)*   001087
Heavy Rare Earth (TbCu2, DyCu2, HoCu2, ErCu2)*   000373
HEM) (Y3Al5O12, melt, slow cooling)*   #Melt Growth   000167
Herring Mechanism in Czochralski Silicon (Si)*   001052
Heteroepitaxial Growth of beta-SiC on Silicon Substrate   000616
Heteroepitaxial Growth of ZnS on GaP by the Close-   000482
Heteroepitaxial Homogeneous CdxHg1-xTe Films (vapor   000707
Heteroepitaxial Layers: Ge On GaAs (vapor transport,   000362
Heteroepitaxial Structures of Si on Sapphire, Obtained   000823
Heteroepitaxial Zinc Oxide Films Grown on Sapphire (000526
Heteroepitaxy of a Deposited Amorphous Germanium Layer   000343
Heteroepitaxy of III-V Compounds (review)*   000728
Heterogeneous Equilibria and Mass Transfer Kinetics in   001003
Heterogeneous Reactions (GaAs, vapor transport, theory)*   000281
Heterostructure Light Modulators* #Liquid Phase Epitaxy   000948
Heterostructures Formed by LPE (GaxAl1-xAs)*   000894
Heterostructures*   #Improved Nucleation   000962
Heusler Alloys, Ni2MnSn and Ni2Mn1-xVxSn, for Neutron   000321
Heusler Cu2.0Mn0.94In1.0 Single Crystal (Bridgman)*   000499
Hexaaluminate Single Crystals (BaAl12O19)*   000611
Hexaborides for Thermionic Emission (LaB6, EuB6, CeB6,   000727
Hexagonal GeO2 Needle Crystals*   000423
Hexagonal Structure Type (flux)*   #LaSRu4O21:   000208
Hexagonal Tungsten Bronze*   #The Thallium   000350
Hexatitanate Fibers by the Hydrothermal Dehydration   000732
HfNi3 (melt)*   000145
HfO2, ThO2 and CeO2 (electrolytic deposition)*   000472
HfSe2, ZrSe2, TiSe2, vapor transport)*   of Transition-   001000
HfS2 (CuxZrS2, CuxHfS2, FexZrS2, FexHfS2, vapor   001090

PERMUTED TITLE INDEX

#The Crystal Structure of Hf3Ni7 (melt)* 000147
#The Crystal Structure of Hf8Ni21 (melt)* 000146
of Spherical Crystals of omega-Phase in the Cd- Hg System* #Growth and Morphology 000512
#Liquid Phase Growth of HgCdTe Epitaxial Layers* 001059
Transport* #Crystallization of HgCdTe Mixed Crystals from the Vapour Phase by Chemical 000342
#Vapor Growth of HgCr2Se4 (epitaxy, theory)* 000332
of Mercuric Iodide Radiation Detectors. Part II. HgI2 (vapor transport, review)*Spectroscopic Evaluation 000834
Crystals from Soluble Complexes. III. Growth of alpha- HgI2 Single Crystals from Dimethylsulfoxide Complexes (000695
Crystals from Soluble Complexes. II. Growth of alpha- HgI2 Single Crystals from Iodomecurate complexes (000694
Mercuric Iodide Crystals in Small Opened Ampoules (HgI2)* #Dynamic Sublimation Technique for Growing 000376
#Modified Chemical Vapor Growth of Cinnabar (HgS) and GaP in Closed Systems* 000279
#Growth of Single Crystals of Cinnabar (alpha- HgS) at Rather Low Temperature (solution, doped)* 000157
#Crystallization of HgTe Crystals from the Vapour Phase* 000341
transport)* #LPE Growth of Hg0.60Cd0.40Te from Te-Rich Solution (film, vapor 000839
#Optically Pumped LPE-Grown Hg1-xCdxTe Lasers (films, flux)* 000368
Homogeneous Mercury Cadmium Telluride Single Crystals (Hg1-xCdxTe, Bridgman)*Possible Method for the Growth of 000292
#Crystal Structure of the Mercury Nitro-Phosphate Hg4PO4NO3.H2O (melt)* 000261
review, doped)* #The History and Present Status of Synthetic Diamond (C, 000685
#A New Organic Low Temperature Conductor: HMTSF-TNAP (solution, organics)* 000087
Rare-Earth Disilicates, R2Si2O7 (R = Tm, Er, Ho, Dy) (Tm2Si2O7, Er2Si2O7, Ho2Si2O7, Dy2Si2O7)* 000595
Compounds, RCu2. I. Heavy Rare Earth (TbCu2, DyCu2, HoCu2, ErCu2)* of Rare Earth Copper Intermetallic 000373
Ho3Fe5O12, Dy3Fe5O12, Er3Fe5O12, Ca3Fe2Ge3O12, HoFeO3, DyFeO3, ErFeO3)* Addition of V2O5 (Y3Fe5O12, 000565
of Some Rare Earth Trifluorides (GdF3, TbF3, DyF3, HoF3, ErF3, Bridgman-Stockbarger, Czochralski)* Growth 000323
of the Rare-Earth Dihydrides (GdH2, TbH2, DyH2, ErH2, HoH2)* #The Growth and Structure of Epitaxial Films 000782
#A New Substrate Holder for Liquid Phase Epitaxy* 000072
#Electrocrystallization of NbO and Hollow NbO2 Needles from Molten Salts (flux)* 000204
#Growth of Hollow Se-Te Whisker Crystals (theory)* 000112
#Growth Mechanism of Hollow ZnO Crystals from ZnSe. II* 000427
and Stimulated Emission of Crystals of Yttrium Holmium Aluminum Garnets ((Y1-xHox)3Al5O12, Czochralski) 000046
#Synthesis and Characterization of Europium(II)- Holoborates, Eu2B5O9Cl and Eu2B5O9Br (melt)* 000580
the Layer Perovskite (C3H7NH3)2MnCl4 Using a Compound Holographic Interference and Conventional Microscope (000640
Silicon-Iodine Transport System (Si)* #Local Selective Homoepitaxy of Silicon at Reduced Temperatures Using a 000136
Growth Conditions (#Dependence of the Optical Homogeneity of Single Germanosillenite Crystals on the 000903
#Heteroepitaxial Homogeneous CdxHg1-xTe Films (vapor transport)* 000707
#Preparation of TiCx Single Crystal with Homogeneous Compositions (float zone melting)* 001092
Hg1-xCdxTe, #A Possible Method for the Growth of Homogeneous Mercury Cadmium Telluride Single Crystals (000292
Tendency by the Bridgman Method - #Growth of Homogeneous Mixed Crystals with a Large Segregation 000244
New Growth Method (vapor transport)* #Properties of Homogeneously Doped ZnSe Single Crystals Obtained by a 000745
Growth Kinetics of Garnet Liquid Phase Epitaxy Using Horizontal Dipping (flux)* #The 001021
TbFe2, Nd, Pr, Ce, cold #Single Crystal Growth by the Horizontal Levitation Zone Melting Method (DyFe2, 000620
#GaSb and InSb Crystals Grown by a Vertical and Horizontal Travelling Heater Method* 000099
Substituted YIG Single Crystals #Interface Shape and Horizontal Variations of Al and Ga Contents in 000483
I. Basic Technique (InSb)* # Horizontal Zone-Melting of Indium Antimonide in Vacuum. 000712
V compounds, evaporation, review, theory)* # Hot Wall Epitaxy (II-VI compounds, IV-VI compounds, III- 000568
#CdTe Thin Films Grown by Hot Wall Epitaxy* 000569
Vapor Growth of Titanium Diboride by a Modified Hot Wire Method (TiB2)* #Chemical 000926
Growth of Titanium Carbide from the Vapor by a Modified Hot Wire Method (TiC)* #Single Crystal 000928
Crystal Growth of Zirconium Carbide from the Vapor by a Modified Hot-Filament Method (de Boer, Van Arkel, vapor 000927
ErKMo2O8, NdKMo2O8, DyVO4, TbVO4, TbPO4, DyPO4, Ho2Ge2O7, Dy2GeMoO8, ZrSiO4, ThGeO4, Er2Si2O7)* 001060
Tb2Ti2O7, Pr2Ti2O7, Tm2Ge2O7, Er2Si2O5, Dy2Si2O5, Ho2SiO5, Pr2MoO6, LiCoPO4)* MnO3, Pb3Mn7O15, Tm2Ti3O7, 001061
R2Si2O7 (R = Tm, Er, Ho, Dy) (Tm2Si2O7, Er2Si2O7, Ho2Si2O7, Dy2Si2O7)*Polymorphic Rare-Earth Disilicates, 000595
and Supercooling in Fluxed Melt Systems (Ho2Si2O7, TbVO 4, DyKMo2O8)* #Relative Supersaturation 000594
Emission of Cubic Bi4Ge3O12 Crystals Doped with Dy3+, Ho3+, Er3+, Tm3+, or Yb3+ Ions (Czochralski)*Stimulated 000458
from Lead Flux with Addition of V2O5 (Y3Fe5O12, Ho3Fe5O12, Dy3Fe5O12, Er3Fe5O12, Ca3Fe2Ge3O12, HoFeO3, 000565
Oxides of Niobium and Rare-Earth Elements (Ln3NbO7) (Ho3NbO7, Y3NbO7, Er3NbO7, Dy3NbO7, La3NbO7, Y3TaO7, 001097
ErKMo2O8, NdKMo2O8, Dy2MoO6, Sm2MoO6, Pr2MoO6, Ho6MoO12)* in the Systems R2O3-K2O-MoO3 (DyKMo2O8, 001062
#Scandium Hydrogenselenite (Sc(HSeO3) 3, evaporation, solution)* 001008
Molecular Volume of Solid Crystalline Lithium Nitrate Hydrates (solution, LiNO3.3H2O, LiNO3.3D2O)*Density and 000023
Kinetics of Lithium and Strontium Nitrate Crystal Hydrates (LiNO3.3H2O, Sr(NO3)2.4H2O)* #Crystallization 000989
#Kinetic Studies of the Intercalation of 2H-NbSe2 with Hydrazine by Transmittance Measurements (theory)* 000085
#Chemical Vapor Deposition of Scandium Hydride (ScH2, epitaxy, film)* 000494
and Quaternary (In,Ga)(As,P) Alloys Grown by the Hydride Vapor Phase Epitaxy Technique (film)* Ternary 000729
Preparation and Characterization of the Group VA Metal Hydrides and Deuterides (V-H, V-D, Nb-H, Nb-D, Ta-H, Ta- 000790
Growth of the Systems NF-AlF3 and NH4F-FeF3 in Hydrofluoric Acid Solution (TlAlF4, KAlF4, NH4AlF4, 000772
#Growth of Thallium Iodide Crystals in Silica Hydrogels (TlI, Tl2I3, gel)* 000881
#Controlled Growth of alpha-Fe Single Crystal Whiskers (hydrogen reduction of ferrous halides, float zone 000318
Crystals of Cubic Boron Nitride in Systems Containing Hydrogen (BN)* #Synthesis of Single 001130
#Kinetics of Silicon Growth Under Low Hydrogen Pressure (vapor transport, Si)* 000255
and Morphology of Copper-Nickel Alloy Crystals by Hydrogen Reduction of a CuI-NiBr2 Mixture (Cu-Ni, 000364
#Scandium Hydrogenselenite (Sc(HSeO3)3, evaporation, solution)* 001008
#Vapor Phase Growth of Alumina Whiskers (Hydrolysis of Aluminum Fluoride (Al2O3)* 001094
84Na2O.0.84MgO.5Al2O3, theory)* # Hydronium Beta" Alumina: A Fast Proton Conductor (0. 000285
#Synthesis and Structure of Cs2S2 (hydrothermal)* 000130
#The Crystal Structure of CuTeO4 (hydrothermal)* 000282
of a Quasi-Two-Dimensional Antiferromagnet Ni(OH)2, hydrothermal)* #Specific Heat 000273
of Grown Synthetic Quartz Inside the Autoclaves (SiO2, hydrothermal) #On the Dissolution 000507
and Twinning Behavior of Ferric Molybdate, Fe2(MoO4)3 (hydrothermal)* #The Crystal Structure 000178
on the Growth Morphology of Cubic Thulium Oxide (Tm2O3, hydrothermal) #Effect of Mineralizers 001039
of Growth Surfaces of Synthetic Quartz (SiO2, hydrothermal) #Scanning Electron Microscopy 000857
Quartz Crystals by Light-Scattering Tomography (SiO2, hydrothermal) of Growth Defects in Synthetic 000652
Nd7(OH)18Br3, Pr7(OH)18Br3, La7(OH)18I3, Pr7(OH18I3, hydrothermal) (Nd(OH)2Br, Pr(OH)2Br, La7(OH)18Br3, 000531
Study of Their Piezoelectric and Elastic Properties (hydrothermal)* #Na2CoGeO4 Single Crystal Growth and 000256
FeV2O6OH0.5: Relation to the Diaspore Structure Type (hydrothermal)* Crystal Structure of a New Mixed Oxide 000662
for Magnetic Applications, Vol. 1 (garnets, review, hydrothermal, epitaxy, flux, Bridgman, film)* #Crystals 000807
Preparative Chemistry of Oxides and Fluorides (review, hydrothermal, theory)* #Advances in 000398
of Phlogopite in Basic Aqueous Solutions Under Hydrothermal Conditions (mica, KMg3AlSi3O10(OH)2)* 000070
OH)2CO3, #Synthesis of Rare-Earth Carbonates Under Hydrothermal Conditions (YOHCO3, GdOHCO3, ErOHCO3, Y2O(000169
#Growth of Single Crystals of Na3Al2Li3F12 Garnet Under Hydrothermal Conditions* 000675

PERMUTED TITLE INDEX

Solubility of KLa(MoO4)2 in Aqueous K2MoO4 Solutions in   Hydrothermal Conditions*   # 000475
Dislocation-Free and Low-Dislocation Quartz Prepared by   Hydrothermal Crystallization (SiO2)*   # 000068
Na2O-H2O (barium germanates, Ba3Ge9O21.H2O, BaGeO3.   Hydrothermal Crystallization in the System BaO-GeO2-   000592
*   Hydrothermal Crystallization of Magnetic Oxides (review)   000232
  Hydrothermal Crystallization of Sodium   000230
Zirconogermanates (Na2ZrGeO5. Na2ZrGe2O7,   Hydrothermal Dehydration Method (K2Ti6O13)*   000732
#Synthesis of Potassium Hexatitanate Fibers by the   Hydrothermal Growth of the Ferroelectric SbNbO4, SbTaO4   000777
and Pb5S2I6 Single Crystals*   #Synthesis and   Hydrothermal Growth of the Systems MF-AlF3 and NH4F-   000702
FeF3 in Hydrofluoric Acid Solution (TlAlF4, KAlF4,   Hydrothermal Growth of Synthetic 11 Angstroms   000366
#Electron Microscopic Characterization of the   Hydrothermal Method on Single Crystal Seeds Coated with   000243
Mechanism of Sodalite Single Crystals Grown by the   Hydrothermal Na2CoGeO4 Single Crystals in Relation to   000233
Their   #X-Ray Topographic Study of Growth Defects in   Hydrothermal Recrystallization (Ca5(PO4)3OH)*   000040
#Preparation of Pure Hydroxyapatite Single Crystals by   Hydrothermal Solutions of H3PO4*   000774
#Solubility of AlPO4 in   Hydrothermal Synthesis of Aluminum Orthophosphate (# 000496
AlPO4)*   Hydrothermal Synthesis of Crystals (review)*   # 000234
  Hydrothermal Synthesis of Several Mixed Oxides of the A(# 000205
6+)B2(3+)O6 Type Under Very High Pressure (UCr2O6,   Hydrothermally Grown Strontium Paracelsian (Sr[Al2Si2O8]   000304
*   #The Morphology of   Hydrous Lutetium Oxide (cubic Lu(OH)3)*   000663
#Aging Studies on   Hydroxide Bromides and Hydroxide Iodides (Nd(OH)2Br, Pr(000531
Equilibria, and Crystal Chemistry of La, Pr, and Nd   Hydroxide Iodides (Nd(OH)2Br, Pr(OH)2Br, La7(OH)18Br3,   000531
Chemistry of La, Pr, and Nd Hydroxide Bromides and   Hydroxyapatite Single Crystals by Hydrothermal   000040
Recrystallization (Ca5(PO4)3OH)*   #Preparation of Pure   hygroscopic)*Characterization of a Europium(II) Bromide-   000200
Chloride Phase EuBr1.5Cl0.5 (melt, slow cooling,   hygroscopic, Bridgman)*   001056
#Crystal Growth of Sodium Sulphide, Na2S (hygroscopic, Bridgman)*   #Optical and   000633
Dilatometric Studies of KCaCl3 and RbCaCl3 Crystals (H2 CVD Method*   000615
#Epitaxial Growth of ZnS by Zn-S-   H2 System (vapor transport, film)*   Growth   000616
of beta-SiC on Silicon Substrate Using SiCl4-C3H8-   H2 System (BP, vapor transport)*   000884
#Crystal Growth of Boron Monophosphide Using a B2H6-PH3-   H2 System (GaAs, vapor transport, kinetics, theory)*   000653
Rate and Purity of Epitaxial Layers in the Ga_AsCl3-   H2 System Based on Flow Graph (vapor transport, theory,   000503
Si)*   #Model of Silicon Epitaxial Growth in SiCl4-HCl-   H2: The Effects of a Bypass Flow (theory)*   Processes   000558
Associated with the Epitaxial Growth of ZnS in   H2O (barium germanates, Ba3Ge9O21.H2O, BaGeO3.   000592
Crystallization in the System BaO-GeO2-Na2O-   H2O (melt)*   #Crystal   000261
Structure of the Mercury Nitro-Phosphate Hg4PO4NO3.   H2O)*   000538
#Are Icicles Single Crystals? (H2O)*   #Morphology   000347
of Ice Droxtals Grown from Supercooled Water Droplets (H2O)*   #Epitaxial Relationships During   000493
the Formation of Three-Dimensional Snow Dendrites (H2O)*   #On the Growth Mechanism of Polycrystalline   000314
Snow Crystals with a Specific Grain Boundary (H2O, theory)*   #Appearance   000689
of Non-Singular Surfaces on Vapour-Grown Ice Crystals (H2O, vapor transport, sublimation, evaporation)*   Field   000028
on Growth and Evaporation of Epitaxial Ice Crystals (H2O, BaGeO3.H2O, Na4Ge9O20)*   Crystallization in the   000592
System BaO-GeO2-Na2O-H2O (barium germanates, Ba3Ge9O21.   H2O, Na4Ge9O020)*   Crystallization in the System BaO-   000592
GeO2-Na2O-H2O (barium germanates, Ba3Ge9O21.H2O, BaGeO3.   H2O2.3H2O)   Growth Kinetics of Different Faces in Pure   000258
Solution and in the Presence of a Surfactant (NaBO2.   H2O2.3H2O, kinetics, theory)*   I. Nucleation Rates   000257
in Pure Solution in the Presence of a Surfactant (NaBO2.   H3PO4*   000774
#Solubility of AlPO4 in Hydrothermal Solutions of   H3PW12O40.21H2O)*   #Dodecatungstophosphoric   000913
Acid-21-Water by Neutron Diffraction (evaporation,   Ice Crystals (H2O, theory)*   000689
#Appearance of Non-Singular Surfaces on Vapour-Grown   Ice Crystals (H2O, vapor transport, sublimation,   000028
Electric Field on Growth and Evaporation of Epitaxial   Ice Droxtals Grown from Supercooled Water Droplets (H2O)   000347
*   #Morphology of   Icicles Single Crystals? (H2O)*   000538
#Are   Ilmenite) Crystals by the Floating-Zone Method*   000954
#Growth of FeTiO3 (image)*   #Growth of Fayalite (Fe2SiO4)   000953
Single Crystals by the Floating Zone Method (arc   image, float zone, Verneuil)*   # 000239
Growth of Oriented Bicrystals of Nickel Oxide (NiO, arc   Imbalance for Ca,Ge-Substituted Garnet Films grown by   000386
Liquid Phase Epitaxy   #Electrical Properties and Charge   Imperfection on the Solubility of Barium Sulphate (000779
BaSO4.   #Effect of S-35 Tracer, Particle Size and   Imperfections in Gadolinium Molybdate and Other   000446
Isotypic Rare-Earth Molybdates (#Crystal Growth and   Implantation and by Antimony Evaporation (Si, film)*   000734
Molecular Beam Epitaxy by Simultaneous Arsenic Ion   Implantation of Pd (electron beam melting, evaporation)*   000177
#Silicide Formation by High-Dose Si+-Ion   Implanted Silicon Layer on Laser Irradiation Energy   000996
Density (Si)*   #Grain Size Dependence in a Self-   Impurities (solution)*   #The   000094
Mechanism of Growth of KDP Crystals in the Presence of   Impurities in Synthetic Quartz Single Crystals (SiO2,   000170
solution)*   #On the Incorporation of Metallic   Impurities on the Crystallization of Supercooled Copper   000485
Melts (Cu)*   #The Influence of   Impurities on the Piezoelectric Properties of Lithium   000800
Germanate (Li2GeO3, Czochralski)*   #The Effect of   Impurities During Zone Recrystallization of Metals and   000994
Preparation of   #Fractionation of Stable Isotopes and   Impurity and Its Effect on the Purity of LPE GaAs (000313
vapor transport)*   #New Gaseous   Impurity Distribution in the Modified Czochralski   000997
Method of Crystal Growth (Si)*   #Variations of   Impurity Distributions in Melt-Grown Crystals, Taking   001026
Account of Periodic Remelt (#Computation of Striated   Impurity Dopant Incorporation and Diffusion During   000898
Molecular Beam Epitaxial Growth of IV-VI   Impurity During Crystallization of Ge by the   000346
Czochralski Method*   #Behavior of Be as an   Impurity Effect on the Formation of Terraces in GaAs   # 000988
LPE Growth (flux)*   Impurity Effect on Grown-In Dislocation Density of InP   # 000856
and GaAs Crystals (Czochralski)*   Impurity Effects on Low-Angle Boundary Formation in   # 000218
Silicon Single Crystals (Si, Czochralski)*   Impurity Redistribution in EFG (Si, ribbon, theory,   000451
kinetics)*   Impurity*   #Morphology and Growth   000375
Mechanism of Vapor Grown Cd Crystals as Affected by Bi   Impurity-Doped Aluminum Nitride Films Produced by   001104
Reactive Evaporation (AlN)*   #Cathodoluminescence of   Inactive Gas (Mg, review)*   001006
#Growth of Polyhedral Metal Crystallites in   InAs*   000312
#Organometallic VPE Growth of   InAs, InSb, Al2O3, melt, theory)*   #Formation of Stress   000406
and Dislocations in Crystal Growth (Si, Ge, GaAs,   InBi Single Crystals (melt, zone melting)*   000110
#Growth & Etching of   Incommensurate Phase Transition of Rb2ZnBr4 (theory,   000226
solution)*   #Neutron-Scattering Study of the   Indates and Thioindates (LiInS2, Bridgman-Stockbarger)*   000509
Principal Physicochemical Properties of Alkali Metal   Indium Antimonide (InSb)*   #The Strain   000637
Configuration Within the Facet of Czochralski Grown   Indium Antimonide in Vacuum. I. Basic Technique (InSb)*   000712
#Horizontal Zone-Melting of   Indium Antimony Selenides (In2Se3:Sb, In2-xSbxSe3,   000911
InSbSe3, Bridgman)*   #Preparation and Properties of Two   Indium Ditelluride Crystals Grown from Near-   000220
#Electrical and Optical Properties of Copper   Indium III: In2Te3O9*   000766
#Crystal Chemistry of a Tellurate IV of   Indium Polytelluride In2Te5(II) (melt)*   001058
  Indium Selenide In2Se3 (vapor transport)*   000557
#Crystal Structure of the   Indium Selenide In2Se3 (vapor transport)*   000557
#Low Temperature Growth of GaP LPE Layers from   Indium Solvent (flux)*   000925
#Solution Growth of   Indium-Doped Silicon (Si)*   000849

PERMUTED TITLE INDEX

of Metals and Preparation of Single Crystals Rich in
Heusler Alloys, Ni2MnSn and Ni2Mn1-xVxSn, for Neutron
Crystal Growth (NaNO3, PbCl2, organics)*   #An
Hydrothermal Method on Single Crystal Seeds Coated with
Materials (review, theory)*   #
slow cooling)*   #Far
transport)*   #Raman and
Sulphate Thin Films Grown by a New Epitaxial Process (#
                       #Investigation of Growth Processes of
transport)*   #Priorities in the
                    #Low-Defect InSb Crystal Growth by
Properties of LPE InxGa1-xAsyP1-y Thin Layers Grown on
  #Impurity Effect on Grown-In Dislocation Density of
               #The Growth of GaxIn1-xAs on (100)
                    #X-Ray Study of LEC-Grown
                                       #Growth of
  #Reproducible Preparation of Twin-Free
1.#Conditions of LPE Growth for Lattice Matched GaInAsP/
                #Control of Substrate Degradation in
Diffusion Method*
Technique (vapor transport)*   #Large-Sized
                          #Growth of p-Type
and Lattice Mismatch Effects in LPE-Grown In1-xGaxAs on
of Single Crystals of Solid Solutions GaxIn1-xP (GaP,
for Liquid Encapsulated Czochralski Crystal Growth (
Supercooling During the Crystal Growth of
                                     #Low-Defect
Travelling Heater Method*   #GaSb and
of Indium Antimonide in Vacuum. I. Basic Technique (
the Facet of Czochralski Grown Indium Antimonide (
Distribution Coefficients for Directionally Solidified
Semiconductor Single Crystals and Films (Si, Ge, GaAs,
and Dislocations in Crystal Growth (Si, Ge, GaAs, InAs,
Two Indium Antimony Selenides (In2Se3:Sb, In2-xSbxSe3,
                    #Growth of Crystalline Slabs of Layered
the Crystal Growth of InSb from the Melt #Morphological
#Analytical Study of the Thermal Transfer Process in an
          #A Simple Analysis of Vapor Phase Growth - Citing an
of Verneuil Sapphire as a Substrate for Silicon on an
407: Electrical, Calorimetric, Magnetic and EPR #Metal-
transport)*   #Structural Aspects of the Metal-
Dichlorosilane as Silicon Source for Epitaxial Films on
#Problems in the Melt- and Vapor Growth of Silicon for
xAlxAs, review, film)*   #
4, Morphological Stability, Convection, Graphite, and
                    #Performance of a New High-
of High Period Polytype of SiC at the Interface of Two
Different Density on Evaporating Crystal Surfaces (#
Crystal Quality During the Crystal Growth of Silicon
) and Its Relationship to the First Stage Graphite/AsF5
                    #Physical Properties and Lithium
Transmittance Measurements (#Kinetic Studies of the
ZrS2 and HfS2 (CuxZrS2, CuxHfS2, FexZrS2, FexHfS2,
and Electrical Properties -Lithium and Molecular
#Structure of the Lennard-Jones (100) Crystal-Liquid
Localized Adsorption in One Layer on a Crystal-Solution
of a Binary Alloy with a Curved Solid-Liquid
Vapour Deposition (theory)*   #Interrelation Between
                    #Improved Nucleation and a Planar
          #Monte Carlo Simulation of the Crystal-Melt
Periodic Closed Growth Lamellae on the Crystal-Melt
#The Occurrence of High Period Polytype of SiC at the
Crystals (alpha-Al2O3, Czochralski)*   #
theory)*   #Effects of
Profiled Substrates (theory)*
Filament, Unconstrained Growth from a Pure Melt (
Single Filament, Unconstrained Growth from a Binary
          Systems. Two Cases of Modelling the Melt Growth
Processes (theory)*   #Correlation of the
C10H16)*
Contents in Substituted YIG Single Crystals Grown by
of Dy3Al5O12*
Crystals (Y3Fe5O12, theory)*
          #On the Analytic Treatment of the Growth of Rough
          #In-Situ Observation of Solidification
Method on Single Crystal Seeds Coated with Informative
transport, flux)*   #On Diffusive-Advective
Perovskite (C3H7NH3)2MnCl4 Using a Compound Holographic
#AlSiP3, A Compound with a Novel Wurtzite-Pyrite
of the Magnetically Ordered Ground State of the
of Single-Crystalline ZnO Film with Smooth Surface on
and Temperature Dependence of the Flow Stress in the
                    #Czochralski Growth of A15 Structure
TbCu2, DyCu2, #Magnetic Properties of Rare Earth Copper
PrCu2, NdCu2)**Magnetic Properties of Rare Earth Copper
review)*   #Crystal Growth (Proceedings of the Third
Slichter Equation as a Diffusion Boundary Layer   #On

Individual Isotopes*   During Zone Recrystallization   000994
Inelastic Scattering Experiments (Czochralski, slow   000321
Inexpensive Zone Melting Furnace for Purification and   000871
Interfacial Layers (Na4Al3Si3O12Cl)* by the   000243
Infra-Red Transmitting Materials. Part 1. Crystalline   000250
Infrared Optical Study of alpha-In2Se3 Compound (melt,   000453
Infrared Spectra of CdIn2S4 and ZnIn2S4 (vapor   001004
Infrared Trichroism of Ferroelectric Triglycine   000329
Ingots of Silicon Carbide Single Crystals (SiC, melt)*   000941
Initial Use of Spacelab for Crystal Growth (vapor   000270
InN Doping (Czochralski)*   000971
InP (film)*   #Characterization of the Optical   001098
InP and GaAs Crystals (Czochralski)*   000856
InP by Liquid-Phase Epitaxy (flux)*   000754
InP Crystals (Czochralski, theory)*   000614
InP Crystals by the Synthesis Solute Diffusion Method*   000603
InP Crystals Using the LEC Technique (Czochralski)*   000125
InP DH Lasers with (100) Substrate in the Range of 1.2-   000035
InP Growth and Properties (Czochralski, doped)*   000384
InP LPE Growth with PH3 Partial Pressure (flux)*   000199
InP Single Crystal Growth by the Synthesis, Solute   001095
InP Single Crystals by the Synthesis, Solute Diffusion   000923
InP Single Crystals by the Temperature Gradient Method*   000422
InP Substrate*   #Composition Latching Phenomenon   000952
InP)*   #Preparation and Properties   000119
InP, CuInS2)*   #Simple Pressurized Chambers   000150
InSb from the Melt Under Stabilizing Thermal Gradient (000478
InSb Crystal Growth by InN Doping (Czochralski)*   000971
InSb Crystals Grown by a Vertical and Horizontal   000099
InSb)*   #Horizontal Zone-Melting   000712
InSb)*   #The Strain Configuration Within   000637
InSb-GaSb (magnetic field effects, eutectics, Bridgman)*   000864
InSb, epitaxy, theory, Czochralski)*   the Growth of   000636
InSb, Al2O3, melt, theory)*   #Formation of Stress   000406
InSbSe3, Bridgman)*   #Preparation and Properties of   000911
InSe by the Czochralski Method*   000187
Instability Under Constitutional Supercooling During   000478
Installation for Preparation of Single Crystals by the   000057
Instance of GaxIn1-xAs (theory)*   000673
Insulator (Czochralski, epitaxy, Al2O3)*   #Evaluation   000793
Insulator Transitions and Phase Diagram of (Ti1-xVx)   000835
Insulator Transitions in (Ti0.9975V0.0025)407 (vapor   000391
Insulators (Si on sapphire, vapor transport)*   #   000254
Integrated Circuits and Solar Cells (Si, review)*   000117
Integrated Optical Devices Fabricated by MBE (GaAs, Ga1-   000152
Integrated Optics (review, theory, C, GaAs)*   Vol.   001081
Intensity Silicon Solar Cell (Si, float zone melting)*   000303
Interacting Spirals*   #The Occurrence   000889
Interaction Between Successive Monatomic Step Trains of   000105
Interaction Between Technological Parameters and   000325
Intercalate (C8AsF6)*   Stage Graphite Salt C8(+)AsF6(-   000076
Intercalates of CrPS4 (slow cooling)*   000571
Intercalation of 2H-NbSe2 with Hydrazine by   000085
Intercalation-Induced Shift of the Absorption Edge in   001090
Intercalations (VOCl)*   Oxychloride: Magnetic, Optical   001031
Interface (theory)*   000528
Interface (theory)*   #   000575
Interface (Ge:Ga, Bridgman, theory)*   Solidification   000206
Interface and Vapour Transport Kinetics in Chemical   000742
Interface in LPE Grown Pb1-xSnxTe Heterostructures*   000962
Interface of a Lennard-Jones Substance (theory)*   000123
Interface of LiCl Doped Rock Salt Crystals (NaCl,   000532
Interface of Two Interacting Spirals*   000889
Interface Growth Feature and Voids in Sapphire Ribbon   001050
Interface Kinetics on the Growth Rate of Dendrites (Sn,   000991
Interface Morphology in Chemical Vapour Deposition on   001011
Interface Morphology During Crystallization - I. Single   000992
Interface Morphology During Crystallization - II.   000993
Interface Motion*   Growth for Metallic One-Component   000059
Interface Reaction Constant with Elementary Surface   000330
Interface Roughness of Vapor-Grown Adamantine Crystals (000753
Interface Shape and Horizontal Variations of Al and Ga   000483
Interface Shape Transitions in the Czochralski Growth   000644
Interface Shape Transitions in Czochralski Grown YAG   000645
Interfaces (theory)*   000764
Interfaces with a Very High Voltage Electron Microscope*   000163
Interfacial Layers (Na4Al3Si3O12Cl)*by the Hydrothermal   000243
Interfacial Mass Transfer (theory, solution, vapor   001074
Interference and Conventional Microscope (solution)*   000640
Intergrowth Structure (flux)*   001044
Intermediate Valence Compound TmSe (slow cooling)*   000031
Intermediately Sputtered ZnO Thin Film on Sapphire (000711
Intermetallic Compound Ni3Ge Single Crystals (Bridgman)*   000033
Intermetallic Compounds (V3Si, V3Ge)*   001073
Intermetallic Compounds, RCu2. I. Heavy Rare Earth (000373
Intermetallic Compounds, RCu2. II. Light Rare Earth (000374
International Summer School on Crystal Growth, 1977) (000066
Interpreting a Quantity in the Bruton, Prim and   001083

PERMUTED TITLE INDEX

Kinetics in Chemical Vapour Deposition (theory)* # Interrelation Between Interface and Vapour Transport 000742
melting, review)* #Properties and Synthesis of Nb-H Interstitial Alloys (electron beam melting, float zone 001072
#An LPE Growth of Thick InxGa1-xAs* 000506
#Characterization of the Optical Properties of LPE InxGa1-xAsyP1-y Thin Layers Grown on InP (film)* 001098
#Molecular Beam Epitaxy of InxGa1-xSb (x = 0 to 1) (GaSb)* 001101
Field on Vertical Bridgman-Stockbarger Growth of InxGa1-xSb* #Influence of Magnetic 000863
#New Ternary Mo(II)-Compounds InxMo15Se19 Containing Mo6Se8 and Mo9Se11 Units* 000186
#Crystal Growth and Characterization of In1.9As0.1Se3 (vapor transport)* 000471
Phenomenon and Lattice Mismatch Effects in LPE-Grown In1-xGaxAs on InP Substrate* #Composition Latching 000952
#Quasi-Grown Layers in Liquid Phase Epitaxial Growth (In1-xGaxAs)* 000951
by MBE and Their Characterization (GaAs, Ga1-xAlxAs, In1-xGaxAs, GaSb1-yAsy)* #Semiconductor Superlattices 000176
Properties of Two Indium Antimony Selenides (In2Se3:Sb, In2-xSbxSe3, InSbSe3, Bridgman)* #Preparation and 000911
#Crystal Structure of the Indium Selenide In2Se3 (vapor transport)* 000557
#Far Infrared Optical Study of alpha- In2Se3 Compound (melt, slow cooling)* 000453
and Properties of Two Indium Antimony Selenides (In2Se3:Sb, In2-xSbxSe3, InSbSe3, Bridgman)*#Preparation 000911
#alpha- In2Te3 (melt)* 000593
#Crystal Chemistry of a Tellurate IV of Indium III: In2Te309* 000766
#Indium Polytelluride In2Te5(II) (melt)* 001058
#Crystal Growth of Antimony Sulphur Iodide (SbSi, flux)* 000109
#Effect of Growth Conditions of Polytypism in Cadmium Iodide Crystals (CdI2, theory, solution, evaporation)* 000432
Dissolution Processes (#Selective Etching of Cesium Iodide Crystals and Comparison of Growth and 000827
#Growth of Thallium Iodide Crystals in Silica Hydrogels (TlI, Tl2I3, gel)* 000881
#Dynamic Sublimation Technique for Growing Mercuric Iodide Crystals in Small Opened Ampoules (HgI2)* 000376
Growth and Nuclear Spectroscopic Evaluation of Mercuric Iodide Radiation Detectors. Part II. HgI2 (vapor 000834
#Crystal Structure of the Zinc Iodide ZnI2 (vapor transport)* 000300
of La, Pr, and Nd Hydroxide Bromides and Hydroxide Iodides (Nd(OH)2Br, Pr(OH)2Br, La7(OH)18Br3, Nd7(OH) 000531
#Transfer of VO2 by Iodine (vapor transport, doped)* 000487
transport, theory, thermodynamics)* #Germanium- Iodine Equilibrium from 700 to 1000 K (Ge, vapor 000191
of Silicon at Reduced Temperatures Using a Silicon- Iodine Transport System (Si)* Selective Homoepitaxy 000136
#Preparation of V3Ga Single Crystals by Iodine Transport* 000217
of Cubic ZnS, ZnSe and ZnSxSe1-x Single Crystals by Iodine Transport* #Growth 000309
of Photoluminescent and Conductive ZnSe:I Crystals by Iodine Vapor Transport* #Preparation 000775
Method for the Growth of Chalcopyrite Crystals by Iodine Vapour Transport (CuInS2, CuGaS2, AgIn5S8)* 000744
the Growth of Gallium Selenide and Gallium Sulphide by Iodine Vapour Transport (GaS, GaSe)* of Conditions for 001078
#Photoluminescence Properties of CuGaSe2 Grown by Iodine Vapour Transport* 000935
than or equal to x less than or equal to 1) Grown by Iodine Vapour Transport* the Series GaSxSe(1-x) (0 less 001077
II. Growth of alpha-HgI2 Single Crystals from Iodomecurate complexes (kinetics)* Soluble Complexes. 000694
Polymorph of KAlSiO4 (flux)* #Alkali Ion Exchange Reactions with RbAlSiO4: A New Metastable 000639
Silicon Molecular Beam Epitaxy by Simultaneous Arsenic Ion Implantation and by Antimony Evaporation (Si, film)* 000734
evaporation)* #Silicide Formation by High-Dose Si+- Ion Implantation of Pd (electron beam melting, 000177
TiO2)* #Anisotropy of Ion Transport in Stoichiometric Rutile Single Crystals (000778
#The Mechanism of Silicon Epitaxial Layer Growth from Ion-Molecular Beams (Si)* 000019
cooling)* #Luminescence and # Ionic Conductivity in N30+ Beta Alumina (solution)* 000286
Superionic Conductors (kinetics, #Crystal Structure and Ionic Conductivity of Barium fluorochloride (BaFCl, slow 001012
#Li3N: A Promising Li Ionic Conductivity of Li24Zn(GeO4)4 and other new Li+ 000396
of Ordered AB Crystals: A Monte Carlo Simulation for Ionic Conductor (electrolytic deposition, review)* 000024
Relations in the Calculation of Heats of Fusion for Ionic Crystal Growth (theory)* #Fast Growth 000182
#Solvent Effects and the Growth Kinetics of Ionic Crystals (review, theory)* #Quantum-mechanical 001087
Transition of #Telluride Halides of Copper with High Ionic Crystals (CaSO4.2H2O, solution, theory)* 000134
#Thin Film Deposition From Beams of Ionic Partial Conductivity. III. On the Phase 000025
#Nitrogen Doping into GaAs1-xPx Using Ionized Atoms and Clusters (Ag on Si, epitaxy)* 000516
Deposition Techniques (vapor #New Developments in Ionized Beam in Molecular Beam Epitaxy* 000617
Ionized-Cluster Beam and Reactive Ionized-Cluster Beam 000944
#New Developments in Ionized-Cluster Beam and Reactive # Ionized-Cluster Beam Deposition and Epitaxy (review)* 000945
GaP and Si Substrates* # Ionized-Cluster Beam Deposition Techniques (vapor 000944
Crystals Doped with Dy3+, Ho3+, Er3+, Tm3+, or Yb3+ Ionized-Cluster Beam Epitaxial Growth of GaP Films on 000651
#The Effects of M(M=Mg, Ca and Ba) Ions (Czochralski)* Emission of Cubic Bi4Ge3012 000458
#The Crystal Structure of Synthetic Lautarite, Ca(Ions on the KCl Flux Growth of MWO4 Needle Crystals* 000715
IO3)2 (flux)* 000331
#Epitaxial Thin Iron Films on Antimony (Fe, vapor transport)* 000875
Y3Fe5012)* #Growth of Bulk Yttrium Iron Garnet Crystals from High Temperature Solutions (000977
Method (Y3Fe5012)* #Evaluation of Yttrium Iron Garnet Single Crystals Grown by the Floating Zone 000480
FeGa) #Magneto-Optical Properties of Some Rare Earth Iron Garnets (flux, (GdPrBi)3(FeGaIn)5012, (YLaBi)3(001107
xGayFe5-yO12, flux)* #A Solubility Model for Rare-Earth Iron Garnets in a PbO/B203 Solution (Y3Fe5012, RExY3- 001022
Method* #Single Crystal Growth of Substituted Yttrium Iron Garnets Y3Fe5-x(Ga,Al)x012 by the Floating Zone 000879
93Gd0.07) #The Flux Growth of Some New Rare Earth and Iron Group Complex Oxides (GdVO3, Gd4Ba6(BO3)9, (BaO. 001063
Pb)MnO3, (Nd,Pb) #The Flux Growth of Some Rare-Earth and Iron Group Complex Oxides (NdBO3, LaBO3/Pr, LaBO3, (La, 001061
#Synthesis and Crystal Structure of Iron Polyphosphide FeP4* 000438
#Morphology of Iron Pyrite Crystals (FeS2, vapor transport)* 001093
#The Preparation of Large Spherical Iron Single Crystals (Fe, strain-anneal)* 000573
#Growth Rate of Iron Whiskers (Fe, vapor transport, theory)* 000462
#Magnetic Properties of beta-Phase Iron-Germanium (FexGe, slow cooling)* 000845
#Properties of VPE-Grown GaN Doped with Al and Some Iron-Group Metals* 000649
Dependence in a Self-Implanted Silicon Layer on Laser Irradiation Energy Density (Si)* #Grain Size 000996
#Unidirectional Solidification of Irregular Silver-Germanium Eutectic Alloys (Ag-Ge)* 000549
Czochralski)* #Observations of Microfacets Near Irregularly Remelted Surfaces in Pulled GaSb Crystals (000517
Dilute Binary Alloy Single Crystal (Fe-Ni, eutectics, # Irreversible Thermodynamics of the Cellular Growth of a 000113
SeO3)2 (solution)* Isotope Effects on the Central Peak Phenomena in KD3(000963
#The Growth of LaAlO3 Crystals Enriched in Isotope 170 (flux)* 000901
of Metals and Preparation of #Fractionation of Stable Isotopes and Impurities During Zone Recrystallization 000994
and Preparation of Single Crystals Rich in Individual Isotopes* During Zone Recrystallization of Metals 000994
Volume of Solid Crystalline Lithium #Influence of Isotopic Composition on the Density and Molecular 000023
and Imperfections in Gadolinium Molybdate and Other Isotypic Rare-Earth Molybdates (Czochralski, melt, Gd2(000446
Temperature Solutions) (review, flux) #Rost Kristallov iz Rastvorov-Rasplavov (Crystal Growth from High- 000976
#Transport Properties of the Systems SnS2-SnI4 and SnS2- I2 (SnS2, vapor transport, kinetics, theory)* 001079
Equilibria and Mass Transfer Kinetics in the System TiP- I2 (TiP, vapor transport)* #Heterogeneous 001003
Velocities of K2PbCu(NO2)6 Near the Cooperative Jahn-Teller Phase Transitions (solution)* #Ultrasonic 000467
#Structure of the Lennard- Jones (100) Crystal-Liquid Interface (theory)* 000528

PERMUTED TITLE INDEX

Simulation of the Crystal-Melt Interface of a Lennard- Jones Substance (theory)* #Monte Carlo 000123
#Structure of Monocrystalline WV2O6 at 298 and 383 K (vapor transport)* 000389
#The Crystal Structure of V4O7 at 120 degrees K (vapor transport)* 000390
#Specific Heat of V6O11 Between 0.4 and 50 K (vapor transport, theory)* 000215
#Germanium-Iodine Equilibrium from 700 to 1000 K (Ge, vapor transport, theory, thermodynamics)* 000191
Preparation, Structure, Heat Capacity from 5 to 300 K and Magnetic Ordering* Vanadium (II) Fluoride, VF2. 000917
Expansion of 2H-MoS2 and 2H-WSe2 Between 10 and 320 K* #The Thermal 000667
in the Solid Electrolytes of the RbAg4I5 Family (KAg4I5, NH4Ag4I5, solution)* #Raman Scattering Studies 000315
of Solution Grown Crystals -- Case of Potash Alum (KAl(SO4)2.12H2O)* of Stirring on Crystalline Quality 000337
#Direct Observation of Secondary Nuclei Production (KAl(SO4)2.12H2O, MgSO4.7H2O, solution)* 000322
of Crystals of Some Transition Metal Fluorides. Part 2 (KAlF4, KMnF3, RbFeF4, Rb2FeF5, Rb3FeF6, RbxFeF3, 001064
and NH4F-FeF3 in Hydrofluoric Acid Solution (TlAlF4, KAlF4, NH4AlF4, RbAlF4, CsAlF4, NH4Fe2F6)* HF-AlF3 000772
Reactions with RbAlSiO4: A New Metastable Polymorph of KAlSiO4 (flux)* #Alkali Ion Exchange 000639
Alkali, Silver, and Europium beta-Alumina (NaAl11O17, KAl11O17, RbAl11O17, AgAl11O17, EuAl11O17, flux, theory) 000621
#Low-Stress Physical Vapor Growth (PVT) (KCl, KBr, KCl:KCN, KCN:KOH, KCl:KOH, theory)* 000809
SnTe, LiF, NaF, KF, LiBr, NaCl, NaBr, LiI, KCl, NaI, KBr, KI)* Pt, Fe, Al, Au, Ag, In, Pb, PbS, PbSe, PbTe, 000949
Scanning Calorimeter. Growth Rate of Crystallite of KBrO3)* Its Quiescent Aqueous Solution by Differential 000367
#Optical and Dilatometric Studies of KCaCl3 and RbCaCl3 Crystals (hygroscopic, Bridgman)* 000633
of Solution-Grown Mixed Crystals of NaCl and KCl (NaCl-KCl eutectic)* Ray Studies and Microstructure 000407
of Solution-Grown Mixed Crystals of NaCl and KCl (NaCl- KCl eutectic)* Growth, X-Ray Studies and Microstructure 000407
Evidence of Spiral Growth on the Surface of Melt-Grown KCl Crystals (Czochralski, theory)* #An 000408
#Effects of Growth Rates on the Defect Generation in KCl Crystals Grown by the Czochralski Method* 000409
#The Effects of M(M=Mg, Ca and Ba) Ions on the KCl Flux Growth of MWO4 Needle Crystals* 000715
#Simultaneous Diffusion of Calcium and Strontium in KCl Single Crystals* 000579
Substrates Using Artificial Surface-Relief Gratings (KCl, films, solution)* Crystal Growth on Amorphous 000899
#Low-Stress Physical Vapor Growth (PVT) (KCl, KBr, KCl:KCN, KCN:KOH, KCl:KOH, theory)* 000809
of Condensation and Evaporation of Alkali Halides (KCl, LiF)* #Molecular (beam) Processes 000628
PbSe, PbTe, SnTe, LiF, NaF, KF, LiBr, NaCl, NaBr, LiI, KCl, NaI, KBr, KI)*Pd, Pt, Fe, Al, Au, Ag, In, Pb, PbS, 000949
#Low-Stress Physical Vapor Growth (PVT) (KCl, KBr, KCl:KCN, KCN:KOH, KCl:KOH, theory)* 000809
Vapor Growth (PVT) (KCl, KBr, KCl:KCN, KCN:KOH, KCl:KOH, theory)* #Low-Stress Physical 000809
#Nucleation and Growth of KClO4 Single Crystals in Silica Gels* 000751
#Modified Gel Technique to Grow Single Crystals of KClO4* 000749
Growth of Potassium Perchlorate in Gelatin Gels (KClO4)* #Crystal 000750
of Orthorhombic Potassium Perchlorate Single Crystals (KClO4)* #Gel Growth and Perfection 000752
#Low-Stress Physical Vapor Growth (PVT) (KCl, KBr, KCl: KCN, KCN:KOH, KCl:KOH, theory)* 000809
Stress Physical Vapor Growth (PVT) (KCl, KBr, KCl:KCN, KCN:KOH, KCl:KOH, theory)* #Low- 000809
#The Mechanism of Growth of KDP Crystals in the Presence of Impurities (solution)* 000094
#X-Ray Topographic Study of Growth Sector Strains in KDP Single Crystals (solution)* 000092
Potassium Dihydrogen Orthophosphate Solutions (KDP)* #Nucleation in Supersaturated 000443
Potassium Dihydrogen Orthophosphate in Silica Gels (KDP)* #Single Crystal Growth of Ferroelectric 000442
#Isotope Effects on the Central Peak Phenomena in KD3(SeO3)2 (solution)* 000963
#Crystal Structure of beta- KEr3F10 (melt, slow cooling)* 000021
Al, Au, Ag, In, Pb, PbS, PbSe, PbTe, SnTe, LiF, NaF, KF, LiBr, NaCl, NaBr, LiI, KCl, NaI, KBr, KI)* Pt, Fe, 000949
ZrMnF6, ZrVF6, RbTiF4, RbVF4, RbVF3, RbV2F6, CsVF4, KFe2F6, KMnF3, RbMnF3)*(ZrFeF6, ZrCrF6, ZrNiF6, ZrCoF6, 000319
LiF, NaF, KF, LiBr, NaCl, NaBr, LiI, KCl, NaI, KBr, KI)* Pt, Fe, Al, Au, Ag, In, Pb, PbS, PbSe, PbTe, SnTe, 000949
Crystals (Si, theory)* #Limitations in Using Kilohertz Radio Frequencies for Float Zone Silicon 000358
Epitaxial Layer Growth from the #Investigations of Kinetic and Thermal Conditions of Silicon Carbide 000942
Crystals from Melt-Solutions on Oriented Seeds (flux) # Kinetic Characteristics of Growth of Y3Al5O12 Single 000978
Hydrazine by Transmittance Measurements (theory)* # Kinetic Studies of the Intercalation of 2H-NbSe2 with 000085
review)* # Kinetic Studies of Nucleation and Growth at Surfaces (001030
#Congruent (Diffusionless) Vapor Transport (theory, kinetics)* 000007
#Impurity Redistribution in EFG (Si, ribbon, theory, kinetics)* 000451
Theory of Normal Growth of Crystals in Binary Systems (kinetics)* #On the 001110
Growth of Bi3B5O12 From its Stoichiometric Melt (kinetics)* #Crystal 000794
of LPE Growth: Layer Thickness Calculation (theory, kinetics)* #New Model 000589
Single Crystals Prepared at High Temperature (theory, kinetics)* #Defects in 000510
Growth of ZnS on GaP by the Close-Spaced Technique (kinetics)* #Heteroepitaxial 000482
Study of Surface Melting (theory, vapor transport, kinetics)* #A Molecular Dynamics 000140
Czochralski Bulk Flow in Microgravity (Al2O3, theory, kinetics)* #Digital Simulation of 000536
Volume Low-Cost Silicon Epitaxy (Si, vapor transport, kinetics)* #Novel Reactor for High 000062
Force for Crystallization from Solution (theory, kinetics)* #The Thermodynamic Driving 000904
Layers Grown by Liquid-Phase Epitaxy (flux, theory, kinetics)* #Surface Morphology of GaAs 000605
Perfection of Single Crystals of Germanium (Ge, theory, kinetics)* #Preferred Growth Directions and 000275
on Czochralski Bulk Flow in Garnet Growth (theory, kinetics)* #Effect of the Buoyancy Parameter 000535
Single Crystals of Quaternary III-V Alloys (review, kinetics)* #Melt and Solution Growth of Bulk 000054
Growth of GaxIn1-xAs and GaxIn1-xP Alloy Films by MBE (kinetics)* #Surface Processes Controlling the 000301
of Crystal Surfaces and Crystal Growth (review, theory, kinetics)* #Survey of Monte Carlo Simulations 001013
I. Growth on Singular (100) GaAs Substrates (film, kinetics)* Deposited by Molecular Beam Epitaxy. 000762
Complexes. I. General Introduction (review, theory, kinetics)* Soluble Single Crystals from Soluble 000693
alpha-HgI2 Single Crystals from Iodomercurate complexes (kinetics)* from Soluble Complexes. II. Growth of 000694
Compounds Along their Polar Axis (ZnO, CdSe, CdS, ZnS, kinetics)* #Crystal Growth and Sublimation in II-VI 000428
FeSi2.33, CoSi2, FeGe2, Mn5Si3, Fe1.76Ge, melt, theory, kinetics)* of Crystals of Certain Metallides (alpha- 000861
Method of Directed Vertical Crystallization (theory, kinetics)* for Preparation of Single Crystals by the 000057
from the Vapour Phase (SiC, film, vapor transport, kinetics)* of Silicon Carbide Epitaxial Layer Growth 000942
(Peltier effect, GaAs, theory, electrolytic deposition, kinetics)**Liquid-Phase Electroepitaxy: Growth Kinetics 000437
Cadmium, Selenium, and Cadmium Selenide (CdSe, epitaxy, kinetics, films, theory)* the Sublimation Properties of 000365
Zone Melting. I. Pure Rotation Driven Flows (Si, Al2O3, kinetics, float zone melting, theory)* in Floating 000173
Li24Zn(GeO4)4 and other new Li+ Superionic Conductors (kinetics, melt)* Structure and Ionic Conductivity of 000396
Force for Crystallization: The Growth Affinity (theory, kinetics, melt, solution)* #On the Driving 001024
Growth of ADP Crystals (NH4H2PO4, theory, solution, kinetics, slow cooling)* #Unstable 000804
Tungstate Melts by Continuous Cooling (BaWO4, flux, kinetics, slow cooling)* from Solutions in Sodium 000812
a Chemical Potential Varying in Space or Time (theory, kinetics, solution)* #Growth Spirals in 000659
#Epitaxial Stratified Growth (kinetics, theory)* 000017
#The Vapour Transport of NbB2 and TaB2 (kinetics, theory)* 000042
Growth Rate of Calcite in Aqueous Solution (CaCO3, kinetics, theory)* #Linear 000627
Metastability During Normal Growth in a Binary Alloy (kinetics, theory)* #Region of 000969
of Solidification in Web-Dendritic Ribbon Growth (Si, kinetics, theory)* #Thermal Analysis 000369
Systems SnS2-SnI4 and SnS2-I2 (SnS2, vapor transport, kinetics, theory)* #Transport Properties of the 001079

PERMUTED TITLE INDEX

of Gadolinium Gallium Garnet (Gd3Ga5012, Czochralski,
Reference to Calcium Sulphate Dihydrate (CaSO4.2H2O,
on the Shape of the Growing Crystal (Czochralski,
Layered Dichalcogenides by Vapour Transport Processes (
with GeBr4 Vapor (Ge, epitaxy, vapor transport,
Kinetics of Garnet Liquid Phase Epitaxy (flux,
in the Ga_AsCl3-H2 System (GaAs, vapor transport,
onto Silicon (vapor transport, electron beam melting,
in the Presence of a Surfactant (NaBO2.H2O2.3H2O,
Melting. II. Combined Free and Rotation Driven Flows (
Crystals with Large Temperature Gradients (CdS,
Field Dynamics in the Synthesis of Diamonds (
Rates of Large-Diameter Silicon Crystals (Si, theory,
at Zero Gravity (Si, Al2O3, theory, float zone melting,
deposition. #Liquid-Phase Electroepitaxy: Growth
Oriented Seeds in #Some Problems of the Crystallization
Single Crystals in Silica Gel (BaMoO4)* #Growth
with Platinum and Gold as #Some Notes on the Growth
Functions of the Mismatch of Crystal Lattice #Growth
Molecular Beam Epitaxy. I. Growth on #Crystal Growth
#Study of Silicon Carbide Epitaxial Growth
#Heterogeneous Equilibria and Mass Transfer
#Interrelation Between Interface and Vapour Transport
Epitaxy* #Growth
benzophenone)* #A New Technique for Determining the
Solutions in Sodium Tungstate Melts by Continuous #
Solutions in Sodium Tungstate Melts by Continuous #
Phenomenological Theories (review)* #
#On the Effect of pH of Solution on the
Constituents (theory)* #
Solutions in Sodium Tungstate Melts by Continuous #
of Sodium Perborate from Aqueous Solutions. II. Growth
#The
Obtained by Sublimation in Vacuum (SiC)* #Growth
Transport Reactions (GaP)* #
#The Liquidus Temperature and Growth-Dissolution
Horizontal Dipping (flux)* #The Growth
in Diffusion and Convection Regimes (solution, NaClO3)*#
Heated Metal Surfaces (solution, CaSO4)* #
#Investigation of the Mechanism and
theory)* #Solvent Effects and the Growth
GaxAl1-xAs)* #Growth and Dissolution
Hydrates (LiNO3.3H2O, Sr(NO3)2.4H2O)* #Crystallization
#Investigation of Structural Formation and Growth
vapor transport, Si)* #
NH2)2, theory)* #The Growth
#Effects of Interface
of Salts from Supersaturated Solutions: Diffusion
of Crucible Shape on a Crystallization Front (Bridgman,
Conditions* #Solubility of
#Epitaxial Growth of KNdP4O12 Laser Waveguides (
Piezoelectric Properties of Potassium Lithium Niobate (
Aqueous Solutions Under Hydrothermal Conditions (mica,
of Some Transition Metal Fluorides. Part 2 (KAlF4,
ZrVF6, RbTiF4, RbVF4, RbVF3, RbV2F6, CsVF4, KFe2F6,
#On the Preparation of Pure, Doped and Reduced
#Epitaxial Growth of
Vapor Growth (PVT) (KCl, KBr, KCl:KCN, KCN:KOH, KCl:
Physical Vapor Growth (PVT) (KCl, KBr, KCl:KCN, KCN:
Trihalogenoplumbates M(PbX3) (CsPbCl3, CsPbBr3, CsPbI3,
High-Temperature Solutions) (review, flux)* #Rost
#A Novel Ruthenium Bronze:
flux)* #The Crystal Structure of La3Ru3O11: A New Cubic
#Magneto-Optical Properties of
Bismuth Niobate Single Crystals by RF Sputtering (
Bismuth Niobate Single Crystals by the BGM Technique (
#Growth and Phase Transitions in Single Crystals of
Susceptibility of Nonmetallic Vanadium Bronze (NaxV2O5,
those of Potassium, Rubidium and Cesium (review,
at High Temperatures (review, melt, flux, Czochralski,
A Modified Bridgman Method (ZnF2, CoF2, BaF2, MgF2,
Properties of Potassium Lithium Niobate (KLN) Crystals
#The Crystal Structure of the Tetragermanate
#Flux Growth of Cr2O3 Crystals from a K2Cr2O7-K2CrO4-
Crystals of Langbeinites Rb2Cd2(SO4)3, Tl2Cd2(SO4)3,
#Flux Growth of Cr2O3 Crystals from a K2Cr2O7-
#Growth of Cr2O3 Crystals by the Decomposition of a
#Flux Growth of Cr2O3 Crystals from a
Bridgman)* #Crystal Growth of
#Crystal Growth of K2CuF4 and Some Solid Solutions of
6, Tl2NiCd(#Crystal Growth of Triple Nitrites in Gels (
The Structure of
Layer- Type Compounds K2Mn1-xMxF4 (M + Fe, Co)
on the Antiferromagnetic Layer- Type Compounds
#Solubility of KLa(MoO4)2 in Aqueous
#Crystal Structure of the Double Molybdate
Triple Nitrites in Gels (K2CuPb(NO2)6, Tl2CuPb(NO2)6,
#Viscosity Measurements in the System Dy2O3-

kinetics, theory)*   Processes and Crystal Growth   000769
kinetics, theory)*   in Aqueous Solution with Special 000193
kinetics, theory)*   with the Melt, and its Influence 001046
kinetics, theory)*   Considerations for the Growth of   000805
kinetics, theory)*   of Germanium (111) Single Crystals 000060
kinetics, theory)*   Temperature and Growth-Dissolution 000333
kinetics, theory)*   Rate and Purity of Epitaxial Layers 000653
kinetics, theory)*   of Films of CdS Evaporated in Vacuum 000073
kinetics, theory)*   I. Nucleation Rates in Pure Solution 000257
kinetics, theory, electron beam melting)* Floating Zone   000174
kinetics, theory, vapor transport)* of Cadmium Sulphide   000289
kinetics, theory, C)*   #Temperature   000862
kinetics, Czochralski)*   #On the Cooling   000165
kinetics, Si)*   of Zone Geometry and Prandtl Number   000175
Kinetics (Peltier effect, GaAs, theory, electrolytic   000437
Kinetics and Growth Mechanism of Y3Al5O12 Crystals on   000093
Kinetics and Habit Modification of Barium Molybdate   000519
Kinetics and Morphology of VLS Silicon Crystals Grown   001075
Kinetics and Structure of Epitaxial Garnet Layers as   000251
Kinetics in (GaAs)n-(AlAs)m Superlattices Deposited by   000762
Kinetics in the SiC-C System (SiC, theory)*   000562
Kinetics in the System TiP-I2 (TiP, vapor transport)*   001003
Kinetics in Chemical Vapour Deposition (theory)*   000742
Kinetics of (111) Si Through an Al Layer by Solid Phase   000585
Kinetics of Crystal Growth from the Melt (organics,   000755
Kinetics of Crystal Growth of Calcium Tungstate from   000810
Kinetics of Crystal Growth of Strontium Tungstate from   000811
Kinetics of Crystal Growth. Microscopic and   000660
Kinetics of Crystallization (theory)*   000525
Kinetics of Crystallization of Alloys of Multiple   000181
Kinetics of Crystallization cf Barium Tungstate from   000812
Kinetics of Different Faces in Pure Solution and in the   000258
Kinetics of Dissolution (theory)*   000498
Kinetics of Epitaxial Layers of Silicon Carbide   000940
Kinetics of Gallium Phosphide Synthesis by Chemical Gas   000231
Kinetics of Garnet Liquid Phase Epitaxy (flux,   000333
Kinetics of Garnet Liquid Phase Epitaxy Using   001021
Kinetics of Growth and Dissolution of Sodium Chlorate   000888
Kinetics of Growth of Calcium Sulfate Crystals at   000334
Kinetics of Growth of LPE GaAs (theory, epitaxy)*   000144
Kinetics of Ionic Crystals (CaSO4.2H2O, solution,   000134
Kinetics of III-V Heterostructures Formed by LPE (000894
Kinetics of Lithium and Strontium Nitrate Crystal   000989
Kinetics of Organic Crystals (solution, melt)*   000773
Kinetics of Silicon Growth Under Low Hydrogen Pressure (000255
Kinetics of Urea Monocrystals from Aqueous Solution (CO(000392
Kinetics on the Growth Rate of Dendrites (Sn, theory)*   000991
Kinetics*   #Crystallization   000623
Kinetics, theory)*   #Effect   000004
KLa(MoO4)2 in Aqueous K2MoO4 Solutions in Hydrothermal   000475
KLaP4O12, solution)*   000643
KLN) Crystals (K2.89Li1.55Nb5.11O15, Czochralski)* and   000010
KMg3AlSi3O10(OH)2)* #Solubility of Phlogopite in Basic   000070
KMnF3, RbFeF4, Rb2FeF5, Rb3FeF6, RbxFeF3, CsFeF4,   001064
KMnF3, RbMnF3)*(ZrFeF6, ZrCrF6, ZrNiF6, ZrCoF6, ZrMnF6,   000319
KNbO3 Single Crystals (Czochralski)*   000295
KNdP4O12 Laser Waveguides (KLaP4O12, solution)*   000643
KOH, theory)*   #Low-Stress Physical   000809
KOH, KCl:KOH, theory)*   #Low-Stress   000809
KPbI3, RbPbI3, vapor transport)* Properties of Alkali-   000377
Kristallov iz Rastvorov-Rasplavov (Crystal Growth from   000976
KRu4O8 (solution)*   001082
KSbO3 Derivative Oxide with No Metal-Metal Bonding (000209
KTb3F10 and LiTbF4 Crystals (Czochralski)*   001068
KxLi1-xNbO3)* Niobate Single-Crystal Films on Potassium   000009
KxLi1-xNbO3)* Niobate Single-Crystal Films on Potassium   000011
KxNa1-x)NbO3 (flux)*   000056
KxV2O5, AgxV2O5, theory)*   Specific Heat and Magnetic   000171
KxWO3, RbxWO3, CsxWO3)* Tungsten Bronzes, in Particular   000402
Kyropoulos, float zone melting, Bridgman)* from Liquids   000139
KY3F10)*the Thermic Screen Translation (TST) Technique: 000541
K2.89Li1.55Nb5.11O15, Czochralski)* and Piezoelectric   000010
K2Ba[Ge4O9]2*   000083
K2B4O7 Mixture*   000860
K2Co2(SO4)3, and their Phase Transitions (solution)*   000138
K2CrO4-K2B4O7 Mixture*   000860
K2Cr2O7 Melt (flux)*   000859
K2Cr2O7-K2CrO4-K2B4O7 Mixture*   000860
K2CuF4 and Some Solid Solutions of K2CuF4-K2ZnF4 (001055
K2CuF4-K2ZnF4 (Bridgman)*   001055
K2CuPb(NO2)6, Tl2CuPb(NO2)6, K2NiPb(NO2)6, Tl2NiPb(NO2)   000550
K2Er(NO3)5 (flux)*   000872
K2Mn1-xFexF4, slow cooling)* on the Antiferromagnetic   000106
K2Mn1-xMxF4 (M + Fe, Co) (K2Mn1-xFexF4, slow cooling)*   000106
K2MoO4 Solutions in Hydrothermal Conditions*   000475
K2Ni(MoO4)2 (slow cooling, flux)*   000486
K2NiPb(NO2)6, Tl2NiPb(NO2)6, Tl2NiCd(NO2)6, Tl2NiHg(NO2) 000550
K2O-MoO3 (DyKMo2O8, melt)*   000320

PERMUTED TITLE INDEX

of RKMo2O8, R2MoO6 and R6MoO12 in the Systems R2O3-	K2O-MoO3 (DyKMo2O8, ErKMo2O8, NdKMo2O8, Dy2MoO6,	001062
Transitions (solution)* #Ultrasonic Velocities of	K2PbCu(NO2)6 Near the Cooperative Jahn-Teller Phase	000467
Fibers by the Hydrothermal Dehydration Method (	K2Ti6O13* #Synthesis of Potassium Hexatitanate	000732
#The Crystal Structure of	K2W4O13*	000720
Growth of K2CuF4 and Some Solid Solutions of K2CuF4-	K2ZnF4 (Bridgman)* #Crystal	001055
#Dielectric and Optical Properties of	K5Nd(MoO4)4 (Czochralski)*	000582
#Lasing and Fluorescence in	K5NdLi2F10 (flux)*	000189
#Crystal Structure of	K5NdLi2F10 (flux)*	000397
#Orientation Dependence of	l-Alanine Incorporation in TGS Crystals (solution)*	000559
#Epitaxial Growth of Ferroelectric PLZT ((Pb, La)	(Zr,Ti)O3) Thin Films*	000415
Emission (LaB6, EuB6, #Single-Crystal Growth of Mixed (	La, Eu, Y, Ce, Ba, Cs) Hexaborides for Thermionic	000727
Preparation, Phase Equilibria, and Crystal Chemistry of	La, Pr, and Nd Hydroxide Bromides and Hydroxide Iodides	000531
LaB6, EuB6, CeB6, BaB6, (La,Eu)B6, La1-xEuB6, (La,Y)B6,	La,Ba)B6, (La,Cs)B6, (Eu,Y)B6, (Eu,Ba)B6, flux)* (	000727
CeB6, BaB6, (La,Eu)B6, La1-xEuB6, (La,Y)B6, La,Ba)B6,	La,Cs)B6, (Eu,Y)B6, (Eu,Ba)B6, flux)* (LaB6, EuB6,	000727
for Thermionic Emission (LaB6, EuB6, CeB6, BaB6,	La,Eu)B6, La1-xEuB6, (La,Y)B6, La,Ba)B6, (La,Cs)B6, (Eu,	000727
and Iron Group Complex Oxides (NdBO3, LaBO3/Pr, LaBO3,	La,Pb)MnO3, (Nd,Pb)MnO3, (Eu,Pb)MnO3, Pb3Mn7O15,	001061
(LaB6, EuB6, CeB6, BaB6, (La,Eu)B6, La1-xEuB6,	La,Y)B6, La,Ba)B6, (La,Cs)B6, (Eu,Y)B6, (Eu,Ba)B6, flux)	000727
#The Growth of	LaAlO3 Crystals Enriched in Isotope 170 (flux)*	000901
#Flux Growth of Lanthanum Borate,	LaBO3*	000508
Some Rare-Earth and Iron Group Complex Oxides (NdBO3-	LaBO3/Pr, LaBO3, (La,Pb)MnO3, (Nd,Pb)MnO3, (Eu,Pb)MnO3,	001061
Earth and Iron Group Complex Oxides (NdBO3, LaBO3/Pr,	LaBO3, (La,Pb)MnO3, (Nd,Pb)MnO3, (Eu,Pb)MnO3,	001061
Starting Compositions for Flux Growth (DyPO4, PrBO3,	LaBO3, DyKMo2O8, ErKMo2O8, NdKMo2O8, DyVO4, TbVO4,	001060
#Chemical Vapor Growth of	LaB6 Whiskers and Crystals Having a Sharp Tip*	000656
Eu, Y, Ce, Ba, Cs) Hexaborides for Thermionic Emission (	LaB6, EuB6, CeB6, BaB6, (La,Eu)B6, La1-xEuB6, (La,Y)B6,	000727
Solution Method and a Study of Their Properties (flux,	LaB6, SmB6, SmB4, EuB6, YbB6)*Rare Earth Borides by the	000359
#Mixed Oxysulfides of Chromium III and Rare Earths (	LaCrOS2, CeCrOS2, PrCrOS2, NdCrOS2, SmCrOS2)*	001048
Rock Salt Crystals (NaCl, #Semi-Periodic Closed Growth	Lamellae on the Crystal-Melt Interface of LiCl Doped	000552
and their Phase #The Growth of Single Crystals of	Langbeinites Rb2Cd2(SO4)3, Tl2Cd2(SO4)3, K2Co2(SO4)3,	000138
#Ceramic Eutectics in the Systems ZrO2-Ln2O3 (Ln =	Lanthanide): Unidirectional Solidification,	000630
Type Structure (phosphides, flux)* #Ternary	Lanthanoid-Transition Metal Pnictides with ThCr2Si2-	000596
#Flux Growth of	Lanthanum Borate, LaBO3*	000508
#Crystal Structure of	Lanthanum Polysulfide LaS2*	000259
#Crystal Structure of Double Oxides of Rhenium. II. The	Lanthanum-Rhenium Oxide La6ReO18 (melt)*	000103
Process Update and Material Characterization (Si,	laser melting, review)* To-Ribbon (RTR) Technique:	000360
cw Operation at 1.07- and 1.35-micrometer Wavelength (	laser melting, Czochralski)* Laser: Room-Temperature	000916
Amorphous Germanium Layer on a Silicon Substrate by	Laser Annealing (Ge film)**Heteroepitaxy of a Deposited	000343
Size Dependence in a Self-Implanted Silicon Layer on	Laser Irradiation Energy Density (Si)* #Grain	000996
#Epitaxial Layers of the	Laser Material Nd(Ga,Cr)3(BO3)4 (flux)*	000578
#The Crystal Structure of a New High -Nd- Concentration	Laser Material: Na3Nd(PO4)2*	000825
#Chemical Synthesis and Crystal Growth of	Laser Quality Praseodymium Pentaphosphate (PrP5O14)*	000127
#Epitaxial Growth of KNdP4O12	Laser Waveguides (KLaP4O12, solution)*	000643
State Chemistry. I. High Temperature Reactions with CO2-	Laser* #A New Preparation Technique in Solid	000657
#Electrical and Structural Characteristics of	Laser-Induced Epitaxial Layers in Silicon (Si)*	001116
#Dynamics of	Laser-Induced Formation of Palladium Silicide (Pd2Si)*	001042
micrometer Wavelength (#Nd:Y2O3 Single-Crystal Fiber	Laser: Room-Temperature cw Operation at 1.07- and 1.35-	000916
#Optically Pumped LPE-Grown Hg1-xCdxTe	Lasers (films, flux)*	000368
#Epitaxy of NdAl3(BO3)4 For Thin Film Miniature	Lasers (flux)*	000577
of LPE Growth for Lattice Matched GaInAsP/InP DH	Lasers with (100) Substrate in the Range of 1.2-1.5	000035
II. High Pressure - High Temperature Reactions with CO2-	Lasers* Preparative Technique in Solid State Chemistry.	000658
#	Lasing and Fluorescence in K5NdLi2F10 (flux)*	000189
#Crystal Structure of Lanthanum Polysulfide	LaS2*	000259
Grown In1-xGaxAs on InP Substrate* #Composition	Latching Phenomenon and Lattice Mismatch Effects in LPE-	000952
Solidification of a Binary Alloy with a Curved Solid- #	Lateral Solute Segregation During Unidirectional	000206
of the Growth of a Metallic Face-Centered Cubic	Lattice in a Crystal from a Molten Mass in the (100)	000995
(theory)* #Determination of the	Lattice Constant of Epitaxial Layers of III-V Compounds	000399
Al, Au, Ag, In, Pb, PbS, PbSe, PbTe, SnTe, #Roles of	Lattice Fitting in Epitaxy (theory, Ni, Cu, Pd, Pt, Pe,	000949
Substrate in the Range of #Conditions of LPE Growth for	Lattice Matched GaInAsP/InP DH Lasers with (100)	000035
of Liquidus Surface in Pb-Sn-Te System by Simplex	Lattice Method (flux, theory, Pb1-xSnxTe)*Investigation	000668
#The Role of	Lattice Misfit in Epitaxy (review, theory)*	001014
Substrate* #Composition Latching Phenomenon and	Lattice Mismatch Effects in LPE-Grown In1-xGaxAs on InP	000952
Utilizing Non-Parallel Setting for Measuring Local	Lattice Mismatches Between Epitaxial Films and	000704
and #Direct Measurement of Temperature Dependence of	Lattice Mismatches Between LPE-Grown Li(Nb,Ta)O3 Film	000922
Coefficients of #Accurate X-ray Determination of the	Lattice Parameters and the Thermal Expansion	000513
Garnet Layers as Functions of the Mismatch of Crystal	Lattice Parameters((Y, Bi, Sm, Ca)3(Fe, Ge)5 O12, flux)*	000251
Controlled Phosphorus #Liquid Phase Epitaxial Growth of	Lattice-Matched GaP-Ga0.5Al0.5P-GaP Layer Under	000460
Capture Centres for Different Two-Dimensional Substrate	Lattices (theory)*Capture of Migrating Adatoms by Point	000873
#The Crystal Structure of Synthetic	Lautarite, Ca(IO3)2 (flux)*	000331
#Evidence for a Universal	Law of Dendritic Growth Rates (theory)*	000534
#Growth Kinetics of (111) Si Through an Al	Layer by Solid Phase Epitaxy*	000585
#Localized Adsorption in One	Layer on a Crystal-Solution Interface (theory)*	000575
film)**Heteroepitaxy of a Deposited Amorphous Germanium	Layer on a Silicon Substrate by Laser Annealing (Ge	000343
#Grain Size Dependence in a Self-Implanted Silicon	Layer on Laser Irradiation Energy Density (Si)*	000996
#Growth Conditions and Crystal Structure Parameters of	Layer Compounds in the Series Mo1-xWxSe2 (vapor	000013
Furnaces for Silicon Carbide Monocrystal and Epitaxial	Layer Growing (SiC, melt)* in High-Temperature	000561
and Thermal Conditions of Silicon Carbide Epitaxial	Layer Growth from the Vapour Phase (SiC, film, vapor	000942
#The Mechanism of Silicon Epitaxial	Layer Growth from Ion-Molecular Beams (Si)*	000019
#Face Stability and Growth Rate Variations of the	Layer Perovskite (C3H7NH3)2CuCl4 (solution, theory)*	000641
Holographic #Observations on Step Bunching for the	Layer Perovskite (C3H7NH3)2MnCl4 Using a Compound	000640
mNH3MX4 Families with M = Cd, Cu, Fe, Mn or Pd and X = #	Layer Perovskites of the (CnH2n+1NH3)2MX4 and NH3(CH2)	000039
Prim and Slichter Equation as a Diffusion Boundary	Layer Thickness (effective distribution coefficient,	001083
#New Model of LPE Growth:	Layer Thickness Calculation (theory, kinetics)*	000589
Epitaxial Growth of Lattice-Matched GaP-Ga0.5Al0.5P-GaP	Layer Under Controlled Phosphorus Vapor Pressure and	000460
Neutron Scattering Experiments on the Antiferromagnetic	Layer- Type Compounds K2Mn1-xMxF4 (M + Fe, Co) (K2Mn1-	000106
#Some Thermodynamic Considerations for the Growth of	Layered Dichalcogenides by Vapour Transport Processes (	000805
#GaAs/AlAs	Layered Films (epitaxy)*	000353
#Growth of Crystalline Slabs of	Layered InSe by the Czochralski Method*	000187
#Dispersion of Acoustic Phonons in	Layered SrGa2 (melt, slow cooling)*	000672
TaxTi1-xSe2 (#Structural and Electrical Properties of	Layered Transition Metal Selenides VxTi1-xSe2 and	000555

PERMUTED TITLE INDEX

#Solar Furnace Annealing of Amorphous Si	Layers (films)* 000537
and Effect of Dislocations in Epitaxial Semiconductor	Layers (III-V, review)* #Nature, Origin 000126
Crystal Seeds Coated with Informative Interfacial	Layers (Na4Al3Si3O12Cl) Hydrothermal Method on Single 000243
#Growth Kinetics and Structure of Epitaxial Garnet	Layers as Functions of the Mismatch of Crystal Lattice 000251
#Preparation and Optical Properties of ZnSe Epitaxial	Layers by a Close-Spaced Technique* 000670
of Barium Lead Hexa-Aluminate Single Crystal	Layers by the Liquid Phase Epitaxy Technique (flux, (Ba, 000071
#Growth and Morphology of 6H-SiC Epitaxial	Layers by CVD (film)* 000699
#Conditions for Stable Growth of Epitaxial GaP	Layers by Molten Salt Electrodeposition (flux)* 000224
#Low Temperature Growth of GaP LPE	Layers from Indium Solvent (flux)* 000925
Influencing the Growth Rate and Purity of Epitaxial	Layers in the Ga_AsCl3-H2 System (GaAs, vapor 000653
theory, #Theoretical Analysis of Equilibrium Adsorption	Layers in CVD Systems (Si-H-Cl, Ga-As-H-Cl) (GaAs, 000184
#Quasi-Grown	Layers in Liquid Phase Epitaxial Growth (In1-xGaxAs)* 000951
Structural Characteristics of Laser-Induced Epitaxial	Layers in Silicon (Si) #Electrical and 001106
#Epitaxial	Layers of the Laser Material Nd(Ga,Cr)3(BO3)4 (flux)* 000578
#Epitaxial	Layers of CuInSe2 on GaAs* 000843
#Determination of the Lattice Constant of Epitaxial	Layers of III-V Compounds (theory)* 000399
Boracite (Ni3B7O13Cl), #Gas Phase Synthesis of Epitaxial	Layers of Nickel-Chlorine Boracite on Chromium-Chlorine 000837
Vacuum (SiC)* #Growth Kinetics of Epitaxial	Layers of Silicon Carbide Obtained by Sublimation in 000940
Geometry (#Epitaxial Growth of Single Crystal Cd1-xZnxS	Layers on (111) GaAs Substrates Using the Close-Spaced 000305
Synthesis and Structure of Epitaxial Cadmium Sulphide	Layers on Cadmium Single Crystals (CdS, evaporation)* 000429
Diffusion Technique* #Growth of ZnSxSe1-x	Layers on ZnS and ZnSe Substrates by a Solid-State- 000045
#In Situ X-Ray Topography of Epitaxial Ge	Layers During Growth* 000363
#Ellipsometric Assessment of (Ga,Al)As/GaAs Epitaxial	Layers During Their Growth in an Organometallic VPE 000974
kinetics)* #Surface Morphology of GaAs	Layers Grown by Liquid-Phase Epitaxy (flux, theory, 000605
of the Optical Properties of LPE InxGa1-xAsyP1-y Thin	Layers Grown on InP (film)* #Characterization 001098
#Improvement of Crystal Composition in Ga1-xAlxAs LPE	Layers Grown under Conditions of Constant Cooling Rate (000249
#Growth and Characterization of Undoped ZnSe Epitaxial	Layers Obtained by Organometallic Chemical Vapour 000114
#Liquid Phase Growth of HgCdTe Epitaxial	Layers* 001059
#Tetragonal Distortion in Heteroepitaxial	Layers: Ge On GaAs (vapor transport, film)* 000362
Thermionic Emission (LaB6, EuB6, CeB6, BaB6, (La,Eu)B6,	La1-xEuB6, (La,Y)B6, La,Ba)B6, (La,Cs)B6, (Eu,Y)B6, (Eu, 000723
#Structure of	La2Fe1.76S5 (melt)* 000102
#Synthesis and Crystal Growing of	La3NbO7* 000759
Elements (Ln3NbO7) (Ho3NbO7, Y3NbO7, Er3NbO7, Dy3NbO7,	La3NbO7, Y3TaO7, Er3TaO7, Gd3TaO7)* and Rare-Earth 001097
#Crystal Structure of	La3ReO8* 000081
Metal-Metal Bonding (flux)* #The Crystal Structure of	La3Ru3O11: A New Cubic KSbO3 Derivative Oxide with No 000209
Oxides of Rhenium. II. The Lanthanum-Rhenium Oxide	La6Re4O18 (melt)* #Crystal Structure of Double 000103
Bromides and Hydroxide Iodides (Nd(OH)2Br, Pr(OH)2Br,	La7(OH)18Br3, Nd7(OH)18Br3, Pr7(OH)18Br3, La7(OH)18I3, 000531
Pr(OH)2Br, La7(OH)18Br3, Nd7(OH)18Br3, Pr7(OH)18Br3,	La7(OH)18I3, Pr7(OH18I3, hydrothermal)* (Nd(OH)2Br, 000531
New Hexagonal Structure Type (flux)* #	La8Ru4O21: A Mixed-Valence Ternary Ruthenium Oxide of a 000208
Method for Growing Large Needles and Single Crystals of	Lead Chloride (PbCl)* #New Gel 000003
#Crystallisation of Some Ferrite Materials from	Lead Flux with Addition of V2O5 (Y3Fe5O12, Ho3Fe5O12, 000565
Phase Epitaxy Technique (flux, (#Preparation of Barium	Lead Hexa-Aluminate Single Crystal Layers by the Liquid 000071
#Growth of Single Crystals of	Lead Monoxide (PbO, flux)* 000717
cooling, evaporation)* #Flux Growth of	Lead Neodymium Chlorapatite (Na2Nd2Pb6(PO4)6Cl, slow 000445
#Phase Diagram and Crystal Growth of	Lead Phosphovanadates (Pb3(PxV1-xO4)2, Pb3(PO4)2)* 000987
#Photoconduction of	Lead Trititanate (PbTi3O7, flux)* 000264
#A Series of	Lead Tungsten Bronzes (PbxWO3)* 000267
Preparation of Twin-Free InP Crystals Using the	LEC Technique (Czochralski)* #Reproducible 000125
#X-Ray Study of	LEC-Grown InP Crystals (Czochralski, theory)* 000614
#Surface Charge and	Ledge Dynamics in Cadmium Sulfide (CdS, evaporation)* 000529
and Silicon Crystals (Al2O3, Si)* #Maximum Stable Zone	Length in Float-Zone Growth of Small-Diameter Sapphire 000479
and Products and Their Application in National Economy,	Leningrad, March 14-17, 1979 (review)*Profiled Crystals 001111
#Structure of the	Lennard-Jones (100) Crystal-Liquid Interface (theory)* 000528
Carlo Simulation of the Crystal-Melt Interface of a	Lennard-Jones Substance (theory)* #Monte 000123
Ce, cold #Single Crystal Growth by the Horizontal	Levitation Zone Melting Method (DyFe2, TbFe2, Nd, Pr, 000620
#Li3N: A Promising	Li Ionic Conductor (electrolytic deposition, review)* 000024
Dependence of Lattice Mismatches Between LPE-Grown	Li(Nb,Ta)O3 Film and LiTaO3 Substrate (flux)* 000922
LiTaO3 Substrates (LiNb1-xTaxO3, flux)* #LPE Growth of	Li(Nb,Ta)O3 Solid-Solution Thin Film Waveguides on 000500
and Ionic Conductivity of Li24Zn(GeO4)4 and other new	Li+ Superionic Conductors (kinetics, melt)* Structure 000396
#Acoustooptic Characteristics of	LiBi(MoO4)2 (Czochralski)* 000016
Au, Ag, In, Pb, PbS, PbSe, PbTe, SnTe, LiF, NaF, KF,	LiBr, NaCl, NaBr, LiI, KCl, NaI, KBr, KI)* Pt, Fe, Al, 000949
Closed Growth Lamellae on the Crystal-Melt Interface of	LiCl Doped Rock Salt Crystals (NaCl, Czochralski)* 000532
Pr2Ti2O7, Tm2Ge2O7, Er2SiO5, Dy2SiO5, Ho2SiO5, Pr2MoO6,	LiCoPO4)* (Eu,Pb)MnO3, Pb3Mn7O15, Tm2Ti3O7, Tb2Ti2O7, 001061
#Thermal and Dielectric Properties of LiKSO4 and	LiCsSO4* 000229
of Condensation and Evaporation of Alkali Halides (KCl,	LiP)* #Molecular (beam) Processes 000628
Pd, Pt, Fe, Al, Au, Ag, In, Eb, PbS, PbSe, PbTe, SnTe,	LiF, NaF, KF, LiBr, NaCl, NaBr, LiI, KCl, NaI, KBr, KI)* 000949
xP on GaP and Its Application to Double Heterostructure	Light Modulators* #Liquid Phase Epitaxy of AlxGa1- 000948
of Rare Earth Copper Intermetallic Compounds, RCu2. II.	Light Rare Earth (PrCu2, NdCu2)* #Magnetic Properties 000374
of Growth Defects in Synthetic Quartz Crystals by	Light-Scattering Tomography (SiO2, hydrothermal)* 000652
PbS, PbSe, PbTe, SnTe, LiF, NaF, KF, LiBr, NaCl, NaBr,	LiI, KCl, NaI, KBr, KI)*Pd, Pt, Fe, Al, Au, Ag, In, Pb, 000949
#Single Crystal Growth of	LiInS2 (directional solidification)* 000456
Properties of Alkali Metal Indates and Thioindates (	LiInS2, Bridgman-Stockbarger)*Principal Physicochemical 000509
#Enantiomorphism of alpha-	LiIO3 Single Crystals (solution)* 000682
#Thermal and Dielectric Properties of	LiKSO4 and LiCsSO4* 000229
Nb, NaxWO3, Czochralski, electrolytic #Growth Rate	Limitations in Electrochemical Crystallization (metals, 000225
Float Zone Silicon Crystals (Si, theory)* #	Limitations in Using Kilohertz Radio Frequencies for 000358
Melts (flux)* #Crystallization of	LiNbO3 from Solution in Borate, Vanadate, and Tungstate 001047
#Second Harmonic Generation in a Sputtered	LiNbO3 Film on MgO (epitaxy)* 000518
Liquid-Phase Epitaxy (flux)* #Thin Films	LiNbO3, Doped with Na+ and Co2+ plus Zr4+, Grown by 000691
Solution Thin Film Waveguides on LiTaO3 Substrates (	LiNb1-xTaxO3, flux)* #LPE Growth of Li(Nb,Ta)O3 Solid- 000500
#Flux Growth of	LiNdP4O12 Single Crystals* 000677
#Top-Seeded Flux Growth of	LiNdP4O12 Single Crystals* 000678
of Precipitates in ZnTe: Shape of the Tellurium Solidus	Line (Bridgman) Electron Microscopy Observations 000098
CaCO3, kinetics, theory)* #	Linear Expansion Anisotropy of CdGa2Se4 (melt)* 000476
Lithium Nitrate Hydrates (solution, LiNO3.3H2O,	Linear Growth Rate of Calcite in Aqueous Solution (000627
Solid Crystalline Lithium Nitrate Hydrates (solution,	LiNO3.3D2O)* and Molecular Volume of Solid Crystalline 000023
of Lithium and Strontium Nitrate Crystal Hydrates (	LiNO3.3H2O, LiNO3.3D2O)*Density and Molecular Volume of 000023
	LiNO3.3H2O, Sr(NO3)2.4H2O)* #Crystallization Kinetics 000989

PERMUTED TITLE INDEX

Phosphate. II. Crystal Growth by Dimer Addition (| LiPO3)*   During Polymerization of Lithium Dihydrogen | 000097
Ternary Component Solution (Ga1-xAlxAs,   #Variation of | Liquid and Solid Compositions During LPE Growth from a | 000970
#Properties of Fe2P Crystals Prepared from a | Liquid Copper Medium* | 000030
CuInS2)*   #Simple Pressurized Chambers for | Liquid Encapsulated Czochralski Crystal Growth (InP, | 000150
melting, vapor transport, sublimation)* | Liquid Encapsulation Zone-Refining of PbS (zone | 000583
#Temperature Gradient Measurement in | Liquid Epitaxial Growth Systems (flux)* | 000574
#Structure of the Lennard-Jones (100) Crystal- | Liquid Interface (theory)* | 000528
Solidification of a Binary Alloy with a Curved Solid- | Liquid Interface (Ge:Ga, Bridgman, theory)* | 000206
Convection of a Solution with #Influence of Density of | Liquid Phase in the Y3Fe5O12-BaO.B2O3 System on | 000048
#Quasi-Grown Layers in | Liquid Phase Epitaxial Deposition of GaP on GaAs* | 000354
System (flux)* | Liquid Phase Epitaxial Growth (In1-xGaxAs) | 000951
CuFe2O4)* | Liquid Phase Epitaxial Growth of CdTe in the CdTe-CdCl2 | 000828
Stepwise Grading (melt)* | Liquid Phase Epitaxial Growth of Copper Ferrite Films (| 001016
Ga0.5Al0.5P-GaP Layer Under Controlled Phosphorus | Liquid Phase Epitaxial Growth of Ga1-xInxSb on GaSb by | 000803
Aluminate Films (flux, Li0.5Fe2.5-xAlxO4)* | Liquid Phase Epitaxial Growth of Lattice-Matched GaP- | 000460
review)* | Liquid Phase Epitaxial Growth of Lithium Ferrite- | 001017
| Liquid Phase Epitaxial Growth of Magnetic Garnets (| 000783
Temperature and Growth-Dissolution Kinetics of Garnet | Liquid Phase Epitaxial Growth of Zn and S Doped GaAs* | 000416
#GaxIn1-xSb Crystals Grown by | Liquid Phase Epitaxy (flux, kinetics, theory)* Liquidus | 000333
Imbalance for Ca,Ge-Substituted Garnet Films grown by | Liquid Phase Epitaxy (melt)* | 000852
Application to Double Heterostructure Light Modulators*# | Liquid Phase Epitaxy from PbO-B2O3 Fluxed Melts* Charge | 000386
| Liquid Phase Epitaxy of AlxGa1-xP on GaP and Its | 000948
Compounds (ZnSnAs2, solution)* | Liquid Phase Epitaxy of Garnets (review)* | 000802
Substrates (flux)#Thin Film Growth of (MnxZn1-x)Fe04 by | Liquid Phase Epitaxy of II-IV-V2, Chalcopyrite | 000089
Barium Lead Hexa-Aluminate Single Crystal Layers by the | Liquid Phase Epitaxy on Bridgman Grown Zn2TiO4 | 000216
#The Growth Kinetics of Garnet | Liquid Phase Epitaxy Technique (flux, (Ba,Pb)Al12O19)* | 000071
#A New Substrate Holder for | Liquid Phase Epitaxy Using Horizontal Dipping (flux)* | 001021
#Growth of GaAs1-xSbx Crystals by Steady-State | Liquid Phase Epitaxy* | 000072
| Liquid Phase Epitaxy* | 000957
of VLS Silicon Crystals Grown with Platinum and Gold as | Liquid Phase Growth of HgCdTe Epitaxial Layers* | 001059
effect, GaAs, theory, electrolytic deposition, | Liquid-Forming Agents (vapor transport, Si)* Morphology | 001075
#The Growth of GaxIn1-xAs on (100) InP by | Liquid-Phase Electroepitaxy: Growth Kinetics (Peltier | 000437
of LiNbO3, Doped with Na+ and Co2+ plus Zr4+, Grown by | Liquid-Phase Epitaxy (flux)* | 000754
#Surface Morphology of GaAs Layers Grown by | Liquid-Phase Epitaxy (flux)*   #Thin Films | 000691
| Liquid-Phase Epitaxy (flux, theory, kinetics)* | 000605
Properties of Er3Al5-xGaxO12 Films Obtained by | Liquid-Phase Epitaxy of ZnSe from Zn-Ga Solution (film) | 000308
#Growth of Whiskers by the Vapor- | Liquid-Phase Epitaxy*   #Spectroscopic | 000245
Czochralski, Kyropoulos, float   #Crystal Growth from | Liquid-Solid Mechanism* | 000338
#Fiber Optic Probe for Thermal Profiling of | Liquids at High Temperatures (review, melt, flux, | 000139
| Liquids During Crystal Growth (solution) | 000395
xSnxTe, PbTe, flux)* | Liquidus of the Ternary System Pb-Tl-Te* | 000101
Method (flux, theory, Pb1-xSnxTe)*   #Investigation of | Liquidus Measurements in the Pb-Sn-Te System (Pb1- | 000047
Garnet Liquid Phase Epitaxy (flux, kinetics,   #The | Liquidus Surface in Pb-Sn-Te System by Simplex Lattice | 000668
#The Crystal Structure of Lithium Fluorosulphate | Liquidus Temperature and Growth-Dissolution Kinetics of | 000333
Czochralski)* | LiSO3F (solution)* | 001108
of #Mismatches Between LPE-Grown Li(Nb,Ta)O3 Film and | LiTaO3 Single Crystals for Saw Device Applications (| 000311
of Li(Nb,Ta)O3 Solid-Solution Thin Film Waveguides on | LiTaO3 Substrate (flux)*   Dependence of Lattice | 000922
#Structural Studies of the Solid Electrolyte High- | LiTaO3 Substrates (LiNb1-xTaxO3, flux)* #LPE Growth | 000500
#Magneto-Optical Properties of KTb3F10 and | LiTa3O8 (electrolytic deposition)* | 000706
Magnetic, Optical and Electrical Properties - | LiTbF4 Crystals (Czochralski)* | 001068
3H2O, Sr(NO3)2.4H2O)*   #Crystallization Kinetics of | Lithium and Molecular Intercalations (VOCl)* | 001031
Dimer   #Crystallization During Polymerization of | Lithium and Strontium Nitrate Crystal Hydrates (LiNO3. | 000989
Li0.5Fe2.5O4)*   #LPE Growth of | Lithium Dihydrogen Phosphate. II. Crystal Growth by | 000097
xAlxO4)*   #Liquid Phase Epitaxial Growth of | Lithium Ferrite on Spinel Substrate Crystals (films, | 000340
#The Crystal Structure of | Lithium Ferrite-Aluminate Films (flux, Li0.5Fe2.5- | 001017
Effect of Impurities on the Piezoelectric Properties of | Lithium Fluorosulphate LiSO3F (solution)* | 001108
#Physical Properties and | Lithium Germanate (Li2GeO3, Czochralski)*   #The | 000800
#Crystal Growth and Characterization of | Lithium Intercalates of CrPS4 (slow cooling)* | 000571
Li2GeO3) Single Crystals (Bridgman)*   #Polar Twins in | Lithium Metasilicate (Li2SiO3, Czochralski)* | 000141
#Elastic and Piezoelectric Properties of Potassium | Lithium Metasilicate and Metagermanate (Li2SiO3, | 000108
Bismuth Niobate Single   #Epitaxial Growth of Potassium | Lithium Niobate (KLN) Crystals (K2.89Li1.55Nb5.11O15, | 000010
Bismuth Niobate Single   #Epitaxial Growth of Potassium | Lithium Niobate Single-Crystal Films on Potassium | 000009
the Density and Molecular Volume of Solid Crystalline | Lithium Niobate Single-Crystal Films on Potassium | 000011
#Low Temperature Method for the Growth of | Lithium Nitrate Hydrates (solution, LiNO3.3H2O, LiNO3. | 000023
#The Crystal Structure of | Lithium Nitride Single Crystals (Li3N)* | 000252
Growth of Lithium Ferrite-Aluminate Films (flux, | Lithium Uranate (Li2UO4, flux)* | 000324
of Lithium Ferrite on Spinel Substrate Crystals (films, | Li0.5Fe2.5-xAlxO4)*   #Liquid Phase Epitaxial | 001017
#Heat-Flow-Controlled Growth During | Li0.5Fe2.504)*   #LPE Growth | 000340
#Crystal Structure of | Li2B4O7 Crystallization (melt)* | 000385
in Lithium Metasilicate and Metagermanate (Li2SiO3, | Li2CaUF8 (flux)* | 001029
on the Piezoelectric Properties of Lithium Germanate (| Li2GeO3) Single Crystals (Bridgman)* #Polar Twins | 000108
Growth and Characterization of Lithium Metasilicate (| Li2GeO3, Czochralski)*   #The Effect of Impurities | 000800
#Polar Twins in Lithium Metasilicate and Metagermanate (| Li2SiO3, Czochralski)*   #Crystal | 000141
#Bridgman-Stockbarger Crystal Growth of | Li2SiO3, Li2GeO3) Single Crystals (Bridgman)* | 000108
#The Crystal Structure of Lithium Uranate (| Li2Ti3O7* | 000634
#Single Crystal Growth of | Li2UO4, flux)* | 000324
kinetics, #Crystal Structure and Ionic Conductivity of | Li2ZnGe (melt)* | 000520
#Czochralski Growth of | Li24Zn(GeO4)4 and other new Li+ Superionic Conductors (| 000986
for the Growth of Lithium Nitride Single Crystals (| Li3N Crystals* | 000842
deposition, #Ceramic Eutectics in the Systems ZrO2-Ln2O3 | Li3N)*   #Low Temperature Method | 000252
Microstructural and Crystallographic Characterization (| Li3N: A Promising Li Ionic Conductor (electrolytic | 000024
#Synthesis of Single Crystal | Ln = Lanthanide): Unidirectional Solidification, | 000630
#Crystal Growth and Crystallographic Data of Some | Ln = Nd, Sm, Dy, skull-melting)*   Solidification, | 000630
Microstructural #Ceramic Eutectics in the Systems ZrO2- | LnPO4 (vapor transport)* | 000733
of Double Oxides of Niobium and Rare-Earth Elements (| Ln2(MoO4)3 Type Mixed Rare Earth Molybdates (RE(MoO4)3, | 000447
Method Utilizing Non-Parallel Setting for Measuring | Ln2O3 (Ln = Lanthanide): Unidirectional Solidification, | 000630
| Ln3NbO7) (Ho3NbO7, Y3NbO7, Er3NbO7, Dy3NbO7, La3NbO7, | 001097
| Local Lattice Mismatches Between Epitaxial Films and | 000704

PERMUTED TITLE INDEX

PERMUTED TITLE INDEX

Mn1-xFexP, x = 0 to 0.15, theory)*   # Magnetic Properties of (Mn,Cr)P and (Mn,Fe)P Compounds (000430
Vanadium Triselenide (BaVSe3)*   #Structural and   # Magnetic Properties of "One-Dimensional" Barium 000474
2 and 4.2K (Stockbarger)*   # Magnetic Properties of a Single Crystal of NdF3 Between 000767
FexGe, slow cooling)*   # Magnetic Properties of beta-Phase Iron-Germanium (000845
Crystals (melt)*   # Magnetic Properties of CdCr2Se4-ZnCr2Se4 Single 000586
#Low Temperature Thermal and   # Magnetic Properties of CeP and CeAs (sublimation)* 000400
#Crystal Growth and   Magnetic Properties of NdFeTiO5 (flux)* 001091
Compounds, RCu2. I. Heavy Rare Earth (TbCu2, DyCu2,   # Magnetic Properties of Rare Earth Copper Intermetallic 000373
Compounds, RCu2. II. Light Rare Earth (PrCu2, NdCu2)*   # Magnetic Properties of Rare Earth Copper Intermetallic 000374
*   #Preparation of Single Crystals of the   Magnetic Semiconductor CdCr2Se4 (vapor transport, doped) 000625
review)*   # Magnetic Spinel Single Crystals by Bridgman Technique (000924
Tb5Ge4. A Neutron Diffraction Study (melt)*   #On the   Magnetic Structure and Magnetic Phase Transitions of 000840
solution)*   # Magnetic Structure of GdCu1-xZnx System (melt)* 000956
#Crystallographic and   Magnetic Structure of RbFeCl3.2D2O and CsFeCl3.2D2O (000077
Ca3Fe2Ge3O12 and Ca3Mn2Ge3O12*   #Flux Growth of Single   Magnetic Sublattice Antiferromagnetic Garnets 000238
NaxV2O5, KxV2O5,   #Low-Temperature Specific Heat and   Magnetic Susceptibility of Nonmetallic Vanadium Bronze (000171
and Molecular   #Vanadium (III) Oxychloride: Magnetic, Optical and Electrical Properties -Lithium 001031
Ferromagnetic EuB6 (review)*   # Magnetic, Transport, and Thermal Properties of 000293
Valence Compound TmSe (slow   #On the Resistivity of the   Magnetically Ordered Ground State of the Intermediate 000031
*   #Skull Melter Growth of   Magnetite (Fe3O4)*
*   Magnetization and Neutron Studies of UTe and USb0.8Te0.2 000154
Magnetostriction and Elastic Constants of a Ruthenium-   Magnetization, Magnetocrystalline Anisotropy, 000631
Crystals (Czochralski)*   Magneto-Optical Properties of KTb3F10 and LiTbF4 001068
#Optical and   Magneto-Optical Properties of Sm0.55Tb0.45FeO3 (flux)* 000914
Garnets (flux, (GdPrBi)3(FeGaIn)5O12, (YLaBi)3(FeGa)   Magneto-Optical Properties of Some Rare Earth Iron 001107
Magneto-Seebeck Effect of Pb-Doped Bi-Sb (Bridgman)* 000034
Elastic Constants of a Ruthenium-Nickel   #Magnetization,   Magnetocrystalline Anisotropy, Magnetostriction and 000631
Nickel   #Magnetization, Magnetocrystalline Anisotropy,   Magnetostriction and Elastic Constants of a Ruthenium- 000631
Epitaxial Growth of ZnO Films on Sapphire by Planar   Magnetron rf Sputtering System*   #High Rate 000882
of Oscillatory   #Some Evidence for the Existence and   Magnitude of a Critical Marangoni Number for the Onset 000847
Man-Made Gemstones (review)* 000271
#Crystal Structure of the Phosphate of Trivalent   Manganese Mn(PO3)3 (flux)* 000058
Evidence for the Existence and Magnitude of a Critical   Marangoni Number for the Onset of Oscillatory Flow in 000847
and Their Application in National Economy, Leningrad,   March 14-17, 1979 (review)*   Crystals and Products 001111
Face-Centered Cubic Lattice in a Crystal from a Molten   Mass in the (100) Orientation (theory)*   of a Metallic 000995
Mass Spectrometric Study on Chemical Transport of VO2* 000063
#On Diffusive-Advective Interfacial   Mass Transfer (theory, solution, vapor transport, flux)* 001074
Boundary with the Melt, and its Influence on the Shape   # Mass Transfer at the Surface of a Crystal Near to its 001046
transport)*   #Heterogeneous Equilibria and   # Mass Transfer Kinetics in the System TiP-I2 (TiP, vapor 001003
Growth of ZnS in H2: The Effects of a Bypass Flow (# Mass Transport Processes Associated with the Epitaxial 000558
Properties*   Metastable Pb1-xCdxS Epitaxial Films. II. Electrical 000910
the Range of 1.2-   #Conditions of LPE Growth for Lattice   Matched GaInAsP/InP DH Lasers with (100) Substrate in 000035
Phosphorus   #Liquid Phase Epitaxial Growth of Lattice-   Matched GaP-Ga0.5Al0.5P-GaP Layer Under Controlled 000040
#Ca12Al14O33: A New Piezoelectric   Material (Czochralski)* 001076
Properties of Cadmium,   #Influence of the Source   Material and the Source Shape on the Sublimation 000365
Ribbon-To-Ribbon (RTR) Technique: Process Update and   Material Characterization (Si, laser melting, review)* 000030
#Epitaxial Layers of the Laser   Material Nd(Ga,Cr)3(BO3)4 (flux)* 000578
#Spacelab and   Material Processing Facilities and Experiments (review)* 000853
Structure of a New High -Nd- Concentration Laser   Material: Na3Nd(PO4)2*   #The Crystal 000825
The Chemical Vapour Transport of Ni-Cl Boracite (no   materials grown - Ni3B7O13Cl theory)* Component System: 000236
#Infra-Red Transmitting Materials. Part 1. Crystalline   Materials (review, theory)* 000250
alpha-   #Synthesis and Crystal Growth of Refractory   Materials by RF Melting in a Cold Container (review, 000020
Y3Fe5O12, Ho3Fe5O12,   #Crystallisation of Some Ferrite   Materials from Lead Flux with Addition of V2O5 (000565
solution, vapor transport, zone melting, doped, review)#   Materials Behaviour in Low Gravity Environment (melt, 001070
#Current Topics in   Materials Science, Volume 1 (review)* 000450
#Current Topics in   Materials Science, Volume 2 (Review)* 000449
theory)*   #Infra-Red Transmitting   Materials. Part 1. Crystalline Materials (review, 000250
Graphite,   #Preparation and Properties of Solid State   Materials, Vol. 4, Morphological Stability, Convection, 001081
theory)*   #The Molten State of   Matter. Melting and Crystal Structure (melt, review, 000999
Small-Diameter Sapphire and Silicon Crystals (Al2O3,   Maximum Stable Zone Length in Float-Zone Growth of 000479
Crystallization of Sodium Zirconogermanates (Na2ZrGeO5.   Na2ZrGe2O7, Na4Zr2Ge3O12, Na3HZr[GeO4]2)* #Hydrothermal 000230
the Growth of GaxIn1-xAs and GaxIn1-xP Alloy Films by   MBE (kinetics)*   #Surface Processes Controlling 000301
Range for Growth of Autoepitaxial GaAs Films by   MBE (molecular beam epitaxy, theory)*   #Temperature 000686
#Semiconductor Surface and Crystal Physics Studied by   MBE (GaAs, Ga1-xAlxAs, review of MBE process)* 000055
#Integrated Optical Devices Fabricated by   MBE (GaAs, Ga1-xAlxAs, review, film)* 000152
xGaxAs, GaSb1-yAsy)*   #Semiconductor Superlattices in   MBE and Their Characterization (GaAs, Ga1-xAlxAs, In1- 000176
#Nonstoichiometry and Carrier Concentration Control in   MBE of Compound Semiconductors (II-VI, IV-VI, III-V, 000897
Physics Studied by MBE (GaAs, Ga1-xAlxAs, review of   MBE process)*   #Semiconductor Surface and Crystal 000055
*   # MBE Techniques for IV-VI Optoelectronic Devices (review) 000394
Barrier Diodes Prepared by Molecular-Beam Epitaxy (MBE) on GaAs (Al, films)*   Crystal-Aluminum Schottky- 000190
#Si-   MBE: Growth and Sb Doping (Si, film)* 000502
#Temperature Gradient   Measurement in Liquid Epitaxial Growth Systems (flux)* 000574
from Its Quiescent Aqueous Solution by Differential   # Measurement of Crystallization of Potassium Bromate 000367
Mismatches Between LPE-Grown Li(Nb,Ta)O3 Film   #Direct   Measurement of Temperature Dependence of Lattice 000922
of 2H-NbSe2 with Hydrazine by Transmittance   Measurements (theory)*   Studies of the Intercalation 000085
flux)*   #Liquidus   Measurements in the Pb-Sn-Te System (Pb1-xSnxTe, PbTe, 000047
melt)*   #Viscosity   Measurements in the System Dy2O3-K2O-MoO3 (DyKMo2O8, 000320
Crystal During Czochralski Growth of Gallium and   # Measurements of the Contact Angle Between Melt and 001017
Growth from the Vapor in Closed Ampules (GeS,   # Measurements of Growth Parameters During the Crystal 000841
#Electronic Bandgap   Measurements of SnS2 Polytypes (vapor transport)* 000008
float zone melting)*   #Fermi Surface   Measurements of ZrB2 by the de Haas van Alphen Effect (000964
Crystal Method Utilizing Non-Parallel Setting for   Measuring Local Lattice Mismatches Between Epitaxial 000704
Fusion for Ionic Crystals (review, theory)*   #Quantum-   Mechanical Relations in the Calculation of Heats of 001087
epitaxy)*   #Investigation of the   Mechanism and Kinetics of Growth of LPE GaAs (theory, 000144
#Growth of Stacking Faults by Bardeen-Herring   Mechanism in Czochralski Silicon (Si) 001052
Impurities (solution)*   #The   Mechanism of Growth of KDP Crystals in the Presence of 000094
Based on Mossbauer Studies (NH4H2PO4, solution)*   # Mechanism of Habit Change of ADP Crystals by Fe3+, 000296
#Growth   Mechanism of Hollow ZnO Crystals from ZnSe. II* 000427

PERMUTED TITLE INDEX

PERMUTED TITLE INDEX

of a Solid Particle Growing from a Binary Alloy
of Cr2O3 Crystals by the Decomposition of a K2Cr2O7
#Crystal Growth of Bi3B5O12 From its Stoichiometric
for Determining the Kinetics of Crystal Growth from the
- I. Single Filament, Unconstrained Growth from a Pure
of the Solid Electrolyte (C5H5NH)2Cu5Br7 From the
Model of Segregation During Crystal Growth from the
During Epitaxial Growth from a Solution in a
Twinned Silicon Crystals Grown in an Al-16 wt.% Si
Filament, Unconstrained Growth from a Binary Alloy
Cubic Zinc Sulphide by Double Decomposition Reaction in
#Growth of Zircon Single Crystals from a Solution in a
and #Measurements of the Contact Angle Between
Quaternary III-V Alloys (review, kinetics)* #
#Evaluation of Temperature Distribution of
#Growth of Niobium Single Crystals by Pulling from a
#Computational Simulation of the
Grown Gadolinium Gallium Garnet (#The Effect of
Bridgman)* #The
Exchange Method (HEM) (Y3Al5O12, melt, slow cooling)* #
One-Component Systems. Two Cases of Modelling the
#Monte Carlo Simulation of the Crystal-
#Semi-Periodic Closed Growth Lamellae on the Crystal-
#Relative Supersaturation and Supercooling in Fluxed
of Defects in Zinc Selenide Crystals Grown from the
Supercooling During the Crystal Growth of InSb from the
Characteristics of CdS Single Crystals Grown from the
Circuits and Solar Cells (Si, review)* #Problems in the
On the Transition from Faceted to Non-Faceted Growth in
Si, #Computation of Striated Impurity Distributions in
#An Evidence of Spiral Growth on the Surface of
#Edge Emission in
Mechanism of Y3Al5O12 Crystals on Oriented Seeds in
of Growth of Y3Al5O12 Single Crystals from
the Surface of a Crystal Near to its Boundary with the
#Skull
#Growth & Etching of InBi Single Crystals (melt, zone
of Chromium Silicide, Cr3Si, Crystals (float zone
Crystal Elastic Constants of Al2Cu (Bridgman, arc
Method for Growing V3Si Single Crystals (electron beam
a New High-Intensity Silicon Solar Cell (Si, float zone
of Bi-Sb Solid Solutions in Capillaries (zone
of ZrB2 by the de Haas van Alphen Effect (float zone
Crystal with Homogeneous Compositions (float zone
and Properties of Beta-Boron Crystals (B, float zone
Cobaltate Single Crystals (SrCoO2.70, float zone
Characterization (Ln = Nd, Sm, Dy, skull-
Niobium Carbide Nitride (NbN, NbN-NbC, NbxNyCz, zone
(hydrogen reduction of ferrous halides, float zone
from Near-Stoichiometric Compositions (doped, zone
Alloys (Nb-Zr, Nb-Hf, electron beam melting, zone
Rotation Driven Flows (kinetics, theory, electron beam
(chemical vapor transport, solution, cold crucible
Environment (melt, solution, vapor transport, zone
Data: Theoretical (directional crystallization, zone
by High-Dose Si+-Ion Implantation of Pd (electron beam
Synthesis of Nb-H Interstitial Alloys (electron beam
in Vacuum onto Silicon (vapor transport, electron beam
Number at Zero Gravity (Si, Al2O3, theory, float zone
Interstitial Alloys (electron beam melting, float zone
Process Update and Material Characterization (Si, laser
and Growth Peculiarities of Tl-Se-TlInSe2 (TlSe, zone
#Single-Crystal Growth of Tantalum Diboride (TaB2, zone
Crystallization Experiments in Spacelab (CdTe, zone
Theory and Crystallization (entropy, specific heat,
Single Crystals with Low Dislocation Density (Al, zone
Rotation Driven Flows (Si, Al2O3, kinetics, float zone
Refined Metallurgical Silicon (Si, Czochralski, zone
#Liquid Encapsulation Zone-Refining of PbS (zone
and Tungsten Alloys (Nb-Zr, Nb-Hf, electron beam
review, melt, flux, Czochralski, Kyropoulos, float zone
at 1.07- and 1.35-micrometer Wavelength (laser
Sheet and Their Effects on Solar Cell Processing (zone
Near the Melting Point and the Vacancy Mechanism of
#A Molecular Dynamics Study of Surface
on Surface Tension Driven Flow in Floating Zone
#The Molten State of Matter.

Molecular Dynamics Simulation (melt, slow   #Study of
in Fe-3wt%Si Single Crystals Grown by Floating Zone
and Crystal Growth of Refractory Materials by RF
Technique (InSb)*   #Horizontal Zone-
RE2O3, eutectics, directional   #Growth from Skull-
NaNO3, PbCl2, organics)*   #An Inexpensive Zone
W, W-C)*   #Growth of Single Crystals of High
Single Crystal Growth by the Horizontal Levitation Zone
theory, compilation,   #Heat Capacity of Metals Near the
Silicon-Germanium Solid Solutions by Electron-Beam Zone

Melt (dendrites, theory)*   #Morphological Stability   000990
Melt (flux)*   #Growth   000859
Melt (kinetics)*   000794
Melt (organics, benzophenone)*   #A New Technique   000755
Melt (theory)*   Morphology During Crystallization   000992
Melt (Bridgman)*   #Crystal Growth   000890
Melt (Czochralski, theory)* Burton, Prim, and Slichter   001084
Melt (GaP, flux)*   #Theory of Crystallization   000591
Melt (Si, flux)*   #Fivefold   000489
Melt (Sn-Pb, theory)*   Crystallization - II. Single   000993
Melt (ZnS, flux)*   #Growth of Single Crystals of 000328
Melt (ZrSiO4)*   000344
Melt and Crystal During Czochralski Growth of Gallium   001071
Melt and Solution Growth of Bulk Single Crystals of   000054
Melt in Silicon Ribbon Growth (EDFG)*   000523
Melt on a Pedestal. III. Characteristic Arrangement of   000684
Melt Flow During Czochralski Growth (theory)*   000491
Melt Flow Phenomena on the Perfection of Czochralski   000638
Melt Growth and Thermal Expansion of Na2CO3 Crystals (000632
Melt Growth of Nd:Y3Al5O12(Nd:YAG) Using the Heat   000167
Melt Growth Interface Motion   Growth for Metallic   000059
Melt Interface of a Lennard-Jones Substance (theory)*   000123
Melt Interface of LiCl Doped Rock Salt Crystals (NaCl,   000532
Melt Systems (Ho2Si2O7, TbVO4, DyKMo2O8)*   000594
Melt Under Argon Pressure (ZnSe, Bridgman)*   #Formation   000477
Melt Under Stabilizing Thermal Gradient (Bridgman,   000478
Melt*   #Improving the Electrical   000588
Melt- and Vapor Growth of Silicon for Integrated   000117
Melt-Grown Crystals (organic)*   # 000355
Melt-Grown Crystals, Taking Account of Periodic Remelt (001026
Melt-Grown KCl Crystals (Czochralski, theory)*   000408
Melt-Grown ZnSexS1-x (Bridgman)*   000710
Melt-Solutions (flux)*   Kinetics and Growth   000093
Melt-Solutions on Oriented Seeds (flux)*Characteristics   000978
Melt, and its Influence on the Shape of the Growing   001046
Melter Growth of Magnetite (Fe3O4)*   000370
melting)*   000110
melting)*   #Growth   000441
melting)*   #Single-   000274
melting)*   #A Simple   000162
melting)*   #Performance of   000303
melting)*   #Crystallization 001035
melting)*   #Fermi Surface Measurements 000964
melting)*   #Preparation of TiCx Single 001092
melting)*   #Some Peculiarities in the Growth 000967
melting)*   #Growth and Properties of Strontium 000955
melting)*   Microstructural and Crystallographic 000630
melting)*   #Crystal Growth of Niobium Nitride and 000192
melting)*   Growth of alpha-Fe Single Crystal Whiskers 000318
melting)*   of Copper Indium Ditelluride Crystals Grown 000220
melting)*   of Niobium-Zirconium, Hafnium, and Tungsten 000307
melting)* Floating Zone Melting. II. Combined Free and   000174
melting)*Characterization of Transition Metal Silicides   000607
melting, doped, review)*   Behaviour in Low Gravity   001070
melting, eutectics)*   Coefficients from Experimental   001080
melting, evaporation)*   #Silicide Formation 000177
melting, float zone melting, review)*   #Properties and 001072
melting, kinetics, theory)* of Films of CdS Evaporated   000073
melting, kinetics, Si)*   of Zone Geometry and Prandtl   000175
melting, review)*   #Properties and Synthesis of Nb-H   001072
melting, review)* the Ribbon-To-Ribbon (RTR) Technique: 000360
melting, slow cooling)*   #Structure   000361
melting, solution)*   000680
melting, sublimation)*Ellipsoid Mirror Furnace for Zone   000276
melting, theory)*   #Percolation   000168
melting, theory)*   Method of High Purity Aluminum   000228
melting, theory)*   in Floating Zone Melting. I. Pure   000173
melting, vapor transport)*   #Solar Cells from Zone- 000195
melting, vapor transport, sublimation)*   000583
melting, zone melting)* of Niobium-Zirconium, Hafnium,   000307
melting, Bridgman)* from Liquids at High Temperatures (000139
melting, Czochralski)*   Room-Temperature cw Operation 000916
melting, Si)* #Properties of Rapidly-Grown RTR Silicon   000546
Melting (theory, compilation, 39 metals)*   of Metals 000351
Melting (theory, vapor transport, kinetics)*   000140
Melting (NaNO3, theory)*   #Experiments   000848
Melting and Crystal Structure (melt, review, theory)*   000999
Melting and Crystallization of Corundum (alpha-Al2O3)*   000563
Melting and Freezing in the Gaussian Core Model by   000915
Melting at Various Growth Rates*   #Growth Striations 001028
Melting in a Cold Container (review, alpha-Al2O3, Al2O3:   000200
Melting of Indium Antimonide in Vacuum. I. Basic   000712
Melting of Zirconia-Rare Earth Oxide Crystals (ZrO2-   000629
Melting Furnace for Purification and Crystal Growth (000871
Melting Metal Alloys and Compounds by Plasma Heating (000829
Melting Method (DyFe2, TbFe2, Nd, Pr, Ce, cold crucible)   000620
Melting Point and the Vacancy Mechanism of Melting (000351
Melting Without a Crucible (Si-Ge)*   Single Crystals of   000820

PERMUTED TITLE INDEX

#Computer Simulation of Convection in Floating Zone   Melting. I. Pure Rotation Driven Flows (Si, Al2O3,   000173
#Computer Simulation of Convection in Floating Zone   Melting. II. Combined Free and Rotation Driven Flows (000174
                                       #Theory of   Melting, Vacancy Model*   001007
at Zero   #Thermocapillary Convection in Floating Zone   Melting: Influence of Zone Geometry and Prandtl Number   000175
LiNbO3 from Solution in Borate, Vanadate, and Tungstate   Melts (flux)*   #Crystallization of   001047
Impurities on the Crystallization of Supercooled Copper   Melts (Cu)*   #The Influence of   000485
for the Onset of Oscillatory Flow in Crystal Growth   Melts (NaNO3, theory)*   of a Critical Marangoni Number   000847
of Barium Tungstate from Solutions in Sodium Tungstate   Melts by Continuous Cooling (BaWO4, flux, kinetics,   000812
of Calcium Tungstate from Solutions in Sodium Tungstate   Melts by Continuous Cooling (CaWO4, flux)*   Growth   000810
Strontium Tungstate from Solutions in Sodium Tungstate   Melts by Continuous Cooling (SrWO4, flux)*   Growth of   000811
Vanadates and   #Solubility Curves in High-Temperature   Melts for the Growth of Single Crystals of Rare Earth   000265
Ferrite Single Crystals from BaO-B2O3-ZnO-NiO-Fe2O3   Melts with a Nonstoichiometric Content of Ferrite-   000107
of CuCr2Se4 Crystals Obtained from Solutions in   Melts*   #Properties   000050
grown by Liquid Phase Epitaxy from PbO-B2O3 Fluxed   Melts*   Imbalance for Ca,Ge-Substituted Garnet Films   000386
*   #Dynamic Sublimation Technique for Growing   Mercuric Iodide Crystals in Small Opened Ampoules (HgI2)   000376
Crystal Growth and Nuclear Spectroscopic Evaluation of   Mercuric Iodide Radiation Detectors. Part II. HgI2 (000834
#A Possible Method for the Growth of Homogeneous   Mercury Cadmium Telluride Single Crystals (Hg1-xCdxTe,   000292
#Crystal Structure of the   Mercury Nitro-Phosphate Hg4PO4NO3.H2O (melt)*   000261
#An X-Ray Topographic Assessment of Cadmium   Mercury Telluride (CdxHg1-xTe, Bridgman)*   000158
on the Compositional Variations in Crystals of Cadmium   Mercury Telluride (Cd1-xHgxTe, slow cooling)*   Speed   000074
Bridgman)*   #Polar Twins in Lithium Metasilicate and   Metagermanate (Li2SiO3, Li2GeO3) Single Crystals (000108
#Growth of Single Crystals of High Melting   Metal Alloys and Compounds by Plasma Heating (W, W-C)*   000829
#A High Temperature High Purity Source for   Metal Beam Epitaxy (Ag, film)*   000287
A New Cubic KSbO3 Derivative Oxide with No Metal-   Metal Bonding (flux)   Crystal Structure of La3Ru3O11:   000209
#Growth of Polyhedral   Metal Crystallites in Inactive Gas (Mg, review)*   001006
Modes and Localized Effective Charges of Transition-   Metal Dichalcogenides (MoS2, MoSe2, WS2, WSe2, HfSe2,   001000
Rb3FeF6,   #Flux Growth of Crystals of Some Transition   Metal Fluorides. Part 2 (KAlF4, KMnF3, RbFeF4, Rb2FeF5,   001064
Preparation and Characterization of the Group VA   Metal Hydrides and Deuterides (V-H, V-D, Nb-H, Nb-D, Ta-   000790
and Principal Physicochemical Properties of Alkali   Metal Indates and Thioindates (LiInS2, Bridgman-   000509
Studies and Structural Systematics of the C2/c Alkali   Metal Metavanadates (NaxLi1-xVO3)*   #Crystallographic   000692
bcc Metals (V, Fe,   #Crystal Structure and Habit of Fine   Metal Particles Formed by Gas-Evaporation Technique: 000821
#Crystal Growth and Characterization of the Transition-   Metal Phosphides CuP2, NiP2, and RhP3 (flux, vapor   000708
phosphides, flux)*   #Ternary Lanthanoid-Transition   Metal Pnictides with ThCr2Si2-Type Structure (000596
and Electrical Properties of Layered Transition   Metal Selenides VxTi1-xSe2 and TaxTi1-xSe2 (TiSe2,   000555
cold   #Growth and Characterization of Transition   Metal Silicides (chemical vapor transport, solution,   000607
of Growth of Calcium Sulfate Crystals at Heated   Metal Surfaces (solution, CaSO4)*   #Kinetics   000334
Potassium, Rubidium and Cesium (review, #On the Alkali   Metal Tungsten Bronzes, in Particular those of   000402
xVx)4O7: Electrical, Calorimetric, Magnetic and EPR #   Metal-Insulator Transitions and Phase Diagram of (Ti1-   000835
vapor transport)*   #Structural Aspects of the   Metal-Insulator Transitions in (Ti0.9975V0.0025)4O7 (000391
La3Ru3O11: A New Cubic KSbO3 Derivative Oxide with No   Metal-Metal Bonding (flux)   #The Crystal Structure of   000209
#Optical Properties of V6O13 Single Crystals in the   Metallic and the Semiconducting Phase*   001023
of Cellular Structures of Monocrystals of Dilute Binary   Metallic Alloys (Cu - 0.5%Si, Cu1-xSbx - x = 0.0015-0.   000210
a Molten Mass in the (#Simulation of the Growth of a   Metallic Face-Centered Cubic Lattice in a Crystal from   000995
(SiO2, solution)*   #On the Incorporation of   Metallic Impurities in Synthetic Quartz Single Crystals   000170
the   #A Fluctuation Theory of Normal Crystal Growth for   Metallic One-Component Systems. Two Cases of Modelling   000059
and Vacancy and Diffusion Characteristics in Binary   Metallic Solid Solutions (theory)*   Quantities   000090
76Ge,   #Growth Characteristics of Crystals of Certain   Metallides (alpha-FeSi2.33, CoSi2, FeGe2, Mn5Si3, Fe1.   000861
#Purification and Characterization of   Metallurgical Silicon (Si, melt, Czochralski)*   000196
vapor transport)*   #Solar Cells from Zone-Refined   Metallurgical Silicon (Si, Czochralski, zone melting,   000195
Vacancy Mechanism of Melting (theory, compilation, 39   metals)*   of Metals Near the Melting Point and the   000351
a Modified Form of Zone Refining-Cascade Purification (metals)*   #Theoretical Solute Redistribution During   000201
Rate Limitations in Electrochemical Crystallization (metals, Nb, NaxWO3, Czochralski, electrolytic   000225
Particles Formed by Gas-Evaporation Technique: bcc   Metals (V, Fe, Cr, Mo and W)*   and Habit of Fine Metal   000821
and Impurities During Zone Recrystallization of   Metals and Preparation of Single Crystals Rich in   000994
Whiskers by Chemical   #Catalytic effects of Various   Metals and Refractory Oxides on the Growth of TiC   000959
of Melting (theory, compilation, 39 #Heat Capacity of   Metals Near the Melting Point and the Vacancy Mechanism   000351
of VPE-Grown GaN Doped with Al and Some Iron-Group   Metals*   #Properties   000649
#Crystal Structure of Scandium   Metaphosphate Sc(PO3)3 (solution)*   000902
#Crystal Growth and Characterization of Lithium   Metasilicate (Li2SiO3, Czochralski)*   000141
Single Crystals (Bridgman)*   #Polar Twins in Lithium   Metasilicate and Metagermanate (Li2SiO3, Li2GeO3)   000108
kinetics, theory)*   #Region of   Metastability During Normal Growth in a Binary Alloy (000969
Physical Properties*   #   Metastable Pb1-xCdxS Epitaxial Films. I. Growth and   000909
#Alkali Ion Exchange Reactions with RbAlSiO4: A New   Metastable Polymorph of KAlSiO4 (flux)*   000639
and Structural Systematics of the C2/c Alkali Metal   Metavanadates (NaxLi1-xVO3)*   #Crystallographic Studies   000692
Crystals and its Influence on Crystal   #Nitrilotri-   Methylenephosphonic Acid) Adsorption on Barium Sulfate (000553
TlAlF4, KAlF4,   #Hydrothermal Growth of the Systems   MF-AlF3 and NH4F-FeF3 in Hydrofluoric Acid Solution (000772
= Cl, Br, and I) (melt, #New Mixed Halide Compounds   MFX of Divalent Rare Earths (M = Sm, Eu, Tm, and Yb - X   000088
of Polyhedral Metal Crystallites in Inactive Gas (Mg, review)*   #Growth   001006
Needle Crystals*   #The Effects of M(M-   Mg, Ca and Ba) Ions on the KCl Flux Growth of MWO4   000715
of Large Diameter Grown by the Verneuil Technique (MgAl2O4, alpha-Al2O3)*   #Powder Feeder for Crystals   000284
A Modified Bridgman Method (ZnF2, CoF2, BaF2,   MgF2, KY3F10)*   Screen Translation (TST) Technique: 000541
Morphology and Crystal Habit of Struvite Crystals (MgNH4PO4.6H2O) (solution, theory)*   #Growth   000002
Harmonic Generation in a Sputtered LiNbO3 Film on   MgO (epitaxy)*   #Second   000518
of Secondary Nuclei Production (KAl(SO4)2.12H2O,   MgSO4.7H2O, solution)*   #Direct Observation   000322
#Crystal Growth and EPR Study of Cr+-Doped   MgTe Single Crystals (sublimation)*   000133
Analysis and Properties of High Purity Zn(1-x)   MgxTe Alloys (distillation)*   #Preparation,   000792
#Luminescence and Bandgap Studies of High Purity   MgxZn1-xTe Ternary Alloys (Bridgman)*   000600
#Growth of   Mg2TiO4 Single Crystals by the Floating Zone Method*   000880
Some Aspects of Polytypism in Crystals (SiC, ZnS, CdI2,   mica, theory, review)*   #   000069
Basic Aqueous Solutions Under Hydrothermal Conditions --   mica, KMg3AlSi3O10(OH)2)*   #Solubility of Phlogopite in   000070
from Gas Phase Under Non-Stoichiometric Conditions --   Microanalytical Evaluation of Composition (vapor   000082
Carbide as Revealed by Field Electron Microscopy (W   microcrystal, crystal growth study)*   Ditungsten   000724
Study of Some Molybdenum Oxide Crystals (MoO3, Mo4O11,   microcrystals*   #Electron Microscopy   000539
Czochralski Silicon Crystals* #The Effect of Doping on   Microdefect Formation in As-Grown Dislocation-Free   000223
Pulled GaSb Crystals (Czochralski)*   #Observations of   Microfacets Near Irregularly Remelted Surfaces in   000517
#Digital Simulation of Czochralski Bulk Flow in   Microgravity (Al2O3, theory, kinetics)*   000536
#Vapour Growth in a   Microgravity Environment (review)*   000091
Laser: Room-Temperature cw Operation at 1.07- and 1.35-   micrometer Wavelength (laser melting, Czochralski)*   000916

PERMUTED TITLE INDEX

DH Lasers with (100) Substrate in the Range of 1.2-1.5
a Compound Holographic Interference and Conventional
Interfaces with a Very High Voltage Electron
#Kinetics of Crystal Growth.
of Synthetic 11 Angstroms Tobermorite (#Electron
Water Solvent (solution)*   #Electron
Ditungsten Carbide as Revealed by Field Electron
#Techniques for Routine UHV in situ Electron
SiO2, hydrothermal)*   #Scanning Electron
of the Tellurium Solidus Line (#Transmission Electron
%oO3, Mo4O11, microcrystals)*   #Electron
Ln2O3 (Ln = Lanthanide): Unidirectional Solidification,
(film, epitaxy, Cu2O)*   #Recent Developments in the
and KCl (NaCl-KCl   #The Growth, X-Ray Studies and
of   #Control of Structure and Properties. Controlling
Different Two-Dimensional Substrate   #Capture of
#Epitaxy of NdAl3(BO3)4 For Thin Film
Oxide (Tm2O3, hydrothermal)*   #Effect of
#Vapor Growth and Epitaxy of
Crystals of Si*   #Influence of Growth Conditions on
Spacelab (CdTe, zone melting,   #A Double-Ellipsoid
#The Role of Lattice
of Epitaxial Garnet Layers as Functions of the
Substrate* #Composition Latching Phenomenon and Lattice
Non-Parallel Setting for Measuring Local Lattice
Direct Measurement of Temperature Dependence of Lattice
Thermionic Emission (LaB6,   #Single-Crystal Growth of
#Preparation, Crystal Growth and Characterization of
#Growth and Polytypism in Vapor-Grown
#Solution Growth of ZnS, ZnSe, CdS and Their
#Growth and Properties of Pb(WO4)1-x(MoO4)x
#Crystal Growth and Assessment of SnxPb1-xTe
Transport*   #Crystallization of HgCdTe
#Formation of
X-Ray Studies and Microstructure of Solution-Grown
Bridgman Method - Pseudobinary #Growth of Homogeneous
#Structural Phase Transition in
Sm, Eu, Tm, and Yb - X = Cl, Br, and I) (melt,   #New
melt, slow   #Synthesis and Crystal Structure of a New
Structure   #Synthesis and Crystal Structure of a New
Pressure (UCr2O6,   #Hydrothermal Synthesis of Several
LaCrOS2, CeCrOS2, PrCrOS2, NdCrOS2, SmCrOS2)*   #
and Crystallographic Data of Some Ln2(MoO4)3 Type
#Growth and Characterization of
#Magnetic
Hexagonal Structure Type (flux)*   #La8Ru4O21: A
Alloy Crystals by Hydrogen Reduction of a CuI-NiBr2
Growth of Cr2O3 Crystals from a K2Cr2O7-K2CrO4-K2B4O7
2HX4 and NH3(CH2)nNH3MX4 Families with M = Cd, Cu, Fe,
Structure of the Phosphate of Trivalent Manganese
transport*   #Formation of Amorphous
Quarternary Chalcogenides Cu2(I)B(II)C(IV)X4 (BII=Zn,
15, theory)*   #Magnetic Properties of (
#Magnetic Properties of (Mn,Cr)P and (
#The Growth Figures of
Properties and Crystal Growth of A(II)B(VI)-
#Layer Perovskites of the (CnH2n+1NH3)2MX4 and NH3(CH2)
Oxides (NdBO3, LaBO3/Pr, LaBO3, (La,Pb)MnO3, (Nd,Pb)
Group Complex Oxides (NdBO3, LaBO3/Pr, LaBO3, (La,Pb)
LaBO3/Pr, LaBO3, (La,Pb)MnO3, (Nd,Pb)MnO3, (Eu,Pb)
Growth and Crystal Chemistry of NiAs-Type Compounds:
Zn2TiO4 Substrates (flux)*   #Thin Film Growth of (
#Twinning and Crystal Structure of
Magnetic Properties of (Mn,Cr)P and (Mn,Fe)P Compounds (
#Crystal Growth of
of Certain Metallides (alpha-FeSi2.33, CoSi2, FeGe2,
by Gas-Evaporation Technique: bcc Metals (V, Fe, Cr,
Mo9Se11 Units*   #New Ternary
Very High Pressure (UCr2O6, UFe2O6, UMn2O6, UMnNiO6,
#Study of Melting and Freezing in the Gaussian Core
and Experimental Investigations (theory)*   #A
Solution (Y3Fe5O12, RExY3-xGayFe5-yO12, #A Solubility
Analysis of Crystallization of the Soft-Core
#Computer
the Vapour Phase (theory)*   #Flow-Graph
#New
theory, kinetics)*   #New
Melt (#A New Look at the Burton, Prim, and Slichter
System Based on Flow Graph (vapor transport, theory, #
#Theory of Melting, Vacancy
Growth for Metallic One-Component Systems. Two Cases of
#The Validity of Steady-State Dendrite Growth
New Technique for Simulating Two-Dimensional Nucleation
on "The Validity of Steady-State Dendrite Growth
Metal Dichalcogenides (MoS2, MoSe2,   #Optical Phonon
Bridgman, theory)*   #Local
#Physical Methods Used for the Characterization of
(theory)*   #

micrometers* LPE Growth for Lattice Matched GaInAsP/InP   000035
Microscope (solution)* Perovskite (C3H7NH3)2MnCl4 Using   000640
Microscope*   #In-Situ Observation of Solidification 000163
Microscopic and Phenomenological Theories (review)*   000660
Microscopic Characterization of the Hydrothermal Growth   000366
Microscopic Characterization of Cubic BN Prepared by   000934
Microscopy (W microcrystal, crystal growth study)*   000724
Microscopy of Growth Processes of Epitaxial Thin Films*   000950
Microscopy of Growth Surfaces of Synthetic Quartz (000857
Microscopy Observations of Precipitates in ZnTe: Shape   000098
Microscopy Study of Some Molybdenum Oxide Crystals (000539
Microstructural and Crystallographic Characterization (000630
Microstructural Characterization of Corrosion Processes   001045
Microstructure of Solution-Grown Mixed Crystals of NaCl   000407
Microstructures and Properties of Superalloys via Use   000671
Migrating Adatoms by Point Capture Centres for   000873
Miniature Lasers (flux)*   000577
Mineralizers on the Growth Morphology of Cubic Thulium   001039
Minerals and Synthetic Crystals (review, theory)*   000932
Mirodefect Distribution in Disloation-Free Single   000687
Mirror Furnace for Zone Crystallization Experiments in   000726
Misfit in Epitaxy (review, theory)*   001014
Mismatch of Crystal Lattice Parameters((Y, Bi, Sm, Ca)3(000251
Mismatch Effects in LPE-Grown In1-xGaxAs on InP   000952
Mismatches Between Epitaxial Films and Substrates (RE   000704
Mismatches Between LPE-Grown Li(Nb,Ta)O3 Film and   000922
Mixed (La, Eu, Y, Ce, Ba, Cs) Hexaborides for   000727
Mixed Bromides of Rare Earths, Sodium, and Cesium (000626
Mixed Cadmium Halide Crystals (CdI2-xBrx, CdI2-xClx)*   000870
Mixed Compounds Using Tellurium as a Solvent*   001065
Mixed Crystals (Czochralski)*   000709
Mixed Crystals (PbxSn1-xTe, review)*   000490
Mixed Crystals from the Vapour Phase by Chemical   000342
Mixed Crystals in the System CuxZn1-xCr2Se4*   000722
Mixed Crystals of NaCl and KCl (NaCl-KCl eutectic)*   000407
Mixed Crystals with a Large Segregation Tendency by the   000244
Mixed Crystals WxMo1-xO3 (melt)*   000824
Mixed Halide Compounds MPX of Divalent Rare Earths (M =   000088
Mixed Oxide "FeV3O8" (FexV1-xO2 - x approx. = 0.25) (000661
Mixed Oxide FeV2O6O0.5: Relation to the Diaspore   000662
Mixed Oxides of the A(6+)B2(3+)O6 Type Under Very High   000205
Mixed Oxysulfides of Chromium III and Rare Earths (001048
Mixed Rare Earth Molybdates (RE(MoO4)3, Czochralski)*   000447
Mixed Sr:Ca Tartrates (Sr1-xCazC4H4O6.4H2O, gel)*   000747
Mixed Valent TaSe (melt)*   000079
Mixed-Valence Ternary Ruthenium Oxide of a New   000208
Mixture (Cu-Ni, dendrites, whiskers)* of Copper-Nickel   000364
Mixture*   000860
Mn or Pd and X = Cl or Br: Importance, Solubilities and   000039
Mn(PO3)3 (flux)*   #Crystal   000058
Mn-Bi Films and Their Crystallization Process (vapor   001103
Mn, Fe, Co - CIV = Si, Ge, Sn - X = S,Se) (Cu2ZnGeS4,   000357
(Mn,Cr)P and (Mn,Fe)P Compounds (Mn1-xFexP, x = 0 to 0.   000430
(Mn,Fe)P Compounds (Mn1-xFexP, x = 0 to 0.15, theory)*   000430
MnAs (whiskers, theory)*   000067
MnB(VI) Systems (review)*   #Physicochemical   000740
mMH3MX4 Families with M = Cd, Cu, Fe, Mn or Pd and X =   000039
MnO3, (Eu,Pb)MnO3, Pb3Mn7O15, Tm2Ti3O7, Tb2Ti2O7,   001061
MnO3, (Nd,Pb)MnO3, (Eu,Pb)MnO3, Pb3Mn7O15, Tm2Ti3O7,   001061
MnO3, Pb3Mn7O15, Tm2Ti3O7, Tb2Ti2O7, Pr2Ti2O7,   001061
MnSb, CoSb, and NiSb (Czochralski)*   #Crystal   000180
MnxZn1-x)Fe2O4 by Liquid Phase Epitaxy on Bridgman Grown 000216
Mn0.75Ga2.17S4 (flux)*   000247
Mn1-xFexP, x = 0 to 0.15, theory)*   # 000430
Mn15Si26 (vapor transport, Bridgman)*   000495
Mn5Si3, Fe1.76Ge, melt, theory, kinetics)* of Crystals   000861
Mo and W)*   and Habit of Fine Metal Particles Formed   000821
Mo(II)-Compounds InxMo15Se19 Containing Mo6Se8 and   000186
MoCr2O6)* Mixed Oxides of the A(6+)B2(3+)O6 Type Under   000205
Model by Molecular Dynamics Simulation (melt, slow   000915
Model for Macroscopic Czochralski Growth: Theoretical   001002
Model for Rare-Earth Iron Garnets in a PbO/B2O3   001022
Model in an FCC Crystal Formation (theory)*#Geometrical   000965
Model of Autoepitaxial Growth of Ge Films (theory)*   000018
Model of Epitaxial Growth of A(III)B(V) Compounds from   000504
Model of LPE Growth: Growth Rate Calculation (theory)*   000590
Model of LPE Growth: Layer Thickness Calculation (000589
Model of Segregation During Crystal Growth from the   001084
Model of Silicon Epitaxial Growth in SiCl4-HCl-H2   000503
Model*   001007
Modelling the Melt Growth Interface Motion*   Crystal   000059
Models (theory)*   000832
Models of Crystal Growth (theory)*   #A   001038
Models" (theory)*   #Reply to Comment of R. Trivedi 000833
Modes and Localized Effective Charges of Transition-   001000
Modes in Rb1-cKc Alloys: A Neutron Scattering Study (000459
Modes of Epitaxial Growth from the Vapor Phase (review,   000542
Modes of Vibrations in Step Trains: Rhythmical Bunching   000933

PERMUTED TITLE INDEX

#(111) Cu2O Growth Modes on (111) Cu Surfaces (epitaxy)* 000388
PrBO3, LaBO3, DyKMo2O8, ErKMo2O8, NdKMo2O8, #Effects of Modifying Starting Compositions for Flux Growth (DyPO4, 001060
GaP and Its Application to Double Heterostructure Light Modulators* #Liquid Phase Epitaxy of AlxGa1-xP on 000948
the Gas Phase. II. Phosphorus Content and Differential Molar Solution Enthalpies (Si, vapor transport, theory)* 000515
#Periodic Doping Structure in GaAs (GaAs, Ga1-xAlxAs, molecular beam epitaxy, review)* 000248
Range for Growth of Autoepitaxial GaAs Films by MBE (molecular beam epitaxy, theory)* #Temperature 000686
Evaporation of Alkali Halides (KCl, LiF)* # Molecular (beam) Processes of Condensation and 000628
Under Ultra-High Vacuum (film)* # Molecular Beam and Solid-Phase Epitaxies of Silicon 000883
(#Impurity Dopant Incorporation and Diffusion During Molecular Beam Epitaxial Growth of IV-VI Semiconductors 000898
Implantation and #n-Type Doping Techniques in Silicon Molecular Beam Epitaxy by Simultaneous Arsenic Ion 000734
#Growth Temperature Dependence in Molecular Beam Epitaxy of Gallium Arsenide (GaAs)* 000666
Molecular Beam Epitaxy of GaSb and GaSbxAs1-x (films)* 001100
Film Superlattices* # Molecular Beam Epitaxy of Ge and Ga1-xAlxAs Thin 000761
* # Molecular Beam Epitaxy of InxGa1-xSb (x = 0 to 1) (GaSb) 001101
# Molecular Beam Epitaxy of ZnSexTe1-x(film)* 001102
#Design Considerations for Molecular Beam Epitaxy Systems (review)* 000576
Kinetics in (GaAs)n-(AlAs)m Superlattices Deposited by Molecular Beam Epitaxy. I. Growth on Singular (100) 000762
#Nitrogen Doping into GaAs1-xPx Using Ionized Beam in Molecular Beam Epitaxy* 000617
Mechanism of Silicon Epitaxial Layer Growth from Ion- Molecular Beams (Si)* #The 000019
of Melting and Freezing in the Gaussian Core Model by Molecular Dynamics Simulation (melt, slow cooling) 000915
vapor transport, kinetics)* #A Molecular Dynamics Study of Surface Melting (theory, 000140
Optical and Electrical Properties -Lithium and Molecular Intercalations (VOCl) Oxychloride: Magnetic, 001031
#Influence of Isotopic Composition on the Density and Molecular Volume of Solid Crystalline Lithium Nitrate 000023
Crystal-Aluminum Schottky-Barrier Diodes Prepared by Molecular-Beam Epitaxy (MBE) on GaAs (Al, films)*Single- 000190
#Preparation of Silicon Plates by Pulling from a Molten Bath According to the Stepanov Procedure (Si)* 000966
Face-Centered Cubic Lattice in a Crystal from a Molten Mass in the (100) Orientation (theory)* Metallic 000995
YAl-Borate Crystals in Preparation from Solutions in Molten Potassium Trimolybdate ((Y,Nd)Al3(BO3), flux)* 000551
Conditions for Stable Growth of Epitaxial GaP Layers by Molten Salt Electrodeposition (flux)* # 000224
of NbO and Hollow NbO2 Needles from Molten Salts (flux)* #Electrocrystallization 000204
melt, review, theory)* #The Molten State of Matter. Melting and Crystal Structure (000999
#A Complexity in the Solidification Behaviour of Molten Y3Al5O12 (Czochralski)* 000202
Peltier Effect (Ge)* #Observations of the Movement of a Molten Zone in Germanium Under the Influence of the 000061
#Crystal Growth and Imperfections in Gadolinium Molybdate and Other Isotypic Rare-Earth Molybdates (000446
#Crystal Structure of the Double Molybdate K2Ni(MoO4)2 (slow cooling, flux)* 000486
#Growth Kinetics and Habit Modification of Barium Molybdate Single Crystals in Silica Gel (BaMoO4)* 000519
#The Crystal Structure and Twinning Behavior of Ferric Molybdate, Fe2(MoO4)3 (hydrothermal)* 000178
in Gadolinium Molybdate and Other Isotypic Rare-Earth Molybdates (Czochralski, melt, Gd2(MoO4)3, Sm2(MoO4)3, 000446
Data of Some Ln2(MoO4)3 Type Mixed Rare Earth Molybdates (RE(MoO4)3, Czochralski)* Crystallographic 000447
Transport and Crystal Growth of Molybdenum Trioxide and Molybdenum Ditelluride (MoO3-x, MoTe2)* #Vapor Phase 000299
Magnetic and Electrical #New Thiochlorides of Molybdenum II: Mo6Cl10Y (Y = S, Se, Te): Structure and 000757
#Electron Microscopy Study of Some Molybdenum Oxide Crystals (MoO3, Mo4O11, microcrystals)* 000539
MoTe2)* #Vapor Phase Transport and Crystal Growth of Molybdenum Trioxide and Molybdenum Ditelluride (MoO3-x, 000299
Evaporating Crystal #Interaction Between Successive Monatomic Step Trains of Different Density on 000105
During Vapor Phase Silicon Epitaxial Growth #In-Process Monitoring and Control of Doping Gas Concentration 000937
and the Crystal Structures of Two Modifications of Gold Monobromide, I-AuBr and P-AuBr (vapor transport)*Growth 000436
Arsenide by Chemical Vapor Deposition from the Complex Monochlorodiethylgallium-Triethylarsine (GaAs)* Gallium 001109
#Crystal Structure of Monophosphate AgCoPO4* 000986
#Si Contamination in Epitaxial Boron Monophosphide (BP)* 000713
transport)* #Crystal Growth of Boron Monophosphide Using a B2H6-PH3-H2 System (BP, vapor 000884
#Growth of Single Crystals of Lead Monoxide (PbO, flux)* 000717
* #Fast Growth of Ordered AB Crystals: A Monte Carlo Simulation for Ionic Crystal Growth (theory) 000182
the Roughening Transition (C, theory)* # Monte Carlo Simulation of a (111) Diamond Face Around 001020
a Lennard-Jones Substance (theory)* # Monte Carlo Simulation of the Crystal-Melt Interface of 000123
Growth (review, theory, kinetics)* #Survey of Monte Carlo Simulations of Crystal Surfaces and Crystal 001013
#Viscosity Measurements in the System Dy2O3-K2O- MoO3 (DyKMo2O8, melt)* 000320
of RKMo2O8, R2Mo06 and R6Mo012 in the Systems R2O3-K2O- MoO3 (DyKMo2O8, ErKMo2O8, NdKMo2O8, Dy2Mo06, Sm2Mo06, 001062
of Molybdenum Trioxide and Molybdenum Ditelluride (MoO3-x, MoTe2)*Vapor Phase Transport and Crystal Growth 000299
MoO4)2 Single Crystals from the Ternary System Nd2O3- MoO3-Na2CO3 (flux, slow cooling)* #Growth of NaNd(000716
Microscopy Study of Some Molybdenum Oxide Crystals (MoO3, Mo4O11, microcrystals)* #Electron 000539
Particles Prepared by a Gas-Evaporation Technique (MoO3, WO3, theory)* #Coalescence Growth of Smoke 000448
#Growth and Properties of Pb(WO4)1-x(MoO4)x Mixed Crystals (Czochralski)* 000709
#Crystal Structure of the Double Molybdate K2Ni(MoO4)2 (slow cooling, flux)* 000486
#Acoustooptic Characteristics of LiBi(MoO4)2 (Czochralski)* 000016
Conditions* #Solubility of KLa(MoO4)2 in Aqueous K2MoO4 Solutions in Hydrothermal 000475
MoO3-Na2CO3 (flux, slow cooling)* #Growth of NaNd(MoO4)2 Single Crystals from the Ternary System Nd2O3- -000716
and Twinning Behavior of Ferric Molybdate, Fe2(MoO4)3 (hydrothermal)* #The Crystal Structure 000178
#Crystal Growth and Crystallographic Data of Some Ln2(MoO4)3 Type Mixed Rare Earth Molybdates (RE(MoO4)3, 000447
(Czochralski, melt, Gd2(MoO4)3, Sm2(MoO4)3, Tb2(MoO4)3)* and Other Isotypic Rare-Earth Molybdates 000446
of Some Ln2(MoO4)3 Type Mixed Rare Earth Molybdates (RE(MoO4)3, Czochralski)* Growth and Crystallographic Data 000447
Isotypic Rare-Earth Molybdates (Czochralski, melt, Gd2(MoO4)3, Sm2(MoO4)3, Tb2(MoO4)3)* Molybdate and Other 000446
Earth Molybdates (Czochralski, melt, Gd2(MoO4)3, Sm2(MoO4)3, Tb2(MoO4)3)* Molybdate and Other Isotypic Rare- 000446
#Dielectric and Optical Properties of K5Nd(MoO4)4 (Czochralski)* 000582
Whiskers to Dendrites in Zinc Reduction Growth (Cu, # Morphological Change from Copper/alpha-copper-Zinc 000481
Supercooling During the Crystal Growth of InSb from # Morphological Instability Under Constitutional 000478
from a Binary Alloy Melt (dendrites, theory)* # Morphological Stability of a Solid Particle Growing 000990
Deposition Processes. II (theory)* # Morphological Stability Analysis in Chemical Vapour 000435
Deposition Processes. I (theory)* # Morphological Stability Analysis in Chemical Vapour 001010
Vapour (theory)* # Morphological Stability Analysis of Growth from the 001009
#Studies of LPE Ripple Based on Morphological Stability Theory* 000698
and Properties of Solid State Materials, Vol. 4, Morphological Stability, Convection, Graphite, and 001081
#Crystals of CTA-DNA: A Growth and Morphological Study (solution)* 000488
MgNH4PO4.6H2O) (solution, theory)* #Growth Morphology and Crystal Habit of Struvite Crystals (000002
Produced by a Solid-Solid Diffusion Method. I. Morphology and Crystallographic Relationships Between 000372
Crystals as Affected by Bi Impurity)* #Growth, Morphology and Growth Mechanism of Vapor Grown Cd 000375
Crystal (vapor transport)* #Growth, Morphology and Slip System of alpha-Si3N4 Single 000696
Phase* # Morphology and Structure of ZnS Grown from the Vapour 000874
Substrates (theory)* #Interface Morphology in Chemical Vapour Deposition on Profiled 001011
Reduction of a CuI-NiBr2 Mixture (Cu-Ni, #Growth and Morphology of Copper-Nickel Alloy Crystals by Hydrogen 000364

PERMUTED TITLE INDEX

Context	Entry	No.
#Effect of Mineralizers on the Growth	Morphology of Cubic Thulium Oxide (Tm203, hydrothermal)*	001039
#Surface	Morphology of GaAs Grown by Vapor Phase Epitaxy*	000705
(flux, theory, kinetics)*	Morphology of GaAs Layers Grown by Liquid-Phase Epitaxy	000605
Paracelsian (Sr[Al2Si208]*	Morphology of Hydrothermally Grown Strontium	000304
#The	Morphology of HCl Etched Silicon Wafers (Si)*	001015
#Surface	Morphology of Ice Droxtals Grown from Supercooled Water	000347
#	Morphology of Iron Pyrite Crystals (FeS2, vapor	001093
Droplets (H20)*	Morphology of Spherical Crystals of omega-Phase in the	000512
transport)*	Morphology of Vapor-Deposited Niobium Nitride (NbN)*	000947
Cd-Hg System*	Morphology of VLS Silicon Crystals Grown with Platinum	001075
#Growth and	Morphology of Weddellite (CaC2O4.2XH2O, gel)*	000302
#Growth Parameters and Crystal	Morphology of 6H-SiC Epitaxial Layers by CVD (film)*	000699
and Gold as #Some Notes on the Growth Kinetics on	Morphology Control of NdP5O14 Single Crystals Grown	000466
#Growth	Morphology During Crystallization - I. Single Filament,	000992
#Growth and	Morphology During Crystallization - II. Single	000993
from Polyphosphoric Acids (solution)*	MoSe2, WS2, WSe2, HfSe2, ZrSe2, TiSe2, vapor transport)*	001000
Unconstrained Growth from a Pure Melt (#Interface	Mossbauer Studies (NH4H2PO4, solution)* #Mechanism	000296
Filament, Unconstrained Growth from a Binary #Interface	Mossbauer Studies of FeNi2BO5 and FeNiBO4 (ferrites,	000005
Charges of Transition-Metal Dichalcogenides (MoS2,	Mossbauer Studies of Sr(Fe0.766W0.234)O3 and its	001001
of Habit Change of ADP Crystals by Fe3+, Based on	Mossbauer Study of FeMo2S4 (vapor transport)*	000006
vapor transport)* #Magnetic and	MoS2 and 2H-WSe2 Between 10 and 320 K*	000667
Related Compounds (#Crystal Growth, and Magnetic and	MoS2, MoSe2, WS2, WSe2, HfSe2, ZrSe2, TiSe2, vapor	001000
#The Thermal Expansion of 2H-	MoTe2)* #Vapor Phase Transport and Crystal Growth of	000299
Effective Charges of Transition-Metal Dichalcogenides (	Motion* Growth for Metallic One-Component Systems.	000059
Molybdenum Trioxide and Molybdenum Ditelluride (MoO3-x,	Movement of a Molten Zone in Germanium Under the	000061
Two Cases of Modelling the Melt Growth Interface	Mo1-xWxSe2 (vapor transport)* Conditions and Crystal	000013
Influence of the Peltier Effect (#Observations of the	Mo4O11, microcrystals)* #Electron Microscopy	000539
Structure Parameters of Layer Compounds in the Series	Mo6C110Se, melt)*of Molybdenum,II: Mo6C110Y (Y = S, Se,	000757
Study of Some Molybdenum Oxide Crystals (MoO3,	Mo6C110Y (Y = S, Se, Te): Structure and Magnetic and	000757
Te): Structure and Magnetic and Electrical Properties (	Mo6Se8 and Mo9Se11 Units*	000186
Electrical #New Thiochlorides of Molybdenum II:	Mo9Se11 Units* #New Ternary	000186
#New Ternary Mo(II)-Compounds InxMo15Se19 Containing	Multi-Steps on the Surface of Vapour Grown FeCr2S4*	000417
Mo(II)-Compounds InxMo15Se19 Containing Mo6Se8 and	Multi-Steps on the Surface of Vapour Grown FeCr2S4*	000419
#Growth Mechanism of	Multilayered Structures in the (GaAl)As System by LPE (	000892
#Step Growth by Elementary Steps and	Multiple Constituents (theory)*	000181
Ga1-xAlxAs, flux)#An Automated System for the Growth of	MWO4 Needle Crystals* #The Effects	000715
#Kinetics of Crystallization of Alloys of	M2M'Ta5O15. Crystal Structure of CaTlTa5O15 (flux)*	000317
of M(M=Mg, Ca and Ba) Ions on the KCl Flux Growth of	M3B7O13X)* #Preparation and Physical	000151
#A New Structural Family	n+ Back-Surface-Field Silicon Solar Cells (Si)*	000298
Properties of Boracite Crystals (Chemical transport,	n-(AlAs)m Superlattices Deposited by Molecular Beam	000762
#High-Efficiency p+-n-	n-n+ Back-Surface-Field Silicon Solar Cells (Si)*	000298
Epitaxy. I. Growth on #Crystal Growth Kinetics in (GaAs)	n-Type CuInSe2 Single Crystals (melt)*	000410
#High-Efficiency p+-	n-Type Doping Techniques in Silicon Molecular Beam	000734
#Electrical Properties of p- and	N(CH3)4)2CoCl4*	000831
Epitaxy by Simultaneous Arsenic Ion Implantation and #	N(CH3)4)2ZnCl4 (evaporation)*	000830
#Ferroelectricity in (	Na+ and Co2+ plus Zr4+, Grown by Liquid-Phase Epitaxy (	000691
#Ferroelectricity in (	NaAl11O17, KAl11O17, RbAl11O17, AgAl11O17, EuAl11O17,	000621
flux)* #Thin Films of LiNbO3, Doped with	NaBO2 (flux, Na2Fe2Ti6O14, Na2NiTi7O16)* #Growth	000602
from Alkali, Silver, and Europium beta-Alumina	NaBO2.H2O2.3H2O)*II. Growth Kinetics of Different Faces	000258
Conditions for Sodium-Titanium Bronze Crystals in	NaBO2.H2O2.3H2O, kinetics, theory)* I. Nucleation	000257
in Pure Solution and in the Presence of a Surfactant (	NaBr, LiI, KCl, NaI, KBr, KI)* Pt, Fe, Al, Au, Ag, In,	000949
Rates in Pure Solution in the Presence of a Surfactant (	NaCl (Au, Ag, vapor transport)*Tilt Boundaries Produced	000207
Pb, PbS, PbSe, PbTe, SnTe, LiF, NaF, KF, LiBr, NaCl,	NaCl and KCl (NaCl-KCl eutectic)* Growth, X-Ray Studies	000407
by Epitaxial Growth on Bicrystalline Substrates of	NaCl Substrates (CdS, CdSe, CdTe, ZnSe)* Growth of Some	000878
and Microstructure of Solution-Grown Mixed Crystals of	NaCl)*	000654
II-VI Compound Films Evaporated Onto Electron-Bombarded	NaCl-KCl eutectic)* X-Ray Studies and Microstructure	000407
#Preparation of Fine Sodium Chloride Crystals (	NaCl, theory)* #Anisotropy	000530
of Solution-Grown Mixed Crystals of NaCl and KCl (	NaCl, theory)* Between Successive Monatomic Step Trains	000105
of Round Evaporation Spirals on Rocksalt Surfaces (	NaCl, theory, melt, Czochralski)* Solidification Front	000128
of Different Density on Evaporating Crystal Surfaces (	NaCl, Czochralski)* Growth Lamellae on the Crystal-	000532
Shape on the Temperature Distribution in the Crystal (	NaCl, NaBr, LiI, KCl, NaI, KBr, KI)*Pd, Pt, Fe, Al, Au,	000949
Melt Interface of LiCl Doped Rock Salt Crystals (	NaClO3)* #Convection Phenomena During	000179
Ag, In, Pb, PbS, PbSe, PbTe, SnTe, LiF, NaF, KF, LiBr,	NaClO3)* #Kinetics of Growth and Dissolution of Sodium	000888
the Growth of Sodium Chlorate Crystals from Solution (	NaClO3, solution)* #Crystal Growth and	000784
Chlorate in Diffusion and Convection Regimes (solution,	NaCrF4, Rb2Cr5F17)* RbxFeF3, CsFeF4, CsFe2F7, Cs3Fe2F9,	001064
Selective Etching Studies on Sodium Chlorate Crystals (	NaF, KF, LiBr, NaCl, NaBr, LiI, KCl, NaI, KBr, KI)* Pd,	000949
CsxFeF3, Na2CoFeF7, Na2NiFeF7, Na2NiAlF7, Na2ZnCrF7,	NaI, KBr, KI)* Pt, Fe, Al, Au, Ag, In, Pb, PbS, PbSe,	000949
Pt, Fe, Al, Au, Ag, In, Pb, PbS, PbSe, PbTe, SnTe, LiF,	NaNbO3 (NaxTh0.25(1-x)NbO3, melt, slow cooling)*	000527
PbTe, SnTe, LiF, NaF, KF, LiBr, NaCl, NaBr, LiI, KCl,	NaNd(MoO4)2 Single Crystals from the Ternary System	000716
of Perovskite Phases of the Systems Th0.25NbO3-	NaNO3, theory)* #Experiments on	000848
Nd2O3-MoO3-Na2CO3 (flux, slow cooling)* #Growth of	NaNO3, theory)* of a Critical Marangoni Number for	000847
Surface Tension Driven Flow in Floating Zone Melting (	NaNO3, PbCl2, organics)* #An Inexpensive Zone	000871
the Onset of Oscillatory Flow in Crystal Growth Melts (	National Economy, Leningrad, March 14-17, 1979 (review)*	001111
Melting Furnace for Purification and Crystal Growth (	Nature, Origin and Effect of Dislocations in Epitaxial	000126
Profiled Crystals and Products and Their Application in	NaxLi1-xVO3)* #Crystallographic Studies and Structural	000692
Semiconductor Layers (III-V, review)*	NaxTh0.25(1-x)NbO3, melt, slow cooling)* Chemistry	000527
Systematics of the C2/c Alkali Metal Metavanadates (	NaV2O5*	000907
of Perovskite Phases of the Systems Th0.25NbO3-NaNbO3 (	NaxV2O5, KxV2O5, AgxV2O5, theory)* Specific Heat and	000171
#Bridgman and Czochralski Growth of beta-Phase	NaxWO3, Czochralski, electrolytic deposition)*	000225
Magnetic Susceptibility of Nonmetallic Vanadium Bronze (	Na0.5AlSi6O14.xH2O)* #Growth of Fibrous	000444
in Electrochemical Crystallization (metals, Nb,	Na14[MnO4]2O*	000135
Sodium Aluminosilicate Crystals in Silica Gel (zeolite,	Na14Mn2O9 = Na14[MnO4]2O*	000135
#The First Oxomanganate(II): Na14Mn2O9	Na2CoFeF7, Na2NiFeF7, Na2NiAlF7, Na2ZnCrF7, NaCrF4,	001064
#The First Oxomanganate(II):	Na2CoGeO4 Single Crystal Growth and Study of Their	000226
Rb3FeF6, RbxFeF3, CsFeF4, CsFe2F7, Cs3Fe2F9, CsxFeF3,	Na2CoGeO4 Single Crystals in Relation to Their Growth	000233
Piezoelectric and Elastic Properties (hydrothermal)*	Na2CO3 (flux, slow cooling)* #Growth of NaNd(MoO4)	000716
Ray Topographic Study of Growth Defects in Hydrothermal	Na2CO3 Crystals (Bridgman)*	000632
2 Single Crystals from the Ternary System Nd2O3-MoO3-	Na2Fe2Ti6O14, Na2NiTi7O16)* #Growth Conditions	000602
#The Melt Growth and Thermal Expansion of		
for Sodium-Titanium Bronze Crystals in NaBO2 (flux,		

PERMUTED TITLE INDEX

Germanate Crystals and Glasses (Na2Ge4O9, Na4Ge9O20,	Na2Ge2O5)* #Structure and Stability of Alkali	000822
and Stability of Alkali Germanate Crystals and Glasses (	Na2Ge4O9, Na4Ge9O20, Na2Ge2O5)* #Structure	000822
#X-Ray and Neutron Diffraction of	Na2MnFeF7: A Variety of Trigonal Weberite (flux)*	001033
#Flux Growth of Lead Neodymium Chlorapatite (	Na2Nd2Pb6(PO4)6Cl, slow cooling, evaporation)*	000445
CsFe2F7, Cs3Fe2F9, CsxFeF3, Na2CoFeF7, Na2NiFeF7,	Na2NiAlF7, Na2ZnCrF7, NaCrF4, Rb2Cr5F17)* CsFeF4,	001064
RbxFeF3, CsFeF4, CsFe2F7, Cs3Fe2F9, CsxFeF3, Na2CoFeF7,	Na2NiFeF7, Na2NiAlF7, Na2ZnCrF7, NaCrF4, Rb2Cr5F17)*	001064
Titanium Bronze Crystals in NaBO2 (flux, Na2Fe2Ti6O14,	Na2NiTi7O16)* #Growth Conditions for Sodium-	000602
#Hydrothermal Crystallization in the System BaO-GeO2-	Na2O-H2O (barium germanates, Ba3Ge9O21.H2O, BaGeO3.H2O,	000592
#Crystal Growth of Sodium Sulphide,	Na2S (hygroscopic, Bridgman)*	001056
Cs3Fe2F9, CsxFeF3, Na2CoFeF7, Na2NiFeF7, Na2NiAlF7,	Na2ZnCrF7, NaCrF4, Rb2Cr5F17)*RbxFeF3, CsFeF4, CsFe2F7,	001064
Crystallization of Sodium Zirconogermanates (	Na2ZrGe2O7, Na4Zr2Ge3O12, Na3HZr[GeO4]2)*	000230
#Growth of Single Crystals of	Na2ZrGeO5. Na2ZrGe2O7, Na4Zr2Ge3O12, Na3Zr[GeO4]2)*	000230
Zirconogermanates (Na2ZrGeO5. Na2ZrGe2O7, Na4Zr2Ge3O12,	Na3Al2Li3F12 Garnet Under Hydrothermal Conditions*	000675
of a New High -Nd- Concentration Laser Material:	Na3HZr[GeO4]2)* #Hydrothermal Crystallization of Sodium	000230
#New Ferroelectric	Na3Nd(PO4)2* #The Crystal Structure	000825
	Na3Sc2(PO4)3 (melt)*	000721
Seeds Coated with Informative Interfacial Layers (	Na4Al3Si3O12Cl)* Hydrothermal Method on Single Crystal	000243
Na2O-H2O (barium germanates, Ba3Ge9O21.H2O, BaGeO3.H2O,	Na4Ge9O20)* Crystallization in the System BaO-GeO2-	000592
of Alkali Germanate Crystals and Glasses (Na2Ge4O9,	Na4Ge9O20, Na2Ge2O5)* #Structure and Stability	000822
of Sodium Zirconogermanates (Na2ZrGeO5. Na2ZrGe2O7,	Na4Zr2Ge3O12, Na3HZr[GeO4]2)* Crystallization	000230
#Preparation, Optical and Dielectric Properties of	Na5W3O9F5 (melt)*	000785
III. Characteristic Arrangement of Dislocations (	Nb)* Crystals by Pulling from a Melt on a Pedestal.	000684
Group VA Metal Hydrides and Deuterides (V-H, V-D, Nb-H,	Nb-D, Ta-H, Ta-D)* and Characterization of the	000790
zone melting, review)* #Properties and Synthesis of	Nb-H Interstitial Alloys (electron beam melting, float	001072
the Group VA Metal Hydrides and Deuterides (V-H, V-D,	Nb-H, Nb-D, Ta-H, Ta-D)* and Characterization of	000790
Niobium-Zirconium, Hafnium, and Tungsten Alloys (Nb-Zr,	Nb-Hf, electron beam melting, zone melting)* of	000307
of Niobium-Zirconium, Hafnium, and Tungsten Alloys (	Nb-Zr, Nb-Hf, electron beam melting, zone melting)*	000307
Niobium Oxides with TeCl4. I. Calculation of the System	Nb/O/Te/Cl and the Transport of NbO2/NbO2.417 (NbO2,	000797
Limitations in Electrochemical Crystallization (metals,	Nb, NaxWO3, Czochralski, electrolytic deposition)* Rate	000225
Dependence of Lattice Mismatches Between LPE-Grown Li	Nb,Ta)O3 Film and LiTaO3 Substrate (flux)* Temperature	000922
Substrates (LiNb1-xTaxO3, flux)* #LPE Growth of Li (	Nb,Ta)O3 Solid-Solution Thin Film Waveguides on LiTaO3	000500
#The Vapour Transport of	NbB2 and TaB2 (kinetics, theory)*	000042
#The Hardness of	NbB2 Single Crystals (flux, slow cooling)*	000679
Niobium Nitride and Niobium Carbide Nitride (NbN, NbN-	NbC, NbxNyCz, zone melting)* #Crystal Growth of	000192
#Superconductivity of Single-Crystal	NbGe2*	000791
Crystal Morphology of Vapor-Deposited Niobium Nitride (	NbN)* #Growth Parameters and	000947
of Niobium Nitride and Niobium Carbide Nitride (NbN,	NbN-NbC, NbxNyCz, zone melting)* #Crystal Growth	000192
Growth of Niobium Nitride and Niobium Carbide Nitride (	NbN, NbN-NbC, NbxNyCz, zone melting)* #Crystal	000192
#Electrocrystallization of	NbO and Hollow NbO2 Needles from Molten Salts (flux)*	000204
of the System Nb/O/Te/Cl and the Transport of NbO2/	NbO2 Needles from Molten Salts (flux)*	000204
TeCl4 (II) The Transport Behaviour of the Equilibrium	NbO2.417 (NbO2, NbO2.417, theory)*TeCl4. I. Calculation	000797
Nb/O/Te/Cl and the Transport of NbO2/NbO2.417 (NbO2,	NbO2.417 to NbO2.5* of Some Niobium Oxides with	000798
The Transport Behaviour of the Equilibrium NbO2.417 to	NbO2.417, theory)* TeCl4. I. Calculation of the System	000797
of the System Nb/O/Te/Cl and the Transport of	NbO2.5*Transport of Some Niobium Oxides with TeCl4 (II)	000798
System Nb/O/Te/Cl and the Transport of NbO2/NbO2.417 (	NbO2/NbO2.417 (NbO2, NbO2.417, theory)* I. Calculation	000797
and Phase Transitions in Single Crystals of (KxNa1-x)	NbO2, NbO2.417, theory)* TeCl4. I. Calculation of the	000797
Phases of the Systems Th0.25NbO3-NaNbO3 (NaxTh0.25(1-x)	NbO3 (flux)* #Growth	000056
theory)* #Kinetic Studies of the Intercalation of 2H-	NbO3, melt, slow cooling)* Chemistry of Perovskite	000527
of Upper Critical Field in Superconducting 2H-	NbSe2 with Hydrazine by Transmittance Measurements (	000085
#Raman Scattering in 3R-	NbS2 (theory)* #Anisotropy	000730
#The Crystal Structure of Niobium Trisulfide,	NbS2 (vapor transport, theory)*	000731
* #Crystal Growth and Phase Transitions in the	NbS3*	000814
Nitride and Niobium Carbide Nitride (NbN, NbN-NbC,	NbxCr1-xO2 (x greater than 0.5) System (vapor transport)	000095
and Properties of Niobium Chalcogenide Halides,	NbxNyCz, zone melting)* #Crystal Growth of Niobium	000192
15 Structure)* #Cold Working	NbX2Y2(X=S, Se - Y=Cl, Br, I) (vapor transport)*	000795
Phase Equilibria, and Crystal Chemistry of La, Pr, and	Nb3Al in the bcc Structure and then Converting to the A-	001067
#Epitaxial Layers of the Laser Material	Nd(Ga,Cr)(BO3)4 (flux)*	000578
Pr, and Nd Hydroxide Bromides and Hydroxide Iodides (	Nd(OH)2Br, Pr(OH)2Br, La7(OH)18Br3, Nd7(OH)18Br3, Pr7(	000531
from Solutions in Molten Potassium Trimolybdate ((Y,	Nd)Al3(BO3), flux)* YAl-Borate Crystals in Preparation	000551
#The Crystal Structure of a New High -	Nd- Concentration Laser Material: Na3Nd(PO4)2*	000825
Levitation Zone Melting Method (DyFe2, TbFe2,	Nd, Pr, Ce, cold crucible)* Growth by the Horizontal	000620
and Crystallographic Characterization (Ln =	Nd, Sm, Dy, skull-melting)* Microstructural	000630
Complex Oxides (NdBO3, LaBO3/Pr, LaBO3, (La,Pb)MnO3,	Nd,Pb)MnO3, (Eu,Pb)MnO3, Pb3Mn7O15, Tm2Ti3O7, Tb2Ti2O7,	001061
melt, slow cooling)* #Melt Growth of Nd:Y3Al5O12 (	Nd:YAG Using the Heat Exchange Method (HEM) (Y3Al5O12,	000167
Operation at 1.07- and 1.35-micrometer Wavelength (	Nd:Y2O3 Single-Crystal Fiber Laser: Room-Temperature cw	000916
HEM) (Y3Al5O12, melt, slow cooling)* #Melt Growth of	Nd:Y3Al5O12 (Nd:YAG) Using the Heat Exchange Method (	000167
#Epitaxy of	NdAl3(BO3)4 For Thin Film Miniature Lasers (flux)*	000577
of Some Rare-Earth and Iron Group Complex Oxides (	NdBO3, LaBO3/Pr, LaBO3, (La,Pb)MnO3, (Nd,Pb)MnO3, (Eu,	001061
III and Rare Earths (LaCrOS2, CeCrOS2, PrCrOS2,	NdCrOS2, SmCrOS2)* #Mixed Oxysulfides of Chromium	001048
Compounds, RCu2. II. Light Rare Earth (PrCu2,	NdCu2)* Properties of Rare Earth Copper Intermetallic	000734
#Crystal Growth and Magnetic Properties of	NdFeTiO5 (flux)*	001091
#Magnetic Properties of a Single Crystal of	NdF3 Between 2 and 4.2K (Stockbarger)*	000767
Flux Growth (DyPO4, PrBO3, LaBO3, DyKMo2O8, ErKMo2O8,	NdKMo2O8, DyVO4, TbVO4, TbPO4, DyPO4, Ho2Ge2O7,	001060
in the Systems R2O3-K2O-MoO3 (DyKMo2O8, ErKMo2O8,	NdKMo2O8, Dy2MoO6, Sm2MoO6, Pr2MoO6, Ho6MoO12)* R6MoO12	001062
(solution)* #Morphology Control of	NdP5O14 Single Crystals Grown from Polyphosphoric Acids	000466
of NaNd(MoO4)2 Single Crystals from the Ternary System	Nd2O3-MoO3-Na2CO3 (flux, slow cooling)* #Growth	000716
and Some New Stimulated Emission Data of Gd3Ga5O12:	Nd3+ Single Crystals (Nd3Ga5O12, doped, Czochralski)*	000457
Emission Data of Gd3Ga5O12:Nd3+ Single Crystals (	Nd3Ga5O12, doped, Czochralski)* and Some New Stimulated	000457
Oxides (GdVO3, Gd4Ba6(BO3)9, (Ba0.93Gd0.07)TiO3,	Nd4Ba6(BO3)9, Pr4Ba6(BO3)9, Ni8Nb3O15, Ni2V2Pb8,	001063
Hydroxide Iodides (Nd(OH)2Br, Pr(OH)2Br, La7(OH)18Br3,	Nd7(OH)18Br3, Pr7(OH)18Br3, La7(OH)18I3, Pr7(OH)18I3,	000531
in Compacted Powders* #Growth of a Crystal	Needle of Si2ON2 by Nitridization in the Si-SiO2 System	000049
#Vapor Growth of Hexagonal GeO2	Needle Crystals*	000423
M(M=Mg, Ca and Ba) Ions on the KCl Flux Growth of MWO4	Needle Crystals* #The Effects of	000715
#New Gel Method for Growing Large	Needles and Single Crystals of Lead Chloride (PbCl)*	000003
#Electrocrystallization of NbO and Hollow NbO2	Needles from Molten Salts (flux)*	000204
cooling, evaporation)* #Flux Growth of Lead	Neodymium Chlorapatite (Na2Nd2Pb6(PO4)6Cl, slow	000445
Preparation from Solutions in Molten Potassium	Neodymium Incorporation into YAl-Borate Crystals in	000551
#Dodecatungstophosphoric Acid-21-Water by	Neutron Diffraction (evaporation, H3PW12O40.21H2O)*	000913

PERMUTED TITLE INDEX

Weberite (flux)*   #X-Ray and
Structure and Magnetic Phase Transitions of Tb5Ge4. A
    of Heusler Alloys, Ni2MnSn and Ni2Mn1-xVxSn, for
Layer- Type Compounds K2Mn1-xMxF4 (M +   #Magnetic and
        #Local Modes in Rb1-cKc Alloys: A
                #Magnetization and
Transition of Rb2ZnBr4 (theory, solution)*   #
Kinetics of Urea Monocrystals from Aqueous Solution (CO (
and X = #Layer Perovskites of the (CnH2n+1NH3) 2MX4 and
    of Deuterium Ammonium Dihydrogen Arsenate (DADA) (
    the Solid Electrolytes of the RbAg4I5 Family (KAg4I5,
NH4F-FeF3 in Hydrofluoric Acid Solution (TlAlF4, KAlF4,
            #On the Growth of
    #Hydrothermal Growth of the Systems MF-AlF3 and
Acid Solution (TlAlF4, KAlF4, NH4AlF4, RbAlF4, CsAlF4,
    Phosphate Crystals from Aqueous Solution (ADP,
of ADP Crystals by Fe3+, Based on Mossbauer Studies (
        #Unstable Growth of ADP Crystals: A
    Heat of a Quasi-Two-Dimensional Antiferromagnet
of Nickel Single Crystals by Recrystallization Method (
Five-Component System: The Chemical Vapour Transport of
    by Hydrogen Reduction of a CuI-NiBr2 Mixture (Cu-
Growth of a Dilute Binary Alloy Single Crystal (Fe-
Perfection in Czochralski Grown Nickel Single Crystals (
PbTe,   #Roles of Lattice Fitting in Epitaxy (theory,
        #Crystal Growth and Crystal Chemistry of
Nickel Alloy Crystals by Hydrogen Reduction of a CuI-
Chemical Vapor Deposition (ZrC)* #Catalytic Effect of
NiBr2 Mixture (Cu-Ni, #Growth and Morphology of Copper-
NiFe2O4, theory)*   #On the Preparation of
    #Growth of Oriented Bicrystals of
    #Crystal Structure of a New Variety of
Magnetostriction and Elastic Constants of a Ruthenium-
    #Crystal Perfection in Czochralski Grown
        #The Growth of
Ni3B7O13Cl, #Gas Phase Synthesis of Epitaxial Layers of
    #The Binary Systems Cobalt-Gallium and
NiO-Fe2O3 Melts with a Nonstoichiometric #Growth of
Nickel Ferrite Single Crystals by Chemical Transport (
Cobalt-Gallium and Nickel-Gallium Compared (CoGa3,
Nickel-Zinc Ferrite Single Crystals from BaO-B2O3-ZnO-
    #Growth of Oriented Bicrystals of Nickel Oxide
    and Piezoelectric Properties of Potassium Lithium
    Niobate Single-Crystal Films on Potassium Bismuth
    Niobate Single-Crystal Films on Potassium Bismuth
Niobate Single #Epitaxial Growth of Potassium Lithium
Niobate Single #Epitaxial Growth of Potassium Lithium
    and Properties of Crystalline Rubidium and Cesium
Y3NbO7, Er3NbO7,   #Flux Growth of Double Oxides of
melting)*   #Crystal Growth of Niobium Nitride and
Br, I) (#Preparation, Structures, and Properties of
    Parameters and Crystal Morphology of Vapor-Deposited
NbC, NbxNyCz, zone melting)*   #Crystal Growth of
of the Equilibrium #On the Chemical Transport of Some
Nb/O/Te/Cl and the #On the Chemical Transport of Some
Pedestal. III. Characteristic Arrangement of #Growth of
    #The Crystal Structure of
Nb-Hf, electron beam melting,   #Elastic Constants of
    of the Transition-Metal Phosphides CuP2,
Chemistry of NiAs-Type Compounds: MnSb, CoSb, and
    #Crystallization Kinetics of Lithium and Strontium
    and Molecular Volume of Solid Crystalline Lithium
    and Crystal Morphology of Vapor-Deposited Niobium
#Crystal Growth of Niobium Nitride and Niobium Carbide
NbxNyCz, zone melting)*   #Crystal Growth of Niobium
    #Synthesis of Single Crystals of Cubic Boron
    #Cathodoluminescence of Impurity-Doped Aluminum
#Low Temperature Method for the Growth of Lithium
transport, flux)*   #
*   #Growth of a Crystal Needle of Si2ON2 by
Barium Sulfate Crystals and its Influence on Crystal #
NO2)6, Tl2NiPb(NO2)6, Tl2NiCd(#Crystal Growth of Triple
    #Crystal Structure of the Mercury
Molecular Beam Epitaxy*
a Nonstoichiometric Content of Ferrite-Forming Oxides (
        #Uniaxial Ferromagnetism in
#Growth of Large Single Crystals of Heusler Alloys,
of Large Single Crystals of Heusler Alloys, Ni2MnSn and
    Structure of a New Variety of Nickel Pyrophosphate:
93Gd0.07)TiO3, Nd4Ba6(BO3)9, Pr4Ba6(BO3)9, Ni8Nb3O15,
    Transport of Ni-Cl Boracite (no materials grown -
Nickel-Chlorine Boracite on Chromium-Chlorine Boracite (
    of the Flow Stress in the Intermetallic Compound
TiO3, Nd4Ba6(BO3)9, Pr4Ba6(BO3)9, Ni8Nb3O15, Ni2V2PbO8,
    #Crystallochemical Study of the System Ni3Pb2S2-
Ni3Pb2S2-Ni3Pb2Se2 and Note on the Shandite Structure (
Ni3Pb2Se2, vapor #Crystallochemical Study of the System
BO3)9, Pr4Ba6(BO3)9, Ni8Nb3O15, Ni2V2PbO8, Ni3Nb2O8,

Neutron Diffraction of Na2MnFeF7: A Variety of Trigonal   001033
Neutron Diffraction Study (melt)*   #On the Magnetic   000840
Neutron Inelastic Scattering Experiments (Czochralski,   000321
Neutron Scattering Experiments on the Antiferromagnetic 000106
Neutron Scattering Study (Bridgman, theory)*   000459
Neutron Studies of UTe and USb0.8Te0.2*   000154
Neutron-Scattering Study of the Incommensurate Phase   000226
NH2)2, theory)*   #The Growth   000392
NH3(CH2)mNH3MX4 Families with M = Cd, Cu, Fe, Mn or Pd   000039
NH4(1-x)D4xH2(1-x)D2xAsO4, solution)*   Phases   000100
NH4Ag4I5, solution)*   #Raman Scattering Studies in   000315
NH4AlF4, RbAlF4, CsAlF4, NH4Fe2F6)* Systems MF-AlF3 and   000772
NH4Cl Dendrites at Very Low Supersaturation (solution)* 000172
NH4F-FeF3 in Hydrofluoric Acid Solution (TlAlF4, KAlF4, 000772
NH4Fe2F6)*Systems MF-AlF3 and NH4F-FeF3 in Hydrofluoric 000772
NH4H2PO4)*   in the Growth of Ammonium Dihydrogen   000219
NH4H2PO4, solution)*   #Mechanism of Habit Change   000296
NH4H2PO4, theory, solution, kinetics, slow cooling)*   000804
Ni(OH)2, hydrothermal)*   #Specific   000273
Ni)*   #The Growth 000548
Ni-Cl Boracite (no materials grown - Ni3B7O13Cl theory)* 000236
Ni, dendrites, whiskers)* Copper-Nickel Alloy Crystals   000364
Ni, eutectics, theory, dendrites)*   of the Cellular 000113
Ni, theory)*   #Crystal   000521
Ni, Cu, Pd, Pt, Fe, Al, Au, Ag, In, Pb, PbS, PbSe,   000049
NiAs-Type Compounds: MnSb, CoSb, and NiSb (Czochralski)* 000180
NiBr2 Mixture (Cu-Ni, dendrites, whiskers)* of Copper- 000364
Nickel on the Growth of Zirconium Carbide Whiskers by   000958
Nickel Alloy Crystals by Hydrogen Reduction of a CuI-   000364
Nickel Ferrite Single Crystals by Chemical Transport (000760
Nickel Oxide (NiO, arc image, float zone, Verneuil)*   000239
Nickel Pyrophosphate: Ni2P2O7 (slow cooling)*   000609
Nickel Single Crystal (RuNi, Bridgman)*   Anisotropy, 000631
Nickel Single Crystals (Ni, theory)*   000521
Nickel Single Crystals by Recrystallization Method (Ni)* 000548
Nickel-Chlorine Boracite on Chromium-Chlorine Boracite (000837
Nickel-Gallium Compared (CoGa3, NiGa4, flux)*   000290
Nickel-Zinc Ferrite Single Crystals from BaO-B2O3-ZnO-   000107
NiFe2O4, theory)*   #On the Preparation of   000760
NiGa4, flux)*   #The Binary Systems   000290
NiO-Fe2O3 Melts with a Nonstoichiometric Content of   000107
NiO, arc image, float zone, Verneuil)*   000239
Niobate (KLN) Crystals (K2.89Li1.55Nb5.11O15,   000010
Niobate Single Crystals by the EGM Technique (KxLi1-   000011
Niobate Single Crystals by RF Sputtering (KxLi1-xNbO3)* 000009
Niobate Single-Crystal Films on Potassium Bismuth   000009
Niobate Single-Crystal Films on Potassium Bismuth   000011
Niobates (Czochralski, Cs8Nb22O59, Rb8Nb22O59)* Growth 000440
Niobium and Rare-Earth Elements (Ln3NbO7) (Ho3NbO7,   001097
Niobium Carbide Nitride (NbN, NbN-NbC, NbxNyCz, zone   000192
Niobium Chalcogenide Halides, NbX2Y2(X=S, Se - Y=Cl,   000795
Niobium Nitride (NbN)*   #Growth   000947
Niobium Nitride and Niobium Carbide Nitride (NbN, NbN-   000192
Niobium Oxides with TeCl4 (II) The Transport Behaviour   000798
Niobium Oxides with TeCl4. I. Calculation of the System 000797
Niobium Single Crystals by Pulling from a Melt on a   000684
Niobium Trisulfide, NbS3*   000814
Niobium-Zirconium, Hafnium, and Tungsten Alloys (Nb-Zr, 000307
NiP2, and RhP3 (flux, vapor transport)*Characterization 000708
NiSb (Czochralski)*   #Crystal Growth and Crystal   000180
Nitrate Crystal Hydrates (LiNO3.3H2O, Sr(NO3)2.4H2O)*   000989
Nitrate Hydrates (solution, LiNO3.3H2O, LiNO3.3D2O)*   000023
Nitride (NbN)*   #Growth Parameters   000947
Nitride (NbN, NbN-NbC, NbxNyCz, zone melting)*   000192
Nitride and Niobium Carbide Nitride (NbN, NbN-NbC,   000192
Nitride in Systems Containing Hydrogen (BN)*   000132
Nitride Films Produced by Reactive Evaporation (AlN)*   001104
Nitride Single Crystals (Li3N)*   000252
Nitrides--Structures and Crystal Growth (review, vapor   000533
Nitridization in the Si-SiO2 System in Compacted Powders 000049
Nitrilotri(Methylenephosphonic Acid) Adsorption on   000553
Nitrites in Gels (K2CuPb(NO2)6, Tl2CuPb(NO2)6, K2NiPb(000550
Nitro-Phosphate Hg4PO4NO3.H2O (melt)*   000261
Nitrogen Doping into GaAs1-xPx Using Ionized Beam in   000617
NixZn1-xFe2O4, flux)* BaO-B2O3-ZnO-NiO-Fe2O3 Melts with 000107
NiZrF6.6H2O (solution)*   000464
Ni2MnSn and Ni2Mn1-xVxSn, for Neutron Inelastic   000321
Ni2Mn1-xVxSn, for Neutron Inelastic Scattering   000321
Ni2P2O7 (slow cooling)*   #Crystal   000609
Ni2V2PbO8, Ni3Nb2O8, Ni3V2O8, Ta2Co4O9, PbCr2.3Ti2.309, 001063
Ni3B7O13Cl theory)*   System: The Chemical Vapour 000236
Ni3B7O13Cl, vapor transport)* of Epitaxial Layers of   000837
Ni3Ge Single Crystals (Bridgman)*Temperature Dependence 000033
Ni3Nb2O8, Ni3V2O8, Ta2Co4O9, PbCr2.3Ti2.309, Pb2CrO5)*   001063
Ni3Pb2Se2 and Note on the Shandite Structure (000198
Ni3Pb2S2-Ni3Pb2Se2 and Note on the Shandite Structure (000198
Ni3Pb2S2-Ni3Pb2Se2 and Note on the Shandite Structure (Study of the System 000198
Ni3V2O8, Ta2Co4O9, PbCr2.3Ti2.309, Pb2CrO5)*   Nd4Ba6(001063

PERMUTED TITLE INDEX

BO3)9, (Ba0.93Gd0.07)TiO3, Nd4Ba6(BO3)9, Pr4Ba6(BO3)9,
   Properties (transport, magnetic susceptibility, and
   The Chemical Vapour Transport of Ni-Cl Boracite (
of La3Ru3O11: A New Cubic KSbO3 Derivative Oxide with
Directionally Solidified InSb-GaSb (magnetic field
Experimental Data: Theoretical (directional
melt, slow cooling)*   #Dynamics of
                  #On the Transition from Faceted to
Mismatches   #An X-Ray Double-Crystal Method Utilizing
on a Quartz Surface (sublimation)*
sublimation)*   #Growth Study of Large
H2O, theory)*   #Appearance of
Evaluation of   #Growth of ZnSiP2 from Gas Phase Under
*   #Physico-Chemical Properties of the
   of Silicon Ribbon Crystals Grown with Wetting and
      Specific Heat and Magnetic Susceptibility of
Crystals from BaO-B2O3-ZnO-NiO-Fe2O3 Melts with a
Chemical Vapor Deposition (ZrC)*   #Synthesis of
MBE of Compound Semiconductors (II-VI, IV-VI, III-V,
Systems. Two Cases of   #A Fluctuation Theory of
                    #Region of Metastability During
                    #On the Theory of
Silicon Crystals Grown with Platinum and Gold as #Some
Si, vapor transport, kinetics)*
                                #A
                          #AlSiP3, A Compound with a
Transitions (solution)**Ultrasonic Velocities of K2PbCu(
6, Tl2NiCd(NO2)6, Tl2NiHg(NO2)6, Tl2CoPb(NO2)6, Tl2CoBa(
   of Triple Nitrites in Gels (K2CuPb(NO2)6, Tl2CuPb(
6, Tl2NiPb(NO2)6, Tl2NiCd(NO2)6, Tl2NiHg(NO2)6, Tl2CoPb(
6, K2NiPb(NO2)6, Tl2NiPb(NO2)6, Tl2NiCd(NO2)6, Tl2NiHg(
   #Crystal Growth of Triple Nitrites in Gels (K2CuPb(
(K2CuPb(NO2)6, Tl2CuPb(NO2)6, K2NiPb(NO2)6, Tl2NiPb(
6, Tl2CuPb(NO2)6, K2NiPb(NO2)6, Tl2NiPb(NO2)6, Tl2NiCd(
Nitrites in Gels (K2CuPb(NO2)6, Tl2CuPb(NO2)6, K2NiPb(
and Strontium Nitrate Crystal Hydrates (LiNO3.3H2O, Sr(
                            #The Structure of K2Er(
Radiation   #State-of-The-Art of Crystal Growth and
xSnxTe Heterostructures*   #Improved
                          #Kinetic Studies of
Silica Gels*
Orthophosphate Solutions (KDP)*
      #A New Technique for Simulating Two-Dimensional
   of Sodium Perborate from Aqueous Solutions. I.
solution)*   #Direct Observation of Secondary
                #Ionic Conductivity in
Oxides with TeCl4. I. Calculation of the System Nb/
(Nd(OH)2Br, Pr(OH)2Br, La7(OH)18Br3, Nd7(OH)18Br3, Pr7(
   and Hydroxide Iodides (Nd(OH)2Br, Pr(OH)2Br, La7(
Iodides (Nd(OH)2Br, Pr(OH)2Br, La7(OH)18Br3, Nd7(
OH)2Br, La7(OH)18Br3, Nd7(OH)18Br3, Pr7(OH)18Br3, La7(
Under Hydrothermal Conditions (mica, KMg3AlSi3O10(
Heat of a Quasi-Two-Dimensional Antiferromagnet Ni(
   of a New Pyrochlore Containing Rhenium: Ca1-xRe2O6(
Hydroxide Bromides and Hydroxide Iodides (Nd(OH)2Br, Pr(
Pr, and Nd Hydroxide Bromides and Hydroxide Iodides (Nd(
Hydrothermal Conditions (YOHCO3, GdOHCO3, ErOHCO3, Y2O(
   #Aging Studies on Hydrous Lutetium Oxide (cubic Lu(
   #Cerium Trihydroxide (Ce(
OH)18Br3, Nd7(OH)18Br3, Pr7(OH)18Br3, La7(OH)18I3, Pr7(
   #Growth and Morphology of Spherical Crystals of
   and Magnitude of a Critical Marangoni Number for the
kinetics, #Growth of Films of CdS Evaporated in Vacuum
                    #The Adsorption of Zr
Growth of Some II-VI Compound Films Evaporated
   #Vapor-Phase Growth of Epitaxial Ga1-xInxSb Using
Technique for Growing Mercuric Iodide Crystals in Small
Nd:Y2O3 Single-Crystal Fiber Laser: Room-Temperature cw
Crystal Growth (solution)*   #Fiber
                              #Preparation,
Crystals (hygroscopic, Bridgman)*
Molecular   #Vanadium (III) Oxychloride: Magnetic,
45Fe03 (flux)*
vapor transport, electrolytic   #The Electrical,
substituted Y3Al5O12, flux)*   #Growing
by a Close-Spaced Vapor Process*   #The
review, film)*   #Integrated
on the Growth Conditions (Bi12GeO20, #Dependence of the
Transition-Metal Dichalcogenides (MoS2, MoSe2, WS2,
PbX3) (CsPbCl3,   #Crystal Structure, Growth, and
Solid Solutions (#Phase Studies, Crystal Growth, and
Crystals Grown from Near-Stoichiometric #Electrical and
                              #Some
Crystals of GeFe2O4 (vapor transport)* #Electrical and
Czochralski)*   #Magneto-
                  #Dielectric and

Ni8Nb3O15, Ni2V2PbO8, Ni3NbO8, Ni3VO8, Ta2CoO409,   001063
NMR) of the Spinel CuV2S4 (vapor transport)* Physical 000544
no materials grown - Ni3B7O13Cl theory)*   System: 000236
No Metal-Metal Bonding (flux)* #The Crystal Structure 000209
Non-Constant Distribution Coefficients for   000864
Non-Constant Distribution Coefficients from   001080
Non-Equilibrium Crystallization of an Alloy (theory, 000826
Non-Faceted Growth in Melt-Grown Crystals (organic)* 000355
Non-Parallel Setting for Measuring Local Lattice   000704
Non-Seeded Growth of Large Single Pb1-xSnxTe Crystals 000960
Non-Seeded Pb1-xSnxTe Single Crystals (vapor transport, 000961
Non-Singular Surfaces on Vapour-Grown Ice Crystals (000689
Non-Stoichiometric Conditions -- Microanalytical   000802
Non-Stoichiometric VS2 and V5S8 Phases (vapor transport) 000470
Non-Wetting Dies (Si, EDFG)*Growth and Characterization 000524
Nonmetallic Vanadium Bronze (NaxV2O5, KxV2O5, AgxV2O5, 000171
Nonstoichiometric Content of Ferrite-Forming Oxides (000107
Nonstoichiometric Zirconium Carbide Whiskers by   000674
Nonstoichiometry and Carrier Concentration Control in 000897
Normal Crystal Growth for Metallic One-Component   000059
Normal Growth in a Binary Alloy (kinetics, theory)*   000969
Normal Growth of Crystals in Binary Systems (kinetics)* 001110
Notes on the Growth Kinetics and Morphology of VLS   001075
Novel Reactor for High Volume Low-Cost Silicon Epitaxy (000062
Novel Ruthenium Bronze: KRu4O8 (solution)*   001082
Novel Wurtzite-Pyrite Intergrowth Structure (flux)*   001044
NO2)6 Near the Cooperative Jahn-Teller Phase   000467
NO2)6)* NO2)6, Tl2CuPb(NO2)6, K2NiPb(NO2)6, Tl2NiPb(NO2) 000550
NO2)6, K2NiPb(NO2)6, Tl2NiPb(NO2)6, Tl2NiCd(NO2)6,   000550
NO2)6, Tl2CoBa(NO2)6)* NO2)6, Tl2CuPb(NO2)6, K2NiPb(NO2) 000550
NO2)6, Tl2CoPb(NO2)6, Tl2CoBa(NO2)6)**NO2)6, Tl2CuPb(NO2) 000550
NO2)6, Tl2CuPb(NO2)6, K2NiPb(NO2)6, Tl2NiPb(NO2)6,   000550
NO2)6, Tl2NiCd(NO2)6, Tl2NiHg(NO2)6, Tl2CoPb(NO2)6,   000550
NO2)6, Tl2NiHg(NO2)6, Tl2CoPb(NO2)6, Tl2CoBa(NO2)6)*NO2 000550
NO2)6, Tl2NiPb(NO2)6, Tl2NiCd(NO2)6, Tl2NiHg(NO2)6,   000550
NO3)2.4H2O)*   #Crystallization Kinetics of Lithium 000989
NO3)5 (flux)*   000872
Nuclear Spectroscopic Evaluation of Mercuric Iodide   000834
Nucleation and a Planar Interface in LPE Grown Pb1-   000962
Nucleation and Growth at Surfaces (review)*   001030
Nucleation and Growth of KClO4 Single Crystals in   000751
Nucleation in Supersaturated Potassium Dihydrogen   000443
Nucleation Models of Crystal Growth (theory)*   001038
Nucleation Rates in Pure Solution in the Presence of a 000257
Nuclei Production (KAl(SO4)2.12H2O, MgSO4.7H2O,   000322
N3O+ Beta Alumina (solution)*   000286
O/Te/Cl and the Transport of NbO2/NbO2.417 (NbO2, NbO2. 000797
OH)18Br3, La7(OH)18I3, Pr7(OH18I3, hydrothermal)*   000531
OH)18Br3, Nd7(OH)18Br3, Pr7(OH)18Br3, La7(OH)18I3, Pr7(000531
OH)18Br3, Pr7(OH)18Br3, La7(OH)18I3, Pr7(OH18I3,   000531
OH)18I3, Pr7(OH18I3, hydrothermal)*   (Nd(OH)2Br, Pr(000531
OH)2)*   of Phlogopite in Basic Aqueous Solutions 000070
OH)2, hydrothermal)*   #Specific   000273
OH)2x (x = 0.3)* #High Pressure Synthesis and Structure 000104
OH)2Br, La7(OH)18Br3, Nd7(OH)18Br3, Pr7(OH)18Br3, La7(000531
OH)2Br, Pr(OH)2Br, La7(OH)18Br3, Nd7(OH)18Br3, Pr7(OH) 000531
OH)2CO3, Er2O2CO3)*   of Rare-Earth Carbonates Under 000169
OH)3)*   000663
OH)3, gel)*   000664
OH18I3, hydrothermal)*   (Nd(OH)2Br, Pr(OH)2Br, La7(000531
omega-Phase in the Cd-Hg System*   000512
Onset of Oscillatory Flow in Crystal Growth Melts (000847
onto Silicon (vapor transport, electron beam melting, 000073
onto W(100) Surfaces*   000221
Onto Electron-Bombarded NaCl Substrates (CdS, CdSe,   000878
Open-Tube Flow System (vapor transport)*   000681
Opened Ampoules (HgI2)*   #Dynamic Sublimation   000376
Operation at 1.07- and 1.35-micrometer Wavelength (000916
Optic Probe for Thermal Profiling of Liquids During   000395
Optical and Dielectric Properties of Na5W3O9F5 (melt)* 000785
Optical and Dilatometric Studies of KCaCl3 and RbCaCl3 000633
Optical and Electrical Properties -Lithium and   001031
Optical and Electrical Properties of CdGeAs2 (Bridgman)* 000412
Optical and Magneto-Optical Properties of Sm0.55Tb0. 000914
Optical and Photoconducting Properties of Fe2-xCrxO3 (000624
Optical films of Aluminum Yttrium Garnet (RE   000044
Optical Absorption Edge of Single-Crystal AlN Prepared 000758
Optical Devices Fabricated by MBE (GaAs, Ga1-xAlxAs, 000152
Optical Homogeneity of Single Germanosillenite Crystals 000903
Optical Phonon Modes and Localized Effective Charges of 001000
Optical Properties of Alkali-Trihalogenoplumbates M(000377
Optical Properties of CdGe(As1-xPx)2 and AgGa(Se1-xSx)2 000635
Optical Properties of Copper Indium Ditelluride   000220
Optical Properties of Cs2NaBiCl6*   000756
Optical Properties of GeS2 Single Crystals (Bridgman)* 000697
Optical Properties of High-Purity p-type Single   000920
Optical Properties of KTb3F10 and LiTbF4 Crystals (001068
Optical Properties of K5Nd(MoO4)4 (Czochralski)*   000582

PERMUTED TITLE INDEX

PERMUTED TITLE INDEX

#High-Efficiency	p+-n-n+ Back-Surface-Field Silicon Solar Cells (Si)*	000298
#Electrical Properties of	p- and n-Type CuInSe2 Single Crystals (melt)*	000410
#Electrical and Optical Properties of High-Purity	p-type Single Crystals of GeFe2O4 (vapor transport)*	000920
Method* #Growth of	p-Type InP Single Crystals by the Temperature Gradient	000422
theory)* #Magnetic Properties of (Mn,Cr)	P and (Mn,Fe)P Compounds (Mn1-xFexP, x = 0 to 0.15,	000430
#Magnetic Properties of (Mn,Cr)P and (Mn,Fe)	P Compounds (Mn1-xFexP, x = 0 to 0.15, theory)*	000430
Properties of Binary, Ternary and Quaternary (In,Ga)(As,	P) Alloys Grown by the Hydride Vapor Phase Epitaxy	000729
of Two Modifications of Gold Monobromide, I-AuBr and	P-AuBr (vapor transport)* and the Crystal Structures	000436
Arsenides by Chemical Vapour Transport (PaAs2, Pa3As4,	PaAs)* and Single-Crystal Growth of Protactinium	000161
of Protactinium Arsenides by Chemical Vapour Transport (	PaAs2, Pa3As4, PaAs)* and Single-Crystal Growth	000161
#Dynamics of Laser-Induced Formation of	Palladium Silicide (Pd2Si)*	001042
#Selected	Papers on Polymer Crystallization (Review)*	001089
Ammonium Dihydrogen Arsenate (#Crystal Growth of the	Para- and Antiferroelectric Phases of Deuterium	000100
#The Morphology of Hydrothermally Grown Strontium	Paracelsian (Sr[Al2Si2O8]*	000304
Between #An X-Ray Double-Crystal Method Utilizing Non-	Parallel Setting for Measuring Local Lattice Mismatches	000704
theory, kinetics)* #Effect of the Buoyancy	Parameter on Czochralski Bulk Flow in Garnet Growth (	000535
VO2 near #Accurate X-ray Determination of the Lattice	Parameters and the Thermal Expansion Coefficients of	000513
Niobium Nitride (NbN)* #Growth	Parameters and Crystal Morphology of Vapor-Deposited	000947
Growth of Silicon #Interaction Between Technological	Parameters and Crystal Quality During the Crystal	000325
vapor #Growth Conditions and Crystal Structure	Parameters of Layer Compounds in the Series Mo1-xWxSe2 (	000013
theory, eutectics, Al-Al3Ni)* #Effect of Some Thermal	Parameters on the Directional Solidification Process (	000268
Pb1-xSnx)1+yTe1-y System* #Influence of the VPE	Parameters on the Epitaxial Film Properties for the (	000433
Closed Ampules (GeS, #Measurements of Growth	Parameters During the Crystal Growth from the Vapor in	000841
Layers as Functions of the Mismatch of Crystal Lattice	Parameters((Y, Bi, Sm, Ca)3(Fe, Ge)5 O12, flux)* Garnet	000251
theory)* #Morphological Stability of a Solid	Particle Growing from a Binary Alloy Melt (dendrites,	000990
Barium Sulphate (BaSO4, #Effect of S-35 Tracer,	Particle Size and Imperfection on the Solubility of	000779
Growth of Co Ferrite on Fine Acicular gamma-Fe2O3	Particles (solution)* #Crystal	000931
Metals (V, #Crystal Structure and Habit of Fine Metal	Particles Formed by Gas-Evaporation Technique: bcc	000821
MoO3, WO3, theory)* #Coalescence Growth of Smoke	Particles Prepared by a Gas-Evaporation Technique (	000448
Preparation of II-IV-V2 Compounds by Gas Phase #	Partition of Principal Components in the Course of	001086
theory)* #Solute	Partitioning During Silicon Dendritic Web Growth (Si,	000855
review)* #Spacelab and	Past and Future Crystal Growth Experiments in Space (	000854
Arsenides by Chemical Vapour Transport (PaAs2,	Pa3As4, PaAs)*and Single-Crystal Growth of Protactinium	000161
of Solidification of Pb-Bi Alloys Rich in Single-Phase	Pb in the Course of Solidification (Czochralski)*Region	000846
#Growth and Properties of	Pb(WO4)1-x(MoO4)x Mixed Crystals (Czochralski)*	000709
Layers by the Liquid Phase Epitaxy Technique (flux, (Ba,	Pb)Al12O19* Barium Lead Hexa-Aluminate Single Crystal	000071
Oxides (NdBO3, LaBO3/Pr, LaBO3, (La,Pb)MnO3, (Nd,	Pb)MnO3, (Eu,Pb)MnO3, Pb3Mn7O15, Tm2Ti3O7, Tb2Ti2O7,	001061
Iron Group Complex Oxides (NdBO3, LaBO3/Pr, LaBO3, (La,	Pb)MnO3, (Nd,Pb)MnO3, (Eu,Pb)MnO3, Pb3Mn7O15, Tm2Ti3O7,	001061
(NdBO3, LaBO3/Pr, LaBO3, (La,Pb)MnO3, (Nd,Pb)MnO3, (Eu,	Pb)MnO3, Pb3Mn7O15, Tm2Ti3O7, Tb2Ti2O7, Pr2Ti2O7,	001061
and Concentration in the Region of Solidification of	Pb-Bi Alloys Rich in Single-Phase Pb in the Course of	000846
#Magneto-Seebeck Effect of	Pb-Doped Bi-Sb (Bridgman)*	000034
#Liquidus Measurements in the	Pb-Sn-Te System (Pb1-xSnxTe, PbTe, flux)*	000047
theory, Pb1- #Investigation of Liquidus Surface in	Pb-Sn-Te System by Simplex Lattice Method (flux,	000668
#Liquidus of the Ternary System	Pb-Tl-Te*	000101
Unconstrained Growth from a Binary Alloy Melt (Sn-	Pb, theory)* Crystallization - II. Single Filament,	000993
in Epitaxy (theory, Ni, Cu, Pd, Pt, Fe, Al, Au, Ag, In,	Pb, PbS, PbSe, PbTe, SnTe, LiF, NaF, KF, LiBr, NaCl,	000949
#Ferroelectric (	Pb,Ba)5Ge3O11 Crystals (flux)*	000155
#Epitaxial Growth of Ferroelectric PLZT ((	Pb,La)(Zr,Ti)O3) Thin Films*	000415
Large Needles and Single Crystals of Lead Chloride (	PbCl)* #New Gel Method for Growing	000003
Furnace for Purification and Crystal Growth (NaNO3,	PbCl2, organics)* #An Inexpensive Zone Melting	000871
9, Ni8Nb3O15, Ni2V2PbO8, Ni3Nb2O8, Ni3V2O8, Ta2Co4O9,	PbCr2.3Ti2.309, Pb2Cr05)*TiO3, Nd4Ba6(BO3)9, Pr4Ba6(BO3)	001063
#The Role of B2O3 in PbO/	PbF2/B2O3 Fluxes (Y3Al15O12, theory)*	000272
#Crystal Data and Crystal Growth of	PbGe3O7 (melt)*	000738
#Low Temperature Thermal Properties of	PbI2 (Bridgman)*	000851
#Polytypism and the Vibrational Properties of	PbI2*	000850
Garnet Films grown by Liquid Phase Epitaxy from	PbO-B2O3 Fluxed Melts* Imbalance for Ca,Ge-Substituted	000386
#A Solubility Model for Rare-Earth Iron Garnets in a	PbO/B2O3 Solution (Y3Fe5O12, RExY3-xGayFe5-yO12, flux)*	001022
#The Role of B2O3 in	PbO/PbF2/B2O3 Fluxes (Y3Al15O12, theory)*	000272
#Growth of Single Crystals of Lead Monoxide (	PbO, flux)*	000717
#Liquid Encapsulation Zone-Refining of	PbS (zone melting, vapor transport, sublimation)*	000583
(theory, Ni, Cu, Pd, Pt, Fe, Al, Au, Ag, In, Pb,	PbS, PbSe, PbTe, SnTe, LiF, NaF, KF, LiBr, NaCl, NaBr,	000949
#Electroless Deposition of Semiconductor Films (	PbS, PbSe, Pb1-xHgxS, solution)*	000868
IV-VI Compounds in a Quasi-Sealed Volume (SnTe, PbTe,	PbSe, evaporation)* #Preparation of Films of	000306
(theory, Ni, Cu, Pd, Pt, Fe, Al, Au, Ag, In, Pb, PbS,	PbSe, PbTe, SnTe, LiF, NaF, KF, LiBr, NaCl, NaBr, LiI,	000949
#Electroless Deposition of Semiconductor Films (PbS,	PbSe, Pb1-xHgxS, solution)*	000868
of IV-VI Compounds in a Quasi-Sealed Volume (SnTe,	PbTe, flux)* #Liquidus	000047
Measurements in the Pb-Sn-Te System (Pb1-xSnxTe,	PbTe, PbSe, evaporation)* #Preparation of Films	000306
Beam Epitaxial Growth of IV-VI Semiconductors (PbTe,	PbTe, Pb1-xSnxSe, Pb1-xSnxTe)* During Molecular	000898
Ni, Cu, Pd, Pt, Fe, Al, Au, Ag, In, Pb, PbS, PbSe,	PbTe, SnTe, LiF, NaF, KF, LiBr, NaCl, NaBr, LiI, KCl,	000949
#Photoconduction of Lead Trititanate (	PbTi3O7, flux)*	000264
Growth and Assessment of SnxPb1-xTe Mixed Crystals (	PbxSn1-xTe, review)* #Crystal	000490
#A Series of Lead Tungsten Bronzes (	PbxWO3)*	000267
and Optical Properties of Alkali-Trihalogenoplumbates M (	PbX3) (CsPbCl3, CsPbBr3, CsPbI3, KPbI3, RbPbI3, vapor	000377
Crystals Obtained by the Traveling Solvent Method from (	Pb0.8Sn0.2) Te* #Semiconductor Single	000064
evaporation)* #Growth of	Pb0.8Sn0.2Te Films in a Quasi-Closed Space (vacuum	000885
#Electrical Properties of	Pb1-xCdxS Epitaxial Films (vapor transport)*	000439
Properties* #Metastable	Pb1-xCdxS Epitaxial Films. I. Growth and Physical	000909
#Metastable	Pb1-xCdxS Epitaxial Films. II. Electrical Properties*	000910
#Electroless Deposition of Epitaxial	Pb1-xHgxS Films*	000869
Deposition of Semiconductor Films (PbS, PbSe,	Pb1-xHgxS, solution)* #Electroless	000868
Parameters on the Epitaxial Film Properties for the (	Pb1-xSnx)1+yTe1-y System* #Influence of the VPE	000433
Beam Epitaxial Growth of IV-VI Semiconductors (PbTe,	Pb1-xSnxSe, Pb1-xSnxTe)* and Diffusion During Molecular	000898
#Non-Seeded Growth of Large Single	Pb1-xSnxTe Crystals on a Quartz Surface (sublimation)*	000960
Improved Nucleation and a Planar Interface in LPE Grown	Pb1-xSnxTe Heterostructures* #	000962
sublimation)* #Growth Study of Large Non-Seeded	Pb1-xSnxTe Single Crystals (vapor transport,	000961
Sn-Te System by Simplex Lattice Method (flux, theory,	Pb1-xSnxTe)* #Investigation of Liquidus Surface in Pb-	000668
Growth of IV-VI Semiconductors (PbTe, Pb1-xSnxSe,	Pb1-xSnxTe)* Diffusion During Molecular Beam Epitaxial	000898
#Liquidus Measurements in the Pb-Sn-Te System (	Pb1-xSnxTe, PbTe, flux)*	000047

PERMUTED TITLE INDEX

Travelling Solvent Method (theory)*   #Growth of
#X-Ray and Dielectric Investigations of
Ni2V2Pb08, Ni3Nb208, Ni3V208, Ta2Co409, PbCr2.3Ti2.309,
Czochralski, flux)*   #High Temperature Form of
Diagram and Crystal Growth of Lead Phosphovanadates (
Growth of Lead Phosphovanadates (Pb3(PxV1-x04)2,
#Crystal Structures of Pb4In9S17 and
LaBO3/Pr, LaBO3, (La,Pb)MnO3, (Nd,Pb)MnO3, (Eu,Pb)MnO3,
#Crystal Structures of
#Crystal Data and Crystal Growth of
Growth of the Ferroelectric SbNbO4, SbTaO4 and
Silicide Formation by High-Dose Si+-Ion Implantation of
and NH3(CH2)mNH3MX4 Families with M = Cd, Cu, Fe, Mn or
#Roles of Lattice Fitting in Epitaxy (theory, Ni, Cu,
of Laser-Induced Formation of Palladium Silicide (
#Isotope Effects on the Central
Boron Crystals (B, float zone melting)*   #Some
slow cooling)*   #Structure and Growth
of Niobium Single Crystals by Pulling from a Melt on a
#High Efficiency GaAs Thin Film Solar Cells by
kinetics)#Liquid-Phase Electroepitaxy: Growth Kinetics of
a Molten Zone in Germanium Under the Influence of the
#Crystal Growth of Some Alkaline-Earth Rare-Earth
and Crystal Growth of Laser Quality Praseodymium
in Pure Solution in the   #Crystallization of Sodium
of Different Faces in Pure   #Crystallization of Sodium
#Crystal Growth of Potassium
#Gel Growth and Perfection of Orthorhombic Potassium
specific heat, melting, theory)*   #
#Profile-Sapphire and Its Structural
the Czochralski Technique and the Assessment of their
#Growth of USb2 Single Crystals and Their Structural
Ni, theory)*   #Crystal
Garnet (#The Effect of Melt Flow Phenomena on the
Crystals (KC104)*   #Gel Growth and
Rate and Direction*   #Dependence of Degree of
kinetics)*   #Preferred Growth Directions and
#Growth of High
Interface of LiCl Doped Rock Salt Crystals (NaCl, #Semi-
molecular beam epitaxy, review)*   #
Distributions in Melt-Grown Crystals, Taking Account of
Method for the Determination of Precipitation Induction
#Face Stability and Growth Rate Variations of the Layer
#Observations on Step Bunching for the Layer
NaxTh0.25(1-x)NbO3, melt, slow   #Crystal Chemistry of
Families with M = Cd, Cu, Fe, Mn or Pd and X =   #Layer
theory)*   #On the Effect of
Method. II. Diffusion-Induced Twins in the alpha-
Solid Diffusion Method. II. Diffusion-Induced   #Two-
Solid Diffusion Method. I. Morphology and   #Two-
Growth of A(III)B(V) Semiconductors from Vapour
of Modes of Epitaxial Growth from the Vapor
Growth of A(III)B(V) Compounds from the Vapour
#Crystallization of WO2 from the Gas
#Growing Monocrystalline Bars of SiC from the Gas
of CuTeBr and the Structure of the High Temperature
Crystals with Phosphorus During Growth from the Gas
Silicon Carbide Epitaxial Layer Growth from the Vapour
of Single Crystals of Zinc Selenide from the Vapour
of HgCdTe Mixed Crystals from the Vapour
#Growth and Morphology of Spherical Crystals of omega-
#Ag5GaS4 -- A New
a Solution with a   #Influence of Density of Liquid
#Crystallography of the fcc
*   #Structural Transitions and
Phosphovanadates (Pb3(PxV1-x04)2, Pb3(PO4)2)*
Calorimetric, Magnetic #Metal-Insulator Transitions and
Bridgman)*   #
GaAs, theory, electrolytic deposition, kinetics)#Liquid-
#Liquid
#Quasi-Grown Layers in Liquid
(flux)*   #Liquid
*   #Liquid
Stepwise Grading (melt)*   #Liquid
5P-GaP Layer Under Controlled Phosphorus Vapor #Liquid
Films (flux, LiO.5Fe2.5-xAlxO4)*   #Liquid
#Liquid
#Solid
#Vapour-
film)*   #Molecular Beam and Solid-
#The Growth of GaxIn1-xAs on (100) InP by Liquid-
Doped with Na+ and Co2+ plus Zr4+, Grown by Liquid-
and Growth-Dissolution Kinetics of Garnet Liquid
#Surface Morphology of GaAs Layers Grown by Liquid-
GaxIn1-xSb Crystals (In Ag solvent)#Liquid
for Ca,Ge-Substituted Garnet Films grown by Liquid
to Double Heterostructure Light Modulators*   #Liquid

Pb1-xSnxTe Single Crystals of Graded Composition by the   000564
Pb2CrO5 (melt)*   000979
Pb2CrO5)*   TiO3, Nd4Ba6(BO3)9, Pr4Ba6(BO3)9, Ni8Nb3O15, 001063
Pb2WO5 and Transformation Phenomena to Its Low Form (000310
Pb3(PxV1-x04)2, Pb3(PO4)2)*   #Phase   000987
Pb3(PO4)2)*   #Phase Diagram and Crystal   000987
Pb3In6.67S13 (melt)*   000335
Pb3Mn7O15, Tm2Ti3O7, Tb2Ti2O7, Pr2Ti2O7, Tm2Ge2O7,   001061
Pb4In9S17 and Pb3In6.67S13 (melt)*   000335
Pb5GeO7 (melt)*   000739
Pb5S2I6 Single Crystals*   #Synthesis and Hydrothermal 000446
Pd (electron beam melting, evaporation)*   #   000177
Pd and X = Cl or Br: Importance, Solubilities and   000039
Pd, Pt, Fe, Al, Au, Ag, In, Eb, PbS, PbSe, PbTe, SnTe,   000949
Pd2Si)*   #Dynamics   001042
Peak Phenomena in KD3(SeO3)2 (solution)*   000963
Peculiarities in the Growth and Properties of Beta-   000967
Peculiarities of Tl-Se-TlInSe2 (TlSe, zone melting,   000361
Pedestal. III. Characteristic Arrangement of   000684
Peeled Film Technology (epitaxy)*   000497
Peltier effect, GaAs, theory, electrolytic deposition,   000437
Peltier Effect (Ge)*   #Observations of the Movement of 000061
Pentaoxometallates (flux, Sr2LaAlO5, Sr2PrAlO5,   000253
Pentaphosphate (PrP5O14)*   #Chemical Synthesis   000127
Perborate from Aqueous Solutions. I. Nucleation Rates   000257
Perborate from Aqueous Solutions. II. Growth Kinetics   000258
Perchlorate in Gelatin Gels (KC104)*   000750
Perchlorate Single Crystals (KC104)*   000752
Percolation Theory and Crystallization (entropy,   000168
Perfection (alpha-Al2O3, melt)*   000246
Perfection (organics)*   of Some Organic Compounds by 000115
Perfection (vapor transport, melt, flux)*   000383
Perfection in Czochralski Grown Nickel Single Crystals (000521
Perfection of Czochralski Grown Gadolinium Gallium   000638
Perfection of Orthorhombic Potassium Perchlorate Single   000752
Perfection of Single Crystals of FeGe2 on Their Growth   000746
Perfection of Single Crystals of Germanium (Ge, theory,   000725
Perfection Ag-Sn Crystals (in Ag solvent)*   000768
Periodic Closed Growth Lamellae on the Crystal-Melt   000532
Periodic Doping Structure in GaAs (GaAs, Ga1-xAlxAs,   000248
Periodic Remelt (Si, theory)*   of Striated Impurity   001026
Periods*   #A   000905
Perovskite (C3H7NH3)2CuCl4 (solution, theory)*   000641
Perovskite (C3H7NH3)2MnCl4 Using a Compound Holographic   000640
Perovskite Phases of the Systems Th0.25NbO3-NaNbO3 (000527
Perovskites of the (CnH2n+1NH3)2MX4 and NH3(CH2)mNH3MX4   000039
pH of Solution on the Kinetics of Crystallization (000525
phase (theory)*   Produced by a Solid-Solid Diffusion 000371
Phase (alpha/beta)-Brass Bicrystals Produced by a Solid- 000371
Phase (alpha/beta)-Brass Bicrystals Produced by a Solid- 000372
Phase (review, theory)*   #Epitaxial   000352
Phase (review, theory)*   Used for the Characterization 000542
Phase (theory)*   #Flow-Graph Model of Epitaxial 000504
Phase (vapor transport)*   000552
Phase (vapor transport)*   000554
Phase (Bridgman)*   III. On the Phase Transition   000025
Phase (Si)*   #Doping of Silicon   000027
Phase (SiC, film, vapor transport, kinetics)*   of   000942
Phase (ZnSe, vapor transport, sublimation)*   #Growth 000214
Phase by Chemical Transport*   #Crystallization   000342
Phase in the Cd-Hg System*   000512
Phase in the System Ag-Ga-S (AgGaS2)*   000688
Phase in the Y3Fe5O12-BaO.B2O3 System on Convection of   000048
Phase in Uni-Directionally Solidified Zn-Al Eutectic*   000421
Phase Composition of Thin Films of Aluminum with Silicon 000473
Phase Diagram and Crystal Growth of Lead   000987
Phase Diagram of (Ti1-xVx)4O7: Electrical,   000835
Phase Diagram of Pseudobinary System CdTe-ZnSe (000725
Phase Electroepitaxy: Growth Kinetics (Peltier effect,   000437
Phase Epitaxial Deposition of GaP on GaAs*   000354
Phase Epitaxial Growth (In1-xGaxAs)*   000951
Phase Epitaxial Growth of CdTe in the CdTe-CdCl2 System   000828
Phase Epitaxial Growth of Copper Ferrite Films (CuFe2O4) 001016
Phase Epitaxial Growth of Ga1-xInxSb on GaSb by   000803
Phase Epitaxial Growth of Lattice-Matched GaP-Ga0.5Al0. 000460
Phase Epitaxial Growth of Lithium Ferrite-Aluminate   001017
Phase Epitaxial Growth of Magnetic Garnets (review)*   000783
Phase Epitaxial Growth of Si Through Al Film*   000584
Phase Epitaxial Growth of Zn and S Doped GaAs*   000416
Phase Epitaxial Growth of ZnS on GaP*   000613
Phase Epitaxies of Silicon Under Ultra-High Vacuum (000883
Phase Epitaxy (flux)*   000754
Phase Epitaxy (flux)*   #Thin Films of LiNbO3, 000691
Phase Epitaxy (flux, kinetics, theory)*   Temperature 000333
Phase Epitaxy (flux, theory, kinetics)*   000605
Phase Epitaxy (melt)*   000333
Phase Epitaxy from PbO-B2O3 Fluxed Melts*   Imbalance 000386
Phase Epitaxy of AlxGa1-xP on GaP and Its Application   000948

PERMUTED TITLE INDEX

#Liquid	Phase Epitaxy of Garnets (review)*	000802
ZnSnAs2, solution)* #Liquid	Phase Epitaxy of II-IV-V2, Chalcopyrite Compounds (	000089
#Liquid-	Phase Epitaxy of ZnSe from Zn-Ga Solution (film)*	000308
flux)* #Thin Film Growth of (MnxZn1-x)FeO4 by Liquid	Phase Epitaxy on Bridgman Grown Zn2TiO4 Substrates (	000216
Temperatures (vapor #Crystalline Defects in Solid	Phase Epitaxy Si Films Deposited at Elevated	000194
(In,Ga)(As,P) Alloys Grown by the Hydride Vapor	Phase Epitaxy Technique (film)* Ternary and Quaternary	000729
Lead Hexa-Aluminate Single Crystal Layers by the Liquid	Phase Epitaxy Technique (flux, (Ba,Pb)Al12O19)* Barium	000071
#GaAs Vapour	Phase Epitaxy Using Computer Control*	001057
#The Growth Kinetics of Garnet Liquid	Phase Epitaxy Using Horizontal Dipping (flux)*	001021
#A New Substrate Holder for Liquid	Phase Epitaxy*	000072
#Chromium Doped GaAs Vapor	Phase Epitaxy*	000469
#Surface Morphology of GaAs Grown by Vapor	Phase Epitaxy*	000705
#Growth of GaAs1-xSbx Crystals by Steady-State Liquid	Phase Epitaxy*	000957
Kinetics of (111) Si Through an Al Layer by Solid	Phase Epitaxy* #Growth	000585
Properties of Er3Al5-xGax012 Films Obtained by Liquid-	Phase Epitaxy* #Spectroscopic	000245
Solid Solutions in the Systems Zn3As2-ZnSe(ZnTe) (#	Phase Equilibria and Properties of Single Crystals of	000726
Y3Al5012, Y3Fe5012) #	Phase Equilibria in Solvent-Garnet Systems (flux,	000159
Nd Hydroxide Bromides and Hydroxide #Preparation,	Phase Equilibria, and Crystal Chemistry of La, Pr, and	000531
and Characterization of a Europium(II) Bromide-Chloride	Phase EuBr1.5Cl0.5 (melt, slow cooling, hygroscopic)*	000200
* #A Simple Analysis of Vapor	Phase Growth - Citing an Instance of GaxIn1-xAs (theory)	000673
Aluminum Fluoride (Al2O3)* #Vapor	Phase Growth of Alumina Whiskers by Hydrolysis of	001094
Flow System (vapor transport)* #Vapor-	Phase Growth of Epitaxial Ga1-xInxSb Using Open-Tube	000681
#Liquid	Phase Growth of HgCdTe Epitaxial Layers*	001059
#Vapor	Phase Growth of Thin Plates of Cr2O3 Single Crystals*	000858
#Magnetic Properties of beta-	Phase Iron-Germanium (FexGe, slow cooling)*	000845
#Bridgman and Czochralski Growth of beta-	Phase NaxV2O5*	000907
Region of Solidification of Pt-Bi Alloys Rich in Single-	Phase Pb in the Course of Solidification (Czochralski)*	000846
#Preparation of ThSiO4 Single Crystals by a Vapour	Phase Reaction*	000454
and Control of Doping Gas Concentration During Vapor	Phase Silicon Epitaxial Growth by Using a Flame	000937
of CdGe(As1-xPx)2 and AgGa(Se1-xSx)2 Solid Solutions (#	Phase Studies, Crystal Growth, and Optical Properties	000635
Boracite on Chromium-Chlorine Boracite (#Gas	Phase Synthesis of Epitaxial Layers of Nickel-Chlorine	000837
Crystals (CdWO4, #The Cadmium Oxide-Tungsten Oxide	Phase System and Growth of Cadmium Tungstate Single	000801
#Ferroelasticity and	Phase Transformation in Rb2Hg(CN)4 Spinel (solution)*	000378
#Electron Diffraction Study of	Phase Transformations of the Compound CuSe*	000866
#Structural	Phase Transition in Mixed Crystals WxMo1-xO3 (melt)*	000824
with High Ionic Partial Conductivity. III. On the	Phase Transition of CuTeBr and the Structure of the	000025
#Neutron-Scattering Study of the Incommensurate	Phase Transition of Rb2ZnBr4 (theory, solution)*	000226
of K2PbCu(NO2)6 Near the Cooperative Jahn-Teller	Phase Transitions (solution)* #Ultrasonic Velocities	000467
Rb2Cd2(SO4)3, Tl2Cd2(SO4)3, K2Co2(SO4)3, and their	Phase Transitions (solution)* Crystals of Langbeinites	000138
5) System (vapor transport)* #Crystal Growth and	Phase Transitions in the NbxCr1-xO2 (x greater than 0.	000095
flux)* #Growth and	Phase Transitions in Single Crystals of (KxNa1-x)NbO3 (	000056
Study (melt)* #On the Magnetic Structure and Magnetic	Phase Transitions of Tb5Ge4. A Neutron Diffraction	000840
Trioxide and Molybdenum Ditelluride (MoO3-x, #Vapor	Phase Transport and Crystal Growth of Molybdenum	000299
Sulfo-Selenide (Cd1+y(SxSe1-x), sublimation)* #Vapor	Phase Transport and Stoichiometry Control of Cadmium	000647
Sulfide (CdS, sublimation)* #Vapor	Phase Transport and Stoichiometry Control of Cadmium	000648
the Course of Preparation of II-IV-V2 Compounds by Gas	Phase Transport, with ZnSiP2 as an Example* in	001086
Microanalytical Evaluation #Growth of ZnSiP2 from Gas	Phase Under Non-Stoichiometric Conditions --	000082
of Doping in Silicon Epitaxial Growth from the Gas	Phase. I. Phosphorus Content and Segregation	000514
of Doping in Growth of Epitaxial Silicon from the Gas	Phase. II. Phosphorus Content and Differential Molar	000515
#Crystallization of HgTe Crystals from the Vapour	Phase*	000341
#Morphology and Structure of ZnS Grown from the Vapour	Phase*	000874
Growth of Large Single Crystals of CdS from the Vapor	Phase* #Seeded	000242
of CdAs2 and ZnAs2 Single Crystals from the Vapor	Phase* #The Growth	000598
Single Crystals in the Metallic and the Semiconducting	Phase* #Optical Properties of V6013	001023
#Crystal-Chemistry, Crystal-Growth, and	Phase-Equilibria of Apatites (review)*	000813
Relationships Between and alpha- and beta-	phases (Cu-Zn, theory)* Morphology and Crystallographic	000372
and Properties of Superalloys via Use of Precipitated	Phases (eutectics, carbides, review)* Microstructures	000671
Properties of the Non-Stoichiometric VS2 and V5S8	Phases (vapor transport)* #Physico-Chemical	000470
NbO3, melt, slow #Crystal Chemistry of Perovskite	Phases of the Systems Th0.25NbO3-NaNbO3 (NaxTh0.25(1-x)	000257
(NH4 #Crystal Growth of the Para- and Antiferroelectric	Phases of Deuterium Ammonium Dihydrogen Arsenate (DADA)	000100
#The Systems RbF-EuF3 and CsF-EuF3. Study of the	Phases RbLnF4 and CsLnF4 (melt, RbEu2F7)*	000037
#Isotope Effects on the Central Peak	Phenomena in KD3(SeO3)2 (solution)*	000963
Gadolinium Gallium Garnet (#The Effect of Melt Flow	Phenomena on the Perfection of Czochralski Grown	000638
#High Temperature Form of Pb2WO5 and Transformation	Phenomena to Its Low Form (Czochralski, flux)*	000310
from Solution (NaClO3)* #Convection	Phenomena During the Growth of Sodium Chlorate Crystals	000179
#Kinetics of Crystal Growth. Microscopic and	Phenomenological Theories (review)*	000660
In1-xGaxAs on InP Substrate* #Composition Latching	Phenomenon and Lattice Mismatch Effects in LPE-Grown	000952
Hydrothermal Conditions (mica, #Solubility of	Phlogopite in Basic Aqueous Solutions Under	000070
Transition-Metal Dichalcogenides (MoS2, MoSe2, #Optical	Phonon Modes and Localized Effective Charges of	001000
#Dispersion of Acoustic	Phonons in Layered SrGa2 (melt, slow cooling)*	000672
#Crystal Structure of the	Phosphate of Trivalent Manganese Mn(PO3)3 (flux)*	000058
of Dislocations in the Growth of Ammonium Dihydrogen	Phosphate Crystals from Aqueous Solution (ADP, NH4H2PO4)	000219
#Crystal Structure of the Mercury Nitro-	Phosphate Hg4PO4NO3.H2O (melt)*	000261
During Polymerization of Lithium Dihydrogen	Phosphate. II. Crystal Growth by Dimer Addition (LiPO3)*	000097
(review, theory, alkaline earth sulfates, calcium	phosphates)* #The Growth of Crystals in Solution	000683
Growth of Single Crystals of Rare Earth Vanadates and	Phosphates (YPO4, YVO4, REPO4, flux)* Melts for the	000265
The Preparation and Properties of Single Crystal Copper	Phosphide (Cu3P, Czochralski)* #	000799
Growth Conditions and Dislocation Structure in Gallium	Phosphide Crystals (GaP, Czochralski, flux)* Between	000723
(GaP)* #Kinetics of Gallium	Phosphide Synthesis by Chemical Gas Transport Reactions	000231
Metal Pnictides with ThCr2Si2-Type Structure (	phosphides, flux)* #Ternary Lanthanoid-Transition	000596
Growth and Characterization of the Transition-Metal	Phosphides CuP2, NiP2, and RhP3 (flux, vapor transport)*	000708
in Growth of Epitaxial Silicon from the Gas Phase. II.	Phosphorus Content and Differential Molar Solution	000515
in Silicon Epitaxial Growth from the Gas Phase. I.	Phosphorus Content and Segregation Coefficient (Si,	000514
#Doping of Silicon Crystals with	Phosphorus During Growth from the Gas Phase (Si)*	000027
Matched GaP-Ga0.5Al0.5P-GaP Layer Under Controlled	Phosphorus Vapor Pressure and Reduced Oxygen	000460
#Phase Diagram and Crystal Growth of Lead	Phosphovanadates (Pb3(PxV1-xO4)2, Pb3(PO4)2)*	000987
transport, electrolytic #The Electrical, Optical and	Photoconducting Properties of Fe2-xCrxO3 (vapor	000624
#	Photoconduction of Lead Trititanate (PbTi3O7, flux)*	000264

PERMUTED TITLE INDEX

vapor transport, melt)* #Photoferromagnetic Effect and   Photoconductivity of Cd1-xHgxCr2Se4 Single Crystals (001037
#Plasma Growth of Rutile Crystals and Their   Photoelectronic Properties (TiO2)*   000387
xHgxCr2Se4 Single Crystals (vapor transport, melt)* #   Photoferromagnetic Effect and Photoconductivity of Cd1-   001037
Vapour Transport*   Photoluminescence Properties of CuGaSe2 Grown by Iodine   000935
Iodine Vapor Transport*   #Preparation of   Photoluminescent and Conductive ZnSe:I Crystals by   000775
Vapor Phase Silicon Epitaxial Growth by Using a Flame   Photometric Detector (Si)*   Gas Concentration During   000937
#Surface   Photovoltage Experiments on SrTiO3 Electrodes (theory)*   000618
organics, vapor transport)*   #Growth of   Phthalic Anhydride in a Closed Crystal-Vapour System (000327
of Epitaxial Growth from the Vapor Phase (review,   #   Physical Methods Used for the Characterization of Modes   000542
susceptibility,   #Preparation, Crystal Structure, and   Physical Properties (transport, magnetic   000544
slow cooling)*   #   Physical Properties and Lithium Intercalates of CrPS4 (000571
Cu2(I)B(II)C(IV) X4 (BII=Zn, Mn, Fe, Co - CIV = Si, Ge,   #   Physical Properties of the Quarternary Chalcogenides   000357
#Preparation and Basic   Physical Properties of BiTeI Single Crystals (Bridgman)*   000240
transport, M3B7O13X)*   #Preparation and   Physical Properties of Boracite Crystals (Chemical   000151
#Crystal Structure and   Physical Properties of CuO0.65VS2 (vapor transport)*   000543
Aqueous TGS Solutions Containing Aniline*   #The   Physical Properties of TGS Single Crystals, Grown from   000266
solution)*   #Growth and   Physical Properties of ZnSiP2 (vapor transport,   000086
#Metastable Pb1-xCdxS Epitaxial Films. I. Growth and   Physical Properties*   000909
KOH, KCl:KOH, theory)*   #Low-Stress   Physical Vapor Growth (PVT) (KCl, KBr, KCl:KCN, KCN: 000809
VS2 and V5S8 Phases (vapor transport)*   Physico-Chemical Properties of the Non-Stoichiometric   000470
vapor   #Synthesis, Preparation of Single Crystals, and   Physicochemical Investigation of CrPr2S4 and CrPr2Se4 (000817
VI)-MnB(VI) Systems (review)*   #   Physicochemical Properties and Crystal Growth of A(II)B(000740
Thioindates (LiInS2,   #Preparation and Principal   Physicochemical Properties of Alkali Metal Indates and   000509
Physics of Magnetic Garnets (review)*   000741
process)*   #Semiconductor Surface and Crystal   Physics Studied by MBE (GaAs, Ga1-xAlxAs, review of MBE   000055
Control of Substrate Degradation in InP LPE Growth with   #   PH3 Partial Pressure (flux)*   000199
#Crystal Growth of Boron Monophosphide Using a B2H6-   PH3-H2 System (BP, vapor transport)*   000884
#Na2CoGeO4 Single Crystal Growth and Study of Their   Piezoelectric and Elastic Properties (hydrothermal)*   000256
#Ca12Al14O33: A New   Piezoelectric Material (Czochralski)*   001076
Czochralski)*   #The Effect of Impurities on the   Piezoelectric Properties of Lithium Germanate (Li2GeO3,   000800
KLN) Crystals (K2.89Li1.55Nb5.11O15,   #Elastic and   Piezoelectric Properties of Potassium Lithium Niobate (000010
Segregation Behavior of Ga-Doped #Application of a Heat   Pipe to Czochralski Growth. Part I. Growth and   000604
(Te)*   #Etch   Pit Orientation in Czochralski-Grown Tellurium Crystals   000877
2H2O, gel)*   #Observations of Etch   Pits on As-Grown Faces of Brushite Crystals (CaHPO4.   000714
Heterostructures*   #Improved Nucleation and a   Planar Interface in LPE Grown Pb1-xSnxTe   000962
#High Rate Epitaxial Growth of ZnO Films on Sapphire by   Planar Magnetron rf Sputtering System*   000882
Carbon (C, film)*   #Reactive Pulse   Plasma Crystallization of Diamond and Diamond-Like   000906
Photoelectronic Properties (TiO2)*   #   Plasma Growth of Rutile Crystals and Their   000387
Crystals of High Melting Metal Alloys and Compounds by   Plasma Heating (W, W-C)*   #Growth of Single   000829
#Spherical Monocrystals for X-Ray Work Obtained by   Plasma Remelting of Alloy Powder (Cu9Al4)*   000043
deposition)*   #Deposition of Crystals from the   Plasmas of ZrO2, HfO2, ThO2 and CeO2 (electrolytic   000472
Stepanov Procedure (Si)*   #Preparation of Silicon   Plates by Pulling from a Molten Bath According to the   000966
#Vapor Phase Growth of Thin   Plates of Cr2O3 Single Crystals*   000858
and Morphology of VLS Silicon Crystals Grown with   Platinum and Gold as Liquid-Forming Agents (vapor   001075
#Thin Films of LiNbO3, Doped with Na+ and Co2+   plus Zr4+, Grown by Liquid-Phase Epitaxy (flux)*   000691
#Epitaxial Growth of Ferroelectric   PLZT ((Pb,La)(Zr,Ti)O3) Thin Films*   000415
#Single Crystal and Film Growth of Actinide   Pnictides by Chemical Vapour Transport (review)*   000160
flux)*   #Ternary Lanthanoid-Transition Metal   Pnictides with ThCr2Si2-Type Structure (phosphides,   000596
of Copper Single Crystals from the Bulk Using Four   Points (Cu, melt)*   #A New Technique for Preparation   000737
Growth and Sublimation in II-VI Compounds Along their   Polar Axis (ZnO, CdSe, CdS, ZnS, kinetics)*   #Crystal   000428
Li2SiO3, Li2GeO3) Single Crystals (Bridgman)*   #   Polar Twins in Lithium Metasilicate and Metagermanate (000108
#Crystal Growth in c Direction and Crystallographic   Polarity in ZnO Crystals (vapor transport, film)*   000426
#Growing and   Polishing Large Thallium Single Crystals (Tl, melt)*   001085
Doping Using   #The Bending of Silicon Wafers by Thin   Polycrystalline Silicon Film Deposition and by Film   000985
Boundary (H2O)*   #On the Growth Mechanism of   Polycrystalline Snow Crystals with a Specific Grain   000314
review)*   #Growth of   Polyhedral Metal Crystallites in Inactive Gas (Mg,   001006
#Selected Papers on   Polymer Crystallization (Review)*   001089
Crystal Growth by Dimer   #Crystallization During   Polymerization of Lithium Dihydrogen Phosphate. II.   000097
Ion Exchange Reactions with KbAlSiO4: A New Metastable   Polymorph of KAlSiO4 (flux)*   #Alkali   000639
Er, Ho, Dy) (Tm2Si2O7, Er2Si2O7)   #Flux Growth of   Polymorphic Rare-Earth Disilicates, R2Si2O7 (R = Tm,   000595
Polymorphism in Bi2Sn2O7 (flux)*   000867
#Synthesis and Crystal Structure of Iron   Polyphosphide FeP4*   000438
Control of NdP5O14 Single Crystals Grown from   Polyphosphoric Acids (solution)*   #Morphology   000466
#Crystal Structure of Lanthanum   Polysulfide LaS2*   000259
#Indium   Polytelluride In2Te5(II) (melt)*   001058
Spirals*   #The Occurrence of High Period   Polytype of SiC at the Interface of Two Interacting   000889
#Electronic Bandgap Measurements of SnS2   Polytypes (vapor transport)*   000008
Polytypism and the Vibrational Properties of PbI2*   000850
vapor transport)*   #   Polytypism and Amorphousness in Silicon Whiskers (Si,   000642
solution, evaporation)* #Effect of Growth Conditions of   Polytypism in Cadmium Iodide Crystals (CdI2, theory,   000432
review)*   #Some Aspects of   Polytypism in Crystals (SiC, ZnS, CdI2, mica, theory,   000069
CdBr2)*   #   Polytypism in Vapor Grown Crystals of Cadmium Bromide (000622
(CdI2-xBrx, CdI2-xClx)*   #Growth and   Polytypism in Vapor-Grown Mixed Cadmium Halide Crystals   000870
Positron Annihilation in ReO3 (vapor transport)*   000015
Quality of Solution Grown Crystals -- Case of   Potash Alum (KAl(SO4)2.12H2O)*   #Stirring on Crystalline   000337
of Potassium Lithium Niobate Single-Crystal Films on   Potassium Bismuth Niobate Single Crystals by the EGM   000011
of Potassium Lithium Niobate Single-Crystal Films on   Potassium Bismuth Niobate Single Crystals by RF   000009
by Differential   #Measurement of Crystallization of   Potassium Bromate from Its Quiescent Aqueous Solution   000367
*   #Single Crystal Growth of Ferroelectric   Potassium Dihydrogen Orthophosphate in Silica Gels (KDP)   000442
#Nucleation in Supersaturated   Potassium Dihydrogen Orthophosphate Solutions (KDP)*   000443
Dehydration Method (K2Ti6O13)*   #Synthesis of   Potassium Hexatitanate Fibers by the Hydrothermal   000732
55Nb5.11O15,   #Elastic and Piezoelectric Properties of   Potassium Lithium Niobate (KLN) Crystals (K2.89Li1.   000010
Potassium Bismuth Niobate Single   #Epitaxial Growth of   Potassium Lithium Niobate Single-Crystal Films on   000009
Potassium Bismuth Niobate Single   #Epitaxial Growth of   Potassium Lithium Niobate Single-Crystal Films on   000011
#Crystal Growth of   Potassium Perchlorate in Gelatin Gels (KClO4)*   000750
#Gel Growth and Perfection of Orthorhombic   Potassium Perchlorate Single Crystals (KClO4)*   000752
Borate Crystals in Preparation from Solutions in Molten   Potassium Trimolybdate ((Y,Nd) Al3(BO3), flux)* into YAl-   000551
Alkali Metal Tungsten Bronzes, in Particular those of   Potassium, Rubidium and Cesium (review, KxWO3, RbxWO3,   000402

PERMUTED TITLE INDEX

solution)*   #Growth Spirals in a Chemical
Conductivity of $Gd_2(SO_4)_3 8H_2O$ (crystal) and $Dy_2Ti_2O_7$ (
for X-Ray Work Obtained by Plasma Remelting of Alloy
the Verneuil Technique ($MgAl_2O_4$, alpha-Al_2O_3)*
by Nitridization in the $Si-SiO_2$ System in Compacted
Pressure on Electrical Resistivity and Thermoelectric
Shape (theory)*
Structure of the Phosphate of Trivalent Manganese Mn(
#Crystal Structure of Scandium Metaphosphate Sc(
#Structure and Luminescence Properties of alpha-CsNd(
of a New High -Nd- Concentration Laser Material: Na_3Nd(
Growth of Lead Phosphovanadates ($Pb_3(P_xV_{1-x}O_4)_2$, Pb_3(
#The Crystal Structure of Cu_4(
#Crystal Structure of Cu_4(
#New Ferroelectric Na_3Sc_2(
Single Crystals by Hydrothermal Recrystallization (Ca_5(
#Flux Growth of Lead Neodymium Chlorapatite ($Na_2Nd_2Pb_6$(
Nd Hydroxide Bromides and Hydroxide Iodides ($Nd(OH)_2Br$,
Phase Equilibria, and Crystal Chemistry of La,
Levitation Zone Melting Method ($DyFe_2$, $TbFe_2$, Nd,
Rare-Earth and Iron Group Complex Oxides ($NdBO_3$, $LaBO_3$/
Floating Zone Melting: Influence of Zone Geometry and
#Chemical Synthesis and Crystal Growth of Laser Quality
melt, Pr_2Ga, Pr_5Ga_3, PrGA)*   #The
Modifying Starting Compositions for Flux Growth ($DyPO_4$,
#Crystal Growth of $SrCl_2$ and Solid Solutions of $SrCl_2$-
of Chromium III and Rare Earths ($LaCrOS_2$, $CeCrOS_2$,
Intermetallic Compounds, RCu_2. II. Light Rare Earth (
and Properties of Superalloys via Use of
Line (#Transmission Electron Microscopy Observations of
#A Method for the Determination of
by the Floating-Zone Method*
Crystals of Germanium (Ge, theory, kinetics)*
#Preparation, Stability, and
#A Gas Handling System for Exacting Crystal Growth
hydrothermal, theory)*   #Advances in
High Pressure - High Temperature Reactions with #A New
theory)*   #On the
the Ternary Compound $ZnGeP_2$ (melt, slow cooling, high
Degradation in InP LPE Growth with PH_3 Partial
#Kinetics of Silicon Growth Under Low Hydrogen
#Overgrowths on Diamond at Atmospheric
Mixed Oxides of the $A(6+)B_2(3+)O_6$ Type Under Very High
Zinc Selenide Crystals Grown from the Melt Under Argon
Technique in Solid State Chemistry. II. High
$Ga_{0.5}Al_{0.5}P-GaP$ Layer Under Controlled Phosphorus Vapor
Power of FeS (Bridgman)*   #Effect of
Containing Rhenium: $Ca_{1-x}Re_2O_6(OH)_{2x}$ (x = 0.3)*   #High
Growth in GaAs VPE at Low Temperature and High Partial
Czochralski Crystal Growth (InP, $CuInS_2$)*   #Simple
System from 0 to 50 At% Gallium (melt, Pr_2Ga, Pr_5Ga_3,
Layer   #On Interpreting a Quantity in the Bruton,
Growth from the Melt (#A New Look at the Burton,
Growth (vapor transport)*
Growth (solution)*   #Fiber Optic
Crystal Growth, 1977) (review)*   #Crystal Growth (
Preparation by Stepanov Method of Profiled Crystals
Preparation by Stepanov Method of Profiled Crystals and
Al_2O_3, melt)*
Union Meeting on the Preparation by Stepanov Method of
#Interface Morphology in Chemical Vapour Deposition on
#Fiber Optic Probe for Thermal
of Solidification of Pb-Bi Alloys Rich in Single-Phase #
review)*   #Li3N: A
Garnet (GGG, theory, $Gd_3Ga_5O_{12}$)*   #Dislocation
$PaAs_2$,   #Preparation and Single-Crystal Growth of

#Hydronium Beta" Alumina: A Fast
Growth of Laser Quality Praseodymium Pentaphosphate (
Gallium System from 0 to 50 At% Gallium (
MoO_3 ($DyKMo_2O_8$, $ErKMo_2O_8$, $NdKMo_2O_8$, Dy_2MoO_6, Sm_2MoO_6,
$Pr_2Ti_2O_7$, $Tm_2Ge_2O_7$, Er_2SiO_5, Dy_2SiO_5, Ho_2SiO_5,
$Nd,Pb)MnO_3$, $(Eu,Pb)MnO_3$, $Pb_3Mn_7O_{15}$, $Tm_2Ti_3O_7$, $Tb_2Ti_2O_7$,
$(GdVO_3$, $Gd_4Ba_6(BO_3)_9$, $(Ba_{0.93}Gd_{0.07})TiO_3$, $Nd_4Ba_6(BO_3)_9$,
Gallium System from 0 to 50 At% Gallium (melt, Pr_2Ga,
$(Nd(OH)_2Br$, $Pr(OH)_2Br$, $La_7(OH)_{18}Br_3$, $Nd_7(OH)_{18}Br_3$,
$La_7(OH)_{18}Br_3$, $Nd_7(OH)_{18}Br_3$, $Pr_7(OH)_{18}Br_3$, $La_7(OH)_{18}I_3$,
a Large Segregation Tendency by the Bridgman Method -
Pyroelectricity and Related Properties in the Fresnoite
#Phase Diagram of
Fe_2TiO_5, electrolytic deposition)#On the Preparation of
of Lattice Fitting in Epitaxy (theory, Ni, Cu, Pd,
of Microfacets Near Irregularly Remelted Surfaces in
Arrangement of   #Growth of Niobium Single Crystals by
Procedure (Si)*   #Preparation of Silicon Plates by
Like Carbon (C, film)*   #Reactive
#Optically
Compounds (SiC, vapor transport, De   #Preparation of

Potential Varying in Space or Time (theory, kinetics,
powder) as a Function of Magnetic Field (solution)*   000659
Powder (Cu_9Al_4)*   #Spherical Monocrystals   000065
Powder Feeder for Crystals of Large Diameter Grown by   000043
Powders*   #Growth of a Crystal Needle of Si_2ON_2 000284
Power of FeS (Bridgman)*   #Effect of   000049
Power Required to Form a Floating Zone and the Zone   000465
$PO_3)_3$ (flux)*   #Crystal   000492
$PO_3)_3$ (solution)*   000058
$PO_3)_4$ Crystals*   000902
$PO_4)_2$*   #The Crystal Structure   000587
$PO_4)_2)$*   #Phase Diagram and Crystal   000825
$PO_4)_2O$ (flux)*   000987
$PO_4)_2O$ (flux)*   000029
$PO_4)_3$ (melt)*   000142
$PO_4)_3OH)$*   #Preparation of Pure Hydroxyapatite   000721
$PO_4)_6Cl$, slow cooling, evaporation)*   000040
$Pr(OH)_2Br$, $La_7(OH)_{18}Br_3$, $Nd_7(OH)_{18}Br_3$, $Pr_7(OH)_{18}Br_3$,   000445
Pr, and Nd Hydroxide Bromides and Hydroxide Iodides (Nd(000531
Pr, Ce, cold crucible)*Crystal Growth by the Horizontal   000531
Pr, $LaBO_3$, $(La,Pb)MnO_3$, $(Nd,Pb)MnO_3$, $(Eu,Pb)MnO_3$,   000620
Prandtl Number at Zero Gravity (Si, Al_2O_3, theory,   001061
Praseodymium Pentaphosphate (PrP_5O_{14})*   000175
Praseodymium-Gallium System from 0 to 50 At% Gallium (000127
$PrBO_3$, $LaBO_3$, $DyKMo_2O_8$, $ErKMo_2O_8$, $NdKMo_2O_8$, $DyVO_4$,   000197
$PrCl_3$ and $SrCl_2-GdCl_3$ (Bridgman-Stockbarger,   001060
$PrCrOS_2$, $NdCrOS_2$, $SmCrOS_2$)*   #Mixed Oxysulfides   001054
$PrCu_2$, $NdCu_2$)*##Magnetic Properties of Rare Earth Copper   001048
Precipitated Phases (eutectics, carbides, review)*   000374
Precipitates in ZnTe: Shape of the Tellurium Solidus   000671
Precipitation Induction Periods*   000098
Preferred Growth Direction of Pure Cu Single Crystals   000905
Preferred Growth Directions and Perfection of Single   000380
Preliminary Crystallographic Study of $Bi_2Sn_2O_7$ (flux)*   000275
Preparations (TiO_2, vapor transport)*   001034
Preparative Chemistry of Oxides and Fluorides (review,   000808
Preparative Technique in Solid State Chemistry. II.   000398
Presentation of Growth Curves for Growth from Solution (000658
pressure, Bridgman)* #Synthesis and Characterization of   001025
Pressure (flux)*   #Control of Substrate   000336
Pressure (vapor transport, Si)*   000199
Pressure (C, flux)*   000255
Pressure (UCr_2O_6, UFe_2O_6, UMn_2O_6, $UMnNiO_6$, $MoCr_2O_6$)*   000748
Pressure (ZnSe, Bridgman)*   #Formation of Defects in   000205
Pressure - High Temperature Reactions with CO_2-Lasers*   000477
Pressure and Reduced Oxygen Contamination (GaP, Ga_0.   000658
Pressure on Electrical Resistivity and Thermoelectric   000460
Pressure Synthesis and Structure of a New Pyrochlore   000465
Pressures (vapor transport)*   #Fast   000104
Pressurized Chambers for Liquid Encapsulated   000393
PrGA)*   #The Praseodymium-Gallium   000150
Prim and Slichter Equation as a Diffusion Boundary   000197
Prim, and Slichter Model of Segregation During Crystal   001083
Priorities in the Initial Use of Spacelab for Crystal   001084
Probe for Thermal Profiling of Liquids During Crystal   000270
Proceedings of the Third International Summer School on   000395
Proceedings of the 8th All Union Meeting on the   000066
Products and Their Application in National Economy,   001111
Profile-Sapphire and Its Structural Perfection (alpha-   011111
Profiled Crystals and Products and Their Application in   011111
Profiled Substrates (theory)*   000246
Profiling of Liquids During Crystal Growth (solution)*   011111
Progress of Temperature and Concentration in the Region   001011
Promising Li Ionic Conductor (electrolytic deposition,   000395
Propagation in Czochralski Grown Gadolinium Gallium   000846
Protactinium Arsenides by Chemical Vapour Transport (000024
Protein Crystallizations (solution, organics)*   000838
Proton Conductor ($0.84Na_2O.0.84MgO.5Al_2O_3$, theory)*   000161
PrP5014)*   #Chemical Synthesis and Crystal   000382
Pr_2Ga, Pr_5Ga_3, PrGA)*   #The Praseodymium-   000285
Pr_2MoO_6, Ho_6MoO_{12})* and R_6MoO_{12} in the Systems $R_2O_3-K_2O$-   000127
Pr_2MoO_6, $LiCoPO_4$)* MnO_3, $Pb_3Mn_7O_{15}$, $Tm_2Ti_3O_7$, $Tb_2Ti_2O_7$,   000197
$Pr_2Ti_2O_7$, $Tm_2Ge_2O_7$, Er_2SiO_5, Dy_2SiO_5, Ho_2SiO_5, Pr_2MoO_6,   001062
$Pr_4Ba_6(BO_3)_9$, $Ni_8Nb_3O_{15}$, $Ni_2V_2PbO_8$, $Ni_3Nb_2O_8$, $Ni_3V_2O_8$,   001061
Pr_5Ga_3, PrGA)*   #The Praseodymium-   001061
$Pr_7(OH)_{18}Br_3$, $La_7(OH)_{18}I_3$, $Pr_7(OH_{18}I_3$, hydrothermal)*   001063
$Pr_7(OH_{18}I_3$, hydrothermal)* $(Nd(OH)_2Br$, $Pr(OH)_2Br$,   000197
Pseudobinary $Cd_xHg_{1-x}Te$*Homogeneous Mixed Crystals with   000531
Pseudobinary System $Ba_2TiGe_2O_8-Ba_2TiSi_2O_8$ (Czochralski)*   000531
Pseudobinary System $CdTe-ZnSe$ (Bridgman)*   000244
Pseudobrookite Single Crystals by Chemical Transport (000836
Pt, Fe, Al, Au, Ag, In, Pb, PbS, PbSe, PbTe, SnTe, LiF,   000725
Pulled GaSb Crystals (Czochralski)*   #Observations   000770
Pulling from a Melt on a Pedestal. III. Characteristic   000949
Pulling from a Molten Bath According to the Stepanov   000517
Pulse Plasma Crystallization of Diamond and Diamond-   000684
Pumped LPE-Grown $Hg_{1-x}Cd_xTe$ Lasers (films, flux)*   000966
Pure and Doped Silicon Carbide by Pyrolysis of Silane   000368
   001043

PERMUTED TITLE INDEX

#Preferred Growth Direction of
Recrystallization (Ca5(PO4)3OH)#   #Preparation of
- I. Single Filament, Unconstrained Growth from a
Simulation of Convection in Floating Zone Melting. I.
Solutions. II. Growth Kinetics of Different Faces in
from Aqueous Solutions. I. Nucleation Rates in
Czochralski)#   #On the Preparation of
During a Modified Form of Zone Refining-Cascade
Silicon (Si, melt, Czochralski)#   #
#An Inexpensive Zone Melting Furnace for
#A Thermodynamic Factor Influencing the Growth Rate and
#New Gaseous Impurity and Its Effect on the
transport)#   #Electrical and Optical Properties of High-
Density (Al, zone melting,   #Preparation Method of High
#Luminescence and Bandgap Studies of High
#A High Temperature High
#Preparation, Analysis and Properties of High
Travelling Heater Method (vapor   #Growth of High-
#Low-Stress Physical Vapor Growth (
and Crystal Growth of Lead Phosphovanadates (Pb3(
#Upper and Lower Critical Fields of TaS2 (
#Morphology of Iron
#AlSiP3, A Compound with a Novel Wurtzite-
3)#   #High Pressure Synthesis and Structure of a New
Pseudobinary System Ba2TiGe2O8-Ba2TiSi2O8 (Czochralski)#
Boer-   #Preparation of Pure and Doped Silicon Carbide by
#Crystal Structure of a New Variety of Nickel
#The Compound Zn4(
#Crystal Structure of Zn4(
#Crystal
Alum (KAl(SO4)2.   #Effect of Stirring on Crystalline
Between Technological Parameters and Crystal
#Chemical Synthesis and Crystal Growth of Laser
Heats of Fusion for Ionic Crystals (review, theory)#   #
Mn, Fe, Co - CIV = Si, Ge,   #Physical Properties of the
Electron Microscopy of Growth Surfaces of Synthetic
#Observation of Growth Defects in Synthetic
#On the Dissolution of Grown Synthetic
#Dislocation-Free and Low-Dislocation
the Incorporation of Metallic Impurities in Synthetic
Seeded Growth of Large Single Pb1-xSnxTe Crystals on a
#Growth of Pb0.8Sn0.2Te Films in a
In1-xGaxAs)#
#Preparation of Films of IV-VI Compounds in a
hydrothermal)#   #Specific Heat of a
#Crystal Growth and Properties of Binary, Ternary and
#LPE Growth of a
#Melt and Solution Growth of Bulk Single Crystals of
of Crystallization of Potassium Bromate from Its
Growth of Polymorphic Rare-Earth Disilicates, R2Si2O7 (
Growth Models" (theory)#   #Reply to Comment of
Crystals (theory, Czochralski, melt, Si)#   #Effects of
and Nuclear Spectroscopic Evaluation of Mercuric Iodide
theory)#   #Limitations in Using Kilohertz
vapor transport)#

the RbAg4I5 Family (KAg4I5, NH4Ag4I5, solution)#

Solar Cell Processing (zone melting, Si)##Properties of
Earth Copper Intermetallic Compounds, RCu2. II. Light
Earth Copper Intermetallic Compounds, RCu2. I. Heavy
BO3)9, (Ba0.93Gd0.07)TiO3,   #The Flux Growth of Some New
Flux Growth and Characterization by X-ray Topography of
of Their   #The Preparation of Single Crystals of the
Heavy Rare Earth (TbCu2, DyCu2,   #Magnetic Properties of
Light Rare Earth (PrCu2, NdCu2)##Magnetic Properties of
YLaBi)3(FeGa)5O12)   #Magneto-Optical Properties of Some
and Crystallographic Data of Some Ln2(MoO4)3 Type Mixed
directional   #Growth from Skull-Melting of Zirconia-
#Czochralski Crystal Growth of
Bridgman-Stockbarger,   #Crystal Growth of Some
Temperature Melts for the Growth of Single Crystals of
SmCrO2S2)#   #Mixed Oxysulfides of Chromium III and
I) (melt,   #New Mixed Halide Compounds MFX of Divalent
Growth and Characterization of Mixed Bromides of
Pr, LaBO3, (La,Pb)MnO3, (Nd,Pb)   #The Flux Growth of Some
YOHCO3, GdOHCO3, ErOHCO3, Y2O(OH)2CO3,   #Synthesis of
#The Growth and Structure of Epitaxial Films of the
Tm2Si2O7, Er2Si2O7,   #Flux Growth of Polymorphic
Er3NbO7,   #Flux Growth of Double Oxides of Niobium and
Y3Fe5O12, RExY3-xGayFe5-yO12,   #A Solubility Model for
in Gadolinium Molybdate and Other Isotypic
Sr2PrAlO5,   #Crystal Growth of Some Alkaline-Earth
Solutions) (review, flux)##Rost Kristallov iz Rastvorov-
Temperature Solutions) (review,   #Rost Kristallov iz
LPE Layers Grown under Conditions of Constant Cooling
Perfection of Single Crystals of FeGe2 on Their Growth
System (#A Thermodynamic Factor Influencing the Growth

Pure Cu Single Crystals by the Floating-Zone Method#   000380
Pure Hydroxyapatite Single Crystals by Hydrothermal   000040
Pure Melt (theory)#   Morphology During Crystallization   000992
Pure Rotation Driven Flows (Si, Al2O3, kinetics, float   000173
Pure Solution and in the Presence of a Surfactant (000258
Pure Solution in the Presence of a Surfactant (NaBO2.   000257
Pure, Doped and Reduced KNbO3 Single Crystals (000295
Purification (metals)#Theoretical Solute Redistribution   000201
Purification and Characterization of Metallurgical   000196
Purification and Crystal Growth (NaWO3, PbCl2, organics)   000871
Purity of Epitaxial Layers in the Ga_AsCl3-H2 System (000653
Purity of LPE GaAs (vapor transport)#   000313
Purity p-type Single Crystals of GeFe2O4 (vapor   000920
Purity Aluminum Single Crystals with Low Dislocation   000228
Purity MgxZn1-xTe Ternary Alloys (Bridgman)#   000600
Purity Source for Metal Beam Epitaxy (Ag, film)#   000287
Purity Zn(1-x)MgxTe Alloys (distillation)#   000792
Purity ZnTe Single Crystals by the Sublimation   000938
PVT) (KCl, KBr, KCl:KCN, KCN:KOH, KCl:KOH, theory)#   000809
PxV1-xO4)2, Pb3(PO4)2)#   #Phase Diagram   000987
Pyridine) 1/2 (vapor transport)#   000468
Pyrite Crystals (FeS2, vapor transport)#   001093
Pyrite Intergrowth Structure (flux)#   001044
Pyrochlore Containing Rhenium: Ca1-xRe2O6(OH)2x (x = 0.   001014
Pyroelectricity and Related Properties in the Fresnoite   000836
Pyrolysis of Silane Compounds (SiC, vapor transport, De   001043
Pyrophosphate: Ni2P2O7 (slow cooling)#   000609
P2S6)3 (melt)#   000131
P2S6)3 (melt)#   000132
Quality of Gel Grown CaHPO4.2H2O Crystals#   000545
Quality of Solution Grown Crystals -- Case of Potash   000337
Quality During the Crystal Growth of Silicon Crystals,   000325
Quality Praseodymium Pentaphosphate (PrP5O14)#   000127
Quantum-Mechanical Relations in the Calculation of   001087
Quarternary Chalcogenides Cu2(I)B(II)C(IV)X4 (BII=Zn,   000357
Quartz (SiO2, hydrothermal)#   #Scanning   000857
Quartz Crystals by Light-Scattering Tomography (SiO2,   000652
Quartz Inside the Autoclaves (SiO2, hydrothermal)#   000507
Quartz Prepared by Hydrothermal Crystallization (SiO2)#   000068
Quartz Single Crystals (SiO2, solution)#   #On   000170
Quartz Surface (sublimation)#   #Non-   000960
Quasi-Closed Space (vacuum evaporation)#   000885
Quasi-Grown Layers in Liquid Phase Epitaxial Growth (000951
Quasi-Sealed Volume (SnTe, PbTe, PbSe, evaporation)#   000306
Quasi-Two-Dimensional Antiferromagnet Ni(OH)2,   000273
Quaternary (In,Ga)(As,P) Alloys Grown by the Hydride   000729
Quaternary GayIn1-yAsxP1-x (film)#   000188
Quaternary III-V Alloys (review, kinetics)#   000054
Quiescent Aqueous Solution by Differential Scanning   000367
R = Tm, Er, Ho, Dy) (Tm2Si2O7, Er2Si2O7, Ho2Si2O7,   000595
R. Trivedi on "The Validity of Steady-State Dendrite   000833
Radial and Axial Temperature Gradients on Tensions in   000326
Radiation Detectors. Part II. HgI2 (vapor transport,   000834
Radio Frequencies for Float Zone Silicon Crystals (Si,   000358
Raman and Infrared Spectra of CdIn2S4 and ZnIn2S4 (001004
Raman Scattering in 3R-NbS2 (vapor transport, theory)#   000731
Raman Scattering Studies in the Solid Electrolytes of   000315
Raman Spectra of ZrS3#   000434
Rapidly-Grown RTR Silicon Sheet and Their Effects on   000546
Rare Earth (PrCu2, NdCu2)#   Magnetic Properties of Rare   000374
Rare Earth (TbCu2, DyCu2, HoCu2, ErCu2)#   of Rare   000373
Rare Earth and Iron Group Complex Oxides (GdVO3, Gd4Ba6(001063
Rare Earth Arsenates (REAsO4, slow cooling)#   #   000900
Rare Earth Borides by the Solution Method and a Study   000359
Rare Earth Copper Intermetallic Compounds, RCu2. I.   000373
Rare Earth Copper Intermetallic Compounds, RCu2. II.   000374
Rare Earth Iron Garnets (flux, (GdPrBi)3(FeGaIn)5O12, (001107
Rare Earth Molybdates (RE(MoO4)3, Czochralski)#   Growth   000447
Rare Earth Oxide Crystals (ZrO2-RE2O3, eutectics,   000269
Rare Earth Tetraborides (YB4, TbB4, ErB4)#   000137
Rare Earth Trifluorides (GdF3, TbF3, DyF3, HoF3, ErF3,   000323
Rare Earth Vanadates and Phosphates (YPO4, YVO4, REPO4,   000265
Rare Earths (LaCrOS2, CeCrOS2, PrCrOS2, NdCrOS2,   001048
Rare Earths (M = Sm, Eu, Tm, and Yb - X = Cl, Br, and   000088
Rare Earths, Sodium, and Cesium (Ca2NaNdBr6,   000626
Rare-Earth and Iron Group Complex Oxides (NdBO3, LaBO3/   001061
Rare-Earth Carbonates Under Hydrothermal Conditions (000169
Rare-Earth Dihydrides (GdH2, TbH2, DyH2, ErH2, HoH2)#   000782
Rare-Earth Disilicates, R2Si2O7 (R = Tm, Er, Ho, Dy)   000595
Rare-Earth Elements (Ln3NbO7) (Ho3NbO7, Y3NbO7,   001097
Rare-Earth Iron Garnets in a PbO/B2O3 Solution (001022
Rare-Earth Molybdates (Czochralski, melt, Gd2(MoO4)3,   000446
Rare-Earth Pentaoxometallates (flux, Sr2LaAlO5,   000253
Rasplavov (Crystal Growth from High-Temperature   000976
Rastvorov-Rasplavov (Crystal Growth from High-   000976
Rate (melt, theory)#   Crystal Composition in Ga1-xAlxAs   000374
Rate and Direction#   #Dependence of Degree of   000746
Rate and Purity of Epitaxial Layers in the Ga_AsCl3-H2   000653

PERMUTED TITLE INDEX

melt)*	#Conditions for Constant Growth	Rate by LPE from a Cooling, Static Solution (theory,	000893
theory)*	#Linear Growth	Rate of Calcite in Aqueous Solution (CaCO3, kinetics,	000627
Solution by Differential Scanning Calorimeter. Growth		Rate of Crystallite (KBrO3)* from Its Quiescent Aqueous	000367
#Effects of Interface Kinetics on the Growth		Rate of Dendrites (Sn, theory)*	000991
Under Constant Conditions (	#The Variability of the	Rate of Growth of Adamantane Crystals from the Vapour	000572
	#Growth	Rate of Iron Whiskers (Fe, vapor transport, theory)*	000462
	#New Model of LPE Growth: Growth	Rate Calculation (theory)*	000590
Planar Magnetron rf Sputtering System*	#High	Rate Epitaxial Growth of ZnO Films on Sapphire by	000882
Assisted Growth of Gallium (Ga, theory)*	#	Rate Equations for Dislocation-Free and Dislocation-	000001
metals, Nb, NaxWO3, Czochralski, electrolytic	#Growth	Rate Limitations in Electrochemical Crystallization (	000225
(solution, theory)*	#Face Stability and Growth	Rate Variations of the Layer Perovskite (C3H7NH3)2CuCl4	000641
#Evidence for a Universal Law of Dendritic Growth		Rates (theory)*	000534
Sodium Perborate from Aqueous Solutions. I. Nucleation		Rates in Pure Solution in the Presence of a Surfactant (	000257
kinetics, Czochralski)*	#On the Cooling	Rates of Large-Diameter Silicon Crystals (Si, theory,	000165
the Czochralski Method*	#Effects of Growth	Rates on the Defect Generation in KCl Crystals Grown by	000409
Grown by Floating Zone Melting at Various Growth		Rates* #Growth Striations in Fe-3wt%Si Single Crystals	001028
#Growth of Single Crystals of Cinnabar (alpha-HgS) at		Rather Low Temperature (solution, doped)*	000157
of Thin Bismuth Single Crystals with High Resistivity		Ratios (Bi, Czochralski)* #The Growth	000887
Thermal Expansion Coefficients of VO2 near #Accurate X-		ray Determination of the Lattice Parameters and the	000513
cooling)* #Flux Growth and Characterization by X-		ray Topography of Rare Earth Arsenates (REAsO4, slow	000900
	#X-	Ray and Dielectric Investigations of Pb2CrO5 (melt)*	000979
Trigonal Weberite (flux)*	#X-	Ray and Neutron Diffraction of Na2MnFe7: A Variety of	001033
(vapor transport)*	#Growth and X-	Ray Analysis of Single Crystals of EuSb4S7 and EuSb2Se4	000022
#Crystal Growth of CdTe for X-		Ray Detectors (review)*	001053
Europium beta-Alumina (NaAl11017, KAl11017,	#X-	Ray Diffuse Scattering from Alkali, Silver, and	000621
Setting for Measuring Local Lattice Mismatches	#An X-	Ray Double-Crystal Method Utilizing Non-Parallel	000704
Crystals of NaCl and KCl (NaCl-KCl	#The Growth, X-	Ray Studies and Microstructure of Solution-Grown Mixed	000407
	#Growth and X-	Ray Study of Cu2SrGeS4 and Cu2BaGeS4 (flux)*	000972
theory)*	#X-	Ray Study of LEC-Grown InP Crystals (Czochralski,	000614
(CdxHg1-xTe, Bridgman)*	#An X-	Ray Topographic Assessment of Cadmium Mercury Telluride	000158
*	#Sublimation Growth and X-	Ray Topographic Characterization of CdTe Single Crystals	000148
Na2CoGeO4 Single Crystals in Relation to Their	#X-	Ray Topographic Study of Growth Defects in Hydrothermal	000233
Single Crystals (solution)*	#X-	Ray Topographic Study of Growth Sector Strains in KDP	000092
of Dilute Binary #Observation by Chemical Attack and X-		Ray Topography of Cellular Structures of Monocrystals	000210
#In Situ X-		Ray Topography of Epitaxial Ge Layers During Growth*	000363
Cu9Al4)* #Spherical Monocrystals for X-		Ray Work Obtained by Plasma Remelting of Alloy Powder (	000043
#Solution Growth of		RbAg4I5 Crystals*	000038
Scattering Studies in the Solid Electrolytes of the		RbAg4I5 Family (KAg4I5, NH4Ag4I5, solution)* #Raman	000315
in Hydrofluoric Acid Solution (TlAlF4, KAlF4, NH4AlF4,		RbAlF4, CsAlF4, NH4Fe2F6)*Systems MF-AlF3 and NH4F-FeF3	000772
#Alkali Ion Exchange Reactions with		RbAlSiO4: A New Metastable Polymorph of KAlSiO4 (flux)*	000639
Silver, and Europium beta-Alumina (NaAl11017, KAl11017,		RbAl11017, AgAl11017, EuAl11017, flux, theory)* Alkali,	000621
#Optical and Dilatometric Studies of KCaCl3 and		RbCaCl3 Crystals (hygroscopic, Bridgman)*	000633
CsF-EuF3. Study of the Phases RbLnF4 and CsLnF4 (melt,		RbEu2F7)* #The Systems RbF-EuF3 and	000037
#Crystal Structure of		RbEu3F10 (melt)*	000036
CsLnF4 (melt, RbEu2F7)* #The Systems		RbF-EuF3 and CsF-EuF3. Study of the Phases RbLnF4 and	000037
#Crystallographic and Magnetic Structure of		RbFeCl3.2D2O and CsFeCl3.2D2O (solution)*	000077
Some Transition Metal Fluorides. Part 2 (KAlF4, KMnF3,		RbFeF4, Rb2FeF5, Rb3FeF6, RbxFeF3, CsFeF4, CsFe2F7,	001064
#The Systems RbF-EuF3 and CsF-EuF3. Study of the Phases		RbLnF4 and CsLnF4 (melt, RbEu2F7)*	000037
RbTiF4, RbVF4, RbVF3, RbV2F6, CsVF4, KFe2F6, KMnF3,		RbMnF3)*(ZrFeF6, ZrCrF6, ZrNiF6, ZrCoF6, ZrMnF6, ZrVF6,	000319
M(PbX3) (CsPbCl3, CsPbBr3, CsPbI3, KPbI3,		RbPbI3, vapor transport)*of Alkali-Trihalogenoplumbates	000377
Halides (CsSnCl3, RbSnCl3, Cs2SnCl6, Rb2SnCl6, CsSnBr3,		RbSnBr3, Rb2SnBr6, Cs2SnBr6, CsSnI3, RbSnI3, CsSn2I5,	000930
of Some Rubidium and Cesium Tin Halides (CsSnCl3,		RbSnCl3, Cs2SnCl6, Rb2SnCl6, CsSnBr3, RbSnBr3,	000930
Rb2SnCl6, CsSnBr3, RbSnBr3, Rb2SnBr6, Cs2SnBr6, CsSnI3,		RbSnI3, CsSn2I5, RbSn2I5)* (CsSnCl3, RbSnCl3, Cs2SnCl6,	000930
RbSnBr3, Rb2SnBr6, Cs2SnBr6, CsSnI3, RbSnI3, CsSn2I5,		RbSn2I5)*CsSnCl3, RbSnCl3, Cs2SnCl6, Rb2SnCl6, CsSnBr3,	000930
IV (ZrFeF6, ZrCrF6, ZrNiF6, ZrCoF6, ZrMnF6, ZrVF6,		RbTiF4, RbVF4, RbVF3, RbV2F6, CsVF4, KFe2F6, KMnF3,	000319
ZrCrF6, ZrNiF6, ZrCoF6, ZrMnF6, ZrVF6, RbTiF4, RbVF4,		RbVF4, RbVF3, RbV2F6, CsVF4, KFe2F6, KMnF3, RbMnF3)* (ZrFeF6,	000319
(ZrFeF6, ZrCrF6, ZrNiF6, ZrCoF6, ZrMnF6, ZrVF6, RbTiF4,		RbVF3, RbV2F6, CsVF4, KFe2F6, KMnF3, RbMnF3)* IV	000319
ZrNiF6, ZrCoF6, ZrMnF6, ZrVF6, RbTiF4, RbVF4, RbVF3,		RbV2F6, CsVF4, KFe2F6, KMnF3, RbMnF3)* (ZrFeF6, ZrCrF6,	000319
Part 2 (KAlF4, KMnF3, RbFeF4, Rb2FeF5, Rb3FeF6,		RbxFeF3, CsFeF4, CsFe2F7, Cs3Fe2F9, CsxFeF3, Na2CoFeF7,	001064
those of Potassium, Rubidium and Cesium (review, KxWO3,		RbxWO3, CsxWO3)* Metal Tungsten Bronzes, in Particular	000402
theory)* #Local Modes in		Rb1-cKc Alloys: A Neutron Scattering Study (Bridgman,	000459
Phase #The Growth of Single Crystals of Langbeinites		Rb2Cd2(SO4)3, Tl2Cd2(SO4)3, K2Co2(SO4)3, and their	000138
Na2CoFeF7, Na2NiFeF7, Na2MnF7, Na2ZnCrF7, NaCrF4,		Rb2Cr5F17)*RbxFeF3, CsFeF4, CsFe2F7, Cs3Fe2F9, CsxFeF3,	001064
Metal Fluorides. Part 2 (KAlF4, KMnF3, RbFeF4,		Rb2FeF5, Rb3FeF6, RbxFeF3, CsFeF4, CsFe2F7, Cs3Fe2F9,	001064
#Ferroelasticity and Phase Transformation in		Rb2Hg(CN)4 Spinel (solution)*	000378
CsSnCl3, RbSnCl3, Cs2SnCl6, RbSnCl6, CsSnBr3, RbSnBr3,		Rb2SnBr6, Cs2SnBr6, CsSnI3, RbSnI3, CsSn2I5, RbSn2I5)* (	000930
and Cesium Tin Halides (CsSnCl3, RbSnCl3, Cs2SnCl6,		Rb2SnCl6, CsSnBr3, RbSnBr3, Rb2SnBr6, Cs2SnBr6, CsSnI3,	000930
Study of the Incommensurate Phase Transition of		Rb2ZnBr4 (theory, solution)* #Neutron-Scattering	000226
Metal Fluorides. Part 2 (KAlF4, KMnF3, RbFeF4, Rb2FeF5,		Rb3FeF6, RbxFeF3, CsFeF4, CsFe2F7, Cs3Fe2F9, CsxFeF3,	001064
Rubidium and Cesium Niobates (Czochralski, Cs8Nb22O59,		Rb8Nb22O59)* #The Growth and Properties of Crystalline	000440
of Rare Earth Copper Intermetallic Compounds,		RCu2. I. Heavy Rare Earth (TbCu2, DyCu2, HoCu2, ErCu2)*	000373
of Rare Earth Copper Intermetallic Compounds,		RCu2. II. Light Rare Earth (PrCu2, NdCu2)* Properties	000374
#Re-Investigation of the		Re-Entrant Corner Effect in Twinned Crystals (theory)*	000484
Twinned Crystals (theory)*	#	Re-Investigation of the Re-Entrant Corner Effect in	000484
Mismatches Between Epitaxial Films and Substrates (		RE garnets)* Setting for Measuring Local Lattice	000704
#Growing Optical films of Aluminum Yttrium Garnet (		RE substituted Y3Al5O12, flux)*	000044
of Some Ln2(MoO4)3 Type Mixed Rare Earth Molybdates (		RE (MoO4)3, Czochralski)* and Crystallographic Data	000447
Crystals of Cubic Zinc Sulphide by Double Decomposition		Reaction in Melt (ZnS, flux)* #Growth of Single	000328
Vapor (Ge, epitaxy, vapor transport, kinetics, theory)**		Reaction of Germanium (111) Single Crystals with GeBr4	000060
theory)* #Correlation of the Interface		Reaction Constant with Elementary Surface Processes (	000330
Preparation of ThSiO4 Single Crystals by a Vapour Phase		Reaction* #	000454
of Single V3O5 Crystals by Chemical Transport		Reactions (vapor transport)* #Preparation	000789
and Its Application to the Study of Heterogeneous		Reactions (GaAs, vapor transport, theory)* Method	000281
Gallium Phosphide Synthesis by Chemical Gas Transport		Reactions (GaP)* #Kinetics of	000231
Characterization (slow cooling)* #Solid State		Reactions to CdTeMoO6 and Its Structural	000297
Technique in Solid State Chemistry. I. High Temperature		Reactions with CO2-Laser* #A New Preparation	000657
State Chemistry. II. High Pressure - High Temperature		Reactions with CO2-Lasers* Technique in Solid	000658
Growth of Anatase (TiO2) Crystals by Chemical Transport		Reactions with HBr and HCl* #	000431

PERMUTED TITLE INDEX

PERMUTED TITLE INDEX

in GaAs (GaAs, Ga1-xAlxAs, molecular beam epitaxy,	review)*
and Crystal Growth of A(II)B(VI)-MnB(VI) Systems (	review)*
Dislocations in Epitaxial Semiconductor Layers (III-V,	review)*
of Actinide Pnictides by Chemical Vapour Transport (	review)*
in MBE of Compound Semiconductors (II-VI, IV-VI, III-V,	review)*
of Silicon for Integrated Circuits and Solar Cells (Si,	review)*
International Summer School on Crystal Growth, 1977) (	review)*
and Material Characterization (Si, laser melting,	review)*
in National Economy, Leningrad, March 14-17, 1979 (	review)*
Radiation Detectors. Part II. HgI2 (vapor transport,	review)*
via Use of Precipitated Phases (eutectics, carbides,	review)*
Alloys (electron beam melting, float zone melting,	review)*
(melt, solution, vapor transport, zone melting, doped,	review)**
Refractory Materials by RF Melting in a Cold Container (	review, alpha-Al2O3, Al2O3:Cr2O3)*
The History and Present Status of Synthetic Diamond (C,	review, doped)*
#Crystal Growth - Theory and Techniques, Volume 2 (	review, epitaxy, Verneuil, III-V compounds, Si)*
Optical Devices Fabricated by MBE (GaAs, Ga1-xAlxAs,	review, film)*
(Crystal Growth from High-Temperature Solutions) (	review, flux)*
#Crystals for Magnetic Applications, Vol. 1 (garnets,	review, hydrothermal, epitaxy, flux, Bridgman, film)*
in Preparative Chemistry of Oxides and Fluorides (	review, hydrothermal, theory)*
of Bulk Single Crystals of Quaternary III-V Alloys (	review, kinetics)*
#Crystal Growth from Liquids at High Temperatures (	review, melt, flux, Czochralski, Kyropoulos, float zone
#Electrocrystallization. Fundamental Aspects (	review, theory)*
#Growth of Crystals, Volume 11 (	review, theory)*
#The Role of Lattice Misfit in Epitaxy (	review, theory)*
Growth and Epitaxy of Minerals and Synthetic Crystals (	review, theory)*
Growth of A(III)B(V) Semiconductors from Vapour Phase (	review, theory)*
Transmitting Materials. Part 1. Crystalline Materials (	review, theory)*
State of Matter. Melting and Crystal Structure (melt,	review, theory)*
the Calculation of Heats of Fusion for Ionic Crystals (	review, theory)*
IV-VI compounds, III-V compounds, evaporation,	review, theory)*
of Modes of Epitaxial Growth from the Vapor Phase (	review, theory)*
phosphates)* #The Growth of Crystals in Solution (	review, theory, alkaline earth sulfates, calcium
Simulations of Crystal Surfaces and Crystal Growth (	review, theory, kinetics)*
from Soluble Complexes. I. General Introduction (	review, theory, kinetics)*
Stability, Convection, Graphite, and Integrated Optics (	review, theory, C, GaAs)*
#Nitrides--Structures and Crystal Growth (	review, vapor transport, flux)*
in Particular those of Potassium, Rubidium and Cesium (	review, KxWO3, RbxWO3, CsxWO3)* Metal Tungsten Bronzes,
#Aspects of Silicon Epitaxy (	review, Si)*
#Current Topics in Materials Science, Volume 2 (	Review)*
#Selected Papers on Polymer Crystallization (	Review)*
Earth Iron Garnets in a PbO/B2O3 Solution (Y3Fe5O12,	RExY3-xGayFe5-yO12, flux)* #A Solubility Model for Rare-
Melting of Zirconia-Rare Earth Oxide Crystals (ZrO2-	RE2O3, eutectics, directional solidification)* Skull-
Growth of ZnO Films on Sapphire by Planar Magnetron	rf Sputtering System*
Synthesis and Crystal Growth of Refractory Materials by	RF Melting in a Cold Container (review, alpha-Al2O3,
Films on Potassium Bismuth Niobate Single Crystals by	RF Sputtering (KxLi1-xNbO3)* Niobate Single-Crystal
6Br2 and Ba10(ReO5)6I2 (vapor #Structure of New	Rhenium Apatites Containing Large Anions: Ba10(ReO5)
of Double Oxides of Rhenium. II. The Lanthanum-	Rhenium Oxide La6Re4O18 (melt)* #Crystal Structure
melt)* #Crystal Structure of Double Oxides of	Rhenium. II. The Lanthanum-Rhenium Oxide La6Re4O18 (
Synthesis and Structure of a New Pyrochlore Containing	Rhenium: Ca1-xReO3(OH)2x (x = 0.3)* #High Pressure
#Growth of Single Crystals of Beta	Rhombohedral Boron by a Modified Verneuil Method (B)*
of the Transition-Metal Phosphides CuP2, NiP2, and	RhP3 (flux, vapor transport)* and Characterization
#Modes of Vibrations in Step Trains:	Rhythmical Bunching (theory)*
#Impurity Redistribution in EFG (Si,	ribbon, theory, kinetics)*
#Silicon Ribbon Growth via the Ribbon-To-	Ribbon (RTR) Technique: Process Update and Material
#Aluminum Redistribution in EFG of Silicon	Ribbon (Si)*
#Interface Growth Feature and Voids in Sapphire	Ribbon Crystals (alpha-Al2O3, Czochralski)*
#Wide Silicon	Ribbon Crystals (EDFG, Si)*
Growth Technique (alpha-Al2O3)#Dislocations in Sapphire	Ribbon Crystals Grown by the Edge-Defined, Film-Fed
(Si, EDFG)* #Growth and Characterization of Silicon	Ribbon Crystals Grown with Wetting and Non-Wetting Dies
of Temperature Distribution of Melt in Silicon	Ribbon Growth (EDFG)* #Evaluation
#Thermal Analysis of Solidification in Web-Dendritic	Ribbon Growth (Si, kinetics, theory)*
Process Update and Material Characterization (#Silicon	Ribbon Growth via the Ribbon-To-Ribbon (RTR) Technique:
of Vapor-Deposited Silicon Carbide Films on Silicon	Ribbon Surfaces (SiC)* #Analysis
#In-Situ Carbon-Silicon	Ribbon Thermocouple (Si, equipment)*
Material #Silicon Ribbon Growth via the	Ribbon-To-Ribbon (RTR) Technique: Process Update and
of Metals and Preparation of Single Crystals	Rich in Individual Isotopes* Zone Recrystallization
in the Region of Solidification of Pb-Bi Alloys	Rich in Single-Phase Pb in the Course of Solidification
Gradient Freeze Method (slow cooling, #The Growth of In-	Rich Bulk Single Crystals of GaxIn1-xPyAs1-y Via the
#LPE Growth of Hg0.60Cd0.40Te from Te-	Rich Solution (film, vapor transport)*
#Studies of LPE	Ripple Based on Morphological Stability Theory*
MoO3 (DyKMo2O8, ErKMo2O8, #Flux Growth of Crystals of	RKMo2O8, R2Mo2O6 and R6Mo6O12 in the Systems R2O3-K2O-
Lamellae on the Crystal-Melt Interface of LiCl Doped	Rock Salt Crystals (NaCl, Czochralski)* Closed Growth
Epitaxial Growth of Reactively Sputtered CuI Films on	Rocksalt Substrates* #The
#Anisotropy of Round Evaporation Spirals on	Rocksalt Surfaces (NaCl, theory)*
Pd, Pt, Fe, Al, Au, Ag, In, Pt, PbS, PbSe, PbTe, SnTe, #	Roles of Lattice Fitting in Epitaxy (theory, Ni, Cu,
micrometer #Nd:Y2O3 Single-Crystal Fiber Laser:	Room-Temperature cw Operation at 1.07- and 1.35-
from High-Temperature Solutions) (review, flux)* #	Rost Kristallov iz Rastvorov-Rasplavov (Crystal Growth
Electrical Properties of Deuterated TGS Produced by the	Rotating Disc Technique (solution)* #Growth and
#Effects of	Rotation on Czochralski Crystal Growth (theory)*
in Floating Zone Melting. II. Combined Free and	Rotation Driven Flows (kinetics, theory, electron beam
of Convection in Floating Zone Melting. I. Pure	Rotation Driven Flows (Si, Al2O3, kinetics, float zone
#On the Analytic Treatment of the Growth of	Rough Interfaces (theory)*
#Interface	Roughness of Vapor-Grown Adamantine Crystals (C10H16)*
Carlo Simulation of a (111) Diamond Face Around the	Roughening Transition (C, theory)* #Monte
theory)* #Anisotropy of	Round Evaporation Spirals on Rocksalt Surfaces (NaCl,
Processing (zone melting, #Properties of Rapidly-Grown	RTR Silicon Sheet and Their Effects on Solar Cell
#Silicon Ribbon Growth via the Ribbon-To-Ribbon (	RTR) Technique: Process Update and Material

#Periodic Doping Structure	000248
#Physicochemical Properties	000740
#Nature, Origin and Effect of	000126
#Single Crystal and Film Growth	000160
and Carrier Concentration Control	000897
#Problems in the Melt- and Vapor Growth	000117
#Crystal Growth (Proceedings of the Third	000066
To-Ribbon (RTR) Technique: Process Update	000360
Crystals and Products and Their Application	001111
Spectroscopic Evaluation of Mercuric Iodide	000834
Microstructures and Properties of Superalloys	000671
#Properties and Synthesis of Nb-H Interstitial	001072
#Materials Behaviour in Low Gravity Environment	001070
#and Crystal Growth of	000020
	000685
	000348
#Integrated	000152
	000976
	000807
#Advances	000398
#Melt and Solution Growth	000054
	000139
	000026
	000183
	001014
#Vapor	000932
#Epitaxial	000352
#Infra-Red	000250
#The Molten	000999
#Quantum-Mechanical Relations in	001087
#Hot Wall Epitaxy (II-VI compounds,	000568
Methods Used for the Characterization	000542
	000683
#Survey of Monte Carlo	001013
Soluble Single Crystals	000693
Vol. 4, Morphological	001081
	000533
	000402
	000701
	000449
	001089
	001022
	000629
	000882
	000020
	000009
	000080
	000103
	000103
	000104
	000213
	000708
	000933
	000451
	000360
	000452
	001050
	000522
	001051
	000524
	000523
	000369
	000360
	001099
	000818
	000360
	000994
	000846
	000053
	000839
	000698
	001062
	000532
	000788
	000530
	000949
	000916
	000976
	000560
	000796
	000174
	000173
	000764
	000753
	001020
	000530
	000546
	000360

PERMUTED TITLE INDEX

Tungsten Bronzes, in Particular those of Potassium, Rubidium and Cesium (review, KxWO3, RbxWO3, CsxWO3)* 000402
Rb8Nb22O59)* #The Growth and Properties of Crystalline Rubidium and Cesium Niobates (Czochralski, Cs8Nb22O59, 000440
Cs2SnCl6, #Gel Growth of Single Crystals of Some Rubidium and Cesium Tin Halides (CsSnCl3, RbSnCl3, 000930
Elastic Constants of a Ruthenium-Nickel Single Crystal (RuNi, Bridgman)* Anisotropy, Magnetostriction and 000631
and Characterization of Ruthenium Dioxide Crystals (RuO2, vapor transport)* #Preparation 000865
#A Novel Ruthenium Bronze: KRu4O8 (solution)* 001082
#Preparation and Characterization of Ruthenium Dioxide Crystals (RuO2, vapor transport)* 000865
* #La8Ru4O21: A Mixed-Valence Ternary Ruthenium Oxide of a New Hexagonal Structure Type (flux) 000208
Anisotropy, Magnetostriction and Elastic Constants of a Ruthenium-Nickel Single Crystal (RuNi, Bridgman)* 000631
TiO2)* #Plasma Growth of Rutile Crystals and Their Photoelectronic Properties (000387
#Anisotropy of Ion Transport in Stoichiometric Rutile Single Crystals (TiO2)* 000778
DyKMo2O8, #Flux Growth of Crystals of RKMo2O8, R2MoO6 and R6MoO12 in the Systems R2O3-K2O-MoO3 (001062
Crystals of RKMo2O8, R2MoO6 and R6MoO12 in the Systems R2O3-K2O-MoO3 (DyKMo2O8, ErKMo2O8, NdKMo2O8, Dy2MoO6, 001062
#Flux Growth of Polymorphic Rare-Earth Disilicates, R2Si2O7 (R = Tm, Er, Ho, Dy) (Tm2Si2O7, Er2Si2O7, 000595
#Flux Growth of Crystals of RKMo2O8, R2MoO6 and R6MoO12 in the Systems R2O3-K2O-MoO3 (DyKMo2O8, 001062
#Growth of Si Strips by Stepanov' s Method* 000116
#Ag5GaS4 -- A New Phase in the System Ag-Ga- S (AgGaS2)* 000688
#Liquid Phase Epitaxial Growth of Zn and S Doped GaAs* 000416
#The Crystal Growth and Atomic Structure of As2(Se, S)3 Compounds (As2S3, As2Se3, vapor transport)* 000896
#Epitaxial Growth of ZnS by Zn- S-H2 CVD Method* 000615
Solubility of Barium Sulphate (BaSO4) #Effect of S-35 Tracer, Particle Size and Imperfection on the 000779
Properties of Niobium Chalcogenide Halides, NbX2Y2(X= S, Se - Y=Cl, Br, I) (vapor transport)* Structures, and 000795
#New Thiochlorides of Molybdenum II: Mo6Cl10Y (Y = S, Se, Te): Structure and Magnetic and Electrical 000757
II)C(IV)X4 (BII=Zn, Mn, Fe, Co - CIV = Si, Ge, Sn - X = S,Se) (Cu2ZnGeS4, Cu2ZnSnS4, Cu2ZnGeSe4, Cu2ZnSnSe4, 000357
Characterisation of Large Single Crystals of ZnS and Zn(S,Se)* #Vapour Growth and Defect 000816
on the Crystal-Melt Interface of LiCl Doped Rock Salt Crystals (NaCl, Czochralski)* Growth Lamellae 000532
Stage #The Synthesis of the First Stage Graphite Salt C8(+)AsF6(-) and Its Relationship to the First 000076
for Stable Growth of Epitaxial GaP Layers by Molten Salt Electrodeposition (flux)* #Conditions 000224
of NbO and Hollow NbO2 Needles from Molten Salts (flux)* #Electrocrystallization 000204
#Crystallization of Salts from Supersaturated Solutions: Diffusion Kinetics* 000623
Silicon Source for Epitaxial Films on Insulators (Si on sapphire, vapor transport)* #Dichlorosilane as 000254
Surface on Intermediately Sputtered ZnO Thin Film on Sapphire (epitaxy)* Crystalline ZnO Film with Smooth 000711
Structure of Heteroepitaxial Zinc Oxide Films Grown on Sapphire (vapor transport, ZnO)* #Real 000526
melt)* #Profile- Sapphire and Its Structural Perfection (alpha-Al2O3, 000246
Zone Length in Float-Zone Growth of Small-Diameter Sapphire and Silicon Crystals (Al2O3, Si)* Stable 000479
Czochralski, epitaxy, Al2O3)* #Evaluation of Verneuil Sapphire as a Substrate for Silicon on an Insulator (000793
#High Rate Epitaxial Growth of ZnO Films on Sapphire by Planar Magnetron rf Sputtering System* 000882
#Epitaxial Growth of CrO2 on Sapphire in Air (vapor transport)* 000413
#Interface Growth Feature and Voids in Sapphire Ribbon Crystals (alpha-Al2O3, Czochralski)* 001050
Film-Fed Growth Technique (alpha-Al2O3) #Dislocations in Sapphire Ribbon Crystals Grown by the Edge-Defined, 001051
#Properties of GaN Grown on Sapphire Substrates (epitaxy)* 000212
#Heteroepitaxial Structures of Si on Sapphire, Obtained by Vacuum Deposition (film, doped)* 000823
of Epitaxial Ice Crystals (#Influence of Environmental Saturation and Electric Field on Growth and Evaporation 000028
#Growth of LiTaO3 Single Crystals for Saw Device Applications (Czochralski)* 000311
#Magneto-Seebeck Effect of Pb-Doped Bi- Sb (Bridgman)* 000034
#Growth of Antimony on Bismuth (Sb film, epitaxy)* 000501
#Si-MBE: Growth and Sb Doping (Si, film)* 000502
#Crystallization of Bi- Sb Solid Solutions in Capillaries (zone melting)* 001035
on Oriented Single Crystal Surfaces of Gold and Silver (Sb)* #Epitaxial Growth of Antimony Electrodeposits 000540
and Properties of Two Indium Antimony Selenides (In2Se3, Sb, In2-xSbxSe3, InSbSe3, Bridgman)* #Preparation 000911
#Synthesis and Hydrothermal Growth of the Ferroelectric SbNbO4, SbTaO4 and Pb5S2I6 Single Crystals* 000777
#Crystal Growth of Antimony Sulphur Iodide (SbSi, flux)* 000109
and Hydrothermal Growth of the Ferroelectric SbNbO4, SbTaO4 and Pb5S2I6 Single Crystals* #Synthesis 000777
#Scandium Hydrogenselenite (Sc(HSeO3)3, evaporation, solution)* 001008
#Crystal Structure of Scandium Metaphosphate Sc(PO3)3 (solution)* 000902
#Chemical Vapor Deposition of Scandium Hydride (ScH2, epitaxy, film)* 000494
solution)* #Scandium Hydrogenselenite (Sc(HSeO3)3, evaporation, 001008
#Crystal Structure of Scandium Metaphosphate Sc(PO3)3 (solution)* 000902
from Its Quiescent Aqueous Solution by Differential Scanning Calorimeter. Growth Rate of Crystallite (KBrO3) 000367
Synthetic Quartz (SiO2, hydrothermal)* #Scanning Electron Microscopy of Growth Surfaces of 000857
Alumina (NaAl11O17, KAl11O17, RbAl11O17, #X-Ray Diffuse Scattering from Alkali, Silver, and Europium beta- 000621
#Raman Scattering in 3R-NbS2 (vapor transport, theory)* 000731
Alloys, Ni2MnSn and Ni2Mn1-xVxSn, for Neutron Inelastic Scattering Experiments (Czochralski, slow cooling)* 000321
Type Compounds K2Mn1-xMxF4 (M + #Magnetic and Neutron Scattering Experiments on the Antiferromagnetic Layer- 000106
RbAg4I5 Family (KAg4I5, NH4Ag4I5, solution)* #Raman Scattering Studies in the Solid Electrolytes of the 000315
#Local Modes in Rb1-cKc Alloys: A Neutron Scattering Study (Bridgman, theory)* 000459
of Rb2ZnBr4 (theory, solution)* #Neutron- Scattering Study of the Incommensurate Phase Transition 000226
of Growth Defects in Synthetic Quartz Crystals by Light- Scattering Tomography (SiO2, hydrothermal)**Observation 000652
Growth (Proceedings of the Third International Summer School on Crystal Growth, 1977) (review)* #Crystal 000066
Epitaxy (MBE) on GaAs (Al, #Single-Crystal-Aluminum Schottky-Barrier Diodes Prepared by Molecular-Beam 000190
#Chemical Vapor Deposition of Scandium Hydride (ScH2, epitaxy, film)* 000494
#Current Topics in Materials Science, Volume 1 (review)* 000450
#Current Topics in Materials Science, Volume 2 (Review)* 000449
Method (ZnF2, CoF2, #Crystal Growth by the Thermic Screen Translation (TST) Technique: A Modified Bridgman 000541
Properties of Niobium Chalcogenide Halides, NbX2Y2(X=S, Se - Y=Cl, Br, I) (vapor transport)* Structures, and 000795
C(IV)X4 (BII=Zn, Mn, Fe, Co - CIV = Si, Ge, Sn - X = S, Se) (Cu2ZnGeS4, Cu2ZnSnS4, Cu2ZnGeSe4, Cu2ZnSnSe4, 000357
of Large Single Crystals of ZnS and Zn(S, Se)* #Vapour Growth and Defect Characterisation 000816
#Growth and Cleavage of Se-Te Alloy Crystals (Bridgman)* 000111
#Growth of Hollow Se-Te Whisker Crystals (theory)* 000112
#Structure and Growth Peculiarities of Tl- Se-TlInSe2 (TlSe, zone melting, slow cooling)* 000361
#New Thiochlorides of Molybdenum II: Mo6Cl10Y (Y = S, Se, Te): Structure and Magnetic and Electrical 000757
#The Crystal Growth and Atomic Structure of As2(Se,S)3 Compounds (As2S3, As2Se3, vapor transport)* 000896
#The Growth of CdS in Sealed Silica Capsules (vapor transport)* 000815
#Preparation of Films of IV-VI Compounds in a Quasi- Sealed Volume (SnTe, PbTe, PbSe, evaporation)* 000306
7H2O, solution)* #Direct Observation of Secondary Nuclei Production (KAl(SO4)2.12H2O, MgSO4. 000322
#X-Ray Topographic Study of Growth Sector Strains in KDP Single Crystals (solution)* 000092
#Magneto- Seebeck Effect of Pb-Doped Bi-Sb (Bridgman)* 000034
#Top- Seeded Flux Growth of LiNdP4O12 Single Crystals* 000678

PERMUTED TITLE INDEX

Vapor Phase*	# Seeded Growth of Large Single Crystals of CdS from the 000242
Quartz Surface (sublimation)*	#Non- Seeded Growth of Large Single Pb1-xSnxTe Crystals on a 000960
sublimation)*	Seeded Pb1-xSnxTe Single Crystals (vapor transport, 000961
#Growth Study of Large Non-	Seeds (flux)* Characteristics of Growth of Y3Al5O12 000978
Single Crystals from Melt-Solutions on Oriented	Seeds (TGS, TBFB)* 000566
#Crystal Growth from Solution Using Cylindrical	Seeds in Melt-Solutions (flux)*Crystallization Kinetics 000093
and Growth Mechanism of Y3Al5O12 Crystals on Oriented	Seeds Coated with Informative Interfacial Layers (000243
Grown by the Hydrothermal Method on Single Crystal	Segregation Behavior of Ga-Doped Ge (theory)* of 000604
a Heat Pipe to Czochralski Growth. Part I. Growth and	Segregation Coefficient (Si, vapor transport, theory)* 000514
Growth from the Gas Phase. I. Phosphorus Content and	Segregation During Crystal Growth from the Melt (001084
#A New Look at the Burton, Prim, and Slichter Model of	Segregation During Unidirectional Solidification of a 000206
Binary Alloy with a Curved Solid-Liquid #Lateral Solute	Segregation Tendency by the Bridgman Method - 000244
#Growth of Homogeneous Mixed Crystals with a Large	Segregation Under Zero Gravity: Ge (Bridgman)* 001088
#Crystal Growth and	# Selected Papers on Polymer Crystallization (Review)* 001089
	# Selective Etching of Cesium Iodide Crystals and 000827
Comparison of Growth and Dissolution Processes (CsI,	Selective Etching Studies on Sodium Chlorate Crystals (000784
NaClO3, solution)* #Crystal Growth and	Selective Homoepitaxy of Silicon at Reduced 000136
Temperatures Using a Silicon-Iodine Transport #Local	Selenides VxTi1-xSe2 and TaxTi1-xSe2 (TiSe2, VSe2, 000555
and Electrical Properties of Layered Transition Metal	Selenide (CdSe, epitaxy, kinetics, films, theory)* 000365
Properties of Cadmium, Selenium, and Cadmium	Selenide (Cd1+y (SxSe1-x), sublimation)* #Vapor Phase 000647
Transport and Stoichiometry Control of Cadmium Sulfo-	Selenide (SnSe, Bridgman)* 000014
#Single Crystal Growth of Stannous	Selenide and Gallium Sulphide by Iodine Vapour 001078
#Optimization of Conditions for the Growth of Gallium	Selenide from the Vapour Phase (ZnSe, vapor transport, 000214
sublimation)* #Growth of Single Crystals of Zinc	Selenide Crystals Grown from the Melt Under Argon 000477
Pressure (ZnSe, Bridgman)**Formation of Defects in Zinc	Selenide In2Se3 (vapor transport)* 000557
#Crystal Structure of the Indium	Selenides (In2Se3:Sb, In2-xSbxSe3, InSbSe3, Bridgman)* 000911
#Preparation and Properties of Two Indium Antimony	Selenium, and Cadmium Selenide (CdSe, epitaxy, 000365
Source Shape on the Sublimation Properties of Cadmium,	Self-Implanted Silicon Layer on Laser Irradiation 000996
Energy Density (Si)* #Grain Size Dependence in a	Semi-Periodic Closed Growth Lamellae on the Crystal- 000532
Melt Interface of LiCl Doped Rock Salt Crystals (NaCl,	Semiconducting Phase* #Optical Properties 001023
of V6O13 Single Crystals in the Metallic and the	Semiconductor CdCr2Se4 (vapor transport, doped)* 000625
#Preparation of Single Crystals of the Magnetic	Semiconductor Films (PbS, PbSe, Pb1-xHgxS, solution)* 000868
#Electroless Deposition of	Semiconductor Layers (III-V, review)* 000126
#Nature, Origin and Effect of Dislocations in Epitaxial	Semiconductor Single Crystals and Films (Si, Ge, GaAs, 000636
InSb, #Creation of Defects During the Growth of	Semiconductor Single Crystals Obtained by the Traveling 000064
Solvent Method from (Pb0.8Sn0.2)Te*	# Semiconductor Superlattices by MBE and Their 000176
Characterization (GaAs, Ga1-xAlxAs, In1-xGaxAs, GaSb1-	Semiconductor Surface and Crystal Physics Studied by 000055
MBE (GaAs, Ga1-xAlxAs, review of MBE process)*	Semiconductors (II-VI, IV-VI, III-V, review)* 000897
and Carrier Concentration Control in MBE of Compound	Semiconductors (PbTe, Pb1-xSnxSe, Pb1-xSnxTe)*Diffusion 000898
During Molecular Beam Epitaxial Growth of A(III)B(V)	Semiconductors from Vapour Phase (review, theory)* 000352
#Epitaxial Growth of A(III)B(V)	Semiconductors: Elements, III-V Compounds and Alloys (000041
Ge, Si, #Some Aspects of the Epitaxial Vapour Growth of	SeO3)2 (solution)* 000963
#Isotope Effects on the Central Peak Phenomena in KD3(	SeO4)3 Single Crystals in Silica Gels* 000121
#Growth of Bi2(	Series of Lead Tungsten Bronzes (PbxWO3)* 000267
#A	Series GaSxSe(1-x) (0 less than or equal to x less than 001077
or equal to 1) #Structural Studies of Compounds in the	Series Mo1-xWxSe2 (vapor transport)* Conditions and 000013
Crystal Structure Parameters of Layer Compounds in the	Setting for Measuring Local Lattice Mismatches Between 000704
#An X-Ray Double-Crystal Method Utilizing Non-Parallel	Se1-xSx)2 Solid Solutions (Bridgman)* Crystal Growth, 000635
and Optical Properties of CdGe(As1-xPx)2 and AgGa(-	Se2TiO6 (slow cooling)* 000547
Study of the System Ni3Pb2S2-Ni3Pb2Se2 and Note on the	Shandite Structure (Ni3Pb2Se2, vapor transport)* 000198
Vapor Growth of LaB6 Whiskers and Crystals Having a	Sharp Tip* #Chemical 000656
melting, Si)* #Properties of Rapidly-Grown RTR Silicon	Sheet and Their Effects on Solar Cell Processing (zone 000654
CuxHfS2, FexZrS2, FexHfS2, vapor #Intercalation-Induced	Shift of the Absorption Edge in ZrS2 and HfS2 (CuxZrS2, 001090
as Silicon Source for Epitaxial Films on Insulators (	Si on sapphire, vapor transport)* #Dichlorosilane 000254
doped)* #Heteroepitaxial Structures of	Si on Sapphire, Obtained by Vacuum Deposition (film, 000823
	# Si Contamination in Epitaxial Boron Monophosphide (BP)* 000713
#Preparation of Ge-	Si Epitaxial Alloys by Sputtering* 000420
transport)* #Crystalline Defects in Solid Phase Epitaxy	Si Films Deposited at Elevated Temperatures (vapor 000194
#Solar Furnace Annealing of Amorphous	Si Layers (films)* 000537
Twinned Silicon Crystals Grown in an Al-16 wt.%	Si Melt (Si, flux)* #Fivefold 000489
Various Growth Rates* #Growth Striations in Fe-3wt%	Si Single Crystals Grown by Floating Zone Melting at 001028
#Growth of	Si Strips by Stepanov's Method* 000116
Cluster Beam Epitaxial Growth of GaP Films on GaP and	Si Substrates* #Ionized- 000651
#Growth Kinetics of (111)	Si Through an Al Layer by Solid Phase Epitaxy* 000585
#Solid Phase Epitaxial Growth of	Si Through Al Film* 000584
evaporation)* #Silicide Formation by High-Dose	Si+-Ion Implantation of Pd (electron beam melting, 000177
Distribution in Dislocation-Free Single Crystals of	Si* #Influence of Growth Conditions on Microdefect 000687
#Aluminum Redistribution in EFG of Silicon Ribbon (	Si)* 000452
#Wide Silicon Ribbon Crystals (EDFG,	Si)* 000522
#Aspects of Silicon Epitaxy (review,	Si)* 000701
#Solution Growth of Indium-Doped Silicon (	Si)* 000849
#Surface Morphology of HCl Etched Silicon Wafers (	Si)* 001015
of the Chemical Vapour Deposition of Silicon (	Si)* #Mechanisms 000118
Free Czochralski Growth of (110) Silicon Crystals (	Si)* #Dislocation- 000262
p+-n+ Back-Surface-Field Silicon Solar Cells (	Si)* #High-Efficiency 000298
Growth Under Low Hydrogen Pressure (vapor transport,	Si)* #Kinetics of Silicon 000255
of Silicon Crystals With Automatic Diameter Control (	Si)* #Growth and Properties 000998
Epitaxial Layer Growth from Ion-Molecular Beams (	Si)* #The Mechanism of Silicon 000019
by Bardeen-Herring Mechanism in Czochralski Silicon (	Si)* #Growth of Stacking Faults 001052
with Phosphorus During Growth from the Gas Phase (	Si)* #Doping of Silicon Crystals 000027
Distributions During Czochralski Crystal Growth (Cu,	Si)* #A Theoretical Study of Temperature 000149
in the Modified Czochralski Method of Crystal Growth (	Si)* #Variations of Impurity Distribution 000997
Volume 2 (review, epitaxy, Verneuil, III-V compounds,	Si)* #Crystal Growth - Theory and Techniques, 000348
from a Molten Bath According to the Stepanov Procedure (	Si)* #Preparation of Silicon Plates by Pulling 000966
of Laser-Induced Epitaxial Layers in Silicon (	Si)* #Electrical and Structural Characteristics 001106
Silicon Layer on Laser Irradiation Energy Density (	Si)* #Grain Size Dependence in a Self-Implanted 000996
and Gold as Liquid-Forming Agents (vapor transport,	Si)* of VLS Silicon Crystals Grown with Platinum 001075
H2 System Based on Flow Graph (vapor transport, theory,	Si)* #Model of Silicon Epitaxial Growth in SiCl4-HCl- 000503

PERMUTED TITLE INDEX

of Small-Diameter Sapphire and Silicon Crystals (Al2O3,	Si)* #Maximum Stable Zone Length in Float-Zone Growth 000479
on Tensions in Crystals (theory, Czochralski, melt,	Si)* #Effects of Radial and Axial Temperature Gradients 000326
Temperatures Using a Silicon-Iodine Transport System (	Si)* #Local Selective Homoepitaxy of Silicon at Reduced 000136
Their Effects on Solar Cell Processing (zone melting,	Si)* #Properties of Rapidly-Grown RTR Silicon Sheet and 000546
(Si, Al2O3, theory, float zone melting, kinetics,	Si)*of Zone Geometry and Prandtl Number at Zero Gravity 000175
Epitaxial Growth by Using a Flame Photometric Detector (	Si)*Doping Gas Concentration During Vapor Phase Silicon 000937
Deposition and by Film Doping Using Boron Diffusion (	Si)*Silicon Wafers by Thin Polycrystalline Silicon Film 000985
by Electron-Beam Zone Melting Without a Crucible (	Si-Ge)* Crystals of Silicon-Germanium Solid Solutions 000820
of Equilibrium Adsorption Layers in CVD Systems (	Si-H-Cl, Ga-As-H-Cl) (GaAs, theory, epitaxy)* Analysis 000184
	Si-MBE: Growth and Sb Doping (Si, film)* 000502
of a Crystal Needle of Si2ON2 by Nitridization in the	Si-SiO2 System in Compacted Powders* #Growth 000049
#Mechanisms of Chemical Vapor Deposition of Silicon (	Si, epitaxy)* 000702
From Beams of Ionized Atoms and Clusters (Ag on	Si, epitaxy)* #Thin Film Deposition 000516
#In-Situ Carbon-Silicon Ribbon Thermocouple (	Si, equipment)* 000818
#Si-MBE: Growth and Sb Doping (	Si, film)* 000502
Arsenic Ion Implantation and by Antimony Evaporation (	Si, film)* Molecular Beam Epitaxy by Simultaneous 000734
Performance of a New High-Intensity Silicon Solar Cell (	Si, float zone melting)* # 000303
Silicon Crystals Grown in an Al-16 wt.% Si Melt (	Si, flux)* #Fivefold Twinned 000489
of Solidification in Web-Dendritic Ribbon Growth (	Si, kinetics, theory)* #Thermal Analysis 000369
Process Update and Material Characterization (	Si, laser melting, review)* To-Ribbon (RTR) Technique: 000360
and Characterization of Metallurgical Silicon (	Si, melt, Czochralski)* #Purification 000196
of Silicon for Integrated Circuits and Solar Cells (	Si, review)* #Problems in the Melt- and Vapor Growth 000117
#Impurity Redistribution in EFG (	Si, ribbon, theory, kinetics)* 000451
Partitioning During Silicon Dendritic Web Growth (	Si, theory)* #Solute 000855
Radio Frequencies for Float Zone Silicon Crystals (	Si, theory)* #Limitations in Using Kilohertz 000358
Melt-Grown Crystals, Taking Account of Periodic Remelt (	Si, theory)* of Striated Impurity Distributions in 001026
the Cooling Rates of Large-Diameter Silicon Crystals (	Si, theory, kinetics, Czochralski)* #On 000165
#Polytypism and Amorphousness in Silicon Whiskers (	Si, vapor transport)* 000642
Novel Reactor for High Volume Low-Cost Silicon Epitaxy (	Si, vapor transport, kinetics)* # 000062
Content and Differential Molar Solution Enthalpies (	Si, vapor transport, theory)* Gas Phase. II. Phosphorus 000515
I. Phosphorus Content and Segregation Coefficient (	Si, vapor transport, theory)*Growth from the Gas Phase. 000514
Floating Zone Melting. I. Pure Rotation Driven Flows (	Si, Al2O3, kinetics, float zone melting, theory)* in 000173
of Zone Geometry and Prandtl Number at Zero Gravity (	Si, Al2O3, theory, float zone melting, kinetics, Si)* 000175
Monocrystals of Dilute Binary Metallic Alloys (Cu – 0.5%	Si, Cu1-xSbx – x = 0.0015-0.01)* Cellular Structures of 000210
Angle Boundary Formation in Silicon Single Crystals (	Si, Czochralski)* #Impurity Effects on Low- 000218
#Solar Cells from Zone-Refined Metallurgical Silicon (	Si, Czochralski, zone melting, vapor transport)* 000195
Crystals Grown with Wetting and Non-Wetting Dies (	Si, EDFG)*Growth and Characterization of Silicon Ribbon 000524
Elements, III-V Compounds and Alloys (Ge,	Si, GaAs, BP)* Vapour Growth of Semiconductors: 000041
Formation of Stress and Dislocations in Crystal Growth (	Si, Ge, GaAs, InAs, InSb, Al2O3, melt, theory)* # 000406
the Growth of Semiconductor Single Crystals and Films (	Si, Ge, GaAs, InSb, epitaxy, theory, Czochralski)* 000636
Cu2(I)B(II)C(IV)X4 (BII=Zn, Mn, Fe, Co – CIV =	Si, Ge, Sn – X = S,Se) (Cu2ZnGeS4, Cu2ZnSnS4, 000357
#The Occurrence of High Period Polytype of	SiC at the Interface of Two Interacting Spirals* 000889
#Growing Monocrystalline Bars of	SiC from the Gas Phase (vapor transport)* 000554
vapor transport, film)* #Heteroepitaxial Growth of beta-	SiC on Silicon Substrate Using SiCl4-C3H8-H2 System (000616
#Growth and Morphology of 6H-	SiC Epitaxial Layers by CVD (film)* 000699
Silicon Carbide Films on Silicon Ribbon Surfaces (	SiC)* #Analysis of Vapor-Deposited 001099
of Silicon Carbide Obtained by Sublimation in Vacuum (	SiC)* #Growth Kinetics of Epitaxial Layers 000940
of Silicon Carbide Epitaxial Growth Kinetics in the	SiC-C System (SiC, theory)* #Study 000562
Carbide Epitaxial Layer Growth from the Vapour Phase (	SiC, film, vapor transport, kinetics)* of Silicon 000942
Processes of Ingots of Silicon Carbide Single Crystals (	SiC, melt)* #Investigation of Growth 000941
Carbide Monocrystal and Epitaxial Layer Growing (	SiC, melt)* in High-Temperature Furnaces for Silicon 000561
Carbide Epitaxial Growth Kinetics in the SiC-C System (	SiC, theory)* #Study of Silicon 000562
Doped Silicon Carbide by Pyrolysis of Silane Compounds (	SiC, vapor transport, De Boer-Van Arkel, doped)* and 001043
#Some Aspects of Polytypism in Crystals (	SiC, ZnS, CdI2, mica, theory, review)* 000069
Growth of beta-SiC on Silicon Substrate Using	SiCl4-C3H8-H2 System (vapor transport, film)* 000616
transport, #Model of Silicon Epitaxial Growth in	SiCl4-HCl-H2 System Based on Flow Graph (vapor 000503
of Pure and Doped Silicon Carbide by Pyrolysis of	Silane Compounds (SiC, vapor transport, De Boer-Van 001043
#The Growth of CdS in Sealed	Silica Capsules (vapor transport)* 000815
#Growth of Fibrous Sodium Aluminosilicate Crystals in	Silica Gel (zeolite, Na0.5AlSi6O14.xH2O)* 000444
Modification of Barium Molybdate Single Crystals in	Silica Gel (BaMoO4)* #Growth Kinetics and Habit 000519
of Ferroelectric Potassium Dihydrogen Orthophosphate in	Silica Gels (KDP)* #Single Crystal Growth 000442
#Growth of Bi2(SeO4)3 Single Crystals in	Silica Gels* 000121
#Growth of Single Crystals of Ag2SeO4 in	Silica Gels* 000122
#Nucleation and Growth of KClO4 Single Crystals in	Silica Gels* 000751
#Growth of Thallium Iodide Crystals in	Silica Hydrogels (TlI, Tl2I3, gel)* 000881
#Dynamics of Laser-Induced Formation of Palladium	Silicide (Pd2Si)* 001042
Pd (electron beam melting, evaporation)* #	Silicide Formation by High-Dose Si+-Ion Implantation of 000177
#Growth of Chromium	Silicide, Cr3Si, Crystals (float zone melting)* 000441
#Growth and Characterization of Transition Metal	Silicides (chemical vapor transport, solution, cold 000607
#Growth of Films of CdS Evaporated in Vacuum onto	Silicon (vapor transport, electron beam melting, 000730
#Mechanisms of the Chemical Vapour Deposition of	Silicon (Si)* 000118
#Solution Growth of Indium-Doped	Silicon (Si)* 000849
Faults by Bardeen-Herring Mechanism in Czochralski	Silicon (Si)* #Growth of Stacking 001052
Characteristics of Laser-Induced Epitaxial Layers in	Silicon (Si)* #Electrical and Structural 001106
#Mechanisms of Chemical Vapor Deposition of	Silicon (Si, epitaxy)* 000702
#Purification and Characterization of Metallurgical	Silicon (Si, melt, Czochralski)* 000196
* #Solar Cells from Zone-Refined Metallurgical	Silicon (Si, Czochralski, zone melting, vapor transport) 000195
Reconstruction on the Growth and Etching Vectors of	Silicon (111) (theory, vapor transport, epitaxy)* Step 001019
Transport System (Si)* #Local Selective Homoepitaxy of	Silicon at Reduced Temperatures Using a Silicon-Iodine 000136
review)* #Problems in the Melt- and Vapor Growth of	Silicon for Integrated Circuits and Solar Cells (Si, 000117
Dependence of Degree of Doping in Growth of Epitaxial	Silicon from the Gas Phase. II. Phosphorus Content and 000515
#Evaluation of Verneuil Sapphire as a Substrate for	Silicon on an Insulator (Czochralski, epitaxy, Al2O3)* 000793
vapor transport, De Boer-#Preparation of Pure and Doped	Silicon Carbide by Pyrolysis of Silane Compounds (SiC, 001043
System (SiC, theory)* #Study of	Silicon Carbide Epitaxial Growth Kinetics in the SiC-C 000562
#Investigations of Kinetic and Thermal Conditions of	Silicon Carbide Epitaxial Layer Growth from the Vapour 000942
#Analysis of Vapor-Deposited	Silicon Carbide Films on Silicon Ribbon Surfaces (SiC)* 001099
Distribution in High-Temperature Furnaces for	Silicon Carbide Monocrystal and Epitaxial Layer Growing 000561
#Growth Kinetics of Epitaxial Layers of	Silicon Carbide Obtained by Sublimation in Vacuum (SiC)* 000940

PERMUTED TITLE INDEX

Context	Index entry	No.
#Investigation of Growth Processes of Ingots of	Silicon Carbide Single Crystals (SiC, melt)*	000941
in Float-Zone Growth of Small-Diameter Sapphire and	Silicon Crystals (Al203, Si) *Maximum Stable Zone Length	000479
#Dislocation-Free Czochralski Growth of (110)	Silicon Crystals (Si)*	000262
#On the Cooling Rates of Large-Diameter	Silicon Crystals (Si, theory)* #Limitations	000358
Gas Phase (Si)* #Doping of	Silicon Crystals (Si, theory, kinetics, Czochralski)	000165
flux)* #Fivefold Twinned	Silicon Crystals with Phosphorus During Growth from the	000027
Some Notes on the Growth Kinetics and Morphology of VLS	Silicon Crystals Grown in an Al-16 wt.% Si Melt (Si,	000489
#Growth and Properties of	Silicon Crystals Grown with Platinum and Gold as Liquid-	001075
Formation in As-Grown Dislocation-Free Czochralski	Silicon Crystals With Automatic Diameter Control (Si)*	000998
and Crystal Quality During the Crystal Growth of	Silicon Crystals* #The Effect of Doping on Microdefect	000223
#Solute Partitioning During	Silicon Crystals, Free from Dislocations (theory)*	000325
Control of Doping Gas Concentration During Vapor Phase	Silicon Dendritic Web Growth (Si, theory)*	000855
#Temperature Dependence of the Degree of Doping in	Silicon Epitaxial Growth by Using a Flame Photometric	000937
on Flow Graph (vapor transport, theory, Si)* #Model of	Silicon Epitaxial Growth from the Gas Phase. I.	000514
(Si)* #The Mechanism of	Silicon Epitaxial Growth in SiCl4-HCl-H2 System Based	000503
#Aspects of	Silicon Epitaxial Layer Growth from Ion-Molecular Beams	000019
#Novel Reactor for High Volume Low-Cost	Silicon Epitaxy (review, Si) *	000701
#The Bending of Silicon Wafers by Thin Polycrystalline	Silicon Epitaxy (Si, vapor transport, kinetics)*	000062
transport, Si)* #Kinetics of	Silicon Film Deposition and by Film Doping Using Boron	000985
#Grain Size Dependence in a Self-Implanted	Silicon Growth Under Low Hydrogen Pressure (vapor	000255
Ion Implantation and by #n-Type Doping Techniques in	Silicon Layer on Laser Irradiation Energy Density (Si)*	000996
to the Stepanov Procedure (Si)* #Preparation of	Silicon Molecular Beam Epitaxy by Simultaneous Arsenic	000734
#Aluminum Redistribution in EFG of	Silicon Plates by Pulling from a Molten Bath According	000966
#Wide	Silicon Ribbon (Si)*	000452
Wetting Dies (Si, EDFG) **Growth and Characterization of	Silicon Ribbon Crystals (EDFG, Si)*	000522
#Evaluation of Temperature Distribution of Melt in	Silicon Ribbon Crystals Grown with Wetting and Non-	000524
Technique: Process Update and Material #	Silicon Ribbon Growth (EDFG)*	000523
#Analysis of Vapor-Deposited Silicon Carbide Films on	Silicon Ribbon Growth via the Ribbon-To-Ribbon (RTR)	000360
#In-Situ Carbon-	Silicon Ribbon Surfaces (SiC)*	001099
Processing (zone #Properties of Rapidly-Grown RTR	Silicon Ribbon Thermocouple (Si, equipment)*	000818
#Impurity Effects on Low-Angle Boundary Formation in	Silicon Sheet and Their Effects on Solar Cell	000546
#Performance of a New High-Intensity	Silicon Single Crystals (Si, Czochralski)*	000218
#New Methods of Making	Silicon Solar Cell (Si, float zone melting)*	000303
#High-Efficiency p+-n-n+ Back-Surface-Field	Silicon Solar Cells (review)*	000886
sapphire, vapor transport)* #Dichlorosilane as	Silicon Solar Cells (review)*	000298
of a Deposited Amorphous Germanium Layer on a	Silicon Source for Epitaxial Films on Insulators (Si on	000254
transport, film)**Heteroepitaxial Growth of beta-SiC on	Silicon Substrate by Laser Annealing (Ge film)*	000343
#An Investigation of Thin Silver Films on Cleaved	Silicon Substrate Using SiCl4-C3H8-H2 System (vapor	000616
#Molecular Beam and Solid-Phase Epitaxies of	Silicon Surfaces (Ag, theory, evaporation)*	000619
#Surface Morphology of HCl Etched	Silicon Under Ultra-High Vacuum (film)*	000883
Deposition and by Film Doping Using #The Bending of	Silicon Wafers (Si)*	001015
#Polytypism and Amorphousness in	Silicon Wafers by Thin Polycrystalline Silicon Film	000985
and Phase Composition of Thin Films of Aluminum with	Silicon Whiskers (Si, vapor transport)*	000642
Melting Without a Crucible (#Growing Single Crystals of	Silicon* #Structural Transitions	000473
Homoepitaxy of Silicon at Reduced Temperatures Using a	Silicon-Germanium Solid Solutions by Electron-Beam Zone	000820
on Oriented Single Crystal Surfaces of Gold and	Silicon-Iodine Transport System (Si)* #Local Selective	000136
#Structure of	Silver (Sb)* Growth of Antimony Electrodeposits	000540
evaporation)* #An Investigation of Thin	Silver Dichromate: Ag2Cr207 (solution)*	000260
#Unidirectional Solidification of Irregular	Silver Films on Cleaved Silicon Surfaces (Ag, theory,	000619
RbAl11017, #X-Ray Diffuse Scattering from Alkali,	Silver-Germanium Eutectic Alloys (Ag-Ge)*	000549
Investigation of Liquidus Surface in Pb-Sn-Te System by	Silver, and Europium beta-Alumina (NaAl11017, KAl11017,	000621
Growth (theory)* #A New Technique for	Simplex Lattice Method (flux, theory, Pb1-xSnxTe)* #	000668
in the Gaussian Core Model by Molecular Dynamics	Simulating Two-Dimensional Nucleation Models of Crystal	001038
#Fast Growth of Ordered AB Crystals: A Monte Carlo	Simulation (melt, slow cooling, theory)* and Freezing	000915
Roughening Transition (C, theory)* #Monte Carlo	Simulation for Ionic Crystal Growth (theory)*	000182
Jones Substance (theory)* #Monte Carlo	Simulation of a (111) Diamond Face Around the	001020
Cubic Lattice in a Crystal from a Molten Mass in the (	Simulation of the Crystal-Melt Interface of a Lennard-	000123
theory)* #Computational	Simulation of the Growth of a Metallic Face-Centered	000995
Pure Rotation Driven Flows (Si, Al203, #Computer	Simulation of the Melt Flow During Czochralski Growth (	000491
Combined Free and Rotation Driven Flows (#Computer	Simulation of Convection in Floating Zone Melting. I.	000173
Al203, theory, kinetics)* #Digital	Simulation of Convection in Floating Zone Melting. II.	000174
review, theory, kinetics)* #Survey of Monte Carlo	Simulation of Czochralski Bulk Flow in Microgravity (	000536
Doping Techniques in Silicon Molecular Beam Epitaxy by	Simulations of Crystal Surfaces and Crystal Growth (	001013
Single Crystals*	Simultaneous Arsenic Ion Implantation and by Antimony	000734
Deposited by Molecular Beam Epitaxy. I. Growth on	Simultaneous Diffusion of Calcium and Strontium in KCl	000579
theory)* #Appearance of Non-	Singular (100) GaAs Substrates (film, kinetics)*	000762
	Singular Surfaces on Vapour-Grown Ice Crystals (H20,	000689
a Crystal Needle of Si2ON2 by Nitridization in the Si-	SiO2 System in Compacted Powders* #Growth of	000049
Quartz Prepared by Hydrothermal Crystallization (	SiO2)* #Dislocation-Free and Low-Dislocation	000068
Microscopy of Growth Surfaces of Synthetic Quartz (	SiO2, hydrothermal)* #Scanning Electron	000857
of Grown Synthetic Quartz Inside the Autoclaves (	SiO2, hydrothermal)* #On the Dissolution	000507
Quartz Crystals by Light-Scattering Tomography (	SiO2, hydrothermal)* of Growth Defects in Synthetic	000652
Impurities in Synthetic Quartz Single Crystals (	SiO2, solution)* #On the Incorporation of Metallic	000170
#Crystal Structure of beta-Eu2(II)	SiO4 (melt)*	000597
Epitaxial Thin Films* #Techniques for Routine UHV in #In-	situ Electron Microscopy of Growth Processes of	000950
#In-	Situ Carbon-Silicon Ribbon Thermocouple (Si, equipment)*	000818
Very High Voltage Electron Microscope* #In-	Situ Observation of Solidification Interfaces with a	000163
Growth* #In-	Situ X-Ray Topography of Epitaxial Ge Layers During	000363
Diffusion Technique (vapor transport)* #Large-	Sized InP Single Crystals by the Synthesis, Solute	000923
Compacted Powders* #Growth of a Crystal Needle of	Si2ON2 by Nitridization in the Si-SiO2 System in	000049
#Growth, Morphology and Slip System of alpha-	Si3N4 Single Crystal (vapor transport)*	000696
#The Growth of beta-	Si3N4 Single Crystals*	000356
and Crystallographic Characterization (Ln = Nd, Sm, Dy,	skull-melting) Solidification, Microstructural	000630
	Skull Melter Growth of Magnetite (Fe304)*	000370
Zr02-RE203, eutectics, directional #Growth from	Skull-Melting of Zirconia-Rare Earth Oxide Crystals (	000629
#Growth of Crystalline	Slabs of Layered InSe by the Czochralski Method*	000187
#On Interpreting a Quantity in the Burton, Prim and	Slichter Equation as a Diffusion Boundary Layer	001083
from the Melt (#A New Look at the Burton, Prim, and	Slichter Model of Segregation During Crystal Growth	001084

PERMUTED TITLE INDEX

transport)*   #Growth, Morphology and
                    #Influence of the CaC03/Ge02 Ratio on Y
of the Mismatch of Crystal Lattice Parameters ((Y, Bi,
          and Crystallographic Characterization (Ln = Nd,
Mixed Halide Compounds MFX of Divalent Rare Earths (M =
               #An Improved Method of Stirring for LPE Growth (
     and a Study of Their Properties (flux, LaB6, SmB6,
     Method and a Study of Their Properties (flux, LaB6,
     and Rare Earths (LaCrOS2, CeCrOS2, PrCrOS2, NdCrOS2,
     = Sm, Eu, Tm, and Yb - X = Cl, Br, and I) (melt, SmFCl,
     (M = Sm, Eu, Tm, and Yb - X = Cl, Br, and I) (melt,
     Tm, and Yb - X = Cl, Br, and I) (melt, SmFCl, SmFBr,
     (MoO3, WO3, theory)*   #Coalescence Growth of
     Vapor Deposition of Single-Crystalline ZnO Film with
               #Optical and Magneto-Optical Properties of
     Rare-Earth Molybdates (Czochralski, melt, Gd2(MoO4)3,
     R203-K20-MoO3 (DyKMo208, ErKMo208, NdKMo208, Dy2Mo06,
     Cu2(I)B(II)C(IV)X4 (BII=Zn, En, Fe, Co - CIV = Si, Ge,
                    #Growth of High Perfection Ag-
     Unconstrained Growth from a Binary Alloy Melt (
                    #Liquidus Measurements in the Pb-
     Pb1-xSnxTe)*   #Investigation of Liquidus Surface in Pb-
     of Interface Kinetics on the Growth Rate of Dendrites (
     Electrical Conductivity and Galvanomagnetic Effects in
     SnF2*   #About
     SnF2 Stannous Flouride. I. Crystallochemistry of alpha-
     theory)*   #Transport Properties of the Systems SnS2-
               #On the Growth Mechanism of Polycrystalline
     Relationships During the Formation of Three-Dimensional
               #Single Crystal Growth of Stannous Selenide (
                    #Electronic Bandgap Measurements in
     #Transport Properties of the Systems SnS2-SnI4 and
     theory)*   #Transport Properties of the Systems
     Properties of the Systems SnS2-SnI4 and SnS2-I2 (
     Cu, Pd, Pt, Fe, Al, Au, Ag, In, Pb, PbS, PbSe, PbTe,
     of Films of IV-VI Compounds in a Quasi-Sealed Volume (
                    #Crystal Growth and Assessment of
     Method on Single Crystal   #On the Growth Mechanism of
     Na0.5AlSi6014.xH2O)*   #Growth of Fibrous
     solution, NaClO3)#Kinetics of Growth and Dissolution of
               #Crystal Growth and Selective Etching Studies on
               #Convection Phenomena During the Growth of
                    #Preparation of Fine
     Rates in Pure Solution in the   #Crystallization of
     Kinetics of Different Faces in Pure #Crystallization of
                    #Crystal Growth of
     Crystallization of Barium Tungstate from Solutions in
     Crystal Growth of Calcium Tungstate from Solutions in
     Crystal Growth of Strontium Tungstate from Solutions in
     Na4Zr2Ge3012,   #Hydrothermal Crystallization of
     Na2Fe2Ti6014, Na2NiTi7016)*   #Growth Conditions for
     and Characterization of Mixed Bromides of Rare Earths,
               #Geometrical Analysis of Crystallization of the
               #Performance of a New High-Intensity Silicon
     of Rapidly-Grown RTR Silicon Sheet and Their Effects on
                    #New Methods of Making Silicon
               #High-Efficiency p+-n-n+ Back-Surface-Field Silicon
     and Vapor Growth of Silicon for Integrated Circuits and
               #High Efficiency GaAs Thin Film
     Si, Czochralski, zone melting, vapor transport)*   #

     Component Solution (Ga1-   #Variation of Liquid and
     Composition on the Density and Molecular Volume of
     Phase (alpha/beta)-Brass Bicrystals Produced by a Solid-
     Phase (alpha/beta)-Brass Bicrystals Produced by a Solid-
     Bridgman)*   #Crystal Growth of the
               #The Crystal Structure of the
     *   #Structural Studies of the
     NH4Ag4I5, solution)*   #Raman Scattering Studies in the
               #Growth of Whiskers by the Vapor-Liquid-
     dendrites, theory)*   #Morphological Stability of a

     Temperatures (vapor transport)* #Crystalline Defects in
               #Growth Kinetics of (111) Si Through an Al Layer by
     and Diffusion Characteristics in Binary Metallic
     Optical Properties of CdGe (As 1-xPx) 2 and AgGa (Se1-xSx) 2
               #Growth of Single Crystals of U1-xThxO2
     Crucible (#Growing Single Crystals of Silicon-Germanium
     #Phase Equilibria and Properties of Single Crystals of
               #Crystallization of Bi-Sb
               #Preparation and Investigation of Crystals of
               #Crystal Growth of K2CuF4 and Some
     Bridgman-Stockbarger,   #Crystal Growth of SrCl2 and
               #Preparation and Properties of Single Crystals of
     with CO2-Laser*   #A New Preparation Technique in
     Temperature Reactions   #A New Preparative Technique in
     Convection, Graphite,   #Preparation and Properties of
     Characterization (slow cooling)*   #

Slip System of alpha-Si3N4 Single Crystal (vapor   000696
Sm Lu Ca Fe Ge Garnet Films Properties (LPE)*   000650
Sm, Ca)3(Fe, Ge)5 012, flux) *Garnet Layers as Functions   000251
Sm, Dy, skull-melting)* Solidification, Microstructural   000630
Sm, Eu, Tm, and Yb - X = Cl, Br, and I) (melt, SmFCl,   000088
Sm,Y) 3 (Fe,Ga) 5012 (flux, theory, film)*   000222
SmB4, EuB6, YbB6)* Earth Borides by the Solution Method   000359
SmB6, SmB4, EuB6, YbB6)* Earth Borides by the Solution   000359
SmCrOS2)*   #Mixed Oxysulfides of Chromium III   001048
SmFBr, SmFI, EuFCl, EuFBr, EuFI, TmFCl, TmFBr, TmFI,   000088
SmFCl, SmFBr, SmFI, EuFCl, EuFBr, EuFI, TmFCl, TmFBr,   000088
SmFI, EuFCl, EuFBr, EuFI, TmFCl, TmFBr, TmFI, YbFCl,   000088
Smoke Particles Prepared by a Gas-Evaporation Technique   000448
Smooth Surface on Intermediately Sputtered ZnO Thin   000711
Sm0.55Tb0.45Fe03 (flux)*   000914
Sm2(MoO4)3, Tb2(MoO4)3)*   Molybdate and Other Isotypic   000446
Sm2Mo06, Pr2Mo06, Ho6Mo012)* and R6Mo012 in the Systems   001062
Sn - X = S,Se) (Cu2ZnGeS4, Cu2ZnSnS4, Cu2ZnGeSe4,   000357
Sn Crystals (in Ag solvent)*   000768
Sn-Pb, theory)*   Crystallization - II. Single Filament,   000993
Sn-Te System (Pb1-xSnxTe, PbTe, flux)*   000047
Sn-Te System by Simplex Lattice Method (flux, theory,   000668
Sn, theory)*   #Effects   000991
SN)x Single Crystal*   # 000463
SnF2 Stannous Flouride. I. Crystallochemistry of alpha-   000235
SnF2*   #About   000235
SnI4 and SnS2-I2 (SnS2, vapor transport, kinetics,   001079
Snow Crystals with a Specific Grain Boundary (H2O)*   000314
Snow Dendrites (H2O)*   #Epitaxial   000493
SnSe, Bridgman)*   000014
SnS2 Polytypes (vapor transport)*   000008
SnS2-I2 (SnS2, vapor transport, kinetics, theory)*   001079
SnS2-SnI4 and SnS2-I2 (SnS2, vapor transport, kinetics,   001079
SnS2, vapor transport, kinetics, theory)*   #Transport   001079
SnTe, LiF, NaF, KF, LiBr, NaCl, NaBr, LiI, KCl, NaI,   000949
SnTe, PbTe, PbSe, evaporation)*   #Preparation   000306
SnxPb1-xTe Mixed Crystals (PtxSn1-xTe, review)*   000490
Sodalite Single Crystals Grown by the Hydrothermal   000243
Sodium Aluminosilicate Crystals in Silica Gel (zeolite,   000444
Sodium Chlorate in Diffusion and Convection Regimes (000888
Sodium Chlorate Crystals (NaClO3, solution)*   000784
Sodium Chlorate Crystals from Solution (NaClO3)*   000179
Sodium Chloride Crystals (NaCl)*   000654
Sodium Perborate from Aqueous Solutions. I. Nucleation   000257
Sodium Perborate from Aqueous Solutions. II. Growth   000258
Sodium Sulphide, Na2S (hygroscopic, Bridgman)*   001056
Sodium Tungstate Melts by Continuous Cooling (BaWO4,   000812
Sodium Tungstate Melts by Continuous Cooling (CaWO4,   000810
Sodium Tungstate Melts by Continuous Cooling (SrWO4,   000811
Sodium Zirconogermanates (Na2ZrGeO5. Na2ZrGe207,   000230
Sodium-Titanium Bronze Crystals in NaBO2 (flux,   000602
Sodium, and Cesium (Ca2NaNdBr6, Cs2NaCeBr6, Cs2NaGdBr6,   000626
Soft-Core Model in an FCC Crystal Formation (theory)*   000965
Solar Cell (Si, float zone melting)*   000330
Solar Cell Processing (zone melting, Si)*   #Properties   000546
Solar Cells (review)*   000886
Solar Cells (Si)*   000298
Solar Cells (Si, review)*   #Problems in the Melt-   000117
Solar Cells by Peeled Film Technology (epitaxy)*   000497
Solar Cells from Zone-Refined Metallurgical Silicon (000195
Solar Furnace Annealing of Amorphous Si Layers (films)*   000537
Solid Compositions During LPE Growth from a Ternary   000970
Solid Crystalline Lithium Nitrate Hydrates (solution,   000023
Solid Diffusion Method. I. Morphology and   000372
Solid Diffusion Method. II. Diffusion-Induced Twins in   000371
Solid Electrolyte (C5H5NH)2Cu5Br7 From the Melt (000890
Solid Electrolyte Ag44I53(C1 1H30W3)3 (solution)*   000973
Solid Electrolyte High-LiTa3O8 (electrolytic deposition)   000706
Solid Electrolytes of the RbAg4I5 Family (KAg4I5,   000315
Solid Mechanism*   000338
Solid Particle Growing from a Binary Alloy Melt (000990
Solid Phase Epitaxial Growth of Si Through Al Film*   000584
Solid Phase Epitaxy Si Films Deposited at Elevated   001094
Solid Phase Epitaxy*   000585
Solid Solutions (theory)*   Quantities and Vacancy   000090
Solid Solutions (Bridgman)*Studies, Crystal Growth, and   000365
Solid Solutions by Chemical Transport Reactions*   000455
Solid Solutions by Electron-Beam Zone Melting Without a   000820
Solid Solutions in the Systems Zn3As2-ZnSe(ZnTe) (000726
Solid Solutions in Capillaries (zone melting)*   001035
Solid Solutions of the System GaAs-ZnSe (Bridgman)*   000185
Solid Solutions of K2CuF4-K2ZnF4 (Bridgman)*   001055
Solid Solutions of SrCl2-PrCl3 and SrCl2-GdCl3 (001054
Solid Solutions GaxIn1-xP (GaP, InP)*   000119
Solid State Chemistry. I. High Temperature Reactions   000657
Solid State Chemistry. II. High Pressure - High   000658
Solid State Materials, Vol. 4, Morphological Stability,   001081
Solid State Reactions to CdTeMo06 and Its Structural   000297

PERMUTED TITLE INDEX

Solidification of a Binary Alloy with a Curved   Solid-Liquid Interface (Ge:Ga, Bridgman, theory)*   000206
Vacuum (film)*   #Molecular Beam and   Solid-Phase Epitaxies of Silicon Under Ultra-High   000883
#Two-Phase (alpha/beta)-Brass Bicrystals Produced by a   Solid-Solid Diffusion Method. I. Morphology and   000372
#Two-Phase (alpha/beta)-Brass Bicrystals Produced by a   Solid-Solid Diffusion Method. II. Diffusion-Induced   000371
Substrates (LiNb1-xTaxO3,   #LPE Growth of Li(Nb,Ta)O3   Solid-Solution Thin Film Waveguides on LiTaO3   000500
of ZnSxSe1-x Layers on ZnS and ZnSe Substrates by a   Solid-State-Diffusion Technique*   #Growth   000045
#Single Crystal Growth of LiInS2 (directional   solidification)*   000456
Oxide Crystals (ZrO2-RE2O3, eutectics, directional   solidification (theory)*   000786
#The Stage II Transient in Off-Eutectic   Solidification (Czochralski)*   of Solidification of   000846
Pb-Bi Alloys Rich in Single-Phase Pb in the Course of   Solidification in Web-Dendritic Ribbon Growth (Si,   000369
kinetics, theory)*   #Thermal Analysis of   Solidification of a Binary Alloy with a Curved Solid-   000206
#Lateral Solute Segregation During Unidirectional   Solidification of Irregular Silver-Germanium Eutectic   000549
Alloys (Ag-Ge)*   #Unidirectional   Solidification of Pb-Bi Alloys Rich in Single-Phase Pb   000846
of Temperature and Concentration in the Region of   Solidification Behaviour of Molten Y3Al5O12 (000202
Czochralski)*   #A Complexity in the   Solidification Front Shape on the Temperature   000128
Distribution in the Crystal (NaCl,   #The Effect of the   Solidification Interfaces with a Very High Voltage   000163
Electron Microscope*   #In-Situ Observation of   Solidification Process (theory, eutectics, Al-Al3Ni)*   000268
#Effect of Some Thermal Parameters on the Directional   Solidification, Microstructural and Crystallographic   000630
Systems ZrO2-Ln2O3 (Ln = Lanthanide): Unidirectional   Solidified InSb-GaSb (magnetic field effects,   000864
Constant Distribution Coefficients for Directionally   Solidified Zn-Al Eutectic*   000421
#Crystallography of the fcc Phase in Uni-Directionally   Solidus Line (Bridgman)*   Microscopy Observations   000098
of Precipitates in ZnTe: Shape of the Tellurium   Solubilities and Simple Growth Techniques (solution)*   000039
M = Cd, Cu, Fe, Mn or Pd and X = Cl or Br: Importance,   # Solubility of AlPO4 in Hydrothermal Solutions of H3PO4*   000774
                                                          Solubility of Barium Sulphate (BaSO4, solution)**Effect   000779
of S-35 Tracer, Particle Size and Imperfection on the   Solubility of KLa(MoO4)2 in Aqueous K2MoO4 Solutions in   000475
Hydrothermal Conditions*   Solubility of Phlogopite in Basic Aqueous Solutions   000070
Under Hydrothermal Conditions (mica, KMg3AlSi3O10(OH)2)*   Solubility Curves in High-Temperature Melts for the   000265
Growth of Single Crystals of Rare Earth Vanadates and   # Solubility Model for Rare-Earth Iron Garnets in a PbO/   001022
B2O3 Solution (Y3Fe5O12, RExY3-xGayFe5-yO12, flux)*   #A   Soluble Complexes. I. General Introduction (review,   000693
Growth of Sparingly Soluble Single Crystals from   Soluble Complexes. II. Growth of alpha-HgI2 Single   000694
Growth of Sparingly Soluble Single Crystals from   Soluble Complexes. III. Growth of alpha-HgI2 Single   000695
Growth of Sparingly Soluble Single Crystals from   Soluble Single Crystals from Soluble Complexes. I.   000693
General Introduction (#Solution Growth of Sparingly   Soluble Single Crystals from Soluble Complexes. II.   000694
Growth of alpha-HgI2   #Solution Growth of Sparingly   Soluble Single Crystals from Soluble Complexes. III.   000695
Growth of alpha-HgI2   #Solution Growth of Sparingly   solute diffusion) Growth of GaP*   000461
#Improvements in SSD (synthesis,   Solute Diffusion Method   000603
#Growth of InP Crystals by the Synthesis   Solute Diffusion Method*   001095
#InP Single Crystal Growth by the Synthesis,   Solute Diffusion Technique (vapor transport)*   000923
#Large-Sized InP Single Crystals by the Synthesis,   # Solute Partitioning During Silicon Dendritic Web Growth   000855
(Si, theory)*   Solute Redistribution During a Modified Form of Zone   000201
Refining-Cascade Purification (metals)*   #Theoretical   Solute Segregation During Unidirectional Solidification   000206
of a Binary Alloy with a Curved Solid-Liquid   (solution)*   000260
#Structure of Silver Dichromate: Ag2Cr2O7   (solution)*   000286
#Ionic Conductivity in N3O+ Beta Alumina   (solution)*   000425
#Birefringence in CdSiP2   (solution)*   000464
#Uniaxial Ferromagnetism in NiZrF6.6H2O   (solution)*   000488
#Crystals of CTA-DNA: A Growth and Morphological Study   (solution)*   000682
#Enantiomorphism of alpha-LiIO3 Single Crystals   (solution)*   000902
#Crystal Structure of Scandium Metaphosphate Sc(PO3)3   (solution)*   001008
#Scandium Hydrogenselenite (Sc(HSeO3)3, evaporation,   (solution)*   001082
#A Novel Ruthenium Bronze: KRu4O8   (solution)*   # 001108
The Crystal Structure of Lithium Fluorosulphate LiSO3F   (solution)*   #Growth   000086
and Physical Properties of ZnSiP2 (vapor transport,   solution)*   #On the   000172
Growth of NH4Cl Dendrites at Very Low Supersaturation   (solution)*   #Crystal   000608
Structure of AgH2PO4. Crystallography of AgH2AsO4   (solution)*   #Isotope   000963
Effects on the Central Peak Phenomena in KD3(SeO3)2   (solution)*   #Epitaxial   000643
Growth of KNdP4O12 Laser Waveguides (KLaP4O12,   solution)*   #Orientation   000559
Dependence of 1-Alanine Incorporation in TGS Crystals   (solution)*   #The Crystal   000973
Structure of the Solid Electrolyte Ag44I53(C11H30N3)3 (solution)*   #Liquid Phase   000089
Epitaxy of II-IV-V2, Chalcopyrite Compounds (ZnSnAs2,   solution)*   #Crystal Growth   000931
of Co Ferrite on Fine Acicular gamma-Fe2O3 Particles   (solution)*   #Single-Crystal   000680
Growth of Tantalum Dibcride (TaB2, zone melting,   solution)*   #Ferroelasticity   000378
and Phase Transformation in Rb2Hg(CN)4 Spinel   (solution)*   #The Mechanism of   000094
Growth of KDP Crystals in the Presence of Impurities   (solution)*   #Fiber Optic Probe   000395
for Thermal Profiling of Liquids During Crystal Growth   (solution)*   #X-Ray Topographic   000092
Study of Growth Sector Strains in KDP Single Crystals   (solution)*   #Crystallographic and   000077
Magnetic Structure of RbFeCl3.2D2O and CsFeCl3.2D2O   (solution)*   #Electron Microscopic   000934
Characterization of Cubic BN Prepared by Water Solvent   (solution)*   #Electroless Deposition   000868
of Semiconductor Films (PbS, PbSe, Pb1-xHgxS,   solution)*   Phases of Deuterium Ammonium   000100
Dihydrogen Arsenate (DADA) (NH4(1-x)D4xH2(1-x)D2xAsO4,   solution)*   #Crystal Growth and Selective   000784
Etching Studies on Sodium Chlorate Crystals (NaClO3,   solution)*   #Growth Spirals in a Chemical   000659
Potential Varying in Space or Time (theory, kinetics,   solution)*   #Morphology Control of NdP5O14   000466
Single Crystals Grown from Polyphosphoric Acids (solution)*   #Direct Observation of Secondary   000322
Nuclei Production (KAl(SO4)2.12H2O, MgSO4.7H2O,   solution)*   #Neutron-Scattering Study of the   000226
Incommensurate Phase Transition of Rb2ZnBr4 (theory,   solution)*   #Mechanism of Habit Change of ADP   000296
Crystals by Fe3+, Based on Mossbauer Studies (NH4H2PO4,   solution)*   #On the Incorporation of Metallic   000170
Impurities in Synthetic Quartz Single Crystals (SiO2,   solution)*   Acid) Adsorption on Barium Sulfate   000553
Crystals and its Influence on Crystal Growth (BaSO4,   solution)*   #Growth and Electrical Properties of   000560
Deuterated TGS Produced by the Rotating Disc Technique (solution)*   #Ultrasonic Velocities of K2PbCu(NO2)   000467
6 Near the Cooperative Jahn-Teller Phase Transitions (solution)*   #Growth and Some Properties of Cerium   000227
Sulphate Enneahydrate Single Crystals (Ce2(SO4)3.9H2O,   solution)*   Etching of Cesium Iodide Crystals and   000827
Comparison of Growth and Dissolution Processes (CsI,   solution)*   #Raman Scattering Studies in the Solid   000315
Electrolytes of the RbAg4I5 Family (KAg4I5, NH4Ag4I5,   solution)*   Crystal Growth on Amorphous Substrates   000899
Using Artificial Surface-Relief Gratings (KCl, films,   solution)*   Conductivity of Gd2(SO4)38H2O (crystal)   000065
and Dy2Ti2O7 (powder) as a Function of Magnetic Field (solution)*   #On the Driving Force for Crystallization: 001024
The Growth Affinity (theory, kinetics, melt,   solution)*   M = Cd, Cu, Fe, Mn or Pd and X = Cl or Br: 000039
Importance, Solubilities and Simple Growth Techniques (solution)**   Crystals of Tetramethyltetrathiafulvalenium   000032
Tetracyanoquinodimethanide (TMTTF-TCNQ) (solution)*

PERMUTED TITLE INDEX

Holographic Interference and Conventional Microscope (solution)* Perovskite (C3H7NH3)2MnCl4 Using a Compound 000640
on the Solubility of Barium Sulphate (BaSO4, solution)* S-35 Tracer, Particle Size and Imperfection 000779
T12Cd2(SO4)3, K2Co2(SO4)3, and their Phase Transitions (solution)*Single Crystals of Langbeinites Rb2Cd2(SO4)3 000138
Transition Metal Silicides (chemical vapor transport, solution, cold crucible melting)* Characterization of 000607
of Cinnabar (alpha-HgS) at Rather Low Temperature (solution, doped)* #Growth of Single Crystals 000157
of Polytypism in Cadmium Iodide Crystals (CdI2, theory, solution, evaporation)* #Effect of Growth Conditions 000432
#Unstable Growth of ADP Crystals (NH4H2PO4, theory, solution, kinetics, slow cooling)* 000804
Formation and Growth Kinetics of Organic Crystals (solution, melt)* #Investigation of Structural 000773
#A New Organic Low Temperature Conductor: HMTSF-TNAP (solution, organics)* 000087
#Protein Crystallizations (solution, organics)* 000382
New Method to Follow Crystal Growth by Coulter Counter solution, organics)* #A 000601
and the Growth Kinetics of Ionic Crystals (CaSO4.2H2O, solution, theory)* #Solvent Effects 000134
and Crystal Habit of Struvite Crystals (MgNH4PO4.6H2O) (solution, theory)* #Growth Morphology 000002
Variations of the Layer Perovskite (C3H7NH3)2CuCl4 (solution, theory)* #Face Stability and Growth Rate 000641
Diffusive-Advective Interfacial Mass Transfer (theory, solution, vapor transport, flux)* #On 001074
#Materials Behaviour in Low Gravity Environment (melt, solution, vapor transport, zone melting, doped, review)* 001070
of Calcium Sulfate Crystals at Heated Metal Surfaces (solution, CaSO4)* #Kinetics of Growth 000334
Volume of Solid Crystalline Lithium Nitrate Hydrates (solution, LiNO3.3H2O, LiNO3.3D2O)*Density and Molecular 000023
of Sodium Chlorate in Diffusion and Convection Regimes (solution, NaClO3)* #Kinetics of Growth and Dissolution 000888
#Liquid-Phase Epitaxy of ZnSe from Zn-Ga Solution (film)* 000308
#LPE Growth of Hg0.60Cd0.40Te from Te-Rich Solution (film, vapor transport)* 000839
calcium phosphates)* #The Growth of Crystals in Solution (review, theory, alkaline earth sulfates, 000683
#On the Presentation of Growth Curves for Growth from Solution (theory)* 001025
Thermodynamic Driving Force for Crystallization from Solution (theory, kinetics)* #The 000904
for Constant Growth Rate by LPE from a Cooling, Static Solution (theory, melt)* #Conditions 000893
#Evaluation of Supersaturation in Crystal Growth from Solution (theory, organics)* 000166
of Ammonium Dihydrogen Phosphate Crystals from Aqueous Solution (ADP, NH4H2PO4)* of Dislocations in the Growth 000219
#Linear Growth Rate of Calcite in Aqueous Solution (CaCO3, kinetics, theory)* 000627
#The Growth Kinetics of Urea Monocrystals from Aqueous Solution (CO(NH2)2, theory)* 000392
Compositions During LPE Growth from a Ternary Component Solution (Ga1-xAlxAs, flux, theory)*of Liquid and Solid 000970
During the Growth of Sodium Chlorate Crystals from Solution (NaClO3)* #Convection Phenomena 000179
the Systems MF-AlF3 and NH4F-FeF3 in Hydrofluoric Acid Solution (TlAlF4, KAlF4, NH4AlF4, RbAlF4, CsAlF4, 000772
Model for Rare-Earth Iron Garnets in a PbO/B2O3 Solution (Y3Fe5O12, RExY3-xGayFe5-yO12, flux)* 001022
II. Growth Kinetics of Different Faces in Pure Solution and in the Presence of a Surfactant (NaBO2. 000258
of Potassium Bromate from Its Quiescent Aqueous Solution by Differential Scanning Calorimeter. Growth 000367
of Crystallization During Epitaxial Growth from a Solution in a Melt (GaP, flux)* #Theory 000591
#Growth of Zircon Single Crystals from a Solution in a Melt (ZrSiO4)* 000344
from Aqueous Solutions. I. Nucleation Rates in Pure Solution in the Presence of a Surfactant (NaBO2.H2O2. 000257
* #Crystallization of LiNbO3 from Solution in Borate, Vanadate, and Tungstate Melts (flux) 001047
#On the Effect of pH of Solution on the Kinetics of Crystallization (theory)* 000525
in the Y3Fe5O12-BaO.B2O3 System on Convection of a Solution with a Temperature Difference* of Liquid Phase 000048
Affinity of Crystal Growth and Dissolution in Aqueous Solution with Special Reference to Calcium Sulphate 000193
Phase. II. Phosphorus Content and Differential Molar Solution Enthalpies (Si, vapor transport, theory)* Gas 000515
2.12H2O)* #Effect of Stirring on Crystalline Quality of Solution Grown Crystals -- Case of Potash Alum (KAl(SO4) 000337
* # Solution Growth and Some Properties of GaP Bulk Crystals 000277
III-V Alloys (review, kinetics)* #Melt and # Solution Growth of Bulk Single Crystals of Quaternary 000054
Solution Growth of Indium-Doped Silicon (Si)* 000849
from Soluble Complexes. I. General Introduction (# Solution Growth of RbAg4I5 Crystals* 000038
from Soluble Complexes. II. Growth of alpha-HgI2 # Solution Growth of Sparingly Soluble Single Crystals 000693
from Soluble Complexes. III. Growth of alpha-HgI2 # Solution Growth of Sparingly Soluble Single Crystals 000694
Compounds Using Tellurium as a Solvent* # Solution Growth of Sparingly Soluble Single Crystals 000695
#Localized Adsorption in One Layer on a Crystal- Solution Growth of ZnS, ZnSe, CdS and Their Mixed 001065
of Single Crystals of the Rare Earth Borides by the Solution Interface (theory)* 000575
LiNb1-xTaxO3, flux)* #LPE Growth of Li(Nb,Ta)O3 Solid- Solution Method and a Study of Their Properties (flux, 000359
#Crystal Growth from Solution Thin Film Waveguides on LiTaO3 Substrates (000500
#The Growth, X-Ray Studies and Microstructure of Solution Using Cylindrical Seeds (TGS, TBFB)* 000566
of Y3Al5O12 Crystals on Oriented Seeds in Melt- Solution-Grown Mixed Crystals of NaCl and KCl (NaCl-KCl 000407
Growth of Magnetic Garnets from High-Temperature Solutions (flux)* Kinetics and Growth Mechanism 000093
and Diffusion Characteristics in Binary Metallic Solid Solutions (review)* #Crystal 000982
Properties of CdGe(As1-xPx)2 and GaGa(Se1-xSx)2 Solid Solutions (theory)*Thermodynamic Quantities and Vacancy 000090
in Supersaturated Potassium Dihydrogen Orthophosphate Solutions (Bridgman)* Crystal Growth, and Optical 000635
Bulk Yttrium Iron Garnet Crystals from High Temperature Solutions (KDP)* #Nucleation 000443
#Growth of Single Crystals of U1-xThxO2 Solid Solutions (Y3Fe5O12)* #Growth of 000977
#Growing Single Crystals of Silicon-Germanium Solid Solutions by Chemical Transport Reactions* 000455
Equilibria and Properties of Single Crystals of Solid Solutions by Electron-Beam Zone Melting Without a 000820
#Crystallization of Bi-Sb Solid Solutions in the Systems Zn3As2-ZnSe(ZnTe) (Zn3As2, 000726
#Solubility of KLa(MoO4)2 in Aqueous K2MoO4 Solutions in Capillaries (zone melting)* 001035
#Properties of CuCr2Se4 Crystals Obtained from Solutions in Hydrothermal Conditions* 000475
into YAl-Borate Crystals in Preparation from Solutions in Melts* 000050
#Kinetics of Crystal Growth of Calcium Tungstate from Solutions in Molten Potassium Trimolybdate ((Y,Nd)Al3(000551
#Kinetics of Crystal Growth of Strontium Tungstate from Solutions in Sodium Tungstate Melts by Continuous 000810
#Kinetics of Crystallization of Barium Tungstate from Solutions in Sodium Tungstate Melts by Continuous 000811
#Preparation and Investigation of Crystals of Solid Solutions in Sodium Tungstate Melts by Continuous 000812
#Solubility of AlPO4 in Hydrothermal Solutions of the System GaAs-ZnSe (Bridgman)* 000185
#Crystal Growth of K2CuF4 and Some Solid Solutions of H3PO4* 000774
Stockbarger, #Crystal Growth of SrCl2 and Solid Solutions of K2CuF4-K2ZnF4 (Bridgman)* 001055
of Growth of Y3Al5O12 Single Crystals from Melt- Solutions of SrCl2-PrCl3 and SrCl2-GdCl3 (Bridgman- 001054
of TGS Single Crystals, Grown from Aqueous TGS Solutions on Oriented Seeds (flux)* Characteristics 000978
#Preparation and Properties of Single Crystals of Solid Solutions Containing Aniline* #The Physical Properties 000266
KMg3AlSi3O10 (#Solubility of Phlogopite in Basic Aqueous Solutions GaxIn1-xP (GaP, InP)* 000119
#Crystallization of Sodium Perborate from Aqueous Solutions Under Hydrothermal Conditions (mica, 000070
Pure #Crystallization of Sodium Perborate from Aqueous Solutions. I. Nucleation Rates in Pure Solution in the 000257
Rasplavov (Crystal Growth from High-Temperature Solutions. II. Growth Kinetics of Different Faces in 000258
#Crystallization of Salts from Supersaturated Solutions) (review, flux)*Rost Kristallov iz Rastvorov- 000976
Growth of High Perfection Ag-Sn Crystals in Ag Solutions: Diffusion Kinetics* 000623
#Low Temperature Growth of GaP LPE Layers from Indium solvent)* 000768
Characterization of Cubic BN Prepared by Water Solvent (flux)* 000925
Solvent (solution)* #Electron Microscopic 000934

PERMUTED TITLE INDEX

Crystals (CaSO4.2H2O, solution, theory)*   #
Single Crystals of Graded Composition by the Travelling
Semiconductor Single Crystals Obtained by the Traveling
CdS and Their Mixed Compounds Using Tellurium as a
                              #Phase Equilibria in
sapphire, vapor transport)*   #Dichlorosilane as Silicon
                     #A High Temperature High Purity
Properties of Cadmium, Selenium, and #Influence of the
Selenium, and #Influence of the Source Material and the
of Solution Grown Crystals -- Case of Potash Alum (KAl(
#Direct Observation of Secondary Nuclei Production (KAl(
of Cerium Sulphate Enneahydrate Single Crystals (Ce2(
of Langbeinites Rb2Cd2(SO4)3, Tl2Cd2(SO4)3, K2Co2(
of Single Crystals of Langbeinites Rb2Cd2(SO4)3, Tl2Cd2(
#The Growth of Single Crystals of Langbeinites Rb2Cd2(
of Magnetic Field (#Low Temperature Conductivity of Gd2(
and Past and Future Crystal Growth Experiments in
#Growth of Pb0.8Sn0.2Te Films in a Quasi-Closed
#Growth Spirals in a Chemical Potential Varying in
xZnxS Layers on (111) GaAs Substrates Using the Close-
#Heteroepitaxial Growth of ZnS on GaP by the Close-
Optical Properties of ZnSe Epitaxial Layers by a Close-
Edge of Single-Crystal AlN Prepared by a Close-
Mirror Furnace for Zone Crystallization Experiments in
Experiments (review)*   #
in Space (review)*   #
                     #Priorities in the Initial Use of
Complexes. I. General Introduction (#Solution Growth of
Complexes. II. Growth of alpha-HgI2 #Solution Growth of
Complexes. III. Growth of alpha-   #Solution Growth of
                          #Raman and Infrared
                               #Raman
                               #Mass
          #State-of-The-Art of Crystal Growth and Nuclear
Emission Data of Gd3Ga5O12:Nd3+ Single   #Growth,
Obtained by Liquid-Phase Epitaxy*   #
                     #Electron Energy-Loss
Yttrium Holmium Aluminum Garnets ((Y1-xHox) #Structure,
Bi4Ge3O12 Crystals Doped with Dy3+, Ho3+, Er3+ #Growth,
Cadmium Mercury Telluride (Cd1- #The Effects of Growth
                     #Growth and Morphology of
                     #The Preparation of Large
Plasma Remelting of Alloy Powder (Cu9Al4)*   #
#Ferroelasticity and Phase Transformation in Rb2Hg(CN)4
                     #Crystal Structure of the
(transport, magnetic susceptibility, and NMR) of the
FeCr2S4, CoCr2S4,   #Preparation of Some Chalcogenide
                     #Magnetic
                     #LPE Growth of Lithium Ferrite on
(Czochralski, theory)*   #An Evidence of
Time (theory, kinetics, solution)*   #Growth
                     #Anisotropy of Round Evaporation
Polytype of SiC at the Interface of Two Interacting
Substrates*   #Growth and Electrical Properties of
                     #The Epitaxial Growth of Reactively
                     #Second Harmonic Generation in a
ZnO Film with Smooth Surface on Intermediately
on Potassium Bismuth Niobate Single Crystals by RF
Growth of ZnO Films on Sapphire by Planar Magnetron rf
                     #Preparation of Ge-Si Epitaxial Alloys by
#Crystal Growth, and Magnetic and Mossbauer Studies of
and Strontium Nitrate Crystal Hydrates (LiNO3.3H2O,
#Growth and Characterization of Mixed
of Hydrothermally Grown Strontium Paracelsian (
GdCl3 (Bridgman-Stockbarger,   #Crystal Growth of
Growth of SrCl2 and Solid Solutions of SrCl2-PrCl3 and
#Crystal Growth of SrCl2 and Solid Solutions of
and Properties of Strontium Cobaltate Single Crystals (
of Sr(Fe0.766W0.234)O3 and its Related Compounds (
                     #Dispersion of Acoustic Phonons in Layered
#Epitaxial Growth of BaTiO3 Vacuum Condensates on
                     #Surface Photovoltage Experiments on
in Sodium Tungstate Melts by Continuous Cooling
#Growth and Characterization of Mixed Sr:Ca Tartrates (
(flux, Sr2LaAlO5, Sr2PrAlO5, Sr2NdAlO5, Sr2SmAlO5,
Sr2GdAlO5, Sr2TbAlO5, Sr2NdAlO5, Sr2NdFeO5, Sr2SmFeO5,
Sr2LaAlO5, Sr2PrAlO5, Sr2NdAlO5, Sr2SmAlO5, Sr2EuAlO5,
Sr2TbAlO5, Sr2NdAlO5, Sr2NdFeO5, Sr2SmFeO5, Sr2EuFeO5,
Alkaline-Earth Rare-Earth Pentaoxometallates (flux,
Sr2NdAlO5, Sr2SmAlO5, Sr2EuAlO5, Sr2GdAlO5, Sr2TbAlO5,
Earth Pentaoxometallates (flux, Sr2LaAlO5, Sr2PrAlO5,
Sr2SmAlO5, Sr2EuAlO5, Sr2GdAlO5, Sr2TbAlO5, Sr2NdAlO5,
Earth Rare-Earth Pentaoxometallates (flux, Sr2LaAlO5,
(flux, Sr2LaAlO5, Sr2PrAlO5, Sr2NdAlO5,
Sr2PrAlO5, Sr2NdAlO5, Sr2SmAlO5, Sr2EuAlO5, Sr2GdAlO5,
                     #Improvements in
Perovskite (C3H7NH3)2CuCl4 (solution, theory)*   #Face

Solvent Effects and the Growth Kinetics of Ionic   000134
Solvent Method (theory)*   #Growth of Pb1-xSnXTe   000564
Solvent Method from (Pb0.8Sn0.2)Te*   #   000064
Solvent*   #Solution Growth of ZnS, ZnSe,   001065
Solvent-Garnet Systems (flux, Y3Al5O12, Y3Fe5O12)*   000159
Source for Epitaxial Films on Insulators (Si on   000254
Source for Metal Beam Epitaxy (Ag, film)*   000287
Source Material and the Source Shape on the Sublimation   000365
Source Shape on the Sublimation Properties of Cadmium,   000365
SO4)2.12H2O)**#Effect of Stirring on Crystalline Quality   000337
SO4)2.12H2O, MgSO4.7H2O, solution)*   000322
SO4)3.9H2O, solution)*   #Growth and Some Properties   000227
SO4)3, and their Phase Transitions (solution)* Crystals   000138
SO4)3, K2Co2(SO4)3, and their Phase Transitions (000138
SO4)3, Tl2Cd2(SO4)3, K2Co2(SO4)3, and their Phase   000138
SO4)38H2O (crystal) and Dy2Ti2O7 (powder) as a Function   000065
Space (review)*   #Spacelab   000854
Space (vacuum evaporation)*   000885
Space or Time (theory, kinetics, solution)*   000659
Spaced Geometry (vapor transport)*of Single Crystal Cd1-   000305
Spaced Technique (kinetics)*   000442
Spaced Technique*   #Preparation and   000670
Spaced Vapor Process*   #The Optical Absorption   000758
Spacelab (CdTe, zone melting, sublimation)*   Ellipsoid   000276
Spacelab and Material Processing Facilities and   000853
Spacelab and Past and Future Crystal Growth Experiments   000854
Spacelab for Crystal Growth (vapor transport)*   000270
Sparingly Soluble Single Crystals from Soluble   000693
Sparingly Soluble Single Crystals from Soluble   000694
Sparingly Soluble Single Crystals from Soluble   000695
Spectra of CdIn2S4 and ZnIn2S4 (vapor transport)*   001004
Spectra of ZrS3*   000434
Spectrometric Study on Chemical Transport of VO2*   000063
Spectroscopic Evaluation of Mercuric Iodide Radiation   000834
Spectroscopic Investigations, and Some New Stimulated   000457
Spectroscopic Properties of Er3Al5-xGax O12 Films   000245
Spectroscopy of TiS2 (vapor transport)*   000288
Spectroscopy, and Stimulated Emission of Crystals of   000046
Spectroscopy, and Stimulated Emission of Cubic   000458
Speed on the Compositional Variations in Crystals of   000074
Spherical Crystals of omega-Phase in the Cd-Hg System*   000512
Spherical Iron Single Crystals (Fe, strain-anneal)*   000573
Spherical Monocrystals for X-Ray Work Obtained by   000043
Spinel (solution)*   000378
Spinel CdErS4 (slow cooling)*   000983
Spinel CuV2S4 (vapor transport)*and Physical Properties   000544
Spinel Single Crystals and Their Electronic Properties (001066
Spinel Single Crystals by Bridgman Technique (review)*   000924
Spinel Substrate Crystals (films, Li0.5Fe2.5O4)*   000340
Spiral Growth on the Surface of Melt-Grown KCl Crystals   000408
Spirals in a Chemical Potential Varying in Space or   000659
Spirals on Rocksalt Surfaces (NaCl, theory)*   000530
Spirals*   #The Occurrence of High Period   000889
Sputter-Deposited Single-Crystal GaSb Films on GaAs   000269
Sputtered CuI Films on Rocksalt Substrates*   000788
Sputtered LiNbO3 Film on MgO (epitaxy)*   000518
Sputtered ZnO Thin Film on Sapphire (epitaxy)*   000711
Sputtering (KxLi1-xNbO3)* Niobate Single-Crystal Films   000009
Sputtering System*   #High Rate Epitaxial   000882
Sputtering*   000420
Sr(Fe0.766W0.234)O3 and its Related Compounds (SrFe1-   001001
Sr(NO3)2.4H2O)*   #Crystallization Kinetics of Lithium   000989
Sr:Ca Tartrates (Sr1-xCazC4H4O6.4H2O, gel)*   000747
Sr[Al2Si2O8]*   #The Morphology   000304
SrCl2 and Solid Solutions of SrCl2-PrCl3 and SrCl2-   001054
SrCl2-GdCl3 (Bridgman-Stockbarger, Czochralski)*Crystal   001054
SrCl2-PrCl3 and SrCl2-GdCl3 (Bridgman-Stockbarger,   001054
SrCoO2.70, float zone melting)*   #Growth   000955
SrFe1-xWxO3, flux)* and Magnetic and Mossbauer Studies   001001
SrGa2 (melt, slow cooling)*   000672
SrTiO3 Crystals*   000984
SrTiO3 Electrodes (theory)*   000618
SrWO4, flux)*   of Strontium Tungstate from Solutions 000811
Sr1-xCazC4H4O6.4H2O, gel)*   000747
Sr2EuAlO5, Sr2GdAlO5, Sr2TbAlO5, Sr2NdAlO5, Sr2NdFeO5,   000253
Sr2EuFeO5, Sr2GdFeO5)* Sr2NdAlO5, Sr2SmAlO5, Sr2EuAlO5,   000253
Sr2GdAlO5, Sr2TbAlO5, Sr2NdAlO5, Sr2NdFeO5, Sr2SmFeO5,   000253
Sr2GdFeO5)* Sr2NdAlO5, Sr2SmAlO5, Sr2EuAlO5, Sr2GdAlO5,   000253
Sr2LaAlO5, Sr2PrAlO5, Sr2NdAlO5, Sr2SmAlO5, Sr2EuAlO5,   000253
Sr2NdAlO5, Sr2NdFeO5, Sr2SmFeO5, Sr2EuFeO5, Sr2GdFeO5)*   000253
Sr2NdFeO5, Sr2SmFeO5, Sr2EuFeO5, Sr2GdFeO5)* Sr2NdAlO5,   000253
Sr2PrAlO5, Sr2NdAlO5, Sr2SmAlO5, Sr2EuAlO5, Sr2GdAlO5,   000253
Sr2SmAlO5, Sr2EuAlO5, Sr2GdAlO5, Sr2TbAlO5, Sr2NdAlO5,   000253
Sr2SmFeO5, Sr2EuFeO5, Sr2GdFeO5)* Sr2NdAlO5, Sr2SmAlO5,   000253
Sr2TbAlO5, Sr2NdAlO5, Sr2NdFeO5, Sr2SmFeO5, Sr2EuFeO5,   000253
SSD (synthesis, solute diffusion) Growth of GaP*   000461
Stability and Growth Rate Variations of the Layer   000641

PERMUTED TITLE INDEX

Alloy Melt (dendrites, theory)*   #Morphological
                    #Elastic Constants and
Na2Ge4O9, Na4Ge9O20, Na2Ge2O5)*   #Structure and
Processes. II (theory)*   #Morphological
Processes. I (theory)*   #Morphological
                    #Morphological
         #Studies of LPE Ripple Based on Morphological
Bi2Sn2O7 (flux)*   #Preparation,
   of Solid State Materials, Vol. 4, Morphological
   During the Crystal Growth of InSb from the Melt Under
Electrodeposition (flux)*   #Conditions for
Recrystallization of Metals and   #Fractionation of
Diameter Sapphire and Silicon Crystals (Al2O3, #Maximum
Czochralski Silicon (Si)*   #Growth of
to the First Stage Graphite/#The Synthesis of the First
   Salt C8(+)AsF6(-) and Its Relationship to the First
theory)*   #The
                    #About SnF2
                    #Single Crystal Growth of
Crystals with Large Temperature   #Analysis of the
   for Constant Growth Rate by LPE from a Cooling,
                    #The History and Present
                    #The Validity of
   #Reply to Comment of R. Trivedi on "The Validity of
         #Growth of GaAs1-xSbx Crystals by
Using a Compound Holographic   #Observations on
Surface of Vapour Grown FeCr2S4*   #
of Silicon (111) (#The Influence of Adsorption and
                    #Unit Cell
Surfaces (#Interaction Between Successive Monatomic
                    #Modes of Vibrations in
   of the 8th All Union Meeting on the Preparation by
Plates by Pulling from a Molten Bath According to the
         #Growth of Si Strips by
         #Thermodynamic Properties of Surface
FeCr2S4*   #Step Growth by Elementary
Crystals. I*   #Growth of Triple Unit Cell
                    #Growth Mechanism of Multi-
         #Step Growth by Elementary Steps and Multi-
#Liquid Phase Epitaxial Growth of Ga1-xInxSb on GaSb by
Aluminum Garnets ((Y1-   #Structure, Spectroscopy, and
with Dy3+, Ho3+, Er3+, Tm3+, #Growth, Spectroscopy, and
   #Growth, Spectroscopic Investigations, and Some New
theory, film)*   #An Improved Method of
Crystals -- Case of Potash Alum (KAl(SO4)2. #Effect of
                    #Bridgman-
   #Influence of Magnetic Field on Vertical Bridgman-
         #Crystal Growth of Tl3VS4 (Bridgman-
   of a Single Crystal of NdF3 Between 2 and 4.2K (
Ca2NaNdF6, Cs2NaCeBr6, Cs2NaGdBr6, Cs2NaYBr6, Bridgman-
Alkali Metal Indates and Thioindates (LiInS2, Bridgman-
   Trifluorides (GdF3, TbF3, DyF3, HoF3, ErF3, Bridgman-
   Solutions of SrCl2-PrCl3 and SrCl2-GdCl3 (Bridgman-
of Copper Indium Ditelluride Crystals Grown from Near-
of   #Growth of ZnSiP2 from Gas Phase Under Non-
         #Crystal Growth of Bi3B5O12 From its
         #Anisotropy of Ion Transport in
   #Physico-Chemical Properties of the Non-
sublimation)*   #Vapor Phase Transport and
SxSe1-x), sublimation)*   #Vapor Phase Transport and
   of Large Spherical Iron Single Crystals (Fe,
Grown Indium Antimonide (InSb)*   #The
         #X-Ray Topographic Study of Growth Sector
                    #Epitaxial
GaAs, InAs, InSb, Al2O3, melt, theory)*   #Formation of
   #Orientation and Temperature Dependence of the Flow
KCN:KOH, KCl:KOH, theory)*   #Low-
Taking Account of Periodic Remelt (Si, #Computation of
Floating Zone Melting at Various Growth Rates* #Growth
                    #Growth of Si
         #Simultaneous Diffusion of Calcium and
zone melting)*   #Growth and Properties of
2.4H2O)*   #Crystallization Kinetics of Lithium and
         #The Morphology of Hydrothermally Grown
Melts by Continuous   #Kinetics of Crystal Growth of
Films on GaAs Substrates (epitaxy)*
Transition Metal Selendies VxTi1-xSe2 and TaxTi1-xSe2 (#
Barium Vanadium Triselenide (BaVSe3)*   #
in (Ti0.9975V0.0025)4O7 (vapor transport)*   #
Layers in Silicon (Si)*   #Electrical and
         #Solid State Reactions to CdTeMoO6 and Its
CaTlTa5O15 (flux)*   #A New
Crystals (solution, melt)*   #Investigation of
         #Profile-Sapphire and Its
         #Growth of USb2 Single Crystals and Their
(melt)*
LiTa3O8 (electrolytic deposition)*
x) (0 less than or equal to x less than or equal to 1) #

Stability of a Solid Particle Growing from a Binary   000990
Stability of bcc In-Tl Alloys (Bridgman)*   000581
Stability of Alkali Germanate Crystals and Glasses (000822
Stability Analysis in Chemical Vapour Deposition   000435
Stability Analysis in Chemical Vapour Deposition   001010
Stability Analysis of Growth from the Vapour (theory)*   001009
Stability Theory*   000698
Stability, and Preliminary Crystallographic Study of   001034
Stability, Convection, Graphite, and Integrated Optics (001081
Stabilizing Thermal Gradient (Bridgman, theory)*   000478
Stable Growth of Epitaxial GaP Layers by Molten Salt   000224
Stable Isotopes and Impurities During Zone   000994
Stable Zone Length in Float-Zone Growth of Small-   000479
Stacking Faults by Bardeen-Herring Mechanism in   001052
Stage Graphite Salt C8(+)AsF6(-) and Its Relationship   000076
Stage Graphite/AsF5 Intercalate (C8AsF6)*Stage Graphite   000076
Stage II Transient in Off-Eutectic Solidification (000786
Stannous Flouride. I. Crystallochemistry of alpha-SnF2*   000235
Stannous Selenide (SnSe, Bridgman)*   000014
Static Method of the Growth of Cadmium Sulphide   000289
Static Solution (theory, melt)*   #Conditions   000893
Status of Synthetic Diamond (C, review, doped)*   000685
Steady-State Dendrite Growth Models (theory)*   000832
Steady-State Dendrite Growth Models" (theory)*   000833
Steady-State Liquid Phase Epitaxy*   000957
Step Bunching for the Layer Perovskite (C3H7NH3)2MnCl4   000640
Step Growth by Elementary Steps and Multi-Steps on the   000419
Step Reconstruction on the Growth and Etching Vectors   001019
Step Structures on Vapour-Grown Cr2S3 Single Crystals*   000676
Step Trains of Different Density on Evaporating Crystal   000105
Step Trains: Rhythmical Bunching (theory)*   000933
Stepanov Method of Profiled Crystals and Products and   001111
Stepanov Procedure (Si)*   #Preparation of Silicon   000966
Stepanov's Method*   000116
Steps (theory)*   001069
Steps and Multi-Steps on the Surface of Vapour Grown   000419
Steps on the Surface of Vapour Grown FeCr2S4 Single   000418
Steps on the Surface of Vapour Grown FeCr2S4*   000417
Steps on the Surface of Vapour Grown FeCr2S4*   000419
Stepwise Grading (melt)*   000803
Stimulated Emission of Crystals of Yttrium Holmium   000046
Stimulated Emission of Cubic Bi4Ge3O12 Crystals Doped   000458
Stimulated Emission Data of Gd3Ga5O12:Nd3+ Single   000457
Stirring for LPE Growth (Sm,Y)3(Fe,Ga)5O12 (flux,   000222
Stirring on Crystalline Quality of Solution Grown   000337
Stockbarger Crystal Growth of Li2Ti3O7*   000634
Stockbarger Growth of InxGa1-xSb*   000863
Stockbarger)*   000411
Stockbarger)*   #Magnetic Properties 000767
Stockbarger)*   of Rare Earths, Sodium, and Cesium (000626
Stockbarger)*   Principal Physicochemical Properties of   000509
Stockbarger, Czochralski)*   Growth of Some Rare Earth   000323
Stockbarger, Czochralski)*   Growth of SrCl2 and Solid   001054
Stoichiometric Compositions (doped, zone melting)*   000220
Stoichiometric Conditions -- Microanalytical Evaluation   000082
Stoichiometric Melt (kinetics)*   000794
Stoichiometric Rutile Single Crystals (TiO2)*   000778
Stoichiometric VS2 and V5S8 Phases (vapor transport)*   000470
Stoichiometry Control of Cadmium Sulfide (CdS,   000648
Stoichiometry Control of Cadmium Sulfo-Selenide (Cd1+y(000647
strain-anneal)*   #The Preparation   000573
Strain Configuration Within the Facet of Czochralski   000637
Strains in KDP Single Crystals (solution)*   000092
Stratified Growth (kinetics, theory)*   000017
Stress and Dislocations in Crystal Growth (Si, Ge,   000406
Stress in the Intermetallic Compound Ni3Ge Single   000033
Stress Physical Vapor Growth (PVT) (KCl, KBr, KCl:KCN,   000809
Striated Impurity Distributions in Melt-Grown Crystals,   001026
Striations in Fe-3wt%Si Single Crystals Grown by   001028
Strips by Stepanov's Method*   000116
Strontium in KCl Single Crystals*   000759
Strontium Cobaltate Single Crystals (SrCoO2.70, float   000955
Strontium Nitrate Crystal Hydrates (LiNO3.3H2O, Sr(NO3)   000989
Strontium Paracelsian (Sr[Al2Si2O8]*   000304
Strontium Tungstate from Solutions in Sodium Tungsten   000811
Structural and Electrical Properties of CuGaSe2 Thin   000844
Structural and Electrical Properties of Layered   000555
Structural and Magnetic Properties of "One-Dimensional"   000474
Structural Aspects of the Metal-Insulator Transitions   000391
Structural Characteristics of Laser-Induced Epitaxial   001106
Structural Characterization (slow cooling)*   000297
Structural Family M2M'Ta5O15. Crystal Structure of   000317
Structural Formation and Growth Kinetics of Organic   000773
Structural Perfection (alpha-Al2O3, melt)*   000246
Structural Perfection (vapor transport, melt, flux)*   000383
Structural Phase Transition in Mixed Crystals WxMo1-xO3   000824
Structural Studies of the Solid Electrolyte High-   000706
Structural Studies of Compounds in the Series GaSxSe(1-   001077

PERMUTED TITLE INDEX

Metavanadates (NaxLi1- #Crystallographic Studies and	Structural Systematics of the C2/c Alkali Metal	000692
Films of Aluminum with Silicon* #	Structural Transitions and Phase Composition of Thin	000473
A Compound with a Novel Wurtzite-Pyrite Intergrowth	Structure (flux)* #AlSiP3,	001044
#The Molten State of Matter. Melting and Crystal	Structure (melt, review, theory)*	000999
Transition Metal Pnictides with ThCr2Si2-Type	Structure (phosphides, flux)* #Ternary Lanthanoid-	000596
the System Ni3Pb2S 2-Ni3Pb2Se2 and Note on the Shandite	Structure (Ni3Pb2Se2, vapor transport)* Study of	000198
#Cold Working Nb3Al in the bcc	Structure and then Converting to the A-15 Structure*	001067
#Crystal	Structure and Antiferromagnetism of EuSb2 (melt)*	000401
TlSe, zone melting, slow cooling)* #	Structure and Growth Peculiarities of Tl-Se-TlInSe2 (	000361
Gas-Evaporation Technique: bcc Metals (V, Fe, #Crystal	Structure and Habit of Fine Metal Particles Formed by	000821
other new Li+ Superionic Conductors (kinetics, #Crystal	Structure and Ionic Conductivity of Li24Zn(GeO4)4 and	000396
4 Crystals* #	Structure and Luminescence Properties of alpha-CsNd(PO3)	000587
of Molybdenum II: Mo6Cl10Y (Y = S, Se, Te):	Structure and Magnetic and Electrical Properties (	000757
Neutron Diffraction Study (melt)* #On the Magnetic	Structure and Magnetic Phase Transitions of Tb5Ge4. A	000840
transport)* #Crystal	Structure and Physical Properties of Cu00.65VS2 (vapor	000543
and Properties of Superalloys via Use of #Control of	Structure and Properties. Controlling Microstructures	000671
and Glasses (Na4Ge4O9, Na4Ge9C20, Na2Ge2O5)* #	Structure and Stability of Alkali Germanate Crystals	000822
Fe2(MoO4)3 (hydrothermal)* #The Crystal	Structure and Twinning Behavior of Ferric Molybdate,	000178
#Relation Between Growth Conditions and Dislocation	Structure in Gallium Phosphide Crystals (GaP,	000723
epitaxy, review)* #Periodic Doping	Structure in GaAs (GaAs, Ga1-xAlxAs, molecular beam	000248
Material: Na3Nd(PO4)2* #The Crystal	Structure of a New High -Nd- Concentration Laser	000825
approx. = 0.25) (melt, slow #Synthesis and Crystal	Structure of a New Mixed Oxide "FeV3O8" (FexV1-xO2 - x	000661
the Diaspore Structure Type (#Synthesis and Crystal	Structure of a New Mixed Oxide FeV2O6OH0.5: Relation to	000662
xRe2O6(OH)2x (x = 0.3)* #High Pressure Synthesis and	Structure of a New Pyrochlore Containing Rhenium: Ca1-	000104
Ni2P2O7 (slow cooling)* #Crystal	Structure of a New Variety of Nickel Pyrophosphate:	000609
#Crystal	Structure of beta-Eu2(II)SiO4 (melt)*	000597
#Crystal	Structure of beta-KEr3F10 (melt, slow cooling)*	000021
cooling, flux)* #Crystal	Structure of the Double Molybdate K2Ni(MoO4)2 (slow	000486
III. On the Phase Transition of CuTeBr and the	Structure of the High Temperature Phase (Bridgman)*	000025
transport)* #Crystal	Structure of the Indium Selenide In2Se3 (vapor	000557
Interface (theory)* #	Structure of the Lennard-Jones (100) Crystal-Liquid	000528
melt)* #Crystal	Structure of the Mercury Nitro-Phosphate Hg4PO4NO3.H2O (	000261
PO3)3 (flux)* #Crystal	Structure of the Phosphate of Trivalent Manganese Mn(	000058
solution)* #The Crystal	Structure of the Solid Electrolyte Ag44I53(C11H30N3)3 (	000973
#Crystal	Structure of the Spinel CdEr2S4 (slow cooling)*	000983
#The Crystal	Structure of the Tetragermanate K2Ba[Ge4O9]2*	000083
solution)* #Crystal	Structure of the Zinc Iodide ZnI2 (vapor transport)*	000300
transport)* #The Crystal Growth and Atomic	Structure of AgH2PO4. Crystallography of AgH2AsO4 (	000608
#The Crystal	Structure of As2(Se,S)3 Compounds (As2S3, As2Se3, vapor	000896
#The Crystal	Structure of Ba2Fe6O11 (flux)*	000120
#The Crystal	Structure of Ba2V2O7 (melt)*	000379
#Crystal Growth and	Structure of BiVO4 (Czochralski)*	000891
#A New Structural Family M2MnTa5O15. Crystal	Structure of CaTlTa5O15 (flux)*	000317
#Preparation and Crystal	Structure of CsAlF4*	000570
#Synthesis and	Structure of Cs2S2 (hydrothermal)*	000130
#The Crystal	Structure of Cs6W11O36*	000719
#Crystal	Structure of CuTa2O6 (flux)*	001036
#The Crystal	Structure of CuTeO4 (hydrothermal)*	000282
#The Crystal	Structure of Cu4(PO4)2O (flux)*	000029
#Crystal	Structure of Cu4(PO4)2O (flux)*	000142
Lanthanum-Rhenium Oxide La6Re4O18 (melt)* #Crystal	Structure of Double Oxides of Rhenium. II. The	000103
Cadmium Single Crystals (CdS, #Direct Synthesis and	Structure of Epitaxial Cadmium Sulphide Layers on	000429
Dihydrides (GdH2, TbH2, DyH2, ErH2, #The Growth and	Structure of Epitaxial Films of the Rare-Earth	000782
the Mismatch of Crystal Lattice #Growth Kinetics and	Structure of Epitaxial Garnet Layers as Functions of	000251
State (Czochralski, #Change in the Energy Band	Structure of Eulytine on Transition to the Amorphous	000294
#Magnetic	Structure of GdCu1-xZnx System (melt)*	000956
Sapphire (vapor transport, ZnC)* #Real	Structure of Heteroepitaxial Zinc Oxide Films Grown on	000526
#The Crystal	Structure of Hf3Ni7 (melt)*	000147
#The Crystal	Structure of Hf8Ni21 (melt)*	000146
#Synthesis and Crystal	Structure of Iron Polyphosphide FeP4*	000438
#The	Structure of K2Er(WO3)5 (flux)*	000872
#The Crystal	Structure of K2W4O13*	000720
#Crystal	Structure of K5NdLi2F10 (flux)*	000397
#Crystal	Structure of Lanthanum Polysulfide LaS2*	000259
#	Structure of La2Fe1.76S5 (melt)*	000102
#Crystal	Structure of La3ReO8*	000081
Oxide with No Metal-Metal Bonding (flux)* #The Crystal	Structure of La3Ru3O11: A New Cubic KSbO3 Derivative	000209
#The Crystal	Structure of Lithium Fluorosulphate LiSO3F (solution)*	001108
#The Crystal	Structure of Lithium Uranate (Li2UO4, flux)*	000324
#Crystal	Structure of Li2CaUF8 (flux)*	001029
#Twinning and Crystal	Structure of Mn0.75Ga2.17S4 (flux)*	000247
vapor transport)* #	Structure of Monocrystalline WV2O6 at 298 and 383 K (	000389
#Crystal	Structure of Monophosphate AgCoPO4*	000986
Anions: Ba10(ReO5)6Br2 and Ba10(ReO5)6I2 (vapor #	Structure of New Rhenium Apatites Containing Large	000080
#The Crystal	Structure of Niobium Trisulfide, NbS3*	000814
#Crystal	Structure of RbEu3F10 (melt)*	000036
#Crystallographic and Magnetic	Structure of RbFeCl3.2D2O and CsFeCl3.2D2O (solution)*	000077
#Crystal	Structure of Scandium Metaphosphate Sc(PO3)3 (solution)*	000902
#Crystal	Structure of Se2Ti06 (slow cooling)*	000547
#	Structure of Silver Dichromate: Ag2Cr2O7 (solution)*	000260
#The Crystal	Structure of Synthetic Lautarite, Ca(IO3)2 (flux)*	000331
#The Crystal	Structure of Te3Nb2O11*	000316
#The Crystal	Structure of ThBC*	000806
#The Crystal	Structure of Triclinic WO3 (sublimation)*	000241
#Morphology and	Structure of V4O7 at 120 degrees K (vapor transport)*	000390
#Crystal	Structure of ZnS Grown from the Vapour Phase*	000874
#Czochralski Growth of A15	Structure of Zn4(P2S6)3 (melt)*	000132
	Structure Intermetallic Compounds (V3Si, V3Ge)*	001073

PERMUTED TITLE INDEX

PERMUTED TITLE INDEX

#Analysis of the Static Method of the Growth of Cadmium | Sulphide Crystals with Large Temperature Gradients (| 000289

#Direct Synthesis and Structure of Epitaxial Cadmium | Sulphide Layers on Cadmium Single Crystals (CdS, | 000429

#Crystal Growth of Sodium | Sulphide, Na2S (hygroscopic, Bridgman)* | 001056

#Crystal Growth of Antimony | Sulphur Iodide (SbSi, flux)* | 000109

#Crystal Growth (Proceedings of the Third International | Summer School on Crystal Growth, 1977) (review)* | 000066

Controlling Microstructures and Properties of | Superalloys via Use of Precipitated Phases (eutectics, | 000671

#Anisotropy of Upper Critical Field in | Superconducting 2H-NbS2 (theory)* | 000730

Superconductivity of Single-Crystal NbGe2* | 000791

#The Influence of Impurities on the Crystallization of | Supercooled Copper Melts (Cu)* | 000485

#Morphology of Ice Droxtals Grown from | Supercooled Water Droplets (H2O)* | 000347

DyKMo2O8)* #Relative Supersaturation and | Supercooling in Fluxed Melt Systems (Ho2Si2O7, TbVO 4, | 000594

Melt #Morphological Instability Under Constitutional | Supercooling During the Crystal Growth of InSb from the | 000478

Ionic Conductivity of Li24Zn(GeO4)4 and other new Li+ | Superionic Conductors (kinetics, melt)* Structure and | 000396

Ga1-xAlxAs, In1-xGaxAs, GaSb1-yAsy)* #Semiconductor | Superlattices by MBE and Their Characterization (GaAs, | 000176

Growth on #Crystal Growth Kinetics in (GaAs)n-(AlAs)m | Superlattices Deposited by Molecular Beam Epitaxy. I. | 000762

Beam Epitaxy of Ge and Ga1-xAlxAs Ultra Thin Film | Superlattices* #Molecular | 000761

Solutions (KDP)* #Nucleation in | Supersaturated Potassium Dihydrogen Orthophosphate | 000443

#Crystallization of Salts from | Supersaturated Solutions: Diffusion Kinetics* | 000623

#On the Growth of NH4Cl Dendrites at Very Low | Supersaturation (solution)* | 000172

(Ho2Si2O7, TbVO 4, DyKMo2O8)* #Relative | Supersaturation and Supercooling in Fluxed Melt Systems | 000594

theory, organics)* #Evaluation of | Supersaturation in Crystal Growth from Solution (| 000166

Growth of Large Single Pb1-xSnxTe Crystals on a Quartz | Surface (sublimation)* #Non-Seeded | 000960

xAlxAs, review of MBE process)* #Semiconductor | Surface and Crystal Physics Studied by MBE (GaAs, Ga1- | 000055

flux, theory, Pb1-xSnxTe)* #Investigation of Liquidus | Surface in Pb-Sn-Te System by Simplex Lattice Method (| 000668

Melt, and its Influence on the #Mass Transfer at the | Surface of a Crystal Near its Boundary with the | 001046

* #An Evidence of Spiral Growth on the | Surface of Melt-Grown KCl Crystals (Czochralski, theory) | 000408

#Growth of Triple Unit Cell Steps on the | Surface of Vapour Grown FeCr2S4 Single Crystals. I* | 000418

#Growth Mechanism of Multi-Steps on the | Surface of Vapour Grown FeCr2S4* | 000417

#Step Growth by Elementary Steps and Multi-Steps on the | Surface of Vapour Grown FeCr2S4* | 000419

Deposition of Single-Crystalline ZnO Film with Smooth | Surface on Intermediately Sputtered ZnO Thin Film on | 000711

CdS, evaporation)* # Surface Charge and Ledge Dynamics in Cadmium Sulfide (| 000529

Effect (float zone melting)* #Fermi | Surface Measurements of ZrB2 by the de Haas van Alphen | 000964

#A Molecular Dynamics Study of | Surface Melting (theory, vapor transport, kinetics)* | 000140

Surface Morphology of GaAs Grown by Vapor Phase Epitaxy* | 000705

Epitaxy (flux, theory, kinetics)* # Surface Morphology of GaAs Layers Grown by Liquid-Phase | 000605

Surface Morphology of HCl Etched Silicon Wafers (Si)* | 001015

theory)* # Surface Photovoltage Experiments on SrTiO3 Electrodes (| 000618

of the Interface Reaction Constant with Elementary | Surface Processes (theory)* #Correlation | 000330

and GaxIn1-xP Alloy Films by MBE (kinetics)* # Surface Processes Controlling the Growth of GaxIn1-xAs | 000301

#Thermodynamic Properties of | Surface Steps (theory)* | 001069

NaNO3, theory)* #Experiments on | Surface Tension Driven Flow in Floating Zone Melting (| 000848

#High-Efficiency p+-n-n+ Back- | Surface-Field Silicon Solar Cells (Si)* | 000298

Crystal Growth on Amorphous Substrates Using Artificial | Surface-Relief Gratings (KCl, films, solution)*Oriented | 000899

#(111) Cu2O Growth Modes on (111)Cu | Surfaces (epitaxy)* | 000388

#Kinetic Studies of Nucleation and Growth at | Surfaces (review)* | 001030

of Growth of Calcium Sulfate Crystals at Heated Metal | Surfaces (solution, CaSO4)* #Kinetics | 000334

Investigation of Thin Silver Films on Cleaved Silicon | Surfaces (Ag, theory, evaporation)* #An | 000619

#Anisotropy of Round Evaporation Spirals on Rocksalt | Surfaces (NaCl, theory)* | 000530

Step Trains of Different Density on Evaporating Crystal | Surfaces (NaCl, theory)* Between Successive Monatomic | 000105

Vapor-Deposited Silicon Carbide Films on Silicon Ribbon | Surfaces (SiC)* #Analysis of | 001099

#Survey of Monte Carlo Simulations of Crystal | Surfaces and Crystal Growth (review, theory, kinetics)* | 001013

#Observations of Microfacets Near Irregularly Remelted | Surfaces in Pulled GaSb Crystals (Czochralski)* | 000517

of Antimony Electrodeposits on Oriented Single Crystal | Surfaces of Gold and Silver (Sb)* #Epitaxial Growth | 000540

#Scanning Electron Microscopy of Growth | Surfaces of Synthetic Quartz (SiO2, hydrothermal)* | 000857

#Appearance of Non-Singular | Surfaces on Vapour-Grown Ice Crystals (H2O, theory)* | 000689

#The Adsorption of Zr onto W(100) | Surfaces* | 000221

Faces in Pure Solution and in the Presence of a | Surfactant (NaBO2.H2O2.3H2O)* Kinetics of Different | 000258

Nucleation Rates in Pure Solution in the Presence of a | Surfactant (NaBO2.H2O2.3H2O, kinetics, theory)* I. | 000257

and Crystal Growth (review, theory, kinetics)* # Survey of Monte Carlo Simulations of Crystal Surfaces | 001013

vapor transport)* #A Quick | Survey Method for the Study of CVD Conditions (TiN, | 000655

Structure, and Physical Properties (transport, magnetic | susceptibility, and NMR) of the Spinel CuV2S4 (vapor | 000544

KxV2O5, #Low-Temperature Specific Heat and Magnetic | Susceptibility of Nonmetallic Vanadium Bronze (NaxV2O5, | 000171

Stoichiometry Control of Cadmium Sulfo-Selenide (Cd1+y(| SxSe1-x), sublimation)* #Vapor Phase Transport and | 000647

#Vapor Growth and Epitaxy of Minerals and | Synthetic Crystals (review, theory)* | 000932

#The History and Present Status of | Synthetic Diamond (C, review, doped)* | 000685

#The Crystal Structure of | Synthetic Lautarite, Ca(IO3)2 (flux)* | 000331

#Scanning Electron Microscopy of Growth Surfaces of | Synthetic Quartz (SiO2, hydrothermal)* | 000857

Tomography (SiO2, #Observation of Growth Defects in | Synthetic Quartz Crystals by Light-Scattering | 000652

hydrothermal)* #On the Dissolution of Grown | Synthetic Quartz Inside the Autoclaves (SiO2, | 000507

#On the Incorporation of Metallic Impurities in | Synthetic Quartz Single Crystals (SiO2, solution)* | 000170

Characterization of the Hydrothermal Growth of | Synthetic 11 Angstrom Tobermorite (Ca6Si6O18.4H2O) | 000366

NaxLi1-xVO3)* #Crystallographic Studies and Structural | Systematics of the C2/c Alkali Metal Metavanadates (| 000692

of Lattice Mismatches Between LPE-Grown Li(Nb, | Ta)O3 Film and LiTaO3 Substrate (flux) Dependence | 000922

Substrates (LiNb1-xTaxO3, flux)* #LPE Growth of Li(Nb, | Ta)O3 Solid-Solution Thin Film Waveguides on LiTaO3 | 000500

Hydrides and Deuterides (V-H, V-D, Nb-H, Nb-D, Ta-H, | Ta-D)* and Characterization of the Group VA Metal | 000790

VA Metal Hydrides and Deuterides (V-H, V-D, Nb-H, Nb-D, | Ta-H, Ta-D)* and Characterization of the Group | 000790

#The Vapour Transport of NbB2 and | TaB2 (kinetics, theory)* | 000042

#Single-Crystal Growth of Tantalum Diboride (| TaB2, zone melting, solution)* | 000680

#Single-Crystal Growth of | Tantalum Diboride (TaB2, zone melting, solution)* | 000680

#Growth and Characterization of Mixed Sr:Ca | Tartrates (Sr1-xCaxC4H4O6.4H2O, gel)* | 000747

Selendies VxTi1-xSe2 and TaxTi1-xSe2 (TiSe2, VSe2, | TaSe2, vapor transport)* of Layered Transition Metal | 000555

#Upper and Lower Critical Fields of | TaS2(Pyridine)1/2 (vapor transport)* | 000468

of Layered Transition Metal Selendies VxTi1-xSe2 and | TaxTi1-xSe2 (TiSe2, VSe2, TaSe2, vapor transport)* | 000555

Pr4Ba6(BO3)9, Ni8Nb3O15, Ni2V2PbO8, Ni3Nb2O8, Ni3V2O8, | Ta2Co4O9, PbCr2.3Ti2.3O9, Pb2CrO5)* TiO3, Nd4Ba6(BO3)9, | 001063

#A New Structural Family M2"M' | Ta5O15. Crystal Structure of CaTlTa5O15 (flux) | 000317

Crystal Growth of Rare Earth Tetraborides (YB4, | TbB4, ErB4)* #Czochralski | 000137

Intermetallic Compounds, RCu2. I. Heavy Rare Earth (| TbCu2, DyCu2, HoCu2, ErCu2)* of Rare Earth Copper | 000373

Growth from Solution Using Cylindrical Seeds (TGS, | TBPB)* #Crystal | 000566

PERMUTED TITLE INDEX

the Horizontal Levitation Zone Melting Method (DyFe2,   TbFe2, Nd, Pr, Ce, cold crucible)*   Crystal Growth by   000620
#Crystal Growth of Some Rare Earth Trifluorides (GdF3,   TbF3, DyF3, HoF3, ErF3, Bridgman-Stockbarger,   000323
of Epitaxial Films of the Rare-Earth Dihydrides (GdH2,   TbH2, DyH2, ErH2, HoH2)*   #The Growth and Structure   000782
LaBO3, DyKMo2O8, ErKMo2O8, NdKMo2O8, DyVO4,   TbPO4, DyPO4, Ho2Ge2O7, Dy2GeMoO8, ZrSiO4, ThGeO4,   001060
and Supercooling in Fluxed Melt Systems (Ho2Si2O7,   TbVO 4, DyKMo2O8)*   #Relative Supersaturation   000594
PrBO3, LaBO3, DyKMo2O8, ErKMo2O8, NdKMo2O8, DyVO4,   TbVO4, TbPO4, DyPO4, Ho2Ge2O7, Dy2GeMoO8, ZrSiO4,   001060
Molybdates (Czochralski, melt, Gd2(MoO4)3, Sm2(MoO4)3,   Tb2(MoO4)3)*   Molybdate and Other Isotypic Rare-Earth   000446
Pb)MnO3, (Nd,Pb)MnO3, (Eu,Pb)MnO3, Pb3Mn7O15, Tm2Ti3O7,   Tb2Ti2O7, Pr2Ti2O7, Tm2Ge2O7, Er2SiO5, Dy2SiO5,   001061
Magnetic Structure and Magnetic Phase Transitions of   Tb5Ge4. A Neutron Diffraction Study (melt)*   #On the   000840
Tetracyanoquinodimethanide (TMTTF-   TCNQ) (solution)*   of Tetramethyltetrathiafulvalenium   000032
#Growth and Cleavage of Se-   Te Alloy Crystals (Bridgman)*   000111
#Liquidus Measurements in the Pb-Sn-   Te System (Pb1-xSnxTe, PbTe, flux)*   000047
xSnxTe)*   #Investigation of Liquidus Surface in Pb-Sn-   Te System by Simplex Lattice Method (flux, theory, Pb1-   000668
#Growth of Hollow Se-   Te Whisker Crystals (theory)*   000112
#Liquidus of the Ternary System Pb-Tl-   Te*   000101
by the Traveling Solvent Method from (Pb0.8Sn0.2)   Te*   000064
#Czochralski Growth of Tellurium Single Crystals (Te) #Semiconductor Single Crystals Obtained   000876
Orientation in Czochralski-Grown Tellurium Crystals (Te)*   #Etch Pit   000877
Thiochlorides of Molybdenum II: Mo6Cl10Y (Y = S, Se,   Te): Structure and Magnetic and Electrical Properties (000757
#LPE Growth of Hg0.60Cd0.40Te from   Te-Rich Solution (film, vapor transport)*   000839
Oxides with TeCl4. I. Calculation of the System Nb/O/   Te/Cl and the Transport of NbO2/NbO2.417 (NbO2, NbO2.   000797
Crystal Growth of Silicon   #Interaction Between   Technological Parameters and Crystal Quality During the   000325
Efficiency GaAs Thin Film Solar Cells by Peeled Film   Technology (epitaxy)*   #High   000497
#On the Chemical Transport of Some Niobium Oxides with   TeCl4 (II) The Transport Behaviour of the Equilibrium   000798
#On the Chemical Transport of Some Niobium Oxides with   TeCl4. I. Calculation of the System Nb/O/Te/Cl and the   000797
Velocities of K2PbCu(NO2)6 Near the Cooperative Jahn-   Teller Phase Transitions (solution)*   #Ultrasonic   000467
#Crystal Chemistry of a   Tellurate IV of Indium III: In2Te3O9*   000766
#An X-Ray Topographic Assessment of Cadmium Mercury   Telluride (CdxHg1-xTe, Bridgman)*   000158
Compositional Variations in Crystals of Cadmium Mercury   Telluride (Cd1-xHgxTe, slow cooling)*   Speed on the   000074
#Vapor Growth of Zinc   Telluride by Closed Tube Method (ZnTe, epitaxy)*   000700
Conductivity. III. On the Phase Transition of CuTeBr   #   Telluride Halides of Copper with High Ionic Partial   000025
Method for the Growth of Homogeneous Mercury Cadmium   Telluride Single Crystals (Hg1-xCdxTe, Bridgman)*   000292
#Bismuth   Tellurides: BiTe and Bi4Te3 (melt)*   001016
of ZnS, ZnSe, CdS and Their Mixed Compounds Using   Tellurium as a Solvent*   #Solution Growth   001065
#Etch Pit Orientation in Czochralski-Grown   Tellurium Crystals (Te)*   000877
#Czochralski Growth of   Tellurium Single Crystals (Te)*   000876
Observations of Precipitates in ZnTe: Shape of the   Tellurium Solidus Line (Bridgman)*   Electron Microscopy   000098
*   #Growth of Uranium Dioxide Crystals Using   Tellurium Tetrachloride and Argon (UO2, vapor transport)   000280
#The Effect of the Solidification Front Shape on the   Temperature Distribution in the Crystal (NaCl, theory,   000128
of Bi2WO6 Single Crystal Below the Transformation   Temperature (slow cooling)*   #Flux Growth   000665
Single Crystals of Cinnabar (alpha-HgS) at Rather Low   Temperature (solution, doped)*   #Growth of   000157
#Defects in Single Crystals Prepared at High   Temperature (theory, kinetics)*   000510
Expansion Coefficients of VO2 near the Transition   Temperature (vapor transport)*   and the Thermal   000513
Solidification of Pb-Bi Alloys Rich in   #Progress of   Temperature and Concentration in the Region of   000846
Liquid Phase Epitaxy (flux, kinetics,   #The Liquidus   Temperature and Growth-Dissolution Kinetics of Garnet   000333
*   #Fast Growth in GaAs VPE at Low   Temperature and High Partial Pressures (vapor transport)   000393
Wavelength (#Nd:Y2O3 Single-Crystal Fiber Laser: Room-   Temperature cw Operation at 1.07- and 1.35-micrometer   000916
Transport Properties of Single Crystals #Effect of High-   Temperature Annealing in Cadmium on the Electrical   000612
Dy2Ti2O7 (powder) as a Function of Magnetic Field (#Low   Temperature Conductivity of Gd2(SO4)3·8H2O (crystal) and   000065
#A New Organic Low   Temperature Conductor: HMTSF-TNAP (solution, organics)*   000087
and (Eu,Yb)3Fe5O12, vapor transport, epitaxy,   #Low   Temperature CVD Garnet Growth (Er3Fe5O12, (Eu,Y)3Fe5O12   000211
Gallium Arsenide (GaAs)*   #Growth   Temperature Dependence in Molecular Beam Epitaxy of   000666
Silicon Epitaxial Growth from the Gas Phase. I.   #   Temperature Dependence of the Degree of Doping in   000514
Intermetallic Compound Ni3Ge Single   #Orientation and   Temperature Dependence of the Flow Stress in the   000033
Epitaxial Silicon from the Gas Phase. II. Phosphorus   #   Temperature Dependence of Degree of Doping in Growth of   000515
LPE-Grown Li(Nb,Ta)O3 Film and   #Direct Measurement of   Temperature Dependence of Lattice Mismatches Between   000922
BaO.B2O3 System on Convection of a Solution with a   Temperature Difference* of Liquid Phase in the Y3Fe5O12-   000048
for Silicon Carbide Monocrystal and Epitaxial #Study of   Temperature Distribution in High-Temperature Furnaces   000561
Growth (EDFG)*   #Evaluation of   Temperature Distribution of Melt in Silicon Ribbon   000523
Growth (Cu, Si)*   #A Theoretical Study of   Temperature Distributions During Czochralski Crystal   000149
#Calcium Orthovanadate, Ca3(VO4)2 - A New High-   Temperature Ferroelectric (Czochralski)*   000339
(kinetics, theory, C)*   #High   Temperature Field Dynamics in the Synthesis of Diamonds   000862
to Its Low Form (Czochralski, flux)*   #High   Temperature Form of Pb2WO5 and Transformation Phenomena   000310
and   #Study of Temperature Distribution in High-   Temperature Furnaces for Silicon Carbide Monocrystal   000561
Growth Systems (flux)*   #   Temperature Gradient Measurement in Liquid Epitaxial   000574
#Growth of p-Type InP Single Crystals by the   Temperature Gradient Method*   000422
of the Growth of Cadmium Sulphide Crystals with Large   Temperature Gradients (CdS, kinetics, theory, vapor   000289
Czochralski, melt, Si)*   #Effects of Radial and Axial   Temperature Gradients on Tensions in Crystals (theory,   000326
Solvent (flux)*   #Low   Temperature Growth of GaP LPE Layers from Indium   000925
Ag, film)*   #A High   Temperature High Purity Source for Metal Beam Epitaxy (000287
Rare Earth Vanadates and   #Solubility Curves in High-   Temperature Melts for the Growth of Single Crystals of   000265
Single Crystals (Li3N)*   #Low   Temperature Method for the Growth of Lithium Nitride   000252
Transition of CuTeBr and the Structure of the High   Temperature Phase (Bridgman)*   III. On the Phase   000025
Films by MBE (molecular beam epitaxy, theory)*   #   Temperature Range for Growth of Autoepitaxial GaAs   000686
Preparation Technique in Solid State Chemistry. I. High   Temperature Reactions with CO2-Laser*   #A New   000657
in Solid State Chemistry. II. High Pressure - High   Temperature Reactions with CO2-Lasers*   Technique   000658
#Crystal Growth of Magnetic Garnets from High-   Temperature Solutions (review)*   000982
#Growth of Bulk Yttrium Iron Garnet Crystals from High   Temperature Solutions (Y3Fe5O12)*   000977
iz Rastvorov-Rasplavov (Crystal Growth from High-   Temperature Solutions) (review, flux)* #Rost Kristallov   000976
of Nonmetallic Vanadium Bronze (NaxV2O5, KxV2O5,   #Low-   Temperature Specific Heat and Magnetic Susceptibility   000171
CeAs (sublimation)*   #Low   Temperature Thermal and Magnetic Properties of CeP and   000400
   #Low   Temperature Thermal Properties of PbI2 (Bridgman)*   000851
Chalcopyrite Crystals by Iodine Vapour Transport (#A   Temperature Variation Method for the Growth of   000744
Kyropoulos, float #Crystal Growth from Liquids at High   Temperatures (review, melt, flux, Czochralski,   000139
in Solid Phase Epitaxy Si Films Deposited at Elevated   Temperatures (vapor transport)*   #Crystalline Defects   000194
Si)* #Local Selective Homoepitaxy of Silicon at Reduced   Temperatures Using a Silicon-Iodine Transport System (000136
cf Homogeneous Mixed Crystals with a Large Segregation   Tendency by the Bridgman Method - Pseudobinary CdxHg1-   000244
theory)*   #Experiments on Surface   Tension Driven Flow in Floating Zone Melting (NaNO3,   000848

PERMUTED TITLE INDEX

PERMUTED TITLE INDEX

Vapor Phase Growth - Citing an Instance of GaxIn1-xAs (theory)*
in Chemical Vapour Deposition on Profiled Substrates (theory)*
in One Layer on a Crystal-Solution Interface (theory)*
Structures During and After Unidirectional Growth (theory)*
Tension Driven Flow in Floating Zone Melting (NaNO3, theory)*
Materials. Part 1. Crystalline Materials (review, theory)*
Analysis in Chemical Vapour Deposition Processes. II (theory)*
Analysis in Chemical Vapour Deposition Processes. I (theory)*
During Normal Growth in a Binary Alloy (kinetics, theory)*
A Fast Proton Conductor (0.84Na20.0.84MgO.5Al2O3, theory)*
Epitaxial Growth Kinetics in the SiC-C System (SiC, theory)*
Growth (PVT) (KCl, KBr, KCl:KCN, KCN:KOH, KCl:KOH, theory)*
Cr)P and (Mn,Fe)P Compounds (Mn1-xFexP, x = 0 to 0.15, theory)*
in Czochralski Grown YAG Crystals (Y3Fe5O12, theory)*
(111) Diamond Face Around the Rougnening Transition (C, theory)*
on the Surface of Melt-Grown KCl Crystals (Czochralski, theory)*
Reaction Constant with Elementary Surface Processes (theory)*
Constant of Epitaxial Layers of III-V Compounds (theory)*
Ferrite Single Crystals by Chemical Transport (NiFe2O4, theory)*
Growth of A(III)B(V) Compounds from the Vapour Phase (theory)*
Habit of Struvite Crystals (MgNH4PO4.6H2O) (solution, theory)*
Ferromagnetic Transition (Bridgman, vapor transport, theory)*
Crystal-Melt Interface of a Lennard-Jones Substance (theory)*
Two-Dimensional Nucleation Models of Crystal Growth (theory)*
Radio Frequencies for Float Zone Silicon Crystals (Si, theory)*
Free and Dislocation-Assisted Growth of Gallium (Ga, theory)*
Kinetics of Ionic Crystals (CaSO4.2H2O, solution, theory)*
of Oxides and Fluorides (review, hydrothermal, theory)*
"The Validity of Steady-State Dendrite Growth Models" (theory)*
IV-VI compounds, III-V compounds, evaporation, review, theory)*
Particle Growing from a Binary Alloy Melt (dendrites, theory)*
in Web-Dendritic Ribbon Growth (Si, kinetics, theory)*
Growth: Theoretical and Experimental Investigations (theory)*
A Monte Carlo Simulation for Ionic Crystal Growth (theory)*
SnS2-SnI4 and SnS2-I2 (SnS2, vapor transport, kinetics, theory)*
Characteristics in Binary Metallic Solid Solutions (theory)*
of Graded Composition by the Travelling Solvent Method (theory)*
Part I. Growth and Segregation Behavior of Ga-Doped Ge (theory)*
Prepared by a Gas-Evaporation Technique (MoO3, WO3, theory)*
Growth of Silicon Crystals, Free from Dislocations (theory)*
the Melt Under Stabilizing Thermal Gradient (Bridgman, theory)*
Single Filament, Unconstrained Growth from a Pure Melt (theory)*
GeBr4 Vapor (Ge, epitaxy, vapor transport, kinetics, theory)*
Shape of the Growing Crystal (Czochralski, kinetics, theory)*
of Heterogeneous Reactions (GaAs, vapor transport, theory)*
Growth of ZnS in H2: The Effects of a Bypass Flow (theory)*
Crystal Growth (Si, Ge, GaAs, InAs, InSb, Al2O3, melt, theory)*
of the Soft-Core Model in an FCC Crystal Formation (theory)*
2H-WbSe2 with Hydrazine by Transmittance Measurements (theory)*
Modes of Epitaxial Growth from the Vapor Phase (review, theory)*
by Vapour Transport Processes (kinetics, theory)*
and Segregation Coefficient (Si, vapor transport, theory)*
of the Layer Perovskite (C3H7NH3)2CuCl4 (solution, theory)*
to Calcium Sulphate Dihydrate (CaSO4.2H2O, kinetics, theory)*
Driven Flows (Si, Al2O3, kinetics, float zone melting, theory)*
Grown Crystals, Taking Account of Periodic Remelt (Si, theory)*
Gallium Garnet (Gd3Ga5O12, Czochralski, kinetics, theory)*
Ga_AsCl3-H2 System (GaAs, vapor transport, kinetics, theory)*
Transport Kinetics in Chemical Vapour Deposition (theory)*
Presence of a Surfactant (NaBO2.H2O2.3H2O, kinetics, theory)*
of Garnet Liquid Phase Epitaxy (flux, kinetics, theory)*
Cl and the Transport of NbO2/NbO2.417 (NbO2, NbO2.417, theory)*
HgI2 Single Crystals from Dimethylsulfoxide Complexes (theory)*
of Oscillatory Flow in Crystal Growth Melts (NaNO3, theory)*
a Crystal from a Molten Mass in the (100) Orientation (theory)*
Nonmetallic Vanadium Bronze (NaxV2O5, KxV2O5, AgxV2O5, theory)*
and Cadmium Selenide (CdSe, epitaxy, kinetics, films, theory)*
for Different Two-Dimensional Substrate Lattices (theory)*
Crystals of Cadmium Sulphide (CdS, vapor transport, theory)*
Czochralski Growth of Gallium and Germanium (Ga, Ge, theory)*
Unconstrained Growth from a Binary Alloy Melt (Sn-Pb, theory)*
Between and alpha- and beta-phases (Cu-Zn, theory)*
(CuxZrS2, CuxHfS2, FexZrS2, FexHfS2, vapor transport, theory)*
KAl11O17, RbAl11O17, AgAl11O17, EuAl11O17, flux, theory)*
Density on Evaporating Crystal Surfaces (NaCl, theory)*
GaAs Films by MBE (molecular beam epitaxy, theory)*
Method. II. Diffusion-Induced Twins in the alpha-phase (theory)*
(vapor transport, electron beam melting, kinetics, theory)*
with Low Dislocation Density (Al, zone melting, theory)*
Molar Solution Enthalpies (Si, vapor transport, theory)*
Grown under Conditions of Constant Cooling Rate (melt, theory)*
During Crystal Growth from the Melt (Czochralski, theory)*
of Ni-Cl Boracite (no materials grown - Ni3B7O13Cl theory)*
from a Ternary Component Solution (Ga1-xAlxAs, flux, theory)*
by Molecular Dynamics Simulation (melt, slow cooling, theory)*
of Heats of Fusion for Ionic Crystals (review, theory)*
Thickness (effective distribution coefficient, melt, theory)*
with a Curved Solid-Liquid Interface (Ge:Ga, Bridgman, theory)*

#A Simple Analysis of 000673
#Interface Morphology 001011
#Localized Adsorption 000575
#Coarsening of Eutectic 001027
#Experiments on Surface 000848
#Infra-Red Transmitting 000250
#Morphological Stability 000435
#Morphological Stability 001010
#Region of Metastability 000969
#Hydronium Beta" Alumina: 000285
#Study of Silicon Carbide 000562
#Low-Stress Physical Vapor 000809
#Magnetic Properties of (Mn, 000430
Interface Shape Transitions 000645
Monte Carlo Simulation of a 001020
#An Evidence of Spiral Growth 000408
#Correlation of the Interface 000330
#Determination of the Lattice 000399
#On the Preparation of Nickel 000760
#Flow-Graph Model of Epitaxial 000504
#Growth Morphology and Crystal 000002
#Heat Capacity of EuS near the 000505
#Monte Carlo Simulation of the 000123
#A New Technique for Simulating 001038
#Limitations in Using Kilohertz 000358
#Rate Equations for Dislocation- 000001
#Solvent Effects and the Growth 000134
#Advances in Preparative Chemistry 000398
#Reply to Comment of R. Trivedi on 000833
#Hot Wall Epitaxy (II-VI compounds, 000568
#Morphological Stability of a Solid 000990
#Thermal Analysis of Solidification 000369
#A Model for Macroscopic Czochralski 001002
#Fast Growth of Ordered AB Crystals: 000182
#Transport Properties of the Systems 001079
Quantities and Vacancy and Diffusion 000090
#Growth of Pb1-xSnTe Single Crystals 000564
of a Heat Pipe to Czochralski Growth. 000604
#Coalescence Growth of Smoke Particles 000448
and Crystal Quality During the Crystal 000325
During the Crystal Growth of InSb from 000478
Morphology During Crystallization - I. 000992
of Germanium (111) Single Crystals with 000060
with the Melt, and its Influence on the 001046
Method and Its Application to the Study 000281
Processes Associated with the Epitaxial 000558
#Formation of Stress and Dislocations in 000406
#Geometrical Analysis of Crystallization 000965
#Kinetic Studies of the Intercalation of 000085
Methods Used for the Characterization of 000542
for the Growth of Layered Dichalcogenides 000805
from the Gas Phase. I. Phosphorus Content 000514
#Face Stability and Growth Rate Variations 000641
in Aqueous Solution with Special Reference 000193
in Floating Zone Melting. I. Pure Rotation 000173
of Striated Impurity Distributions in Melt- 001026
Processes and Crystal Growth of Gadolinium 000769
Rate and Purity of Epitaxial Layers in the 000653
#Interrelation Between Interface and Vapour 000742
I. Nucleation Rates in Pure Solution in the 000257
Temperature and Growth-Dissolution Kinetics 000333
TeCl4. I. Calculation of the System Nb/O/Te/ 000797
from Soluble Complexes. III. Growth of alpha- 000695
of a Critical Marangoni Number for the Onset 000847
of a Metallic Face-Centered Cubic Lattice in 000995
Specific Heat and Magnetic Susceptibility of 000171
Sublimation Properties of Cadmium, Selenium, 000365
of Migrating Adatoms by Point Capture Centres 000873
the Electrical Transport Properties of Single 000612
Contact Angle Between Melt and Crystal During 001071
During Crystallization - II. Single Filament, 000993
Morphology and Crystallographic Relationships 000372
Shift of the Absorption Edge in ZrS2 and HfS2 001090
Silver, and Europium beta-Alumina (NaAl11O17, 000621
Successive Monatomic Step Trains of Different 000105
#Temperature Range for Growth of Autoepitaxial 000686
Bicrystals Produced by a Solid-Solid Diffusion 000371
Films of CdS Evaporated in Vacuum onto Silicon 000073
Method of High Purity Aluminum Single Crystals 000228
Phase. II. Phosphorus Content and Differential 000515
*of Crystal Composition in Ga1-xAlxAs LPE Layers 000249
*Burton, Prim, and Slichter Model of Segregation 001084
*Component System: The Chemical Vapour Transport 000236
*Liquid and Solid Compositions During LPE Growth 000970
*Melting and Freezing in the Gaussian Core Model 000915
*Quantum-Mechanical Relations in the Calculation 001087
*Slichter Equation as a Diffusion Boundary Layer 001083
*Unidirectional Solidification of a Binary Alloy 000206

PERMUTED TITLE INDEX

#The Growth of Crystals in Solution (review,
the Melting Point and the Vacancy Mechanism of Melting (
a Dilute Binary Alloy Single Crystal (Fe-Ni, eutectics,
Electroepitaxy: Growth Kinetics (Peltier effect, GaAs,
II. Combined Free and Rotation Driven Flows (kinetics,
of the Mechanism and Kinetics of Growth of LPE GaAs
Layers in CVD Systems (Si-H-Cl, Ga-As-H-Cl) (GaAs,
Parameters on the Directional Solidification Process
of Thin Silver Films on Cleaved Silicon Surfaces (Ag,
of Stirring for LPE Growth (Sm,Y)3(Fe,Ga)5O12 (flux,
Geometry and Prandtl Number at Zero Gravity (Si, Al2O3,
#Congruent (Diffusionless) Vapor Transport
#Impurity Redistribution in EFG (Si, ribbon,
#New Model of LPE Growth: Layer Thickness Calculation
in Single Crystals Prepared at High Temperature
Driving Force for Crystallization from Solution
of Czochralski Bulk Flow in Microgravity (Al2O3,
of GaAs Layers Grown by Liquid-Phase Epitaxy (flux,
Parameter on Czochralski Bulk Flow in Garnet Growth (
and Perfection of Single Crystals of Germanium (Ge,
Soluble Complexes. I. General Introduction (review,
(alpha-FeSi2.33, CoSi2, FeGe2, Mn5Si3, Fe1.76Ge, melt,
of Crystal Surfaces and Crystal Growth (review,
by the Method of Directed Vertical Crystallization (
Driving Force for Crystallization: The Growth Affinity
in a Chemical Potential Varying in Space or Time (
Cooling Rates of Large-Diameter Silicon Crystals (Si,
Growth Rate by LPE from a Cooling, Static Solution (
of Non-Equilibrium Crystallization of an Alloy (
on the Temperature Distribution in the Crystal (NaCl,
of Supersaturation in Crystal Growth from Solution (
of Polytypism in Crystals (SiC, ZnS, CdI2, mica,
of the Incommensurate Phase Transition of Rb2ZnBr4 (
of Polytypism in Cadmium Iodide Crystals (CdI2,
#Unstable Growth of ADP Crystals (NH4H2PO4,
#On Diffusive-Advective Interfacial Mass Transfer (
Equilibrium from 700 to 1000 K (Ge, vapor transport,
with Large Temperature Gradients (CdS, kinetics,
from the Vapour Under Constant Conditions (C10O16,
on the Growth and Etching Vectors of Silicon (111) (
#A Molecular Dynamics Study of Surface Melting
Field Dynamics in the Synthesis of Diamonds (kinetics,
Convection, Graphite, and Integrated Optics (review,
Single Crystals and Films (Si, Ge, GaAs, InSb, epitaxy,
Axial Temperature Gradients on Tensions in Crystals (
in Czochralski Grown Gadolinium Gallium Garnet (GGG,
PtSe, PbTe, SnTe, #Roles of Lattice Fitting in Epitaxy (
in Pb-Sn-Te System by Simplex Lattice Method (flux,
HCl-H2 System Based on Flow Graph (vapor transport,
melting, theory)*   #Percolation
Verneuil, III-V compounds, Si)*   #Crystal Growth -
Basic Features and Equilibrium Properties (#Analytical
a Solution in a Melt (GaP, flux)*

Component Systems. Two Cases of   #A Fluctuation
kinetics)*   #On the
                #Basic Consideration for an Analytic
#Studies of LPE Ripple Based on Morphological Stability

sublimation)*   #Low Temperature
Ribbon Growth (Si, kinetics, theory)*
Growth from the Vapour #Investigations of Kinetic and
                          #The Melt Growth and
320 K*   #The
X-ray Determination of the Lattice Parameters and the
Crystal Growth of InSb from the Melt Under Stabilizing
                                              #The
Process (theory, eutectics, Al-Al3Ni)* #Effect of Some
solution)*   #Fiber Optic Probe for
                          #Magnetic, Transport, and
                                        #Low Temperature
Preparation of Single Crystals #Analytical Study of the
Bridgman Method (ZnF2, CoF2,   #Crystal Growth by the
Growth of Mixed (La, Eu, Y, Ce, Ba, Cs) Hexaborides for
Influence of Zone Geometry and Prandtl Number at Zero #
                          #In-Situ Carbon-Silicon Ribbon
Dichalcogenides by Vapour Transport Processes (#Some
Solution (theory, kinetics)*   #The
Purity of Epitaxial Layers in the Ga_AsCl3-H2 System #A

Characteristics in Binary Metallic #Connections Between
from 700 to 1000 K (Ge, vapor transport, theory,
Binary Alloy Single Crystal (Fe-Ni,   #Irreversible
#Effect of Pressure on Electrical Resistivity and
TbVO4, TbPO4, DyPO4, Ho2Ge2O7, Dy2Ge2O7, ZrSiO4,
                                        #An LPE Growth of
and Slichter Equation as a Diffusion Boundary Layer
                          #New Model of LPE Growth: Layer

theory, alkaline earth sulfates, calcium phosphates)*   000683
theory, compilation, 39 metals)*Capacity of Metals Near 000351
theory, dendrites)*   of the Cellular Growth of 000113
theory, electrolytic deposition, kinetics)*Liquid-Phase 000437
theory, electron beam melting)*   Floating Zone Melting. 000174
theory, epitaxy)*   #Investigation 000144
theory, epitaxy)*   Analysis of Equilibrium Adsorption 000184
theory, eutectics, Al-Al3Ni)* #Effect of Some Thermal 000268
theory, evaporation)*   #An Investigation 000619
theory, film)*   #An Improved Method 000222
theory, float zone melting, kinetics, Si)*   of Zone 000175
theory, kinetics)*   000007
theory, kinetics)*   000451
theory, kinetics)*   000589
theory, kinetics)*   #Defects 000510
theory, kinetics)*   #The Thermodynamic 000904
theory, kinetics)*   #Digital Simulation 000536
theory, kinetics)*   #Surface Morphology 000605
theory, kinetics)*   #Effect of the Buoyancy 000535
theory, kinetics)*   #Preferred Growth Directions 000275
theory, kinetics)*   Soluble Single Crystals from 000693
theory, kinetics)*   of Crystals of Certain Metallides 000861
theory, kinetics)* #Survey of Monte Carlo Simulations 001013
theory, kinetics)*   for Preparation of Single Crystals 000057
theory, kinetics, melt, solution)*   #On the 001024
theory, kinetics, solution)*   #Growth Spirals 000659
theory, kinetics, Czochralski)*   #On the 000165
theory, melt)*   #Conditions for Constant 000893
theory, melt, slow cooling)*   #Dynamics 000826
theory, melt, Czochralski)* Solidification Front Shape 000128
theory, organics)*   #Evaluation 000166
theory, review)*   #Some Aspects 000069
theory, solution)*   #Neutron-Scattering Study 000226
theory, solution, evaporation)*   of Growth Conditions 000432
theory, solution, kinetics, slow cooling)*   000804
theory, solution, vapor transport, flux)*   001074
theory, thermodynamics)*   #Germanium-Iodine 000191
theory, vapor transport)*   of Cadmium Sulphide Crystals 000289
theory, vapor transport)* Growth of Adamantane Crystals 000572
theory, vapor transport, epitaxy)* Step Reconstruction 001019
theory, vapor transport, kinetics)*   000140
theory, C)*   #Temperature 000862
theory, C, GaAs)*   Vol. 4, Morphological Stability, 001081
theory, Czochralski)*During the Growth of Semiconductor 000636
theory, Czochralski, melt, Si)* #Effects of Radial and 000326
theory, Gd3Ga5O12)*   #Dislocation Propagation 000838
theory, Ni, Cu, Pd, Pt, Fe, Al, Au, Ag, In, Pb, PbS, 000949
theory, Pb1-xSnxTe)* #Investigation of Liquidus Surface 000668
theory, Si)**Model of Silicon Epitaxial Growth in SiCl4- 000503
Theory and Crystallization (entropy, specific heat, 000168
Theory and Techniques, Volume 2 (review, epitaxy, 000348
Theory of Crystal Growth of Two-Component Systems I. 000765
Theory of Crystallization During Epitaxial Growth from 000591
Theory of Melting, Vacancy Model* 001007
Theory of Normal Crystal Growth for Metallic One- 000059
Theory of Normal Growth of Crystals in Binary Systems (001110
Theory of Two-Component Crystal Growth* 000763
Theory* 000698
Thermal and Dielectric Properties of LiKSO4 and LiCsSO4* 000229
Thermal and Magnetic Properties of CeP and CeAs (000400
Thermal Analysis of Solidification in Web-Dendritic 000369
Thermal Conditions of Silicon Carbide Epitaxial Layer 000942
Thermal Expansion of Na2CO3 Crystals (Bridgman)* 000632
Thermal Expansion of 2H-MoS2 and 2H-WSe2 Between 10 and 000667
Thermal Expansion Coefficients of VO2 near the 000513
Thermal Gradient (Bridgman, theory)*   During the 000478
Thermal Oxidation of GaAs* 000156
Thermal Parameters on the Directional Solidification 000268
Thermal Profiling of Liquids During Crystal Growth (000395
Thermal Properties of Ferromagnetic EuB6 (review)* 000293
Thermal Properties of PbI2 (Bridgman)* 000851
Thermal Transfer Process in an Installation for 000507
Thermic Screen Translation (TST) Technique: A Modified 000541
Thermionic Emission (LaB6, EuB6, CeB6, BaB6, (La,Eu)B6, 000727
Thermocapillary Convection in Floating Zone Melting: 000175
Thermocouple (Si, equipment)* 000818
Thermodynamic Considerations for the Growth of Layered 000805
Thermodynamic Driving Force for Crystallization from 000904
Thermodynamic Factor Influencing the Growth Rate and 000653
Thermodynamic Properties of Surface Steps (theory)* 001069
Thermodynamic Quantities and Vacancy and Diffusion 000090
thermodynamics)*   #Germanium-Iodine Equilibrium 000191
Thermodynamics of the Cellular Growth of a Dilute 000113
Thermoelectric Power of FeS (Bridgman)* 000465
ThGeO4, Er2Si2O7)* DyKMo2O8, ErKMo2O8, NdKMo2O8, DyVO4, 001060
Thick InxGa1-xAs* 000506
Thickness (effective distribution coefficient, melt, 001083
Thickness Calculation (theory, kinetics)* 000589

PERMUTED TITLE INDEX

Ratios (Bi, Czochralski)*   #The Growth of
with Smooth Surface on Intermediately Sputtered ZnO
Clusters (Ag on Si, epitaxy)*   #
Epitaxy on Bridgman Grown Zn2TiO4 Substrates (flux)*
    #Epitaxy of NdAl3(BO3)4 For
epitaxy)*   #High Efficiency GaAs
    #Molecular Beam Epitaxy of Ge and Ga1-xAlxAs Ultra
xTaxO3, flux)**LPE Growth of Li(Nb,Ta)O3 Solid-Solution
    #Structural Transitions and Phase Composition of
, Grown by Liquid-Phase Epitaxy (flux)*
    #Structural and Electrical Properties of CuGaSe2
#Growth and Electrical Properties of Epitaxial CuInS2
Trichroism of Ferroelectric Triglycine Sulphate
    #CdTe
Growth of Ferroelectric PLZT ((Pb,La)(Zr,Ti)O3)
Electron Microscopy of Growth Processes of Epitaxial
of the Optical Properties of LPE InxGa1-xAsyP1-y
    #Vapor Phase Growth of
Film Doping Using   #The Bending of Silicon Wafers by
theory, evaporation)*   #An Investigation of
    #A New Technique for Preparation of
: Structure and Magnetic and Electrical Properties #New
Physicochemical Properties of Alkali Metal Indates and
1977) (review)*   #Crystal Growth (Proceedings of the
#Deposition of Crystals from the Plasmas of ZrO2, HfO2,
    #Preparation of
of Mineralizers on the Growth Morphology of Cubic
#Crystal Chemistry of Perovskite Phases of the Systems
#Epitaxial Growth of Ferroelectric PLZT ((Pb,La)(Zr
    #Single Crystals with Compositions in the Range
Crystals with Compositions in the Range TiB1.89 -
of Titanium Diboride by a Modified Hot Wire Method (
Range TiB1.89 - TiB1.96 Prepared by Chemical Transport (
Various Metals and Refractory Oxides on the Growth of
Carbide from the Vapor by a Modified Hot Wire Method (
float zone melting)*   #Preparation of
by Vapor Transport of Monocrystals of AlF3, FeF3, and
Bicrystalline Substrates of #Characterization of (001)
Spirals in a Chemical Potential Varying in Space or
Growth of Single Crystals of Some Rubidium and Cesium
#A Quick Survey Method for the Study of CVD Conditions (
and HCl*   #Growth of Anatase
Ion Transport in Stoichiometric Rutile Single Crystals (
Rutile Crystals and Their Photoelectronic Properties (
System for Exacting Crystal Growth Preparations (
Complex Oxides (GdVO3, Gd4Ba6(BO3)9, (Ba0.93Gd0.07)
Growth of LaB6 Whiskers and Crystals Having a Sharp
#On the Relations Between the Number of Gold Drops and
Equilibria and Mass Transfer Kinetics in the System
and Mass Transfer Kinetics in the System TiP-I2 (
Dichalcogenides (MoS2, MoSe2, WS2, WSe2, HfSe2, ZrSe2,
Transition Metal Selendies VxTi1-xSe2 and TaxTi1-xSe2 (
of Addition of Vanadium on the Chemical Transport of
    #Electron Energy-Loss Spectroscopy of
Na2NiTi7O16)*   #Growth Conditions for Sodium-
Method (TiC)*   #Single Crystal Growth of
    #Chemical Vapor Growth of
on the Chemical Transport of TiSx, x = 1.40-1.70 (
Aspects of the Metal-Insulator Transitions in (
#Metal-Insulator Transitions and Phase Diagram of (
    #Elastic Constants and Stability of bcc In-
    #Structure and Growth Peculiarities of
    #Liquidus of the Ternary System Pb-
#Growing and Polishing Large Thallium Single Crystals (
HF-AlF3 and NH4F-FeF3 in Hydrofluoric Acid Solution (
Growth of Thallium Iodide Crystals in Silica Hydrogels (
    #Structure and Growth Peculiarities of Tl-Se-
#Structure and Growth Peculiarities of Tl-Se-TlInSe2 (
    #Electrical Properties of
Growth of Single Crystals of Langbeinites Rb2Cd2(SO4)3,
NO2)6, Tl2NiCd(NO2)6, Tl2NiHg(NO2)6, Tl2CoPb(NO2)6,
NO2)6, Tl2NiPb(NO2)6, Tl2NiCd(NO2)6, Tl2NiHg(NO2)6,
Growth of Triple Nitrites in Gels (K2CuPb(NO2)6,
of Thallium Iodide Crystals in Silica Hydrogels (TlI,
NO2)6, Tl2CuPb(NO2)6, K2NiPb(NO2)6, Tl2NiPb(NO2)6,
NO2)6, K2NiPb(NO2)6, Tl2NiPb(NO2)6, Tl2NiCd(NO2)6,
in Gels (K2CuPb(NO2)6, Tl2CuPb(NO2)6, K2NiPb(NO2)6,
Tungsten Bronze*   #The Thallium Tungstate
    #Crystal Growth of
Compounds MFX of Divalent Rare Earths (M = Sm, Eu,
of Polymorphic Rare-Earth Disilicates, R2Si2O7 (R =
(melt, SmFCl, SmFBr, SmFI, EuFCl, EuFBr, EuFI, TmFCl,
and I) (melt, SmFCl, SmFBr, SmFI, EuFCl, EuFBr, EuFI,
SmFCl, SmFBr, SmFI, EuFCl, EuFBr, EuFI, TmFCl, TmFBr,
    #Magnetic Mixed Valent
Ground State of the Intermediate Valence Compound
Tetracyanoquinodimethanide (

Thin Bismuth Single Crystals with High Resistivity   000887
Thin Film on Sapphire (epitaxy)*   Crystalline ZnO Film   000711
Thin Film Deposition From Beams of Ionized Atoms and   000516
Thin Film Growth of (MnxZn1-x)Fe2O4 by Liquid Phase   000216
Thin Film Miniature Lasers (flux)*   000577
Thin Film Solar Cells by Peeled Film Technology (000497
Thin Film Superlattices*   000761
Thin Film Waveguides on LiTaO3 Substrates (LiNb1-   000500
Thin Films of Aluminum with Silicon*   000473
Thin Films of LiNbO3, Doped with Na+ and Co2+ plus Zr4+   000691
Thin Films on GaAs Substrates (epitaxy)*   000844
Thin Films on GaAs Substrates*   000690
Thin Films Grown by a New Epitaxial Process (TGS)*   000329
Thin Films Grown by Hot Wall Epitaxy*   000569
Thin Films*   #Epitaxial   000415
Thin Films*   #Techniques for Routine UHV in situ   000950
Thin Iron Films on Antimony (Fe, vapor transport)*   000875
Thin Layers Grown on InP (film)*   #Characterization   001098
Thin Plates of Cr2O3 Single Crystals*   000858
Thin Polycrystalline Silicon Film Deposition and by   000985
Thin Silver Films on Cleaved Silicon Surfaces (Ag,   000619
Thin Single Crystal Films of Copper (Cu, evaporation)*   000736
Thiochlorides of Molybdenum II: Mo6Cl10Y (Y = S, Se, Te)   000757
Thioindates (LiInS2, Bridgman-Stockbarger)*   Principal   000509
Third International Summer School on Crystal Growth,   000066
ThO2 and CeO2 (electrolytic deposition)*   000472
ThSiO4 Single Crystals by a Vapour Phase Reaction*   000454
Thulium Oxide (Tm2O3, hydrothermal)*   #Effect   001039
Th0.25NbO3-NaNbO3 (NaxTh0.25(1-x)NbO3, melt, slow   000527
Ti)O3) Thin Films*   000415
TiB1.89 - TiB1.96 Prepared by Chemical Transport (TiB2)*   000291
TiB1.96 Prepared by Chemical Transport (TiB2)*   #Single   000291
TiB2)*   #Chemical Vapor Growth   000926
TiB2)*   #Single Crystals with Compositions in the   000291
TiC Whiskers by Chemical Vapor Deposition*   effects of   000959
TiC)   #Single Crystal Growth of Titanium   000928
TiCx Single Crystal with Homogeneous Compositions (001092
TiF3*   #Preparation   000124
Tilt Boundaries Produced by Epitaxial Growth on   000207
Time (theory, kinetics, solution)*   #Growth   000659
Tin Halides (CsSnCl3, RbSnCl3, Cs2SnCl6, Rb2SnCl6,   000930
TiN, vapor transport)*   000655
TiO2) Crystals by Chemical Transport Reactions with HBr   000431
TiO2)*   #Anisotropy of   000778
TiO2)*   #Plasma Growth of   000387
TiO2, vapor transport)*   #A Gas Handling   000808
TiO3, Nd4Ba6(BO3)9, Pr4Ba6(BO3)9, Ni8Nb3O15, Ni2V2PbO8,   001063
TiP*   #Chemical Vapor   000656
TiP Whiskers in VLS Growth*   000929
TiP-I2 (TiP, vapor transport)*   #Heterogeneous   001003
TiP, vapor transport)*   #Heterogeneous Equilibria   001003
TiSe2, vapor transport)*   Charges of Transition-Metal   001000
TiSe2, VSe2, TaSe2, vapor transport)*   of Layered   000555
TiSx, x = 1.40-1.70 (TiyV1-ySx)*   #Effect   000819
TiS2 (vapor transport)*   000288
Titanium Bronze Crystals in NaBO2 (flux, Na2Fe2Ti6O14,   000602
Titanium Carbide from the Vapor by a Modified Hot Wire   000928
Titanium Diboride by a Modified Hot Wire Method (TiB2)*   000926
TiyV1-ySx)*   #Effect of Addition of Vanadium   000819
Ti0.9975V0.0025)407 (vapor transport)*   #Structural   000391
Ti1-xVx)407: Electrical, Calorimetric, Magnetic and EPR   000835
Tl Alloys (Bridgman)*   000581
Tl-Se-TlInSe2 (TlSe, zone melting, slow cooling)*   000361
Tl-Te*   000101
Tl, melt)*   001085
TlAlF4, KAlF4, NH4AlF4, RbAlF4, CsAlF4, NH4Fe2F6)*   000772
TlI, Tl2I3, gel)*   #   000881
TlInSe2 (TlSe, zone melting, slow cooling)*   000361
TlSe, zone melting, slow cooling)*   000361
TlTe Single Crystals (melt, theory)*   000405
Tl2Cd2(SO4)3, K2Co2(SO4)3, and their Phase Transitions (000138
Tl2CoBa(NO2)6)* 6, Tl2CuPb(NO2)6, K2NiPb(NO2)6, Tl2NiPb(000550
Tl2CoPb(NO2)6, Tl2CoBa(NO2)6)* 6, Tl2CuPb(NO2)6, K2NiPb(000550
Tl2CuPb(NO2)6, K2NiPb(NO2)6, Tl2NiPb(NO2)6, Tl2NiCd(NO2)   000550
Tl2I3, gel)*   #Growth   000881
Tl2NiCd(NO2)6, Tl2NiHg(NO2)6, Tl2CoPb(NO2)6, Tl2CoBa(000550
Tl2NiHg(NO2)6, Tl2CoPb(NO2)6, Tl2CoBa(NO2)6)*6, Tl2CuPb(000550
Tl2NiPb(NO2)6, Tl2NiCd(NO2)6, Tl2NiHg(NO2)6, Tl2CoPb(000550
Tl2W4O13: A Tunnel Structure Related to the Hexagonal   000350
Tl3VS4 (Bridgman-Stockbarger)*   000411
Tm, and Yb - X = Cl, Br, and I) (melt, SmFCl, SmFBr,   000088
Tm, Er, Ho, Dy) (Tm2Si2O7, Er2Si2O7, Ho2Si2O7, Dy2Si2O7)   000595
TmFBr, TmFI, YbFCl, YbFBr, YbFI)*   - X = Cl, Br, and I)   000088
TmFCl, TmFBr, TmFI, YbFCl, YbFBr, YbFI)*   - X = Cl, Br,   000088
TmFI, YbFBr, YbFI)*   - X = Cl, Br, and I) (melt,   000088
TmSe (melt)*   000079
TmSe (slow cooling)*   of the Magnetically Ordered   000031
TMTTF-TCNQ) (solution)* Tetramethyltetrathiafulvalenium   000032

PERMUTED TITLE INDEX

(Eu,Pb) MnO3, Pb3Mn7O15, Tm2Ti3O7, Tb2Ti2O7, Pr2Ti2O7, | Tm2Ge2O7, Er2SiO5, Dy2SiO5, Ho2SiO5, Pr2MoO6, LiCoPO4)* 001061
on the Growth Morphology of Cubic Thulium Oxide (| Tm2O3, hydrothermal)*   #Effect of Mineralizers 001039
Rare-Earth Disilicates, R2Si2O7 (R = Tm, Er, Ho, Dy) | Tm2Si2O7, Er2Si2O7, Ho2Si2O7, Dy2Si2O7)* of Polymorphic 000595
(La,Pb) MnO3, (Nd,Pb) MnO3, (Eu,Pb) MnO3, Pb3Mn7O15, | Tm2Ti3O7, Tb2Ti2O7, Pr2Ti2O7, Tm2Ge2O7, Er2SiO5, 001061
Cubic Bi4Ge3O12 Crystals Doped with Dy3+, Ho3+, Er3+, | Tm3+, or Yb3+ Ions (Czochralski)*Stimulated Emission of 000458
#A New Organic Low Temperature Conductor: HMTSF- | TNAP (solution, organics)* 000087
of the Hydrothermal Growth of Synthetic 11 Angstroms | Tobermorite (Ca6Si6O18.4H2O) Crystals* Characterization 000366
in Synthetic Quartz Crystals by Light-Scattering | Tomography (SiO2, hydrothermal)*   of Growth Defects 000652
Top-Seeded Flux Growth of LiNdP4O12 Single Crystals* 000678
#Current | Topics in Materials Science, Volume 1 (review)* 000450
#Current | Topics in Materials Science, Volume 2 (Review)* 000449
CdxHg1-xTe, Bridgman)*   #An X-Ray | Topographic Assessment of Cadmium Mercury Telluride (000158
#Sublimation Growth and X-Ray | Topographic Characterization of CdTe Single Crystals* 000148
Na2CoGeO4 Single Crystals in Relation to Their   #X-Ray | Topographic Study of Growth Defects in Hydrothermal 000233
Single Crystals (solution)*   #X-Ray | Topographic Study of Growth Sector Strains in KDP 000092
Dilute Binary #Observation by Chemical Attack and X-Ray | Topography of Cellular Structures of Monocrystals of 000210
#In Situ X-Ray | Topography of Epitaxial Ge Layers During Growth* 000363
cooling)*   #Flux Growth and Characterization by X-ray | Topography of Rare Earth Arsenates (REAsO4, slow 000900
System for the Bridgman Growth of Crystals with | Toxic and/or Highly Volatile Components (equipment)* 000164
Solubility of Barium Sulphate (BaSO4,   #Effect of S-35 | Tracer, Particle Size and Imperfection on the 000779
Surfaces #Interaction Between Successive Monatomic Step | Trains of Different Density on Evaporating Crystal 000105
#Modes of Vibrations in Step | Trains: Rhythmical Bunching (theory)* 000933
#On Diffusive-Advective Interfacial Mass | Transfer (theory, solution, vapor transport, flux)* 001074
Boundary with the Melt, and its Influence on the #Mass | Transfer at the Surface of a Crystal Near to its 001046
Transfer of VO2 by Iodine (vapor transport, doped)* 000487
transport)*   #Heterogeneous Equilibria and Mass | Transfer Kinetics in the System TiP-I2 (TiP, vapor 001003
Single Crystals by the #Analytical Study of the Thermal | Transfer Process in an Installation for Preparation of 000057
#Ferroelasticity and Phase | Transformation in Rb2Hg(CN)4 Spinel (solution)* 000378
flux)*   #High Temperature Form of Pb2WO5 and | Transformation Phenomena to Its Low Form (Czochralski, 000310
#Flux Growth of Bi2WO6 Single Crystal Below the | Transformation Temperature (slow cooling)* 000665
#Electron Diffraction Study of Phase | Transformations of the Compound CuSe* 000866
#The Stage II | Transient in Off-Eutectic Solidification (theory)* 000786
#Heat Capacity of EuS near the Ferromagnetic | Transition (Bridgman, vapor transport, theory)* 000505
of a (111) Diamond Face Around the Roughening | Transition (C, theory)*   #Monte Carlo Simulation 001020
Grown Crystals (organic)*   #On the | Transition from Faceted to Non-Faceted Growth in Melt- 000355
#Structural Phase | Transition in Mixed Crystals WxMo1-xO3 (melt)* 000824
with High Ionic Partial Conductivity. III. On the Phase | Transition of CuTeBr and the Structure of the High 000025
#Neutron-Scattering Study of the Incommensurate Phase | Transition of Rb2ZnBr4 (theory, solution)* 000226
#Change in the Energy Band Structure of Eulytine on | Transition to the Amorphous State (Czochralski, 000294
RbFeP4, Rb2FeP5,   #Flux Growth of Crystals of Some | Transition Metal Fluorides. Part 2 (KAlF4, KMnF3, 001064
(phosphides, flux)*   #Ternary Lanthanoid- | Transition Metal Pnictides with ThCr2Si2-Type Structure 000596
TiSe2, #Structural and Electrical Properties of Layered | Transition Metal Selenides VxTi1-xSe2 and TaxTi1-xSe2 (000555
solution, cold crucible #Growth and Characterization of | Transition Metal Silicides (chemical vapor transport, 000607
and the Thermal Expansion Coefficients of VO2 near the | Transition Temperature (vapor transport)*   Parameters 000513
Optical Phonon Modes and Localized Effective Charges of | Transition-Metal Dichalcogenides (MoS2, MoSe2, WS2, 001000
vapor   #Crystal Growth and Characterization of the | Transition-Metal Phosphides CuP2, NiP2, and RhP3 (flux, 000708
of K2PbCu(NO2)6 Near the Cooperative Jahn-Teller Phase | Transitions (solution)*   #Ultrasonic Velocities 000467
SO4)3, Tl2Cd2(SO4)3, K2Co2(SO4)3, and their Phase | Transitions (solution)* Crystals of Langbeinites Rb2Cd2(000138
Aluminum with Silicon*   #Structural | Transitions and Phase Composition of Thin Films of 000473
Electrical, Calorimetric, Magnetic and #Metal-Insulator | Transitions and Phase Diagram of (Ti1-xVx)4O7; 000835
#Structural Aspects of the Metal-Insulator | Transitions in (Ti0.9975V0.0025)4O7 (vapor transport)* 000391
#Interface Shape | Transitions in the Czochralski Growth of Dy3Al5O12* 000644
System (vapor transport)*   #Crystal Growth and Phase | Transitions in the NbxCr1-xO2 (x greater than 0.5) 000095
Y3Fe5O12, theory)*   #Interface Shape | Transitions in Czochralski Grown YAG Crystals (000645
#Growth and Phase | Transitions in Single Crystals of (KxNa1-x)NbO3 (flux)* 000056
melt)*   #On the Magnetic Structure and Magnetic Phase | Transitions of Tb5Ge4. A Neutron Diffraction Study (000840
(ZnF2, CoF2,   #Crystal Growth by the Thermic Screen | Translation (TST) Technique: A Modified Bridgman Method 000541
Precipitates in ZnTe: Shape of the Tellurium Solidus # | Transmission Electron Microscopy Observations of 000098
of the Intercalation of 2H-NbSe2 with Hydrazine by | Transmittance Measurements (theory)*   #Kinetic Studies 000085
review, theory)*   #Infra-Red | Transmitting Materials. Part 1. Crystalline Materials (000250
#Mossbauer Study of FeMo2S4 (vapor | transport)* 000006
#Positron Annihilation in ReO3 (vapor | transport)* 000015
#Crystallization of WO2 from the Gas Phase (vapor | transport)* 000052
#On the Ordering of Fe Atoms in FexNbS2 (vapor | transport)* 000129
#Electron Energy-Loss Spectroscopy of TiS2 (vapor | transport)* 000288
#Crystal Structure of the Zinc Iodide ZnI2 (vapor | transport)* 000300
#The Crystal Structure of V4O7 at 120 degrees K (vapor | transport)* 000390
#Growth of CuInS2 and Its Characterization (vapor | transport)* 000403
#Epitaxial Growth of CrO2 on Sapphire in Air (vapor | transport)* 000413
#Crystal Structure of the Indium Selenide In2Se3 (vapor | transport)* 000557
#Growth of CdP4 Single Crystals (evaporation, vapor | transport)* 000599
#Heteroepitaxial Homogeneous CdxHg1-xTe Films (vapor | transport)* 000707
#Growth of alpha-Ag2S Whiskers in a VLS System (vapor | transport)* 000718
#Synthesis of Single Crystal LnPO4 (vapor | transport)* 000733
#The Growth of CdS in Sealed Silica Capsules (vapor | transport)* 000815
#Epitaxial Thin Iron Films on Antimony (Fe, vapor | transport)* 000875
#Organometallic VPE Growth of AlxGa1-xAs (vapor | transport)* 000919
#Synthesis and Crystal Growth of CdGeP2 (vapor | transport)* 001041
#Morphology of Iron Pyrite Crystals (FeS2, vapor | transport)* 001093
and Infrared Spectra of CdIn2S4 and ZnIn2S4 (vapor | transport)*   #Raman 001004
and Lower Critical Fields of TaS2(Pyridine)1/2 (vapor | transport)*   #Upper 000468
Growth and Characterization of In1.9As0.1Se3 (vapor | transport)*   #Crystal 000471
Structure and Physical Properties of Cu0.65VS2 (vapor | transport)*   #Crystal 000543
Monocrystalline Bars of SiC from the Gas Phase (vapor | transport)*   #Growing 000554
of Monocrystalline WV2O6 at 298 and 383 K (vapor | transport)*   #Structure 000389
Properties of Pb1-xCdxS Epitaxial Films (vapor | transport)*   #Electrical 000439
Bandgap Measurements of SnS2 Polytypes (vapor | transport)*   #Electronic 000008
of Hg0.60Cd0.40Te from Te-Rich Solution (film, vapor | transport)*   #LPE Growth 000839

PERMUTED TITLE INDEX

and Amorphousness in Silicon Whiskers (Si, vapor transport)* #Polytypism 000642
the Initial Use of Spacelab for Crystal Growth (vapor transport)* #Priorities in 000270
Method for the Study of CVD Conditions (TiN, vapor transport)* #A Quick Survey 000655
and Crystal Structures of BiReO4 and BiRe2O6 (vapor transport)* #The Preparation 000781
Tetrabromide Crystals for Optical Studies (CBr4, vapor transport)* #Growth of Carbon 000263
and Slip System of alpha-Si3N4 Single Crystal (TiN, vapor transport)* #Growth, Morphology 000696
and Its Effect on the Purity of LPE GaAs (vapor transport)* #New Gaseous Impurity 000313
for Exacting Crystal Growth Preparations (TiO2, vapor transport)* #A Gas Handling System 000808
V3O5 Crystals by Chemical Transport Reactions (vapor transport)* #Preparation of Single 000789
Epitaxial Ga1-xInxSb Using Open-Tube Flow System (vapor transport)* #Vapor-Phase Growth of 000681
Mn-Bi Films and Their Crystallization Process (vapor transport)* #Formation of Amorphous 001103
Studies of FeNi2BO5 and FeNiBO4 (ferrites, vapor transport)* #Magnetic and Mossbauer 000005
Monophosphide Using a B2H6-PH3-H2 System (BP, vapor transport)* #Crystal Growth of Boron 000884
at Low Temperature and High Partial Pressures (vapor transport)* #Fast Growth in GaAs VPE 000393
Electrical Properties of CdCr2S4 Single Crystals (vapor transport)* #Preparation and Certain 000780
of Single Crystals of EuSb4S7 and EuSb2Se4 (vapor transport)* #Growth and X-Ray Analysis 000022
of the Non-Stoichiometric VS2 and V5S8 Phases (vapor transport)* #Physico-Chemical Properties 000470
in a Closed Crystal-Vapour System (organics, vapor transport)* #Growth of Phthalic Anhydride 000327
Mass Transfer Kinetics in the System TiP-I2 (TiP, vapor transport)* #Heterogeneous Equilibria and 001003
of CuGaxIn1-xSe2 Chalcopyrite Compound (vapor transport)* #Crystal Growth and Properties 000743
Structure of As2(Se,S)3 Compounds (As2S3, As2Se3, vapor transport)* #The Crystal Growth and Atomic 000896
by the Synthesis, Solute Diffusion Technique (vapor transport)* #Large-Sized InP Single Crystals 000923
Insulator Transitions in (Ti0.9975V0.0025)4O7 (vapor transport)* #Structural Aspects of the Metal- 000391
of Ruthenium Dioxide Crystals (RuO2, vapor transport)* #Preparation and Characterization 000865
of High-Purity p-type Single Crystals of GeFe2O4 (vapor transport)* #Electrical and Optical Properties 000920
Using Tellurium Tetrachloride and Argon (UO2, vapor transport)* #Growth of Uranium Dioxide Crystals 000280
Ionized-Cluster Beam Deposition Techniques (vapor transport)* in Ionized-Cluster Beam and Reactive 000944
in the NbxCr1-xO2 (x greater than 0.5) System (vapor transport)* #Crystal Growth and Phase Transitions 000095
Epitaxial Films on Insulators (Si on sapphire, vapor transport)* #Dichlorosilane as Silicon Source for 000254
VxTi1-xSe2 and TaxTi1-xSe2 (TiSe2, VSe2, TaSe2, vapor transport)* of Layered Transition Metal Selendies 000555
of VO2 near the Transition Temperature (vapor transport)* and the Thermal Expansion Coefficients 000513
Metal Phosphides CuP2, NiP2, and RhP3 (flux, vapor transport)* and Characterization of the Transition- 000708
Halides, NbX2Y2(X=S, Se - Y=Cl, Br, I) (vapor transport)* and Properties of Niobium Chalcogenide 000795
and Note on the Shandite Structure (Ni3Pb2Se2, vapor transport)* Study of the System Ni3Pb2S2-Ni3Pb2Se2 000198
Single Crystals Obtained by a New Growth Method (vapor transport)* #Properties of Homogeneously Doped ZnSe 000745
Temperature Gradients (CdS, kinetics, theory, vapor transport)* of Cadmium Sulphide Crystals with Large 000289
Investigation of CrPr2S4 and CrPr2Se4 (vapor transport)* of Single Crystals, and Physicochemical 000817
Vapour Under Constant Conditions (C10O16, theory, vapor transport)* of Growth of Adamantane Crystals from the 000572
-- Microanalytical Evaluation of Composition (vapor transport)* Phase Under Non-Stoichiometric Conditions 000082
M(PbX3) (CsPbCl3, CsPbBr3, CsPbI3, KPbI3, RbPbI3, vapor transport)* Properties of Alkali-Trihalogenoplumbates 000377
Calorimetric, Magnetic and EPR Studies (vapor transport)* Phase Diagram of (Ti1-xVx)4O7: Electrical, 000835
Si Films Deposited at Elevated Temperatures (vapor transport)* #Crystalline Defects in Solid Phase Epitaxy 000194
by the Sublimation Travelling Heater Method (vapor transport)* #Growth of High-Purity ZnTe Single Crystals 000938
GaAs Substrates Using the Close-Spaced Geometry (vapor transport)* of Single Crystal Cd1-xZnxS Layers on (111) 000305
of Gold Monobromide, I-AuBr and P-AuBr (vapor transport)* the Crystal Structures of Two Modifications 000436
(MoS2, MoSe2, WS2, WSe2, HfSe2, ZrSe2, TiSe2, vapor transport)* Charges of Transition-Metal Dichalcogenides 001000
of Layer Compounds in the Series Mo1-xWxSe2 (vapor transport)* Conditions and Crystal Structure Parameters 000013
Silicon (Si, Czochralski, zone melting, vapor transport)* #Solar Cells from Zone-Refined Metallurgical 000195
susceptibility, and NMR) of the Spinel CuV2S4 (vapor transport)* and Physical Properties (transport, magnetic 000544
Cu2MnGeSe4, Cu2MnSnSe4, Cu2FeGeS4, Cu2CoGeS4, (vapor transport)* Cu2ZnSnSe4, Cu2MnSiS4, Cu2MnGeS4, Cu2MnSnS4, 000357
on Chromium-Chlorine Boracite (Ni3B7O13Cl, vapor transport)* #Epitaxial Layers of Nickel-Chlorine Boracite 000837
an Organometallic VPE System (Ga1-xAlxAs, GaAs, vapor transport)* GaAs Epitaxial Layers During Their Growth in 000974
Their Electronic Properties (FeCr2S4, CoCr2S4, vapor transport)* #Some Chalcogenide Spinel Single Crystals and 001066
Large Anions: Ba10(ReO5)6Br2 and Ba10(ReO5)6I2 (vapor transport)* #Structure of New Rhenium Apatites Containing 000080
on Bicrystalline Substrates of NaCl (Au, Ag, vapor transport)* #Tilt Boundaries Produced by Epitaxial Growth 000207
#Transfer of VO2 by Iodine (vapor transport, doped)* 000487
Crystals of the Magnetic Semiconductor CdCr2Se4 (vapor transport, doped)* #Preparation of Single 000625
Photoconducting Properties of Fe2-xCrxO3 (vapor transport, electrolytic deposition)* #Electrical, Optical 000624
Films of CdS Evaporated in Vacuum onto Silicon (vapor transport, electron beam melting, kinetics, theory)* of 000073
and Etching Vectors of Silicon (111) (theory, vapor transport, epitaxy)* Step Reconstruction on the Growth 001019
(Er3Fe5O12, (Eu,Y)3Fe5O12 and (Eu,Yb)3Fe5O12, vapor transport, epitaxy, films)* CVD Garnet Growth 000211
Distortion in Heteroepitaxial Layers: Ge On GaAs (vapor transport, film)* #Tetragonal 000362
and Crystallographic Polarity in ZnO Crystals (vapor transport, film)* #Crystal Growth in c Direction 000426
on Silicon Substrate Using SiCl4-C3H8-H2 System (vapor transport, film)* #Heteroepitaxial Growth of beta-SiC 000616
#Nitrides--Structures and Crystal Growth (review, vapor transport, flux)* 000533
Interfacial Mass Transfer (theory, solution, vapor transport, flux)* #On Diffusive-Advective 001074
Dynamics Study of Surface Melting (theory, vapor transport, kinetics)* #A Molecular 000140
for High Volume Low-Cost Silicon Epitaxy (Si, vapor transport, kinetics)* #Novel Reactor 000062
Layer Growth from the Vapour Phase (SiC, film, vapor transport, kinetics)* of Silicon Carbide Epitaxial 000942
Single Crystals with GeBr4 Vapor (Ge, epitaxy, vapor transport, kinetics, theory)* of Germanium (111) 000600
of the Systems SnS2-SnI4 and SnS2-I2 (SnS2, vapor transport, kinetics, theory)* #Transport Properties 001079
Epitaxial Layers in the Ga_AsCl3-H2 System (GaAs, vapor transport, kinetics, theory)* Growth Rate and Purity of 000653
Crystal Structure, and Physical Properties (transport, magnetic susceptibility, and NMR) of the 000454
of Cd1-xHgxCr2Se4 Single Crystals (vapor transport, melt)* Effect and Photoconductivity 001037
Single Crystals and Their Structural Perfection (vapor transport, melt, flux)* #Growth of USb2 000383
Iodide Radiation Detectors. Part II. HgI2 (vapor transport, review)* Spectroscopic Evaluation of Mercuric 000834
#Growth and Physical Properties of ZnSiP2 (vapor transport, solution)* 000086
of Transition Metal Silicides (chemical vapor transport, solution, cold crucible melting)* 000607
Encapsulation Zone-Refining of PbS (zone melting, vapor transport, sublimation)* #Liquid 000583
of Large Non-Seeded Pb1-xSnxTe Single Crystals (vapor transport, sublimation)* #Growth Study 000961
of Zinc Selenide from the Vapour Phase (ZnSe, vapor transport, sublimation)* #Growth of Single Crystals 000214
and Evaporation of Epitaxial Ice Crystals (H2O, vapor transport, sublimation, evaporation)* Field on Growth 000028
#Specific Heat of V6O11 Between 0.4 and 50 K (vapor transport, theory)* 000215
#Growth Rate of Iron Whiskers (Fe, vapor transport, theory)* 000462
#Raman Scattering in 3R-NbS2 (vapor transport, theory)* 000731
EuS near the Ferromagnetic Transition (Bridgman, vapor transport, theory)* #Heat Capacity of 000505
to the Study of Heterogeneous Reactions (GaAs, vapor transport, theory)* Method and Its Application 000281
Content and Segregation Coefficient (Si, vapor transport, theory)* from the Gas Phase. I. Phosphorus 000514

PERMUTED TITLE INDEX

and Differential Molar Solution Enthalpies (Si, vapor
of Single Crystals of Cadmium Sulphide (CdS, vapor
and HfS2 (CuxZrS2, CuxHfS2, FexZrS2, FexHfS2, vapor
Iodine Equilibrium from 700 to 1000 K (Ge, vapor
in SiCl4-HCl-H2 System Based on Flow Graph (vapor
in Low Gravity Environment (melt, solution, vapor
#Crystal Growth of Mn15Si26 (vapor
Carbide by Pyrolysis of Silane Compounds (SiC, vapor
and Physical Properties of Boracite Crystals (Chemical
of Silicon Growth Under Low Hydrogen Pressure (vapor
with Platinum and Gold as Liquid-Forming Agents (vapor
Zinc Oxide Films Grown on Sapphire (vapor
Modified Hot-Filament Method (de Boer, Van Arkel, vapor
Film Growth of Actinide Pnictides by Chemical Vapour
#Congruent (Diffusionless) Vapor
the Growth of Chalcopyrite Crystals by Iodine Vapour
of Pseudobrookite Single Crystals by Chemical
Gallium Selenide and Gallium Sulphide by Iodine Vapour
of Nickel Ferrite Single Crystals by Chemical
Growth of Protactinium Arsenides by Chemical Vapour
in the Range TiB1.89 - TiB1.96 Prepared by Chemical
Molybdenum Ditelluride (MoO3-x, MoTe2)*   #Vapor Phase
CdS, sublimation)*   #Vapor Phase
Selenide (Cd1+y(SxSe1-x), sublimation)*   #Vapor Phase
TiO2)*   #Anisotropy of Ion
#New Observations in the Chemical
#Preparation by Vapor
#The Vapour
TeCl4. I. Calculation of the System Nb/O/Te/Cl and the
Method to a Five-Component System: The Chemical Vapour
Transport Behaviour of the Equilibrium #On the Chemical
Calculation of the System Nb/O/Te/Cl   #On the Chemical
#Effect of Addition of Vanadium on the Chemical
#Mass Spectrometric Study on Chemical
Transport of Some Niobium Oxides with TeCl4 (II) The
theory)*   #Interrelation Between Interface and Vapour
#Growth of Tungstenite Single Crystals by Direct Vapour
for the Growth of Layered Dichalcogenides by Vapour
Growth of ZnS in H2: The Effects of a Bypass Flow #Mass
I2 (SnS2, vapor transport, kinetics, theory)*   #
High-Temperature Annealing in Cadmium on the Electrical
#Preparation of Single V3O5 Crystals by Chemical
Kinetics of Gallium Phosphide Synthesis by Chemical Gas
#Growth of Anatase (TiO2) Crystals by Chemical
Crystals of U1-xThxO2 Solid Solutions by Chemical
Silicon at Reduced Temperatures Using a Silicon-Iodine
#Preparation of V3Ga Single Crystals by Iodine
#Crystal Growth of Actinide Dioxides by Chemical
Cubic ZnS, ZnSe and ZnSxSe1-x Single Crystals by Iodine
Properties of CuGaSe2 Grown by Iodine Vapour
HgCdTe Mixed Crystals from the Vapour Phase by Chemical
and Conductive ZnSe:I Crystals by Iodine Vapor
to x less than or equal to 1) Grown by Iodine Vapour
(review)*   #Magnetic,
of Preparation of II-IV-V2 Compounds by Gas Phase
#New Observations in the Chemical Transport of GeO2. I.
#Semiconductor Single Crystals Obtained by the
of High-Purity ZnTe Single Crystals by the Sublimation
High-Resistivity CdTe*   #Growth by
and InSb Crystals Grown by a Vertical and Horizontal
Pb1-xSnxTe Single Crystals of Graded Composition by the
Films Grown by a New Epitaxial Process (TGS)* #Infrared
#The Crystal Structure of
Deposition from the Complex Monochlorodiethylgallium-
Stockbarger,   #Crystal Growth of Some Rare Earth
#Single Crystal Growth and Properties of Deuterated
Process (TGS)* #Infrared Trichroism of Ferroelectric
Ray and Neutron Diffraction of Na2MnFeF7: A Variety of
Structure, Growth, and Optical Properties of Alkali-
#Cerium
in Preparation from Solutions in Molten Potassium
#Vapor Phase Transport and Crystal Growth of Molybdenum
K2NiPb(NO2)6, Tl2NiPb(NO2)6, Tl2NiCd(#Crystal Growth of
FeCr2S4 Single Crystals. I*   #Growth of
Properties of "One-Dimensional" Barium Vanadium
#The Crystal Structure of Niobium
#Photoconduction of Lead
#Crystal Structure of the Phosphate of
Growth Models" (theory)*   #Reply to Comment of R.
#Crystal Growth by the Thermic Screen Translation (
#Vapor-Phase Growth of Epitaxial Ga1-xInxSb Using Open-
#Vapor Growth of Zinc Telluride by Closed
Continuous   #Kinetics of Crystal Growth of Calcium
Continuous   #Kinetics of Crystal Growth of Strontium
Continuous   #Kinetics of Crystallization of Barium
of LiNbO3 from Solution in Borate, Vanadate, and
of Barium Tungstate from Solutions in Sodium
Growth of Calcium Tungstate from Solutions in Sodium

transport, theory)*   Gas Phase. II. Phosphorus Content   000515
transport, theory)* the Electrical Transport Properties   000612
transport, theory)*Shift of the Absorption Edge in ZrS2   001090
transport, theory, thermodynamics)*   #Germanium-   000191
transport, theory, Si)*   of Silicon Epitaxial Growth   000503
transport, zone melting, doped, review)*   Behaviour   001070
transport, Bridgman)*   000495
transport, De Boer-Van Arkel, doped)* and Doped Silicon   001043
transport, M3B7O13X)*   #Preparation   000151
transport, Si)*   #Kinetics   000255
transport, Si)*Morphology of VLS Silicon Crystals Grown   001075
transport, ZnO)*   #Real Structure of Heteroepitaxial 000526
transport, ZrC)*   Growth of Zirconium Carbide by a   000927
Transport (review)*   #Single Crystal and   000160
Transport (theory, kinetics)*   000007
Transport (CuInS2, CuGaS2, AgIn5S8)*   Method for   000744
Transport (Fe2TiO5, electrolytic deposition)*   000770
Transport (GaS, GaSe)* of Conditions for the Growth of   001078
Transport (NiFe2O4, theory)*   #On the Preparation   000760
Transport (PaAs2, Pa3As4, PaAs)*   and Single-Crystal   000161
Transport (TiB2)*   #Single Crystals with Compositions   000291
Transport and Crystal Growth of Molybdenum Trioxide and   000299
Transport and Stoichiometry Control of Cadmium Sulfide (000648
Transport and Stoichiometry Control of Cadmium Sulfo-   000647
Transport in Stoichiometric Rutile Single Crystals (000778
Transport of GeO2. I. Transporting Agent Chlorine*   000787
Transport of Monocrystals of AlF3, FeF3, and TiF3*   000124
Transport of NbB2 and TaB2 (kinetics, theory)*   000042
Transport of NbO2/NbO2.417 (NbO2, NbO2.417, theory)*   000797
Transport of Ni-Cl Boracite (no materials grown -   000236
Transport of Some Niobium Oxides with TeCl4 (II) The   000798
Transport of Some Niobium Oxides with TeCl4. I.   000797
Transport of TiSx, x = 1.40-1.70 (TiyV1-ySx)*   000819
Transport of VO2*   000063
Transport Behaviour of the Equilibrium NbO2.417 to NbO2.   000798
Transport Kinetics in Chemical Vapour Deposition (000742
Transport Method (WS2, sublimation)*   000012
Transport Processes (kinetics, theory)* Considerations   000805
Transport Processes Associated with the Epitaxial   000558
Transport Properties of the Systems SnS2-SnI4 and SnS2-   001079
Transport Properties of Single Crystals of Cadmium   000612
Transport Reactions (vapor transport)*   000789
Transport Reactions (GaP)*   #   000231
Transport Reactions with HBr and HCl*   000431
Transport Reactions*   #Growth of Single   000455
Transport System (Si)* #Local Selective Homoepitaxy of   000136
Transport*   000217
Transport*   000912
Transport*   #Growth of   000309
Transport*   #Photoluminescence   000935
Transport*   #Crystallization of   000342
Transport*   #Preparation of Photoluminescent   000775
Transport* the Series GaSxSe(1-x) (0 less than or equal   001077
Transport, and Thermal Properties of Ferromagnetic EuB6   000293
Transport, with ZnSiP2 as an Example*   in the Course   001086
Transporting Agent Chlorine*   000787
Traveling Solvent Method from (Pb0.8Sn0.2)Te*   000064
Travelling Heater Method (vapor transport)*   #Growth   000938
Travelling Heater Method and Characteristic of Undoped   000939
Travelling Heater Method*   #GaSb   000099
Travelling Solvent Method (theory)*   #Growth of   000564
Trichroism of Ferroelectric Triglycine Sulphate Thin   000329
Triclinic WO3 (sublimation)*   000441
Triethylarsine (GaAs)*   Arsenide by Chemical Vapor   001109
Trifluorides (GdF3, TbF3, DyF3, HoF3, ErF3, Bridgman-   000323
Triglycine Fluoroberyllate (TGFB)*   000567
Triglycine Sulphate Thin Films Grown by a New Epitaxial   000329
Trigonal Weberite (flux)*   #X- 001033
Trihalogenoplumbates M(PbX3) (CsPbCl3, CsPbBr3, CsPbI3,   000377
Trihydroxide (Ce(OH)3, gel)*   000664
Trimolybdate ((Y,Nd)Al3(BO3), flux)*YAl-Borate Crystals   000551
Trioxide and Molybdenum Ditelluride (MoO3-x, MoTe2)*   000299
Triple Nitrites in Gels (K2CuPb(NO2)6, Tl2CuPb(NO2)6,   000550
Triple Unit Cell Steps on the Surface of Vapour Grown   000418
Triselenide (BaVSe3)*   #Structural and Magnetic   000474
Trisulfide, NbS3*   000484
Trititanate (PbTi3O7, flux)*   000264
Trivalent Manganese Mn(PO3)3 (flux)*   000058
Trivedi on "The Validity of Steady-State Dendrite   000833
TST) Technique: A Modified Bridgman Method (ZnF2, CoF2,   000541
Tube Flow System (vapor transport)*   000681
Tube Method (ZnTe, epitaxy)*   000700
Tungstate from Solutions in Sodium Tungstate Melts by   000810
Tungstate from Solutions in Sodium Tungstate Melts by   000811
Tungstate from Solutions in Sodium Tungstate Melts by   000812
Tungstate Melts (flux)*   #Crystallization   001047
Tungstate Melts by Continuous Cooling (BaWO4, flux,   000812
Tungstate Melts by Continuous Cooling (CaWO4, flux)*   000810

PERMUTED TITLE INDEX

Growth of Strontium Tungstate from Solutions in Sodium
Oxide-Tungsten Oxide Phase System and Growth of Cadmium
Hexagonal Tungsten Bronze*   #The Thallium
#Elastic Constants of Niobium-Zirconium, Hafnium, and
Tl2W4013: A Tunnel Structure Related to the Hexagonal
                                    #A Series of Lead
Rubidium and Cesium (review,   #On the Alkali Metal
Tungstate Single Crystals (CdWO4,   #The Cadmium Oxide-
Method (WS2, sublimation)*   #Growth of
Bronze*   #The Thallium Tungstate Tl2W4013: A
Czochralski)*   #Reproducible Preparation of
#Re-Investigation of the Re-Entrant Corner Effect in
(Si, flux)*   #Fivefold

hydrothermal)*   #The Crystal Structure and
a Solid-Solid Diffusion Method. II. Diffusion-Induced
Li2SiO3, Li2GeO3) Single Crystals (Bridgman)*   #Polar
of the A(6+)B2(3+)06 Type Under Very High Pressure (
A(6+)B2(3+)06 Type Under Very High Pressure (UCr2O6,
Epitaxial Thin Films*   #Techniques for Routine
          #Molecular Beam Epitaxy of Ge and Ga1-xAlxAs
Beam and Solid-Phase Epitaxies of Silicon Under
Cooperative Jahn-Teller Phase Transitions (solution)*   #
Type Under Very High Pressure (UCr2O6, UFe2O6, UMn2O6,
B2(3+)06 Type Under Very High Pressure (UCr2O6, UFe2O6,
During Crystallization - II. Single Filament,
Morphology During Crystallization - I. Single Filament,
by Travelling Heater Method and Characteristic of
Organometallic Chemical #Growth and Characterization of
          #Crystallography of the fcc Phase in

#Coarsening of Eutectic Structures During and After
Curved Solid-Liquid #Lateral Solute Segregation During
Germanium Eutectic Alloys (Ag-Ge)*
Eutectics in the Systems ZrO2-Ln2O3 (Ln = Lanthanide):
Profiled Crystals and #Proceedings of the 8th All
Crystals*
Single Crystals. I*   #Growth of Triple
II)-Compounds InxMo15Se19 Containing Mo6Se8 and Mo9Se11
          #Evidence for a
solution, kinetics, slow cooling)*
Crystals Using Tellurium Tetrachloride and Argon (
via the Ribbon-To-Ribbon (RTR) Technique: Process
vapor transport)*
*   #Anisotropy of
          #The Crystal Structure of Lithium
and Argon (UO2, vapor transport)*   #Growth of
theory)*   #The Growth Kinetics of
#Magnetization and Neutron Studies of UTe and
vapor transport, melt, flux)*   #Growth of
Electrical Heating for Verneuil Furnaces in the
          #Magnetization and Neutron Studies of
Reactions*   #Growth of Single Crystals of
Hot Wall Epitaxy (II-VI compounds, IV-VI compounds, III-
Techniques, Volume 2 (review, epitaxy, Verneuil, III-
Growth of Bulk Single Crystals of Quaternary III-
#Growth Effects in the Heteroepitaxy of III-
of the Lattice Constant of Epitaxial Layers of III-
Vapour Growth of Semiconductors: Elements, III-
#Growth and Dissolution Kinetics of III-
#Flow-Graph Model of Epitaxial Growth of A(III)B(
#Epitaxial Growth of A(III)B(
of the Group VA Metal Hydrides and Deuterides (V-H,
of the Group VA Metal Hydrides and Deuterides (
of Dislocations in Epitaxial Semiconductor Layers (III-
in MBE of Compound Semiconductors (II-VI, IV-VI, III-
Formed by Gas-Evaporation Technique: bcc Metals (
Crystal Preparation and Characterization of the Group
#Connections Between Thermodynamic Quantities and
#Heat Capacity of Metals Near the Melting Point and the
          #Theory of Melting,
#Growth of Pb0.8Sn0.2Te Films in a Quasi-Closed Space #
and Solid-Phase Epitaxies of Silicon Under Ultra-High
Layers of Silicon Carbide Obtained by Sublimation in
melting,   #Growth of Films of CdS Evaporated in
          #Epitaxial Growth of BaTiO3
Structures of Si on Sapphire, Obtained by
#Horizontal Zone-Melting of Indium Antimonide in
Magnetically Ordered Ground State of the Intermediate
Structure Type (flux)*   #La8Ru4O21: A Mixed-
          #Magnetic Mixed
*   #The
theory)*   #Reply to Comment of R. Trivedi on "The
#Fermi Surface Measurements of ZrB2 by the de Haas
of Silane Compounds (SiC, vapor transport, De Boer-
Carbide by a Modified Hot-Filament Method (de Boer,
#Crystallization of LiNbO3 from Solution in Borate,
Melts for the Growth of Single Crystals of Rare Earth

Tungstate Melts by Continuous Cooling (SrWO4, flux)*   000811
Tungstate Single Crystals (CdWO4, Czochralski)* Cadmium   000801
Tungstate Tl2W4013: A Tunnel Structure Related to the   000350
Tungsten Alloys (Nb-Zr, Nb-Hf, electron beam melting,   000307
Tungsten Bronze*   #The Thallium Tungstate   000350
Tungsten Bronzes (PbxWO3)*   000267
Tungsten Bronzes, in Particular those of Potassium,   000402
Tungsten Oxide Phase System and Growth of Cadmium   000801
Tungstenite Single Crystals by Direct Vapour Transport   000012
Tunnel Structure Related to the Hexagonal Tungsten   000350
Twin-Free InP Crystals Using the LEC Technique (000125
Twinned Crystals (theory)*   000484
Twinned Silicon Crystals Grown in an Al-16 wt.% Si Melt   000489
Twinning and Crystal Structure of Mn0.75Ga2.17S4 (flux)*   000247
Twinning Behavior of Ferric Molybdate, Fe2(MoO4)3 (000178
Twins in the alpha-phase (theory)*   Produced by   000371
Twins in Lithium Metasilicate and Metagermanate (000108
UCr2O6, UFe2O6, UMn2O6, UMnNiO6, MoCr2O6)* Mixed Oxides   000205
UFe2O6, UMn2O6, UMnNiO6, MoCr2O6)* Mixed Oxides of the   000205
UHV in situ Electron Microscopy of Growth Processes of   000950
Ultra Thin Film Superlattices*   000761
Ultra-High Vacuum (film)*   #Molecular   000883
Ultrasonic Velocities of K2PbCu(NO2)6 Near the   000467
UMnNiO6, MoCr2O6)* Mixed Oxides of the A(6+)B2(3+)06   000205
UMn2O6, UMnNiO6, MoCr2O6)*   Mixed Oxides of the A(6+)   000205
Unconstrained Growth from a Binary Alloy Melt (Sn-Pb,   000993
Unconstrained Growth from a Pure Melt (theory)*   000992
Undoped High-Resistivity CdTe*   #Growth   000939
Undoped ZnSe Epitaxial Layers Obtained by   000114
Uni-Directionally Solidified Zn-Al Eutectic*   000421
Uniaxial Ferromagnetism in NiZrF6.6H2O (solution)*   000464
Unidirectional Growth (theory)*   001027
Unidirectional Solidification of a Binary Alloy with a   000206
Unidirectional Solidification of Irregular Silver-   000549
Unidirectional Solidification, Microstructural and   000630
Union Meeting on the Preparation by Stepanov Method of   001111
Unit Cell Step Structures on Vapour-Grown Cr2S3 Single   000676
Unit Cell Steps on the Surface of Vapour Grown FeCr2S4   000418
Units*   #New Ternary Mo(000186
Universal Law of Dendritic Growth Rates (theory)*   000534
Unstable Growth of ADP Crystals (NH4H2PO4, theory,   000804
UO2, vapor transport)*   #Growth of Uranium Dioxide   000280
Update and Material Characterization (Si, laser   000360
Upper and Lower Critical Fields of TaS2(Pyridine)1/2 (000468
Upper Critical Field in Superconducting 2H-NbS2 (theory)   000730
Uranate (Li2UO4, flux)*   000324
Uranium Dioxide Crystals Using Tellurium Tetrachloride   000280
Urea Monocrystals from Aqueous Solution (CO(NH2)2,   000392
USb0.8Te0.2*   000154
USb2 Single Crystals and Their Structural Perfection (000383
USSR (review)*   #Subsidiary   000349
UTe and USb0.8Te0.2*   000154
U1-xThxO2 Solid Solutions by Chemical Transport   000455
V compounds, evaporation, review, theory)*   #   000568
V compounds, Si)*   #Crystal Growth - Theory and   000348
V Alloys (review, kinetics)*   #Melt and Solution   000054
V Compounds (review)*   000728
V Compounds (theory)*   #Determination   000399
V Compounds and Alloys (Ge, Si, GaAs, BP)*the Epitaxial   000041
V Heterostructures Formed by LPE (GaxAl1-xAs)   000894
V) Compounds from the Vapour Phase (theory)*   000504
V) Semiconductors from Vapour Phase (review, theory)*   000352
V-D, Nb-H, Nb-D, Ta-H, Ta-D)*   and Characterization   000790
V-H, V-D, Nb-H, Nb-D, Ta-H, Ta-D)* and Characterization   000790
V, review)*   #Nature, Origin and Effect   000126
V, review)*   and Carrier Concentration Control   000897
V, Fe, Cr, Mo and W)* and Habit of Fine Metal Particles   000821
VA Metal Hydrides and Deuterides (V-H, V-D, Nb-H, Nb-D,   000790
Vacancy and Diffusion Characteristics in Binary   000090
Vacancy Mechanism of Melting (theory, compilation, 39   000351
Vacancy Model*   001007
vacuum evaporation)*   000885
Vacuum (film)*   #Molecular Beam   000883
Vacuum (SiC)*   #Growth Kinetics of Epitaxial   000940
Vacuum onto Silicon (vapor transport, electron beam   000073
Vacuum Condensates on SrTiO3 Crystals*   000984
Vacuum Deposition (film, doped)*   #Heteroepitaxial   000823
Vacuum. I. Basic Technique (InSb)*   000712
Valence Compound TmSe (slow cooling)*Resistivity of the   000031
Valence Ternary Ruthenium Oxide of a New Hexagonal   000208
Valent TmSe (melt)*   000079
Validity of Steady-State Dendrite Growth Models (theory)   000832
Validity of Steady-State Dendrite Growth Models" (000833
van Alphen Effect (float zone melting)*   000964
Van Arkel, doped)* Doped Silicon Carbide by Pyrolysis   001043
Van Arkel, vapor transport, ZrC)* Growth of Zirconium   000927
Vanadate, and Tungstate Melts (flux)*   001047
Vanadates and Phosphates (YPO4, YVO4, REPO4, flux)*   000265

PERMUTED TITLE INDEX

Heat Capacity from 5 to 300 K and Magnetic #Crystalline   Vanadium (II) Fluoride, VF2. Preparation, Structure,   000917
Electrical Properties -Lithium and Molecular   # Vanadium (III) Oxychloride: Magnetic, Optical and   001031
70 (TiyV1-ySx)*   #Effect of Addition of   Vanadium on the Chemical Transport of TiSx, x = 1.40-1.   000819
Heat and Magnetic Susceptibility of Nonmetallic   Vanadium Bronze (NaxV2O5, KxV2O5, AgxV2O5, theory)*   000171
and Magnetic Properties of "One-Dimensional" Barium   Vanadium Triselenide (BaVSe3)*   #Structural   000474
#Mossbauer Study of FeMo2S4 (vapor transport)*   000006
#Electronic Bandgap Measurements of SnS2 Polytypes (vapor transport)*   000008
#Positron Annihilation in ReO3 (vapor transport)*   000015
#Crystallization of WO2 from the Gas Phase (vapor transport)*   000052
#On the Ordering of Fe Atoms in FexNbS2 (vapor transport)*   000129
#Electron Energy-Loss Spectroscopy of TiS2 (vapor transport)*   000288
#Crystal Structure of the Zinc Iodide ZnI2 (vapor transport)*   000300
#Structure of Monocrystalline WV2O6 at 298 and 383 K (vapor transport)*   000389
#The Crystal Structure of V4O7 at 120 degrees K (vapor transport)*   000390
#Growth of CuInS2 and Its Characterization (vapor transport)*   000403
#Epitaxial Growth of CrO2 on Sapphire in Air (vapor transport)*   000413
#Electrical Properties of Pb1-xCdxS Epitaxial Films (vapor transport)*   000439
#Upper and Lower Critical Fields of TaS2(Pyridine)1/2 (vapor transport)*   000468
#Crystal Growth and Characterization of In1.9As0.1Se3 (vapor transport)*   000471
#Crystal Structure of the Indium Selenide In2Se3 (vapor transport)*   000557
#Growth of CdP4 Single Crystals (evaporation, vapor transport)*   000599
#Polytypism and Amorphousness in Silicon Whiskers (Si, vapor transport)*   000642
#Heteroepitaxial Homogeneous CdxHg1-xTe Films (vapor transport)*   000707
#Growth of alpha-Ag2S Whiskers in a VLS System (vapor transport)*   000718
#Synthesis of Single Crystal LnPO4 (vapor transport)*   000733
#The Growth of CdS in Sealed Silica Capsules (vapor transport)*   000815
#Epitaxial Thin Iron Films on Antimony (Fe, vapor transport)*   000875
#Organometallic VPE Growth of AlxGa1-xAs (vapor transport)*   000919
#Raman and Infrared Spectra of CdIn2S4 and ZnIn2S4 (vapor transport)*   001004
#Synthesis and Crystal Growth of CdGeP2 (vapor transport)*   001041
#Morphology of Iron Pyrite Crystals (FeS2, vapor transport)*   001093
Growing Monocrystalline Bars of SiC from the Gas Phase (vapor transport) #   000554
Growth of Hg0.60Cd0.40Te from Te-Rich Solution (film, vapor transport)#LPE   000839
Survey Method for the Study of CVD Conditions (TiN, vapor transport)#A Quick   000655
Structure and Physical Properties of CuO0.65VS2 (vapor transport)#Crystal   000543
Carbon Tetrabromide Crystals for Optical Studies (CBr4, vapor transport)#Growth of   000263
in the Initial Use of Spacelab for Crystal Growth (vapor transport)#Priorities   000270
Impurity and Its Effect on the Purity of LPE GaAs (vapor transport)#New Gaseous   000313
Mossbauer Studies of FeNi2BO5 and FeNiBO4 (ferrites, vapor transport)#Magnetic and   000005
System for Exacting Crystal Growth Preparations (TiO2, vapor transport)#A Gas Handling   000808
and Properties of CuGaxIn1-xSe2 Chalcopyrite Compound (vapor transport)#Crystal Growth   000743
of Boron Monophosphide Using a B2H6-PH3-H2 System (BP, vapor transport)#Crystal Growth   000884
GaAs VPE at Low Temperature and High Partial Pressures (vapor transport)#Fast Growth in   000393
Single V3O5 Crystals by Chemical Transport Reactions (vapor transport)#Preparation of   000789
Characterization of Ruthenium Dioxide Crystals (RuO2, vapor transport)#Preparation and   000865
and Crystal Structures of BiReO4 and BiRe2O6 (vapor transport)#The Preparation   000781
Analysis of Single Crystals of EuSb4S7 and EuSb2Se4 (vapor transport)#Growth and X-Ray   000022
Anhydride in a Closed Crystal-Vapour System (organics, vapor transport)#Growth of Phthalic   000327
and Slip System of alpha-Si3N4 Single Crystal (vapor transport)#Growth, Morphology   000696
of Epitaxial Ga1-xInxSb Using Open-Tube Flow System (vapor transport)#Vapor-Phase Growth   000681
Mn-Bi Films and Their Crystallization Process (vapor transport)#Formation of Amorphous   001103
Crystals by the Synthesis, Solute Diffusion Technique (vapor transport)#Large-Sized InP Single   000923
Electrical Properties of CdCr2S4 Single Crystals (vapor transport)#Preparation and Certain   000780
and Mass Transfer Kinetics in the System TiP-I2 (TiP, vapor transport)#Heterogeneous Equilibria   001003
Crystals Using Tellurium Tetrachloride and Argon (UO2, vapor transport)#Growth of Uranium Dioxide   000280
Metal-Insulator Transitions in (Ti0.9975V0.0025)4O7 (vapor transport)#Structural Aspects of the   000391
Reactive Ionized-Cluster Beam Deposition Techniques (vapor transport)#in Ionized-Cluster Beam and   000944
of the Non-Stoichiometric VS2 and V5S8 Phases (vapor transport)#Physico-Chemical Properties   000470
Metallurgical Silicon (Si, Czochralski, zone melting, vapor transport)#Solar Cells from Zone-Refined   000195
Structure of As2(Se,S)3 Compounds (As2S3, As2Se3, vapor transport)#The Crystal Growth and Atomic   000896
M(PbX3) (CsPbCl3, CsPbBr3, CsPbI3, KPbI3, RbPbI3, vapor transport)#of Alkali-Trihalogenoplumbates   000377
Parameters of Layer Compounds in the Series Mo1-xWxSe2 (vapor transport)#Conditions and Crystal Structure   000013
Cu2MnGeSe4, Cu2MnSnSe4, Cu2FeGeS4, Cu2CoGeS4, vapor transport)#Cu2MnSiS4, Cu2MnGeS4, Cu2MnSnS4,   000357
for Epitaxial Films on Insulators (Si on sapphire, vapor transport)#Dichlorosilane as Silicon Source   000254
Modifications of Gold Monobromide, I-AuBr and P-AuBr (vapor transport)#and the Crystal Structures of Two   000436
Large Temperature Gradients (CdS, kinetics, theory, vapor transport)#of Cadmium Sulphide Crystals with   000289
of High-Purity p-type Single Crystals of GeFe2O4 (vapor transport)#Electrical and Optical Properties   000920
Crystals by the Sublimation Travelling Heater Method (vapor transport)#Growth of High-Purity ZnTe Single   000938
ZnSe Single Crystals Obtained by a New Growth Method (vapor transport)#Properties of Homogeneously Doped   000745
Electrical, Calorimetric, Magnetic and EPR Studies (vapor transport)#and Phase Diagram of (Ti1-xVx)4O7:   000835
Large Anions: Ba10(ReO5)6Br2 and Ba10(ReO5)6I2 (vapor transport)#of New Rhenium Apatites Containing   000080
Transition-Metal Phosphides CuP2, NiP2, and RhP3 (flux, vapor transport)#Growth and Characterization of the   000708
Epitaxy Si Films Deposited at Elevated Temperatures (vapor transport)#Crystalline Defects in Solid Phase   000194
magnetic susceptibility, and NMR) of the Spinel CuV2S4 (vapor transport)#and Physical Properties (transport,   000544
(MoS2, MoSe2, WS2, WSe2, HfSe2, ZrSe2, TiSe2, vapor transport)#of Transition-Metal Dichalcogenides   001000
and Their Electronic Properties (FeCr2S4, CoCr2S4, vapor transport)#Chalcogenide Spinel Single Crystals   001066
Physicochemical Investigation of CrPr2S4 and CrPr2Se4 (vapor transport)#Preparation of Single Crystals, and   000817
-- Microanalytical Evaluation of Composition (vapor transport)#Under Non-Stoichiometric Conditions   000082
in an Organometallic VPE System (Ga1-xAlxAs, GaAs, vapor transport)#Epitaxial Layers During Their Growth   000974
Coefficients of VO2 near the Transition Temperature (vapor transport)#Parameters and the Thermal Expansion   000513
in the NbxCr1-xO2 (x greater than 0.5) System (vapor transport)#Crystal Growth and Phase Transitions   000095
the Vapour Under Constant Conditions (C10O16, theory, vapor transport)#of Growth of Adamantane Crystals from   000572
VxTi1-xSe2 and TaxTi1-xSe2 (TiSe2, VSe2, TaSe2, vapor transport)#of Layered Transition Metal Selendies   000555
(111) GaAs Substrates Using the Close-Spaced Geometry (vapor transport)#of Single Crystal Cd1-xZnxS Layers on   000305
Chalcogenide Halides, NbX2Y2(X=S, Se - Y=Cl, Br, I) (vapor transport)#Structures, and Properties of Niobium   000795
Growth on Bicrystalline Substrates of NaCl (Au, Ag, vapor transport)#Tilt Boundaries Produced by Epitaxial   000207
Boracite on Chromium-Chlorine Boracite (Ni3B7O13Cl, vapor transport)#of Epitaxial Layers of Nickel-Chlorine   000837
and Note on the Shandite Structure (Ni3Pb2Se2, vapor transport)#Study of the System Ni3Pb2S2-Ni3Pb2Se2   000198

PERMUTED TITLE INDEX

#Transfer of VO2 by Iodine (vapor transport, doped)*		000487
Single Crystals of the Magnetic Semiconductor CdCr2Se4 (vapor transport, doped)* #Preparation of		000625
Optical and Photoconducting Properties of Fe2-xCrxO3 (vapor transport, electrolytic deposition)* Electrical,		000624
of Films of CdS Evaporated in Vacuum onto Silicon (vapor transport, electron beam melting, kinetics,		000073
Growth and Etching Vectors of Silicon (111) (theory, vapor transport, epitaxy)* Step Reconstruction on the		001019
Growth (Er3Fe5O12, (Eu,Y)3Fe5O12 and (Eu,Yb)3Fe5O12, vapor transport, epitaxy, films)*Temperature CVD Garnet		000211
Distortion in Heteroepitaxial Layers: Ge On GaAs (vapor transport, film)* #Tetragonal		000362
and Crystallographic Polarity in ZnO Crystals (vapor transport, film)* #Crystal Growth in c Direction		000426
SiC on Silicon Substrate Using SiCl4-C3H8-H2 System (vapor transport, film)* #Heteroepitaxial Growth of beta-		000616
#Nitrides--Structures and Crystal Growth (review, vapor transport, flux)*		000533
Advective Interfacial Mass Transfer (theory, solution, vapor transport, flux)* #On Diffusive-		001074
#A Molecular Dynamics Study of Surface Melting (theory, vapor transport, kinetics)*		000140
Reactor for High Volume Low-Cost Silicon Epitaxy (Si, vapor transport, kinetics)* #Novel		000062
Layer Growth from the Vapour Phase (SiC, film, vapor transport, kinetics)*of Silicon Carbide Epitaxial		000942
Properties of the Systems SnS2-SnI4 and SnS2-I2 (SnS2, vapor transport, kinetics, theory)* #Transport		001079
(111) Single Crystals with GeBr4 Vapor (Ge, epitaxy, vapor transport, kinetics, theory)* of Germanium		000060
of Epitaxial Layers in the Ga_AsCl3-H2 System (GaAs, vapor transport, kinetics, theory)* Rate and Purity		000653
Photoconductivity of Cd1-xHgxCr2Se4 Single Crystals (vapor transport, melt)* #Photoferromagnetic Effect and		001037
USb2 Single Crystals and Their Structural Perfection (vapor transport, melt, flux)* #Growth of		000383
of Mercuric Iodide Radiation Detectors. Part II. HgI2 (vapor transport, review)* Spectroscopic Evaluation		000834
#Growth and Physical Properties of ZnSiP2 (vapor transport, solution)*		000086
of Transition Metal Silicides (chemical vapor transport, solution, cold crucible melting)*		000067
Study of Large Non-Seeded Pt1-xSnxTe Single Crystals (vapor transport, sublimation)* #Growth		000961
Encapsulation Zone-Refining of PbS (zone melting, vapor transport, sublimation)* #Liquid		000583
Crystals of Zinc Selenide from the Vapour Phase (ZnSe, vapor transport, sublimation)* #Growth of Single		000214
Growth and Evaporation of Epitaxial Ice Crystals (H2O, vapor transport, sublimation, evaporation)* Field on		000028
#Specific Heat of V6O11 Between 0.4 and 50 K (vapor transport, theory)*		000215
#Growth Rate of Iron Whiskers (Fe, vapor transport, theory)*		000462
#Raman Scattering in 3R-NbS2 (vapor transport, theory)*		000731
of EuS near the Ferromagnetic Transition (Bridgman, vapor transport, theory)* #Heat Capacity		000505
I. Phosphorus Content and Segregation Coefficient (Si, vapor transport, theory)* Growth from the Gas Phase.		000514
to the Study of Heterogeneous Reactions (GaAs, vapor transport, theory)* Method and Its Application		000281
Properties of Single Crystals of Cadmium Sulphide (CdS, vapor transport, theory)* on the Electrical Transport		000612
in ZrS2 and HfS2 (CuxZrS2, CuxHfS2, FexZrS2, FexHfS2, vapor transport, theory)* Shift of the Absorption Edge		001090
Content and Differential Molar Solution Enthalpies (Si, vapor transport, theory)* the Gas Phase. II. Phosphorus		000515
#Germanium-Iodine Equilibrium from 700 to 1000 K (Ge, vapor transport, theory, thermodynamics)*		000191
Growth in SiCl4-HCl-H2 System Based on Flow Graph (vapor transport, theory, Si)*Model of Silicon Epitaxial		000503
Behaviour in Low Gravity Environment (melt, solution, vapor transport, zone melting, doped, review)*Materials		001070
#Crystal Growth of Mn15Si26 (vapor transport, Bridgman)*		000495
Silicon Carbide by Pyrolysis of Silane Compounds (SiC, vapor transport, De Boer-Van Arkel, doped)* and Doped		001043
Kinetics of Silicon Growth Under Low Hydrogen Pressure (vapor transport, Si)* #		000255
Grown with Platinum and Gold as Liquid-Forming Agents (vapor transport, Si)*Morphology of VLS Silicon Crystals		001075
of Heteroepitaxial Zinc Oxide Films Grown on Sapphire (vapor transport, ZnO)* #Real Structure		000526
by a Modified Hot-Filament Method (de Boer, Van Arkel, vapor transport, ZrC)* Growth of Zirconium Carbide		000927
#Reaction of Germanium (111) Single Crystals with GeBr4 Vapor (Ge, epitaxy, vapor transport, kinetics, theory)*		000060
#Single Crystal Growth of Titanium Carbide from the Vapor by a Modified Hot Wire Method (TiC)*		000928
of Growth Parameters During the Crystal Growth from the Vapor in Closed Ampules (GeS, sublimation)*Measurements		000841
on the Growth of Zirconium Carbide Whiskers by Chemical Vapor Deposition (ZrC)* #Catalytic Effect of Nickel		000958
Zirconium Carbide Whiskers by Chemical Vapor Deposition (ZrC)* #Synthesis of Nonstoichiometric		000674
#Crystal Growth of Gallium Arsenide by Chemical Vapor Deposition from the Complex		001109
Substrates* #Organometallic Vapor Deposition of Epitaxial ZnSe Films on GaAs		000921
film)* #Chemical Vapor Deposition of Scandium Hydride (ScH2, epitaxy,		000494
#Mechanism of Chemical Vapor Deposition of Silicon (Si, epitaxy)*		000702
Smooth Surface on Intermediately Sputtered #Chemical Vapor Deposition of Single-Crystalline ZnO Film with		000711
Oxides on the Growth of TiC Whiskers by Chemical Vapor Deposition* of Various Metals and Refractory		000959
#Epitaxial Growth of ZnSe by Vapor Flow Method*		000669
#Morphology and Growth Mechanism of Vapor Grown Cd Crystals as Affected by Bi Impurity*		000375
#Polytypism in Vapor Grown Crystals of Cadmium Bromide (CdBr2)*		000662
KOH, theory)* #Low-Stress Physical Vapor Growth (PVT) (KCl, KBr, KCl:KCN, KCN:KOH, KCl:		000809
Crystals (review, theory)* # Vapor Growth and Epitaxy of Minerals and Synthetic		000932
* #Modified Chemical Vapor Growth of Cinnabar (HgS) and GaP in Closed Systems		000279
# Vapor Growth of Hexagonal GeO2 Needle Crystals*		000423
# Vapor Growth of HgCr2Se4 (epitaxy, theory)*		000332
Sharp Tip* #Chemical Vapor Growth of LaB6 Whiskers and Crystals Having a		000656
Solar Cells (Si, review)* #Problems in the Melt- and Vapor Growth of Silicon for Integrated Circuits and		000117
Wire Method (TiB2)* #Chemical Vapor Growth of Titanium Diboride by a Modified Hot		000926
ZnTe, epitaxy)* # Vapor Growth of Zinc Telluride by Closed Tube Method (		000700
# Vapor Growth of Zn2GeO4 Single Crystals*		000424
Characterization of Modes of Epitaxial Growth from the Vapor Phase (review, theory)* Methods Used for the		000542
Quaternary (In,Ga)(As,P) Alloys Grown by the Hydride Vapor Phase Epitaxy Technique (film)* Ternary and		000729
#Chromium Doped GaAs Vapor Phase Epitaxy*		000469
#Surface Morphology of GaAs Grown by Vapor Phase Epitaxy*		000705
theory)* #A Simple Analysis of Vapor Phase Growth - Citing an Instance of GaxIn1-xAs (		000673
Aluminum Fluoride (Al2O3)* # Vapor Phase Growth of Alumina Whiskers by Hydrolysis of		001094
Crystals* # Vapor Phase Growth of Thin Plates of Cr2O3 Single		000858
and Control of Doping Gas Concentration During Vapor Phase Silicon Epitaxial Growth by Using a Plane		000937
Trioxide and Molybdenum Ditelluride (MoO3-x, MoTe2)* # Vapor Phase Transport and Crystal Growth of Molybdenum		000299
Cadmium Sulfo-Selenide (Cd1+y(SxSe1-x), sublimation)* # Vapor Phase Transport and Stoichiometry Control of		000647
Cadmium Sulfide (CdS, sublimation)* # Vapor Phase Transport and Stoichiometry Control of		000648
#Seeded Growth of Large Single Crystals of CdS from the Vapor Phase*		000242
#The Growth of CdAs2 and ZnAs2 Single Crystals from the Vapor Phase*		000598
GaP-Ga0.5Al0.5P-GaP Layer Under Controlled Phosphorus Vapor Pressure and Reduced Oxygen Contamination (GaP,		000460
Edge of Single-Crystal AlN Prepared by a Close-Spaced Vapor Process* #The Optical Absorption		000758
#Congruent (Diffusionless) Vapor Transport (theory, kinetics)*		000007
#Preparation by Vapor Transport of Monocrystals of AlF3, FeF3, and TiF3*		000124
and Conductive ZnSe:I Crystals by Iodine Vapor Transport* #Preparation of Photoluminescent		000775
#Growth Parameters and Crystal Morphology of Vapor-Deposited Niobium Nitride (NbN)*		000947
Surfaces (SiC)* #Analysis of Vapor-Deposited Silicon Carbide Films on Silicon Ribbon		001099

PERMUTED TITLE INDEX

#Interface Roughness of	Vapor-Grown Adamantine Crystals (C10H16)*	000753
#Growth and Polytypism in	Vapor-Grown Mixed Cadmium Halide Crystals (CdI2-xBrx,	000870
CdI2-xClx)*	Vapor-Liquid-Solid Mechanism*	000338
#Growth of Whiskers by the	Vapor-Phase Growth of Epitaxial Ga1-xInxSb Using Open-	000681
Tube Flow System (vapor transport)* #	Vapour (theory)*	001009
#Morphological Stability Analysis of Growth from the	Vapour Deposition (theory)* #Interrelation Between	000742
Interface and Vapour Transport Kinetics in Chemical	Vapour Deposition of Silicon (Si)*	000118
#Mechanisms of the Chemical	Vapour Deposition on Profiled Substrates (theory)*	001011
#Interface Morphology in Chemical	Vapour Deposition Processes. I (theory)*	001010
#Morphological Stability Analysis in Chemical	Vapour Deposition Processes. II (theory)*	000435
Epitaxial Layers Obtained by Organometallic Chemical	Vapour Deposition* and Characterization of Undoped ZnSe	000114
#Growth of Triple Unit Cell Steps on the Surface of	Vapour Grown FeCr2S4 Single Crystals. I*	000418
#Growth Mechanism of Multi-Steps on the Surface of	Vapour Grown FeCr2S4*	000417
by Elementary Steps and Multi-Steps on the Surface of	Vapour Grown FeCr2S4* #Step Growth	000419
Single Crystals of ZnS and Zn(S,Se)* #	Vapour Growth and Defect Characterisation of Large	000816
#	Vapour Growth and Epitaxy (review)*	000946
#	Vapour Growth in a Microgravity Environment (review)*	000091
Compounds and Alloys (#Some Aspects of the Epitaxial	Vapour Growth of Semiconductors: Elements, III-V	000041
#Epitaxial Growth of A(III)B(V) Semiconductors from	Vapour Phase (review, theory)*	000352
of Epitaxial Growth of A(III)B(V) Compounds from the	Vapour Phase (theory)* #Flow-Graph Model	000504
of Silicon Carbide Epitaxial Layer Growth from the	Vapour Phase (SiC, film, vapor transport, kinetics)*	000942
#Growth of Single Crystals of Zinc Selenide from the	Vapour Phase (ZnSe, vapor transport, sublimation)*	000214
#Crystallization of HgCdTe Mixed Crystals from the	Vapour Phase by Chemical Transport*	000342
#GaAs	Vapour Phase Epitaxy Using Computer Control*	001057
#Preparation of ThSiO4 Single Crystals by a	Vapour Phase Reaction*	000454
#Crystallization of HgTe Crystals from the	Vapour Phase*	000341
#Morphology and Structure of ZnS Grown from the	Vapour Phase*	000874
#Growth of Phthalic Anhydride in a Closed Crystal-	Vapour System (organics, vapor transport)*	000327
and Film Growth of Actinide Pnictides by Chemical	Vapour Transport (review)* #Single Crystal	000160
for the Growth of Chalcopyrite Crystals by Iodine	Vapour Transport (CuInS2, CuGaS2, AgInSS8)* Method	000744
of Gallium Selenide and Gallium Sulphide by Iodine	Vapour Transport (GaS, GaSe)* Conditions for the Growth	001078
Crystal Growth of Protactinium Arsenides by Chemical	Vapour Transport (PaAs2, Pa3As4, PaAs)* and Single-	000161
#The	Vapour Transport of NbB2 and TaB2 (kinetics, theory)*	000042
Method to a Five-Component System: The Chemical	Vapour Transport of Ni-Cl Boracite (no materials grown	000236
(theory)* #Interrelation Between Interface and	Vapour Transport Kinetics in Chemical Vapour Deposition	000742
#Growth of Tungstenite Single Crystals by Direct	Vapour Transport Method (WS2, sublimation)*	000012
for the Growth of Layered Dichalcogenides by	Vapour Transport Processes (kinetics, theory)*	000805
Photoluminescence Properties of CuGaSe2 Grown by Iodine	Vapour Transport* #	000935
or equal to x less than or equal to 1) Grown by Iodine	Vapour Transport*in the Series GaSxSe(1-x) (0 less than	001077
of the Rate of Growth of Adamantane Crystals from the	Vapour Under Constant Conditions (C10016, theory, vapor	000572
#Unit Cell Step Structures on	Vapour-Grown Cr2S3 Single Crystals*	000676
#Appearance of Non-Singular Surfaces on	Vapour-Grown Ice Crystals (H2O, theory)*	000689
#	Vapour-Phase Epitaxial Growth of ZnS on GaP*	000613
Crystals from the Vapour Under Constant Conditions #The	Variability of the Rate of Growth of Adamantane	000572
#Crystal Structure of a New	Variety of Nickel Pyrophosphate: Ni2P2O7 (slow cooling)*	000609
#X-Ray and Neutron Diffraction of Na2MnFeF7: A	Variety of Trigonal Weberite (flux)*	001033
and Step Reconstruction on the Growth and Etching	Vectors of Silicon (111) (theory, vapor transport,	001019
Teller Phase Transitions (solution)* #Ultrasonic	Velocities of K2PbCu(NO2)6 Near the Cooperative Jahn-	000467
#Subsidiary Electrical Heating for	Verneuil Furnaces in the USSR (review)*	000349
Crystals of Beta Rhombohedral Boron by a Modified	Verneuil Method (B)* #Growth of Single	000213
#The	Verneuil Process (review)*	000283
Insulator (Czochralski, epitaxy, Al2O3)* #Evaluation of	Verneuil Sapphire as a Substrate for Silicon on an	000793
Feeder for Crystals of Large Diameter Grown by the	Verneuil Technique (MgAl2O4, alpha-Al2O3)* #Powder	000284
Bicrystals of Nickel Oxide (NiO, arc image, float zone,	Verneuil)* #Growth of Oriented	000239
- Theory and Techniques, Volume 2 (review, epitaxy,	Verneuil, III-V compounds, Si)* #Crystal Growth	000348
#GaSb and InSb Crystals Grown by a	Vertical and Horizontal Travelling Heater Method*	000099
#Influence of Magnetic Field on	Vertical Bridgman-Stockbarger Growth of InxGa1-xSb*	000863
of Single Crystals by the Method of Directed	Vertical Crystallization (theory, kinetics)*Preparation	000057
300 K and Magnetic #Crystalline Vanadium (II) Fluoride,	VF2. Preparation, Structure, Heat Capacity from 5 to	000917
#Polytypism and	Vibrational Properties of PbI2*	000850
#Modes of	Vibrations in Step Trains: Rhythmical Bunching (theory)*	000933
DyKMo2O8, melt)* #	Viscosity Measurements in the System Dy2O3-K2O-MoO3 (	000320
Between the Number of Gold Drops and TiP Whiskers in	VLS Growth* #On the Relations	000929
#Some Notes on the Growth Kinetics and Morphology of	VLS Silicon Crystals Grown with Platinum and Gold as	001075
#Growth of alpha-Ag2S Whiskers in a	VLS System (vapor transport)*	000718
Properties -Lithium and Molecular Intercalations (	VOCl)* Oxychloride; Magnetic, Optical and Electrical	001031
Czochralski)* #Interface Growth Feature and	Voids in Sapphire Ribbon Crystals (alpha-Al2O3,	001050
#Development of Depressions and	Voids During LPE Growth of GaAs (film)*	000084
Bridgman Growth of Crystals with Toxic and/or Highly	Volatile Components (equipment)*Improved System for the	000164
of Solidification Interfaces with a Very High	Voltage Electron Microscope* #In-Situ Observation	000163
of Films of IV-VI Compounds in a Quasi-Sealed	Volume (SnTe, PbTe, PbSe, evaporation)* #Preparation	000306
of Isotopic Composition on the Density and Molecular	Volume of Solid Crystalline Lithium Nitrate Hydrates (	000023
kinetics)* #Novel Reactor for High	Volume Low-Cost Silicon Epitaxy (Si, vapor transport,	000062
#Current Topics in Materials Science,	Volume 1 (review)*	000450
#Growth of Crystals,	Volume 11 (review, theory)*	000183
Si)* #Crystal Growth - Theory and Techniques,	Volume 2 (review, epitaxy, Verneuil, III-V compounds,	000348
#Current Topics in Materials Science,	Volume 2 (Review)*	000449
#Transfer of	VO2 by Iodine (vapor transport, doped)*	000487
Parameters and the Thermal Expansion Coefficients of	VO2 near the Transition Temperature (vapor transport)*	000513
#Mass Spectrometric Study on Chemical Transport of	VO2*	000063
Czochralski)* #Calcium Orthovanadate, Ca3(	VO4)2 - A New High-Temperature Ferroelectric (	000339
vapor transport)* #Fast Growth in GaAs	VPE at Low Temperature and High Partial Pressures (	000393
#Organometallic	VPE Growth of AlxGa1-xAs (theory)*	000918
#Organometallic	VPE Growth of AlxGa1-xAs (vapor transport)*	000919
(Pb1-xSnx)1+yTe1-y System*	VPE Growth of InAs*	000312
#Influence of the	VPE Parameters on the Epitaxial Film Properties for the	000433
Layers During Their Growth in an Organometallic	VPE System (Ga1-xAlxAs, GaAs, vapor transport)*	000974
#Properties of	VPE-Grown GaN Doped with Al and Some Iron-Group Metals*	000649

PERMUTED TITLE INDEX

Metal Selendies VxTi1-xSe2 and TaxTi1-xSe2 (TiSe2,   VSe2, TaSe2, vapor transport)*   of Layered Transition 000555

#Physico-Chemical Properties of the Non-Stoichiometric VS2 and V5S8 Phases (vapor transport)* 000470

Properties of Layered Transition Metal Selendies VxTi1-xSe2 and TaxTi1-xSe2 (TiSe2, VSe2, TaSe2, vapor 000555

Components in the Course of Preparation of II-IV- V2 Compounds by Gas Phase Transport, with ZnSiP2 as an 001086

#Liquid Phase Epitaxy of II-IV- V2, Chalcopyrite Compounds (ZnSnAs2, solution)* 000089

Some Ferrite Materials from Lead Flux with Addition of V2O5 (Y3Fe5O12, Ho3Fe5O12, Dy3Fe5O12, Er3Fe5O12, 000565

#Single Crystal Growth of Alexandrite in a V2O5 Flux (BeAl2O4)* 001005

#Preparation of V3Ga Single Crystals by Iodine Transport* 000217

Growth of A15 Structure Intermetallic Compounds (V3Si, V3Ge)*   #Czochralski 001073

transport)* #Preparation of Single V3O5 Crystals by Chemical Transport Reactions (vapor 000789

#A Simple Method for Growing V3Si Single Crystals (electron beam melting)* 000162

Growth of A15 Structure Intermetallic Compounds (V3Si, V3Ge)*   #Czochralski 001073

#The Crystal Structure of V4O7 at 120 degrees K (vapor transport)* 000390

Chemical Properties of the Non-Stoichiometric VS2 and V5S8 Phases (vapor transport)* #Physico- 000470

#Specific Heat of V6O11 Between 0.4 and 50 K (vapor transport, theory)* 000215

Semiconducting Phase* #Optical Properties of V6O13 Single Crystals in the Metallic and the 001023

Carbide as Revealed by Field Electron Microscopy (W microcrystal, crystal growth study)*   Ditungsten 000724

#The Adsorption of Zr onto W(100) Surfaces)* 000221

#Wolfram (W)* 000096

Evaporation Technique: bcc Metals (V, Fe, Cr, Mo and W)* and Habit of Fine Metal Particles Formed by Gas- 000821

Metal Alloys and Compounds by Plasma Heating (W, W-C)* #Growth of Single Crystals of High Melting 000829

Melting Metal Alloys and Compounds by Plasma Heating (W, W-C)* #Growth of Single Crystals of High 000829

#Surface Morphology of HCl Etched Silicon Wafers (Si)* 001015

and by Film Doping Using Boron #The Bending of Silicon Wafers by Thin Polycrystalline Silicon Film Deposition 000985

compounds, evaporation, review, theory)* #Hot Wall Epitaxy (II-VI compounds, IV-VI compounds, III-V 000568

#CdTe Thin Films Grown by Hot Wall Epitaxy* 000569

#Asymmetric 180 degree Bloch Walls in Fe Single Crystal Films (theory)* 000936

21H2O)* #Dodecatungstophosphoric Acid-21- Water by Neutron Diffraction (evaporation, H3PW12O40. 000913

#Morphology of Ice Droxtals Grown from Supercooled Water Droplets (H2O)* 000347

Microscopic Characterization of Cubic BN Prepared by Water Solvent (solution)*   #Electron 000934

#Epitaxial Growth of KNdP4O12 Laser Waveguides (KLaP4O12, solution)* 000643

#LPE Growth of Li(Nb,Ta)O3 Solid-Solution Thin Film Waveguides on LiTaO3 Substrates (LiNb1-xTaxO3, flux)* 000500

Temperature cw Operation at 1.07- and 1.35-micrometer Wavelength (laser melting, Czochralski)*   Laser: Room- 000916

#Solute Partitioning During Silicon Dendritic Web Growth (Si, theory)* 000855

#Thermal Analysis of Solidification in Web-Dendritic Ribbon Growth (Si, kinetics, theory)* 000369

Neutron Diffraction of Na2MnFeF7: A Variety of Trigonal Weberite (flux)*   #X-Ray and 001033

#Growth Morphology of Weddellite (CaC2O4.2XH2O, gel)* 000302

Characterization of Silicon Ribbon Crystals Grown with Wetting and Non-Wetting Dies (Si, EDFG)* #Growth and 000524

of Silicon Ribbon Crystals Grown with Wetting and Non- Wetting Dies (Si, EDFG)* #Growth and Characterization 000524

#Growth of Hollow Se-Te Whisker Crystals (theory)* 000112

Reduction of a CuI-NiBr2 Mixture (Cu-Ni, dendrites, whiskers)* of Copper-Nickel Alloy Crystals by Hydrogen 000364

#The Growth Figures of MnAs (whiskers, theory)* 000067

zone #Controlled Growth of alpha-Fe Single Crystal Whiskers (hydrogen reduction of ferrous halides, float 000318

#Growth Rate of Iron Whiskers (Fe, vapor transport, theory)* 000462

#Polytypism and Amorphousness in Silicon Whiskers (Si, vapor transport)* 000642

#Chemical Vapor Growth of LaB6 Whiskers and Crystals Having a Sharp Tip* 000656

#Growth of Whiskers by the Vapor-Liquid-Solid Mechanism* 000338

#Synthesis of Nonstoichiometric Zirconium Carbide Whiskers by Chemical Vapor Deposition (ZrC)* 000674

Effect of Nickel on the Growth of Zirconium Carbide Whiskers by Chemical Vapor Deposition (ZrC)* #Catalytic 000958

Metals and Refractory Oxides on the Growth of TiC Whiskers by Chemical Vapor Deposition*   of Various 000959

#Vapor Phase Growth of Alumina Whiskers by Hydrolysis of Aluminum Fluoride (Al2O3)* 001094

#Growth of alpha-Ag2S Whiskers in a VLS System (vapor transport)* 000718

the Relations Between the Number of Gold Drops and TiP Whiskers in VLS Growth*   #On 000929

#Morphological Change from Copper/alpha-copper-Zinc Whiskers to Dendrites in Zinc Reduction Growth (Cu, 000481

Vapor Growth of Titanium Diboride by a Modified Hot Wire Method (TiB2)*   #Chemical 000926

of Titanium Carbide from the Vapor by a Modified Hot Wire Method (TiC)*   #Single Crystal Growth 000926

Wolfram (W)* 000096

Cu9Al4)* #Spherical Monocrystals for X-Ray Work Obtained by Plasma Remelting of Alloy Powder (000043

to the A-15 Structure* #Cold Working Nb3Al in the bcc Structure and then Converting 001067

#Crystallization of WO2 from the Gas Phase (vapor transport)* 000052

#The Crystal Structure of Triclinic WO3 (sublimation)* 000241

Prepared by a Gas-Evaporation Technique (MoO3, WO3, theory)* #Coalescence Growth of Smoke Particles 000448

#Growth and Properties of Pb(WO4)1-x(MoO4)x Mixed Crystals (Czochralski)* 000709

#The Thermal Expansion of 2H-MoS2 and 2H- WSe2 Between 10 and 320 K* 000667

of Transition-Metal Dichalcogenides (MoS2, MoSe2, WS2, WSe2, HfSe2, ZrSe2, TiSe2, vapor transport)*   Charges 001000

Single Crystals by Direct Vapour Transport Method (WS2, sublimation)*   #Growth of Tungstenite 000012

of Transition-Metal Dichalcogenides (MoS2, MoSe2, WS2, WSe2, HfSe2, ZrSe2, TiSe2, vapor transport)* 001000

#Fivefold Twinned Silicon Crystals Grown in an Al-16 wt.% Si Melt (Si, flux)* 000489

#AlSiP3, A Compound with a Novel Wurtzite-Pyrite Intergrowth Structure (flux)* 001044

#Structure of Monocrystalline WV2O6 at 298 and 383 K (vapor transport)* 000389

#Structural Phase Transition in Mixed Crystals WxMo1-xO3 (melt)* 000824

#LPE Growth of Quaternary GayIn1-yAsxP1- x (film)* 000188

#Molecular Beam Epitaxy of GaSb and GaSbxAs1- x (films)* 001100

#Edge Emission in Melt-Grown ZnSexS1- x (Bridgman)* 000710

of (Mn,Cr)P and (Mn,Fe)P Compounds (Mn1-xFexP, x = 0 to 0.15, theory)*   #Magnetic Properties 000430

#Molecular Beam Epitaxy of InxGa1-xSb (x = 0 to 1) (GaSb)* 001101

Dilute Binary Metallic Alloys (Cu - 0.5%Si, Cu1-xSbx x = 0.0015-0.01)*Cellular Structures of Monocrystals of 000210

a New Pyrochlore Containing Rhenium: Ca1-xRe2O6(OH)2x (x = 0.3)* #High Pressure Synthesis and Structure of 000104

Addition of Vanadium on the Chemical Transport of TiSx, x = 1.40-1.70 (TiyV1-ySx)*   #Effect of 000819

Structure of a New Mixed Oxide "FeV3O8" (FexV1-xO2- x approx. = 0.25) (melt, slow cooling)*   and Crystal 000661

Crystal Growth and Phase Transitions in the NbxCr1-xO2 (x greater than 0.5) System (vapor transport)* # 000005

in the Series GaSxSe(1-x) (0 less than or equal to x less than or equal to 1) Grown by Iodine Vapour 001077

Diffusion Technique* #Growth of ZnSxSe1- x Layers on ZnS and ZnSe Substrates by a Solid-State- 000045

#Growth and Properties of Pb(WO4)1-x(MoO4) x Mixed Crystals (Czochralski)* 000709

Conductivity and Galvanomagnetic Effects in (SN) x Single Crystal*   #Electrical 000463

#Growth of Cubic ZnS, ZnSe and ZnSxSe1- x Single Crystals by Iodine Transport* 000309

#Molecular Beam Epitaxy of ZnSexTe1- x(film)* 001102

Growth of Substituted Yttrium Iron Garnets Y3Fe5- x(Ga,Al)xO12 by the Floating Zone Method*Single Crystal 000879

PERMUTED TITLE INDEX

#Growth and Properties of Pb(WO4)1- x(MoO4)x Mixed Crystals (Czochralski)* 000709
#Americium Ditelluride (AmTe2- x* 000153
#Structural Studies of Compounds in the Series GaSxSe(1- x) (0 less than or equal to x less than or equal to 1) 001077
Control of Cadmium Sulfo-Selenide (Cd1+y(SxSe1- x), sublimation)*   Phase Transport and Stoichiometry 000647
Ammonium Dihydrogen Arsenate (DADA) (NH4(1-x) D4xH2(1- x)D2xAsO4, solution)*   Phases of Deuterium 000100
of Deuterium Ammonium Dihydrogen Arsenate (DADA) (NH4(1- x) D4xH2(1-x) D2xAsO4, solution)*Antiferroelectric Phases 000100
Zn2TiO4 Substrates (flux)* #Thin Film Growth of (MnxZn1- x)Fe04 by Liquid Phase Epitaxy on Bridgman Grown 000216
Analysis and Properties of High Purity Zn(1- x)HgxTe Alloys (distillation)*   #Preparation, 000792
and Phase Transitions in Single Crystals of (KxNa1- x)NbO3 (flux)*   #Growth 000056
Phases of the Systems Th0.25NbO3-NaNbO3 (NaxTh0.25(1- x)NbO3, melt, slow cooling)*   Chemistry of Perovskite 000527
of Molybdenum Trioxide and Molybdenum Ditelluride (MoO3- x, MoTe2)* #Vapor Phase Transport and Crystal Growth 000299
CH2)mNH3MX4 Families with M = Cd, Cu, Fe, Mn or Pd and X = Cl or Br: Importance, Solubilities and Simple 000039
MFX of Divalent Rare Earths (M = Sm, Eu, Tm, and Yb - X = Cl, Br, and I) (melt, SmFCl, SmFBr, SmFI, EuFCl, 000088
I)B(II)C(IV)X4 (BII=Zn, Mn, Fe, Co - CIV = Si, Ge, Sn - X = S, Se) (Cu2ZnGeS4, Cu2ZnSnS4, Cu2ZnGeSe4, 000357
Thermal Expansion Coefficients of VO2 near   #Accurate X-ray Determination of the Lattice Parameters and the 000513
cooling)*   #Flux Growth and Characterization by X-ray Topography of Rare Earth Arsenates (REAsO4, slow 000900
X-Ray and Dielectric Investigations of Pb2CrO5 (melt) 000979
of Trigonal Weberite (flux)* # X-Ray and Neutron Diffraction of Na2MnFeF7: A Variety 001033
EuSb2Se4 (vapor transport)* #Growth and X-Ray Analysis of Single Crystals of EuSb4S7 and 000022
#Crystal Growth of CdTe for X-Ray Detectors (review)* 001053
Europium beta-Alumina (NaAl11O17, KAl11O17, RbAl11O17, X-Ray Diffuse Scattering from Alkali, Silver, and 000621
Setting for Measuring Local Lattice Mismatches   #An X-Ray Double-Crystal Method Utilizing Non-Parallel 000704
Mixed Crystals of NaCl and KCl (NaCl-KCl   #The Growth, X-Ray Studies and Microstructure of Solution-Grown 000407
#Growth and X-Ray Study of Cu2SrGeS4 and Cu2BaGeS4 (flux)* 000972
theory)* # X-Ray Study of LEC-Grown InP Crystals (Czochralski, 000614
Telluride (CdxHg1-xTe, Bridgman)*   #An X-Ray Topographic Assessment of Cadmium Mercury 000158
Crystals*   #Sublimation Growth and X-Ray Topographic Characterization of CdTe Single 000148
Hydrothermal Na2CoGeO4 Single Crystals in Relation to # X-Ray Topographic Study of Growth Defects in 000233
Single Crystals (solution)* X-Ray Topographic Study of Growth Sector Strains in KDP 000092
of Dilute Binary   #Observation by Chemical Attack and X-Ray Topography of Cellular Structures of Monocrystals 000210
#In Situ X-Ray Topography of Epitaxial Ge Layers During Growth* 000363
(Cu9Al4)*   #Spherical Monocrystals for X-Ray Work Obtained by Plasma Remelting of Alloy Powder 000043
and Properties of Niobium Chalcogenide Halides, NbX2Y2(X=S, Se - Y=Cl, Br, I) (vapor transport)*   Structures, 000795
Cooling Rate #Improvement of Crystal Composition in Ga1- xAlxAs LPE Layers Grown under Conditions of Constant 000249
#Molecular Beam Epitaxy of Ge and Ga1- xAlxAs Ultra Thin Film Superlattices* 000761
Structures in the (GaAl)As System by LPE (Ga1- xAlxAs, flux)*   System for the Growth of Multilayered 000892
LPE Growth from a Ternary Component Solution (Ga1- xAlxAs, flux, theory)*   and Solid Compositions During 000970
#Periodic Doping Structure in GaAs (GaAs, Ga1- xAlxAs, molecular beam epitaxy, review)* 000248
Surface and Crystal Physics Studied by MBE (GaAs, Ga1- xAlxAs, review of MBE process)*   #Semiconductor 000055
Integrated Optical Devices Fabricated by MBE (GaAs, Ga1- xAlxAs, review, film)* # 000152
Their Growth in an Organometallic VPE System (Ga1- xAlxAs, GaAs, vapor transport)* Epitaxial Layers During 000974
by MBE and Their Characterization (GaAs, Ga1- xAlxAs, In1-xGaxAs, GaSb1-yAsy)*   Superlattices 000176
of Lithium Ferrite-Aluminate Films (flux, Li0.5Fe2.5- xAlxO4)*   #Liquid Phase Epitaxial Growth 001017
#VPE Growth of AlxGa1- xAs (theory)* 000918
of Vapor Phase Growth - Citing an Instance of GaxIn1- xAs (theory)*   #A Simple Analysis 000673
#Organometallic VPE Growth of AlxGa1- xAs (vapor transport)* 000919
#Surface Processes Controlling the Growth of GaxIn1- xAs and GaxIn1-xP Alloy Films by MBE (kinetics)* 000301
#The Growth of GaxIn1- xAs on (100) InP by Liquid-Phase Epitaxy (flux)* 000754
#An LPE Growth of Thick InxGa1- xAs* 000506
#Preparation and Properties of Bulk GaxIn1- xAs* 000552
of III-V Heterostructures Formed by LPE (GaxAl1- xAs)*   #Growth and Dissolution Kinetics 000894
of the Optical Properties of LPE InxGa1- xAsyP1-y Thin Layers Grown on InP (film)* 001098
in Vapor-Grown Mixed Cadmium Halide Crystals (CdI2- xBrx, CdI2-xClx)*   #Growth and Polytypism 000870
and Characterization of Mixed Sr:Ca Tartrates (Sr1- xCazC4H4O6.4H2O, gel)*   #Growth 000747
#Electrical Properties of Pb1- xCdxS Epitaxial Films (vapor transport)* 000439
#Metastable Pb1- xCdxS Epitaxial Films. I. Growth and Physical Properties 000909
#Metastable Pb1- xCdxS Epitaxial Films. II. Electrical Properties* 000910
#Optically Pumped LPE-Grown Hg1- xCdxTe Lasers (films, flux)* 000368
Mercury Cadmium Telluride Single Crystals (Hg1- xCdxTe, Bridgman)* Method for the Growth of Homogeneous 000292
Grown Mixed Cadmium Halide Crystals (CdI2-xBrx, CdI2- xClx)*   #Growth and Polytypism in Vapor- 000870
Optical and Photoconducting Properties of Fe2- xCrxO3 (vapor transport, electrolytic deposition)* 000624
#Formation of Mixed Crystals in the System CuxZn1- xCr2Se4* 000722
Emission (LaB6, EuB6, CeB6, BaB6, (La,Eu)B6, La1- xEuB6, (La,Y)B6, La,Ba)B6, (La,Cs)B6, (Eu,Y)B6, (Eu,Ba) 000727
Layer- Type Compounds K2Mn1-xMxF4 (M + Fe, Co) (K2Mn1- xFexF4, slow cooling)*   on the Antiferromagnetic 000106
Properties of (Mn,Cr)P and (Mn,Fe)P Compounds (Mn1- xFexP, x = 0 to 0.15, theory)*   #Magnetic 000430
Content of Ferrite-Forming Oxides (NixZn1- xFe2O4, flux)* NiO-Fe2O3 Melts with a Nonstoichiometric 000107
and Lattice Mismatch Effects in LPE-Grown In1- xGaxAs on InP Substrate*Composition Latching Phenomenon 000952
Grown Layers in Liquid Phase Epitaxial Growth (In1- xGaxAs)*   #Quasi- 000951
MBE and Their Characterization (GaAs, Ga1-xAlxAs, In1- xGaxAs, GaSb1-yAsy)*   #Semiconductor Superlattices by 000176
#Spectroscopic Properties of Er3Al5- xGaxO12 Films Obtained by Liquid-Phase Epitaxy* 000245
Iron Garnets in a PbO/B2O3 Solution (Y3Fe5O12, RExY3- xGayFe5-yO12, flux)* #A Solubility Model for Rare-Earth 001022
#Photoferromagnetic Effect and Photoconductivity of Cd1- xHgxCr2Se4 Single Crystals (vapor transport, melt)* 001037
#Electroless Deposition of Epitaxial Pb1- xHgxS Films* 000869
Deposition of Semiconductor Films (PbS, PbSe, Pb1- xHgxS, solution)*   #Electroless 000868
in Crystals of Cadmium Mercury Telluride (Cd1- xHgxTe, slow cooling)*   on the Compositional Variations 000074
of Crystals of Yttrium Holmium Aluminum Garnets ((Y1- xHox)3Al5O12, Czochralski)*   and Stimulated Emission 000046
Crystals in Silica Gel (zeolite, Na0.5Al5Si6O14. xH2O)*   #Growth of Fibrous Sodium Aluminosilicate 000444
#Liquid Phase Epitaxial Growth of Ga1- xInxSb on GaSb by Stepwise Grading (melt)* 000803
#Vapor-Phase Growth of Epitaxial Ga1- xInxSb Using Open-Tube Flow System (vapor transport)* 000681
on the Antiferromagnetic Layer- Type Compounds K2Mn1- xMxF4 (M + Fe, Co) (K2Mn1-xFexF4, slow cooling)* 000106
Bismuth Niobate Single Crystals by RF Sputtering (KxLi1- xNbO3)*   Niobate Single Crystals on Potassium 000009
Niobate Single Crystals by the EGM Technique (KxLi1- xNbO3)*   Single-Crystal Films on Potassium Bismuth 000011
of Substituted Yttrium Iron Garnets Y3Fe5-x(Ga,Al) xO12 by the Floating Zone Method*#Single Crystal Growth 000879
#Crystal Growth and Phase Transitions in the NbxCr1- xO2 (x greater than 0.5) System (vapor transport)* 000095
Crystal Structure of a New Mixed Oxide "FeV3O8" (FexV1- xO2 - x approx. = 0.25) (melt, slow cooling)*   and 000661
#Structural Phase Transition in Mixed Crystals WxMo1- xO3 (melt)* 000824
and Crystal Growth of Lead Phosphovanadates (Pb3(PxV1- xO4)2, Pb3(PO4)2)*   #Phase Diagram 000987
Properties of Single Crystals of Solid Solutions GaxIn1- xP (GaP, InP)*   #Preparation and 000119

PERMUTED TITLE INDEX

PERMUTED TITLE INDEX

of the Crystallization Kinetics and Growth Mechanism of | Y3Al5O12 Crystals on Oriented Seeds in Melt-Solutions (| 000093
Oriented Seeds (#Kinetic Characteristics of Growth of | Y3Al5O12 Single Crystals from Melt-Solutions on | 000978
Y3Al5O12, melt, slow cooling)* #Melt Growth of Nd: | Y3Al5O12(Nd:YAG) Using the Heat Exchange Method (HEM) (| 000167
films of Aluminum Yttrium Garnet (RE substituted | Y3Al5O12, flux)* #Growing Optical | 000044
Y3Al5O12(Nd:YAG) Using the Heat Exchange Method (HEM) (| Y3Al5O12, melt, slow cooling)* #Melt Growth of Nd: | 000167
#Phase Equilibria in Solvent-Garnet Systems (flux, | Y3Al5O12, Y3Fe5O12)* | 000159
by the Floating Zone Method (Y3Fe5O12, Y3Fe4Ga5O12, | Y3Fe4Al5O12)* in Substituted YIG Single Crystals Grown | 000483
Crystals Grown by the Floating Zone Method (Y3Fe5O12, | Y3Fe4Ga5O12, Y3Fe4Al5O12)* in Substituted YIG Single | 000483
Crystal Growth of Substituted Yttrium Iron Garnets | Y3Fe5-x(Ga,Al)xO12 by the Floating Zone Method* #Single | 000879
Equilibria in Solvent-Garnet Systems (flux, Y3Al5O12, | Y3Fe5O12)* #Phase | 000159
Iron Garnet Crystals from High Temperature Solutions (| Y3Fe5O12)* #Growth of Bulk Yttrium | 000977
Single Crystals Grown by the Floating Zone Method (| Y3Fe5O12)* #Evaluation of Yttrium Iron Garnet | 000480
with a #Influence of Density of Liquid Phase in the | Y3Fe5O12-BaO.B2O3 System on Convection of a Solution | 000048
Shape Transitions in Czochralski Grown YAG Crystals (| Y3Fe5O12, theory)* #Interface | 000645
Ferrite Materials from Lead Flux with Addition of V2O5 | Y3Fe5O12, Ho3Fe5O12, Dy3Fe5O12, Er3Fe5O12, | 000565
for Rare-Earth Iron Garnets in a PbO/B2O3 Solution (| Y3Fe5O12, RExY3-xGayFe5-yO12, flux)**A Solubility Model | 001022
YIG Single Crystals Grown by the Floating Zone Method (| Y3Fe5O12, Y3Fe4Ga5O12, Y3Fe4Al5O12)* in Substituted | 000483
of Niobium and Rare-Earth Elements (Ln3NbO7) (Ho3NbO7, | Y3NbO7, Er3NbO7, Dy3NbO7, La3NbO7, Y3TaO7, Er3TaO7, | 001097
(Ln3NbO7) (Ho3NbO7, Y3NbO7, Er3NbO7, Dy3NbO7, La3NbO7, | Y3TaO7, Er3TaO7, Gd3TaO7)* and Rare-Earth Elements | 001097
Fibrous Sodium Aluminosilicate Crystals in Silica Gel (| zeolite, Na0.5Al5i6O14.xH2O)* #Growth of | 000444
Fe2O3 Melts with a Nonstoichiometric #Growth of Nickel- | Zinc Ferrite Single Crystals from BaO-B2O3-ZnO-NiO- | 000107
#Crystal Structure of the | Zinc Iodide ZnI2 (vapor transport)* | 000300
ZnO)* #Real Structure of Heteroepitaxial | Zinc Oxide Films Grown on Sapphire (vapor transport, | 000526
from Copper/alpha-copper-Zinc Whiskers to Dendrites in | Zinc Reduction Growth (Cu, alpha-Cu-Zn)* Change | 000481
transport, sublimation)* #Growth of Single Crystals of | Zinc Selenide from the Vapour Phase (ZnSe, vapor | 000214
Pressure (ZnSe, Bridgman)* #Formation of Defects in | Zinc Selenide Crystals Grown from the Melt Under Argon | 000477
ZnS, flux)* #Growth of Single Crystals of Cubic | Zinc Sulphide by Double Decomposition Reaction in Melt (| 000328
#Vapor Growth of | Zinc Telluride by Closed Tube Method (ZnTe, epitaxy)* | 000700
Cu, #Morphological Change from Copper/alpha-copper- | Zinc Whiskers to Dendrites in Zinc Reduction Growth (| 000481
ZrSiO4)* #Growth of | Zircon Single Crystals from a Solution in a Melt (| 000344
eutectics, directional #Growth from Skull-Melting of | Zirconia-Rare Earth Oxide Crystals (ZrO2-RE2O3, | 000629
Boer, Van Arkel, vapor #Single Crystal Growth of | Zirconium Carbide by a Modified Hot-Filament Method (de | 000927
(ZrC)* #Synthesis of Nonstoichiometric | Zirconium Carbide Whiskers by Chemical Vapor Deposition | 000674
(ZrC)* #Catalytic Effect of Nickel on the Growth of | Zirconium Carbide Whiskers by Chemical Vapor Deposition | 000958
electron beam melting, #Elastic Constants of Niobium- | Zirconium, Hafnium, and Tungsten Alloys (Nb-Zr, Nb-Hf, | 000307
Na3HZr[GeO4]2)* #Hydrothermal Crystallization of Sodium | Zirconogermanates (Na2ZrGeO5, Na2ZrGe2O7, Na4Zr2Ge3O12, | 000230
#Liquid Phase Epitaxial Growth of | Zn and S Doped GaAs* | 000416
Characterisation of Large Single Crystals of ZnS and | Zn(S,Se)* #Vapour Growth and Defect | 000816
#Preparation, Analysis and Properties of High Purity | Zn(1-x)HgxTe Alloys (distillation)* | 000792
to Dendrites in Zinc Reduction Growth (Cu, alpha-Cu- | Zn)* Change from Copper/alpha-copper-Zinc Whiskers | 000481
of the fcc Phase in Uni-Directionally Solidified | Zn-Al Eutectic* #Crystallography | 000421
#Liquid-Phase Epitaxy of ZnSe from | Zn-Ga Solution (film)* | 000308
#Epitaxial Growth of ZnS by | Zn-S-H2 CVD Method* | 000615
Relationships Between and alpha- and beta-phases (Cu- | Zn, theory)* Method. I. Morphology and Crystallographic | 000372
the Quarternary Chalcogenides Cu2(I)B(II)C(IV)X4 (BII= | Zn, Mn, Fe, Co - CIV = Si, Ge, Sn - X = S,Se) (| 000357
#The Growth of CdAs2 and | ZnAs2 Single Crystals from the Vapor Phase* | 000598
#Magnetic Properties of CdCr2Se4- | ZnCr2Se4 Single Crystals (melt)* | 000586
(TST) Technique: A Modified Bridgman Method (| ZnF2, CoF2, BaF2, MgF2, KY3F10)* Screen Translation | 000541
#Flux Growth of | ZnGa2O4 Single Crystals* | 001018
#Synthesis and Characterization of the Ternary Compound | ZnGeP2 (melt, slow cooling, high pressure, Bridgman)* | 000336
#Raman and Infrared Spectra of CdIn2S4 and | ZnIn2S4 (vapor transport)* | 001004
#Crystal Structure of the Zinc Iodide | ZnI2 (vapor transport)* | 000300
Growth in c Direction and Crystallographic Polarity in | ZnO Crystals (vapor transport, film)* #Crystal | 000426
#Growth Mechanism of Hollow | ZnO Crystals from ZnSe. II* | 000427
#Chemical Vapor Deposition of Single-Crystalline | ZnO Film with Smooth Surface on Intermediately | 000711
System* #High Rate Epitaxial Growth of | ZnO Films on Sapphire by Planar Magnetron rf Sputtering | 000882
Film with Smooth Surface on Intermediately Sputtered | ZnO Thin Film on Sapphire (epitaxy)* Crystalline ZnO | 000711
Zinc Oxide Films Grown on Sapphire (vapor transport, | ZnO)* #Real Structure of Heteroepitaxial | 000526
of Nickel-Zinc Ferrite Single Crystals from BaO-B2O3- | ZnO-NiO-Fe2O3 Melts with a Nonstoichiometric Content of | 000107
Sublimation in II-VI Compounds Along their Polar Axis (| ZnO, CdSe, CdS, ZnS, kinetics)* #Crystal Growth and | 000428
and Defect Characterisation of Large Single Crystals of | ZnS and Zn(S,Se)* #Vapour Growth | 000816
Technique* #Growth of ZnSxSe1-x Layers on | ZnS and ZnSe Substrates by a Solid-State-Diffusion | 000045
#Epitaxial Growth of | ZnS by Zn-S-H2 CVD Method* | 000615
Processes Associated with the Epitaxial Growth of | ZnS in H2: The Effects of a Bypass Flow (theory)* | 000558
#Heteroepitaxial Growth of | ZnS on GaP by the Close-Spaced Technique (kinetics)* | 000482
#Vapour-Phase Epitaxial Growth of | ZnS on GaP* | 000613
#Morphology and Structure of | ZnS Grown from the Vapour Phase* | 000874
Zinc Sulphide by Double Decomposition Reaction in Melt (| ZnS, flux)* #Growth of Single Crystals of Cubic | 000328
II-VI Compounds Along their Polar Axis (ZnO, CdSe, CdS, | ZnS, kinetics)* #Crystal Growth and Sublimation in | 000428
#Some Aspects of Polytypism in Crystals (SiC, | ZnS, CdI2, mica, theory, review)* | 000069
Transport* #Growth of Cubic | ZnS, ZnSe and ZnSxSe1-x Single Crystals by Iodine | 000309
Tellurium as a Solvent* #Solution Growth of | ZnS, ZnSe, CdS and Their Mixed Compounds Using | 001065
#Phase Diagram of Pseudobinary System CdTe- | ZnSe (Bridgman)* | 000725
of Crystals of Solid Solutions of the System GaAs- | ZnSe (Bridgman)* #Preparation and Investigation | 000185
#Growth of Cubic ZnS, | ZnSe and ZnSxSe1-x Single Crystals by Iodine Transport* | 000309
#Epitaxial Growth of | ZnSe by Vapor Flow Method* | 000669
#Liquid-Phase Epitaxy of | ZnSe from Zn-Ga Solution (film)* | 000308
#Preparation and Optical Properties of | ZnSe Epitaxial Layers by a Close-Spaced Technique* | 000670
Chemical Vapour #Growth and Characterization of Undoped | ZnSe Epitaxial Layers Obtained by Organometallic | 000114
#Organometallic Vapor Deposition of Epitaxial | ZnSe Films on GaAs Substrates* | 000921
vapor transport)* #Properties of Homogeneously Doped | ZnSe Single Crystals Obtained by a New Growth Method (| 000745
#Growth of ZnSxSe1-x Layers on ZnS and | ZnSe Substrates by a Solid-State-Diffusion Technique* | 000045
#Growth Mechanism of Hollow ZnO Crystals from | ZnSe. II* | 000427
Crystals of Solid Solutions in the Systems Zn3As2- | ZnSe(ZnTe) (Zn3As2, ZnSe, ZnTe, Zn3As2-ZnSe, Zn3As2- | 000726
Electron-Bombarded NaCl Substrates (CdS, CdSe, CdTe, | ZnSe)* of Some II-VI Compound Films Evaporated Onto | 000878
Single Crystals of Zinc Selenide from the Vapour Phase (| ZnSe, vapor transport, sublimation)* #Growth of | 000214
Crystals Grown from the Melt Under Argon Pressure (| ZnSe, Bridgman)* #Formation of Defects in Zinc Selenide | 000477
a Solvent* #Solution Growth of ZnS, | ZnSe, CdS and Their Mixed Compounds Using Tellurium as | 001065

PERMUTED TITLE INDEX

Solutions in the Systems Zn3As2-ZnSe(ZnTe) (Zn3As2, | ZnSe, ZnTe, Zn3As2-ZnSe, Zn3As2-ZnTe)*Crystals of Solid | 000726
Systems Zn3As2-ZnSe(ZnTe) (Zn3As2, ZnSe, ZnTe, Zn3As2- | ZnSe, Zn3As2-ZnTe)* Crystals of Solid Solutions in the | 000726
#Preparation of Photoluminescent and Conductive | ZnSe:I Crystals by Iodine Vapor Transport* | 000775
#Edge Emission in Melt-Grown | ZnSeS1-x (Bridgman)* | 000710
#Molecular Beam Epitaxy of | ZnSexTe1-x (film)* | 001102
#Growth and Physical Properties of | ZnSiP2 (vapor transport, solution)* | 000086
of II-IV-V2 Compounds by Gas Phase Transport, with | ZnSiP2 as an Example* in the Course of Preparation | 001086
Conditions -- Microanalytical Evaluation of #Growth of | ZnSiP2 from Gas Phase Under Non-Stoichiometric | 000082
Phase Epitaxy of II-IV-V2, Chalcopyrite Compounds (| ZnSnAs2, solution)* #Liquid | 000089
State-Diffusion Technique* #Growth of | ZnSxSe1-x Layers on ZnS and ZnSe Substrates by a Solid- | 000045
#Growth of Cubic ZnS, ZnSe and | ZnSxSe1-x Single Crystals by Iodine Transport* | 000309
Heater Method (vapor transport)* #Growth of High-Purity | ZnTe Single Crystals by the Sublimation Travelling | 000938
Crystals of Solid Solutions in the Systems Zn3As2-ZnSe(| ZnTe) (Zn3As2, ZnSe, ZnTe, Zn3As2-ZnSe, Zn3As2-ZnTe)* | 000726
ZnSe(ZnTe) (Zn3As2, ZnSe, ZnTe, Zn3As2-ZnSe, Zn3As2- | ZnTe)*Crystals of Solid Solutions in the Systems Zn3As2- | 000726
#Vapor Growth of Zinc Telluride by Closed Tube Method (| ZnTe, epitaxy)* | 000700
in the Systems Zn3As2-ZnSe(ZnTe) (Zn3As2, ZnSe, | ZnTe, Zn3As2-ZnSe, Zn3As2-ZnTe)* of Solid Solutions | 000726
Electron Microscopy Observations of Precipitates in | ZnTe: Shape of the Tellurium Solidus Line (Bridgman)* | 000098
#Vapor Growth of | Zn2GeO4 Single Crystals* | 000424
MnxZn1-x)FeO4 by Liquid Phase Epitaxy on Bridgman Grown | Zn2TiO4 Substrates (flux)* #Thin Film Growth of (| 000216
of Single Crystals of Solid Solutions in the Systems | Zn3As2-ZnSe(ZnTe) (Zn3As2, ZnSe, ZnTe, Zn3As2-ZnSe, | 000726
in the Systems Zn3As2-ZnSe(ZnTe) (Zn3As2, ZnSe, ZnTe, | Zn3As2-ZnSe, Zn3As2-ZnTe)* Crystals of Solid Solutions | 000726
Zn3As2-ZnSe(ZnTe) (Zn3As2, ZnSe, ZnTe, Zn3As2-ZnSe, | Zn3As2-ZnTe)*Crystals of Solid Solutions in the Systems | 000726
of Solid Solutions in the Systems Zn3As2-ZnSe(ZnTe) (| Zn3As2, ZnSe, ZnTe, Zn3As2-ZnSe, Zn3As2-ZnTe)* Crystals | 000726
#Optical Properties of | Zn3P2 (evaporation)* | 000278
#The Compound | Zn4(P2S6)3 (melt)* | 000131
#Crystal Structure of | Zn4(P2S6)3 (melt)* | 000132
#Growth & Etching of InBi Single Crystals (melt, | zone melting)* | 000110
#Growth of Chromium Silicide, Cr3Si, Crystals (float | zone melting)* | 000441
of a New High-Intensity Silicon Solar Cell (Si, float | zone melting)* #Performance | 000303
of Bi-Sb Solid Solutions in Capillaries (| zone melting)* #Crystallization | 001035
Single Crystal with Homogeneous Compositions (float | zone melting)* #Preparation of TiCx | 001092
Strontium Cobaltate Single Crystals (SrCoO2.70, float | zone melting)* #Growth and Properties of | 000955
Growth and Properties of Beta-Boron Crystals (B, float | zone melting)* #Some Peculiarities in the | 000967
of ZrB2 by the de Haas van Alphen Effect (float | zone melting)* #Fermi Surface Measurements | 000964
Whiskers (hydrogen reduction of ferrous halides, float | zone melting)* Growth of alpha-Fe Single Crystal | 000318
and Niobium Carbide Nitride (NbN, NbN-NbC, NbxNyCz, | zone melting)* #Crystal Growth of Niobium Nitride | 000192
Tungsten Alloys (Nb-Zr, Nb-Hf, electron beam melting, | zone melting)* of Niobium-Zirconium, Hafnium, and | 000307
Grown from Near-Stoichiometric Compositions (doped, | zone melting)* of Copper Indium Ditelluride Crystals | 000220
Gravity Environment (melt, solution, vapor transport, | zone melting, doped, review)*Materials Behaviour in Low | 001070
Data: Theoretical (directional crystallization, | zone melting, eutectics)*Coefficients from Experimental | 001080
Number at Zero Gravity (Si, Al2O3, theory, float | zone melting, kinetics, Si) Zone Geometry and Prandtl | 000175
Nb-H Interstitial Alloys (electron beam melting, float | zone melting, review)* #Properties and Synthesis of | 001072
and Growth Peculiarities of Tl-Se-TlInSe2 (TlSe, | zone melting, slow cooling)* #Structure | 000361
#Single-Crystal Growth of Tantalum Diboride (TaB2, | zone melting, solution)* | 000680
for Zone Crystallization Experiments in Spacelab (CdTe, | zone melting, sublimation)* Ellipsoid Mirror Furnace | 000276
Pure Rotation Driven Flows (Si, Al2O3, kinetics, float | zone melting, theory)* in Floating Zone Melting. I. | 000173
Single Crystals with Low Dislocation Density (Al, | zone melting, theory)* Method of High Purity Aluminum | 000228
Zone-Refined Metallurgical Silicon (Si, Czochralski, | zone melting, vapor transport)* #Solar Cells from | 000195
#Liquid Encapsulation Zone-Refining of PbS (| zone melting, vapor transport, sublimation)* | 000583
(review, melt, flux, Czochralski, Kyropoulos, float | zone melting, Bridgman)* Liquids at High Temperatures | 000139
Sheet and Their Effects on Solar Cell Processing (| zone melting, Si)* of Rapidly-Grown RTR Silicon | 000546
Bicrystals of Nickel Oxide (NiO, arc image, float | zone, Verneuil)* #Growth of Oriented | 000239
#Power Required to Form a Floating | Zone and the Zone Shape (theory)* | 000492
Effect (Ge)* #Observations of the Movement of a Molten | Zone in Germanium Under the Influence of the Peltier | 000061
zone melting, #A Double-Ellipsoid Mirror Furnace for | Zone Crystallization Experiments in Spacelab (CdTe, | 000276
Convection in Floating Zone Melting: Influence of | Zone Geometry and Prandtl Number at Zero Gravity (Si, | 000175
Crystals (Al2O3, #Maximum Stable Zone Length in Float- | Zone Growth of Small-Diameter Sapphire and Silicon | 000479
Sapphire and Silicon Crystals (Al2O3, #Maximum Stable | Zone Length in Float-Zone Growth of Small-Diameter | 000479
#Experiments on Surface Tension Driven Flow in Floating | Zone Melting (NaNO3, theory)* | 000848
in Fe-3wt%Si Single Crystals Grown by Floating | Zone Melting at Various Growth Rates*#Growth Striations | 001028
Growth (NaNO3, PbCl2, organics)* #An Inexpensive | Zone Melting Furnace for Purification and Crystal | 000871
#Single Crystal Growth by the Horizontal Levitation | Zone Melting Method (DyFe2, TbFe2, Nd, Pr, Ce, cold | 000620
of Silicon-Germanium Solid Solutions by Electron-Beam | Zone Melting Without a Crucible (Si-Ge)*Single Crystals | 000820
#Computer Simulation of Convection in Floating | Zone Melting. I. Pure Rotation Driven Flows (Si, Al2O3, | 000173
Flows (#Computer Simulation of Convection in Floating | Zone Melting. II. Combined Free and Rotation Driven | 000174
Number at Zero #Thermocapillary Convection in Floating | Zone Melting: Influence of Zone Geometry and Prandtl | 000175
of Fayalite (Fe2SiO4) Single Crystals by the Floating | Zone Method (arc image)* #Growth | 000953
of Cobalt Single Crystals by the Electron Beam Floating- | Zone Method (Co)* #Growth | 000381
Iron Garnet Single Crystals Grown by the Floating | Zone Method (Y3Fe5O12)* #Evaluation of Yttrium | 000480
Substituted YIG Single Crystals Grown by the Floating | Zone Method (Y3Fe5O12, Y3Fe4GaO12, Y3Fe4AlO12)* in | 000483
#Growth of Mg2TiO4 Single Crystals by the Floating- | Zone Method* | 000880
#Growth of FeTiO3 (Ilmenite) Crystals by the Floating- | Zone Method* | 000954
Direction of Pure Cu Single Crystals by the Floating- | Zone Method* #Preferred Growth | 000380
(Ca2MgSi2O7) and Gehlenite (Ca2Al2SiO7) by the Floating | Zone Method* #Single Crystal Growth of Akermanite | 000404
Yttrium Iron Garnets Y3Fe5-x(Ga,Al)xO12 by the Floating | Zone Method* #Single Crystal Growth of Substituted | 000879
#Fractionation of Stable Isotopes and Impurities During | Zone Recrystallization of Metals and Preparation of | 000994
Shape of the Crystallization Front During Electron-Beam | Zone Refining of beta-B Without a Crucible* # | 000968
Solute Redistribution During a Modified Form of | Zone Refining-Cascade Purification (metals)*Theoretical | 000201
#Power Required to Form a Floating Zone and the | Zone Shape (theory)* | 000492
in Using Kilohertz Radio Frequencies for Float | Zone Silicon Crystals (Si, theory)* #Limitations | 000358
Technique (InSb)* #Horizontal | Zone-Melting of Indium Antimonide in Vacuum. I. Basic | 000712
zone melting, vapor transport)* #Solar Cells from | Zone-Refined Metallurgical Silicon (Si, Czochralski, | 000195
sublimation)* #Liquid Encapsulation | Zone-Refining of PbS (zone melting, vapor transport, | 000583
#The Adsorption of | Zr onto W(100) Surfaces* | 000221
of Niobium-Zirconium, Hafnium, and Tungsten Alloys (Nb- | Zr, Nb-Hf, electron beam melting, zone melting)* | 000307
#Epitaxial Growth of Ferroelectric PLZT ((Pb,La)(| Zr,Ti)O3) Thin Films* | 000415
melting)* #Fermi Surface Measurements of | ZrB2 by the de Haas van Alphen Effect (float zone | 000964
Carbide Whiskers by Chemical Vapor Deposition (| ZrC)* #Synthesis of Nonstoichiometric Zirconium | 000674

PERMUTED TITLE INDEX

Abbaschian, G.J.	1	Rate Equations for Dislocation-Free and Dislocation-Assisted Grow
Abbona, F.	2	Growth Morphology and Crystal Habit of Struvite Crystals (MgNH4PO
Abdulkhadar, M.	3	New Gel Method for Growing Large Needles and Single Crystals of L
Abduragimov, G.A.	4	Effect of Crucible Shape on a Crystallization Front (Bridgman, Ki
Abe, M.	5	Magnetic and Mossbauer Studies of FeNi2BO5 and FeNiBO4 (ferrites,
	6	Mossbauer Study of FeMo2S4 (vapor transport)
Abernathy, J.R.	7	Congruent (Diffusionless) Vapor Transport (theory, kinetics)
Abrahams, S.C.	339	Calcium Orthovanadate, Ca3(VO4)2 - A New High-Temperature Ferroel
Abrasimov, N.V.	966	Preparation of Silicon Plates by Pulling from a Molten Bath Accor
Acharya, S.	8	Electronic Bandgap Measurements of SnS2 Polytypes (vapor transpor
Acrivos, J.V.	85	Kinetic Studies of the Intercalation of 2H-NbSe2 with Hydrazine b
Adachi, G.	580	Synthesis and Characterization of Europium(II)-Holoborates, Eu2B5
Adachi, M.	9	Epitaxial Growth of Potassium Lithium Niobate Single-Crystal Film
	10	Elastic and Piezoelectric Properties of Potassium Lithium Niobate
	11	Epitaxial Growth of Potassium Lithium Niobate Single-Crystal Film
Agarwal, M.K.	12	Growth of Tungstenite Single Crystals by Direct Vapour Transport
	13	Growth Conditions and Crystal Structure Parameters of Layer Compo
Agnihotri, O.P.	14	Single Crystal Growth of Stannous Selenide (SnSe, Bridgman)
Agrawal, V.K.	870	Growth and Polytypism in Vapor-Grown Mixed Cadmium Halide Crystal
Aguilo, M.	804	Unstable Growth of ADP Crystals (NH4H2PO4, theory, solution, kine
Ahlers, G.	505	Heat Capacity of EuS near the Ferromagnetic Transition (Bridgman,
Ahmed, S.	835	Metal-Insulator Transitions and Phase Diagram of (Ti1-xVx)407: E
Aikami, T.	874	Morphology and Structure of ZnS Grown from the Vapour Phase
Ainger, F.W.	800	The Effect of Impurities on the Piezoelectric Properties of Lithi
Akahane, T.	15	Positron Annihilation in ReO3 (vapor transport)
Akasaki, I.	613	Vapour-Phase Epitaxial Growth of ZnS on GaP
Akhin'ko, A.T.	723	Relation Between Growth Conditions and Dislocation Structure in G
Akimov, S.V.	16	Acoustooptic Characteristics of LiBi(MoO4)2 (Czochralski)
Akita, K.	506	An LPE Growth of Thick InxGa1-xAs
Al-Jassim, M.	126	Nature, Origin and Effect of Dislocations in Epitaxial Semiconduc
Al-Saddawi, S.D.	558	Mass Transport Processes Associated with the Epitaxial Growth of
Aldred, A.T.	474	Structural and Magnetic Properties of "One-Dimensional" Barium Va
Aleksandrov, L.N.	17	Epitaxial Stratified Growth (kinetics, theory)
	18	Computer Model of Autoepitaxial Growth of Ge Films (theory)
	19	The Mechanism of Silicon Epitaxial Layer Growth from Ion-Molecula
Aleksandrov, V.I.	20	Synthesis and Crystal Growth of Refractory Materials by RF Meltin
Aleonard, S.	21	Crystal Structure of beta-KEr3F10 (melt, slow cooling)
Aliev, O.M.	22	Growth and X-Ray Analysis of Single Crystals of EuSb4S7 and EuSb2
	817	Synthesis, Preparation of Single Crystals, and Physicochemical In
Alimova, I.A.	23	Influence of Isotopic Composition on the Density and Molecular Vo
Alpen, U.v.	24	Li3N: A Promising Li Ionic Conductor (electrolytic deposition, r
	25	Telluride Halides of Copper with High Ionic Partial Conductivity.
Amblard, J.	26	Electrocrystallization. Fundamental Aspects (review, theory)
Ammer, S.A.	27	Doping of Silicon Crystals with Phosphorus During Growth from the
Amouric, M.	70	Solubility of Phlogopite in Basic Aqueous Solutions Under Hydroth
Andersen, N.H.	87	A New Organic Low Temperature Conductor: HMTSF-TNAP (solution, or
Anderson, B.J.	28	Influence of Environmental Saturation and Electric Field on Growt
Anderson, J.B.	29	The Crystal Structure of Cu4(PO4)2O (flux)
Andersson, Y.	30	Properties of Fe2P Crystals Prepared from a Liquid Copper Medium
Andreev, A.A.	778	Anisotropy of Ion Transport in Stoichiometric Rutile Single Cryst
Andres, K.	31	On the Resistivity of the Magnetically Ordered Ground State of th
Andrews, A.M.	803	Liquid Phase Epitaxial Growth of Ga1-xInxSb on GaSb by Stepwise G
Angilello, J.	865	Preparation and Characterization of Ruthenium Dioxide Crystals (R
Ansel, R.	754	The Growth of GaxIn1-xAs on (100) InP by Liquid-Phase Epitaxy (fl
Antony, A.V.	442	Single Crystal Growth of Ferroelectric Potassium Dihydrogen Ortho
	443	Nucleation in Supersaturated Potassium Dihydrogen Orthophosphate
Anzai, H.	32	Growth of Large Crystals of Tetramethyltetrathiafulvalenium Tetra
Aoki, K.	33	Orientation and Temperature Dependence of the Flow Stress in the
Aoki, M.	1065	Solution Growth of ZnS, ZnSe, CdS and Their Mixed Compounds Using
Aoki, R.	731	Raman Scattering in 3R-NbS2 (vapor transport, theory)
Aono, T.	34	Magneto-Seebeck Effect of Pb-Doped Bi-Sb (Bridgman)
Appalasami, S.	810	Kinetics of Crystal Growth of Calcium Tungstate from Solutions in
	811	Kinetics of Crystal Growth of Strontium Tungstate from Solutions
Aquilano, D.	302	Growth Morphology of Weddellite (CaC2O4.2XH2O, gel)
Aragon, R.	370	Skull Melter Growth of Magnetite (Fe3O4)
Arai, S.	35	Conditions of LPE Growth for Lattice Matched GaInAsP/InP DH Laser
Arai, T.	731	Raman Scattering in 3R-NbS2 (vapor transport, theory)
Arbus, A.	36	Crystal Structure of RbEu3F10 (melt)
	37	The Systems RbF-EuF3 and CsF-EuF3. Study of the Phases RbLnF4 an
Arend, H.	38	Solution Growth of RbAg4I5 Crystals
	39	Layer Perovskites of the (CnH2n+1NH3)2MX4 and NH3(CH2)mNH3MX4 Fam
	295	On the Preparation of Pure, Doped and Reduced KNbO3 Single Crysta
	545	Crystal Quality of Gel Grown CaHPO4.2H2O Crystals
Arends, J.	40	Preparation of Pure Hydroxyapatite Single Crystals by Hydrotherma
Ariumi, T. (ed.)	946	Vapour Growth and Epitaxy (review)
Arizumi, T.	41	Some Aspects of the Epitaxial Vapour Growth of Semiconductors: El

Armas, B.   42   The Vapour Transport of NbB2 and TaB2 (kinetics, theory)
Arnberg, L.   43   Spherical Monocrystals for X-Ray Work Obtained by Plasma Remeltin
Arora, S.K.   747   Growth and Characterization of Mixed Sr:Ca Tartrates (Sr1-xCazC4H
Arsen'ev, P.A.   44   Growing Optical films of Aluminum Yttrium Garnet (RE substituted
Asami, S.   45   Growth of ZnSxSe1-x Layers on ZnS and ZnSe Substrates by a Solid-
Ashurov, M.Kh.   46   Structure, Spectroscopy, and Stimulated Emission of Crystals of Y
Astier, R.   766   Crystal Chemistry of a Tellurate IV of Indium III: In2Te309
Astles, M.G.   47   Liquidus Measurements in the Pb-Sn-Te System (Pb1-xSnxTe, PbTe, f
Aubert, J.J.   1036 Crystal Structure of CuTa206 (flux)
Aubree, J.   987   Phase Diagram and Crystal Growth of Lead Phosphovanadates (Pb3(Px
Authier, A.   233   X-Ray Topographic Study of Growth Defects in Hydrothermal Na2CoGe
Averbuch-Pouchot, M.T.   260   Structure of Silver Dichromate: Ag2Cr207 (solution)
Averbuch, M.T.   986   Crystal Structure of Monophosphate AgCoPO4
Avgustimov, V.L.   306   Preparation of Films of IV-VI Compounds in a Quasi-Sealed Volume
Avignon, M.   610   Magnetic and Electrical Properties of BaVS3 and BaVxTi1-xS3 (flux
Avvakumova, L.A.   48   Influence of Density of Liquid Phase in the Y3Fe5012-Ba0.B203 Sys
Ayabe, M.   461   Improvements in SSD (synthesis, solute diffusion) Growth of GaP
Ayers, J.D.   217   Preparation of V3Ga Single Crystals by Iodine Transport
Aznarez, J.A.   539   Electron Microscopy Study of Some Molybdenum Oxide Crystals (MoO3
Azuma, N.   49   Growth of a Crystal Needle of Si2ON2 by Nitridization in the Si-S
Babel, D.   1033 X-Ray and Neutron Diffraction of Na2MnFeF7: A Variety of Trigona
Babitsyna, A.A.   50   Properties of CuCr2Se4 Crystals Obtained from Solutions in Melts
                          51   Crystal Growth of CuCr2Se4
Babushkin, A.V.   52   Crystallization of WO2 from the Gas Phase (vapor transport)
Bachmann, K.J.   53   The Growth of In-Rich Bulk Single Crystals of GaxIn1-xPyAs1-y Via
                          54   Melt and Solution Growth of Bulk Single Crystals of Quaternary II
                          552   Preparation and Properties of Bulk GaxIn1-xAs
Bachrach, R.Z.   55   Semiconductor Surface and Crystal Physics Studied by MBE (GaAs, G
Badurski, M.   56   Growth and Phase Transitions in Single Crystals of (KxNa1-x) NbO3
Bagaev, V.N.   746   Dependence of Degree of Perfection of Single Crystals of FeGe2 on
Bagamadova, A.M.   526   Real Structure of Heteroepitaxial Zinc Oxide Films Grown on Sapph
Bagdasarov, Kh.S.   44   Growing Optical films of Aluminum Yttrium Garnet (RE substituted
                          57   Analytical Study of the Thermal Transfer Process in an Installati
Baghdadi, A.   360   Silicon Ribbon Growth via the Ribbon-To-Ribbon (RTR) Technique:
                          546   Properties of Rapidly-Grown RTR Silicon Sheet and Their Effects o
Bagieu-Beucher, M.   58   Crystal Structure of the Phosphate of Trivalent Manganese Mn(PO3)
Bagirov, S.B.   593   alpha-In2Te3 (melt)
Baikov, Yu.A.   59   A Fluctuation Theory of Normal Crystal Growth for Metallic One-Co
                          1110 On the Theory of Normal Growth of Crystals in Binary Systems (kin
Bakhiya, T.N.   823   Heteroepitaxial Structures of Si on Sapphire, Obtained by Vacuum
Balchin, A.A.   805   Some Thermodynamic Considerations for the Growth of Layered Dicha
                          1077 Structural Studies of Compounds in the Series GaSxSe(1-x) (0 less
                          1078 Optimization of Conditions for the Growth of Gallium Selenide and
Balkanov, M.R.   60   Reaction of Germanium (111) Single Crystals with GeBr4 Vapor (Ge,
Ballman, A.A.   339   Calcium Orthovanadate, Ca3(VO4)2 - A New High-Temperature Ferroel
Balooch, M.   61   Observations of the Movement of a Molten Zone in Germanium Under
Ban, V.S.   62   Novel Reactor for High Volume Low-Cost Silicon Epitaxy (Si, vapor
Bando, Y.   63   Mass Spectrometric Study on Chemical Transport of VO2
Bannai, E.   964   Fermi Surface Measurements of ZrB2 by the de Haas van Alphen Effe
                          1092 Preparation of TiCx Single Crystal with Homogeneous Compositions
Bansaragtschin, B.   64   Semiconductor Single Crystals Obtained by the Traveling Solvent M
Barbier, D.   600   Luminescence and Bandgap Studies of High Purity MgxZn1-xTe Ternar
Barclay, J.A.   65   Low Temperature Conductivity of Gd2(SO4)38H2O (crystal) and Dy2Ti
Bardsley, W. (ed.)   66   Crystal Growth (Proceedings of the Third International Summer Sch
Barner, K.   67   The Growth Figures of MnAs (whiskers, theory)
Barns, R.L.   68   Dislocation-Free and Low-Dislocation Quartz Prepared by Hydrother
                          538   Are Icicles Single Crystals? (H20)
Baronnet, A.   69   Some Aspects of Polytypism in Crystals (SiC, ZnS, CdI2, mica, the
                          70   Solubility of Phlogopite in Basic Aqueous Solutions Under Hydroth
Bartels, G.   71   Preparation of Barium Lead Hexa-Aluminate Single Crystal Layers b
                          72   A New Substrate Holder for Liquid Phase Epitaxy
Bartels, W.J.   399   Determination of the Lattice Constant of Epitaxial Layers of III-
Barter, B.H.   73   Growth of Films of CdS Evaporated in Vacuum onto Silicon (vapor t
Bartlett, B.E.   74   The Effects of Growth Speed on the Compositional Variations in Cr
                          75   A Study of Casting in the CdxHg1-xTe System
Bartlett, N.   76   The Synthesis of the First Stage Graphite Salt C8(+)AsF6(-) and I
Basten, J.A.J.   77   Crystallographic and Magnetic Structure of RbFeCl3.2D20 and CsFeC
Bastin, G.F.   78   Crystallography of Aligned Fe-Al Eutectoid (FeAl, FeAl2)
Batalin, G.I.   473   Structural Transitions and Phase Composition of Thin Films of Alu
Batlogg, B.   79   Magnetic Mixed Valent TmSe (melt)
Baud, G.   80   Structure of New Rhenium Apatites Containing Large Anions: Ba10(
                          81   Crystal Structure of La3Re08
                          103   Crystal Structure of Double Oxides of Rhenium. II. The Lanthanum
                          104   High Pressure Synthesis and Structure of a New Pyrochlore Contain
Baudelet, B.   371   Two-Phase (alpha/beta)-Brass Bicrystals Produced by a Solid-Solid
                          372   Two-Phase (alpha/beta)-Brass Bicrystals Produced by a Solid-Solid

Bauer, C.L.	207	Characterization of (001) Tilt Boundaries Produced by Epitaxial G
Baum, H.	82	Growth of ZnSiP2 from Gas Phase Under Non-Stoichiometric Conditio
Baumgartner, O.	83	The Crystal Structure of the Tetragermanate K2Ba[Ge4O9]2
Bauser, E.	84	Development of Depressions and Voids During LPE Growth of GaAs (f
Bayard, M.	189	Lasing and Fluorescence in K5NdLi2F10 (flux)
Beal, A.R.	85	Kinetic Studies of the Intercalation of 2H-NbSe2 with Hydrazine b
Becherer, H.	86	Growth and Physical Properties of ZnSiP2 (vapor transport, soluti
Bechgaard, K.	87	A New Organic Low Temperature Conductor: HMTSF-TNAP (solution, or
Beck, H.P.	88	New Mixed Halide Compounds MFX of Divalent Rare Earths (M = Sm, E
Beckman, O.	30	Properties of Fe2P Crystals Prepared from a Liquid Copper Medium
Bedair, S.M.	89	Liquid Phase Epitaxy of II-IV-V2, Chalcopyrite Compounds (ZnSnAs2
Bednarczyk, E.	912	Crystal Growth of Actinide Dioxides by Chemical Transport
Beinglass, I.	834	State-of-The-Art of Crystal Growth and Nuclear Spectroscopic Eval
Beke, D.L.	90	Connections Between Thermodynamic Quantities and Vacancy and Diff
Bell, M.I.	229	Thermal and Dielectric Properties of LiKSO4 and LiCsSO4
Belorusets, E.D.	19	The Mechanism of Silicon Epitaxial Layer Growth from Ion-Molecula
Belouet, C.	91	Vapour Growth in a Microgravity Environment (review)
	92	X-Ray Topographic Study of Growth Sector Strains in KDP Single Cr
Belov, N.V.	551	Neodymium Incorporation into YAl-Borate Crystals in Preparation f
	592	Hydrothermal Crystallization in the System BaO-GeO2-Na2O-H2O (bar
	902	Crystal Structure of Scandium Metaphosphate Sc(PO3)3 (solution)
Belyaev, L.M.	93	Some Problems of the Crystallization Kinetics and Growth Mechanis
Belyaev, L.M.(ed.)	976	Rost Kristallov iz Rastvorov-Rasplavov (Crystal Growth from High-
Belyaevskii, T.V.	151	Preparation and Physical Properties of Boracite Crystals (Chemica
Belyskiy, N.K.	598	The Growth of CdAs2 and ZnAs2 Single Crystals from the Vapor Pha
Belyustin, A.V.	94	The Mechanism of Growth of KDP Crystals in the Presence of Impuri
Ben-Dor, L.	95	Crystal Growth and Phase Transitions in the NbxCr1-xO2 (x greater
Benesovsky, F.	96	Wolfram (W)
Benkhoucha, R.	97	Crystallization During Polymerization of Lithium Dihydrogen Phosp
Bennema, P.	40	Preparation of Pure Hydroxyapatite Single Crystals by Hydrotherma
	105	Interaction Between Successive Monatomic Step Trains of Different
	182	Fast Growth of Ordered AB Crystals: A Monte Carlo Simulation for
	933	Modes of Vibrations in Step Trains: Rhythmical Bunching (theory)
	1013	Survey of Monte Carlo Simulations of Crystal Surfaces and Crystal
Bensahel, D.	98	Transmission Electron Microscopy Observations of Precipitates in
Benz, K.W.	99	GaSb and InSb Crystals Grown by a Vertical and Horizontal Travell
Berdowski, J.	100	Crystal Growth of the Para- and Antiferroelectric Phases of Deute
Berg, H.	67	The Growth Figures of MnAs (whiskers, theory)
Berg, L.G.	101	Liquidus of the Ternary System Pb-Tl-Te
Berger, R.S.	141	Crystal Growth and Characterization of Lithium Metasilicate (Li2S
Bergeron, C.G.	385	Heat-Flow-Controlled Growth During Li2B4O7 Crystallization (melt)
	794	Crystal Growth of Bi3B5O12 From its Stoichiometric Melt (kinetics
Berliner, L.B.	623	Crystallization of Salts from Supersaturated Solutions: Diffusio
Bernardini, J.	768	Growth of High Perfection Ag-Sn Crystals (in Ag solvent)
Bernardini, J.E.	1073	Czochralski Growth of A15 Structure Intermetallic Compounds (V3Si
Besrest, F.	102	Structure of La2Fe1.76S5 (melt)
Besse, J.-P.	80	Structure of New Rhenium Apatites Containing Large Anions: Ba10(
	103	Crystal Structure of Double Oxides of Rhenium. II. The Lanthanum
Besse, J.P.	81	Crystal Structure of La3ReO8
	104	High Pressure Synthesis and Structure of a New Pyrochlore Contain
Besson, J.M.	187	Growth of Crystalline Slabs of Layered InSe by the Czochralski Me
Bethge, H.	105	Interaction Between Successive Monatomic Step Trains of Different
Bevaart, L.	106	Magnetic and Neutron Scattering Experiments on the Antiferromagne
Bezmaternykh, L.N.	107	Growth of Nickel-Zinc Ferrite Single Crystals from BaO-B2O3-ZnO-N
Bhalla, A.S.	108	Polar Twins in Lithium Metasilicate and Metagermanate (Li2SiO3, L
	109	Crystal Growth of Antimony Sulphur Iodide (SbSi, flux)
	141	Crystal Growth and Characterization of Lithium Metasilicate (Li2S
Bhatt, V.P.	110	Growth & Etching of InBi Single Crystals (melt, zone melting)
	111	Growth and Cleavage of Se-Te Alloy Crystals (Bridgman)
	112	Growth of Hollow Se-Te Whisker Crystals (theory)
Bholagir, A.	871	An Inexpensive Zone Melting Furnace for Purification and Crystal
Bhoskar, B.T.	444	Growth of Fibrous Sodium Aluminosilicate Crystals in Silica Gel (
Bierlein, J.D.	867	Polymorphism in Bi2Sn2O7 (flux)
Billia, B.	113	Irreversible Thermodynamics of the Cellular Growth of a Dilute Bi
	210	Observation by Chemical Attack and X-Ray Topography of Cellular S
Bingham, D.	65	Low Temperature Conductivity of Gd2(SO4)38H2O (crystal) and Dy2Ti
Bisaro, R.	754	The Growth of GaxIn1-xAs on (100) InP by Liquid-Phase Epitaxy (fl
Blackwell, J.C.	892	An Automated System for the Growth of Multilayered Structures in
Blanconnier, P.	114	Growth and Characterization of Undoped ZnSe Epitaxial Layers Obta
Blanzat, B.	756	Optical Properties of Cs2NaBiCl6
Blasse, G.	1012	Luminescence and Ionic Conductivity of Bariumfluorochloride (BaFC
Bleay, J.	115	The Growth of Single Crystals of Some Organic Compounds by the Cz
Bletskan, N.I.	116	Growth of Si Strips by Stepanov's Method
	723	Relation Between Growth Conditions and Dislocation Structure in G
Bloem, J.	117	Problems in the Melt- and Vapor Growth of Silicon for Integrated
	118	Mechanisms of the Chemical Vapour Deposition of Silicon (Si)

	884	Crystal Growth of Boron Monophosphide Using a B2H6-PH3-H2 System
	1015	Surface Morphology of HCl Etched Silicon Wafers (Si)
Blom, G.M. (ed.)	946	Vapour Growth and Epitaxy (review)
Blomen, L.J.M.J.	1025	On the Presentation of Growth Curves for Growth from Solution (th
Bobrov, Yu.A.	151	Preparation and Physical Properties of Boracite Crystals (Chemica
Bochkarev, E.P.	636	Creation of Defects During the Growth of Semiconductor Single Cry
Bochkova, T.M.	294	Change in the Energy Band Structure of Eulytine on Transition to
Bochu, B.	1036	Crystal Structure of CuTa2O6 (flux)
Bodnar', I.V.	119	Preparation and Properties of Single Crystals of Solid Solutions
Bodyachevskii, S.V.	48	Influence of Density of Liquid Phase in the Y3Fe5O12-BaO.B2O3 Sys
	563	Melting and Crystallization of Corundum (alpha-A12O3)
Boettinger, W.J.	521	Crystal Perfection in Czochralski Grown Nickel Single Crystals (N
Bogdanova, V.I.	688	Ag5GaS4 -- A New Phase in the System Ag-Ga-S (AgGaS2)
Bohm, J.	457	Growth, Spectroscopic Investigations, and Some New Stimulated Emi
	458	Growth, Spectroscopy, and Stimulated Emission of Cubic Bi4Ge3O12
Boistelle, R.	2	Growth Morphology and Crystal Habit of Struvite Crystals (MgNH4PO
Boivin, J.-C.	120	Crystal Structure of Ba2Fe6O11 (flux)
Bokii, G.B.	862	Temperature Field Dynamics in the Synthesis of Diamonds (kinetics
Boncheva-Mladenova, Z.	121	Growth of Bi2(SeO4)3 Single Crystals in Silica Gels
	122	Growth of Single Crystals of Ag2SeO4 in Silica Gels
Bonissent, A.	123	Monte Carlo Simulation of the Crystal-Melt Interface of a Lennard
Bonnamy, C.	124	Preparation by Vapor Transport of Monocrystals of AlF3, FeF3, and
Bonner, W.A.	125	Reproducible Preparation of Twin-Free InP Crystals Using the LEC
Bonnerot, J.	171	Low-Temperature Specific Heat and Magnetic Susceptibility of Nonm
Bonnet, M.M.	255	Kinetics of Silicon Growth Under Low Hydrogen Pressure (vapor tra
Boo, W.O.J.	917	Crystalline Vanadium (II) Fluoride, VF2. Preparation, Structure,
Booker, G.R.	126	Nature, Origin and Effect of Dislocations in Epitaxial Semiconduc
Borkowski, B.	127	Chemical Synthesis and Crystal Growth of Laser Quality Praseodymi
Borodin, V.A.	128	The Effect of the Solidification Front Shape on the Temperature
Boswarva, I.M.	73	Growth of Films of CdS Evaporated in Vacuum onto Silicon (vapor t
Boswell, F.W.	129	On the Ordering of Fe Atoms in FexNbS2 (vapor transport)
	1090	Intercalation-Induced Shift of the Absorption Edge in ZrS2 and Hf
Botsaris, G.D.	1038	A New Technique for Simulating Two-Dimensional Nucleation Models
Bottcher, P.	130	Synthesis and Structure of Cs2S2 (hydrothermal)
Bouchetiere, M.	131	The Compound Zn4(P2S6)3 (melt)
	132	Crystal Structure of Zn4(P2S6)3 (melt)
Bougnot, J.	336	Synthesis and Characterization of the Ternary Compound ZnGeP2 (me
Boulanger, D.	133	Crystal Growth and EPR Study of Cr+-Doped MgTe Single Crystals (s
Boulon, G.	36	Crystal Structure of RbEu3F10 (melt)
Bourne, J.R.	134	Solvent Effects and the Growth Kinetics of Ionic Crystals (CaSO4.
Bowers, J.E.	839	LPE Growth of Hg0.60Cd0.40Te from Te-Rich Solution (film, vapor t
Brachtel, G.	135	The First Oxomanganate(II): Na14Mn2O9 = Na14[MnO4]2O
Brandt, G.	241	The Crystal Structure of Triclinic WO3 (sublimation)
Brantov, S.K.	966	Preparation of Silicon Plates by Pulling from a Molten Bath Accor
Braun, D.J.	438	Synthesis and Crystal Structure of Iron Polyphosphide FeP4
Braun, P.D.	136	Local Selective Homoepitaxy of Silicon at Reduced Temperatures Us
Brec, R.	571	Physical Properties and Lithium Intercalates of CrPS4 (slow cooli
Breiter, M.W.	286	Ionic Conductivity in N3O+ Beta Alumina (solution)
Bressel, B.	137	Czochralski Crystal Growth of Rare Earth Tetraborides (YB4, TbB4,
Brezina, B.	138	The Growth of Single Crystals of Langbeinites Rb2Cd2(SO4)3, T12Cd
Briant, J.L.	285	Hydronium Beta" Alumina: A Fast Proton Conductor (0.84Na2O.0.84H
	286	Ionic Conductivity in N3O+ Beta Alumina (solution)
Brice, J.C.	139	Crystal Growth from Liquids at High Temperatures (review, melt, f
Bridenbaugh, P.M.	220	Electrical and Optical Properties of Copper Indium Ditelluride Cr
Broughton, J.Q.	140	A Molecular Dynamics Study of Surface Melting (theory, vapor tran
Brower, W.S.	639	Alkali Ion Exchange Reactions with RbAlSiO4: A New Metastable Pol
Brun, M.K.	141	Crystal Growth and Characterization of Lithium Metasilicate (Li2S
Brunel-Laugt, M.	142	Crystal Structure of Cu4(PO4)2O (flux)
Bruni, F.J.	143	Gadolinium Gallium Garnet (Gd3Ga5O12, review)
Bryskiewicz, T.	144	Investigation of the Mechanism and Kinetics of Growth of LPE GaAs
Bsenko, L.	145	The Crystal Structures of Two Modifications of HfNi3 (melt)
	146	The Crystal Structure of Hf8Ni21 (melt)
	147	The Crystal Structure of Hf3Ni7 (melt)
Buck, P.	148	Sublimation Growth and X-Ray Topographic Characterization of CdTe
Buckley-Golder, I.M.	149	A Theoretical Study of Temperature Distributions During Czochrals
Buder, R.	610	Magnetic and Electrical Properties of BaVS3 and BaVxTi1-xS3 (flux
	835	Metal-Insulator Transitions and Phase Diagram of (Ti1-xVx)4O7: E
Buehler, E.	150	Simple Pressurized Chambers for Liquid Encapsulated Czochralski C
	505	Heat Capacity of EuS near the Ferromagnetic Transition (Bridgman,
	1073	Czochralski Growth of A15 Structure Intermetallic Compounds (V3Si
Bugakov, V.I.	151	Preparation and Physical Properties of Boracite Crystals (Chemica
	733	Synthesis of Single Crystal LnPO4 (vapor transport)
Buhrig, E.	86	Growth and Physical Properties of ZnSiP2 (vapor transport, soluti
Burdette, H.E.	521	Crystal Perfection in Czochralski Grown Nickel Single Crystals (N
Burgess, E.L.	298	High-Efficiency p+-n-n+ Back-Surface-Field Silicon Solar Cells (S
Burkhanov, G.S.	829	Growth of Single Crystals of High Melting Metal Alloys and Compou

Burnham, R.D.	152	Integrated Optical Devices Fabricated by MBE (GaAs, Ga1-xAlxAs, r
Burns, J.H.	153	Americium Ditelluride (AmTe2-x
Burrus, C.A.	916	Nd:Y2O3 Single-Crystal Fiber Laser: Room-Temperature cw Operatio
Busch, G.	154	Magnetization and Neutron Studies of UTe and USb0.8Te0.2
Bush, A.A.	155	Ferroelectric (Pb,Ba)5Ge3O11 Crystals (flux)
	979	X-Ray and Dielectric Investigations of Pb2CrO5 (melt)
Busing, W.R.	913	Dodecatungstophosphoric Acid-21-Water by Neutron Diffraction (eva
Butaeva, T.I.	458	Growth, Spectroscopy, and Stimulated Emission of Cubic Bi4Ge3O12
Butcher, D.N.	156	The Thermal Oxidation of GaAs
Butter, E.	354	Liquid Phase Epitaxial Deposition of GaP on GaAs
Butti, C.	157	Growth of Single Crystals of Cinnabar (alpha-HgS) at Rather Low T
Buzaneva, E.V.	473	Structural Transitions and Phase Composition of Thin Films of Alu
Buzynin, A.N.	116	Growth of Si Strips by Stepanov's Method
	723	Relation Between Growth Conditions and Dislocation Structure in G
Bye, K.L.	158	An X-Ray Topographic Assessment of Cadmium Mercury Telluride (Cdx
Bykov, A.B.	93	Some Problems of the Crystallization Kinetics and Growth Mechanis
	159	Phase Equilibria in Solvent-Garnet Systems (flux, Y3Al5O12, Y3Fe5
	977	Growth of Bulk Yttrium Iron Garnet Crystals from High Temperature
	978	Kinetic Characteristics of Growth of Y3Al5O12 Single Crystals fro
Cafiero, A.V.	727	Single-Crystal Growth of Mixed (La, Eu, Y, Ce, Ba, Cs) Hexaboride
Calestani, G.	160	Single Crystal and Film Growth of Actinide Pnictides by Chemical
	161	Preparation and Single-Crystal Growth of Protactinium Arsenides b
Callaghan, T.	162	A Simple Method for Growing V3Si Single Crystals (electron beam m
Calvo, C.	379	The Crystal Structure of Ba2V2O7 (melt)
	692	Crystallographic Studies and Structural Systematics of the C2/c A
Calvo, R.	293	Magnetic, Transport, and Thermal Properties of Ferromagnetic EuB6
Camel, D.	163	In-Situ Observation of Solidification Interfaces with a Very High
Campisano, S.U.	996	Grain Size Dependence in a Self-Implanted Silicon Layer on Laser
Capella, L.	113	Irreversible Thermodynamics of the Cellular Growth of a Dilute Bi
	210	Observation by Chemical Attack and X-Ray Topography of Cellular S
Capper, P.	74	The Effects of Growth Speed on the Compositional Variations in Cr
	75	A Study of Casting in the CdxHg1-xTe System
	164	An Improved System for the Bridgman Growth of Crystals with Toxic
	165	On the Cooling Rates of Large-Diameter Silicon Crystals (Si, theo
Capponi, J.J.	205	Hydrothermal Synthesis of Several Mixed Oxides of the A(6+)B2(3+)
Cardew, P.T.	166	Evaluation of Supersaturation in Crystal Growth from Solution (th
Carre, D.	259	Crystal Structure of Lanthanum Polysulfide LaS2
	300	Crystal Structure of the Zinc Iodide ZnI2 (vapor transport)
	557	Crystal Structure of the Indium Selenide In2Se3 (vapor transport)
Carruthers, J.R.	604	Application of a Heat Pipe to Czochralski Growth. Part I. Growth
Caslavsky, J.L.	167	Melt Growth of Nd:Y3Al5O12(Nd:YAG) Using the Heat Exchange Method
Castro, C.A.	471	Crystal Growth and Characterization of In1.9As0.1Se3 (vapor trans
	708	Crystal Growth and Characterization of the Transition-Metal Phosp
	911	Preparation and Properties of Two Indium Antimony Selenides (In2S
Cecchi, J.H.	602	Growth Conditions for Sodium-Titanium Bronze Crystals in NaBO2 (f
Cerclet, M.	114	Growth and Characterization of Undoped ZnSe Epitaxial Layers Obta
Chaban, I.A.	168	Percolation Theory and Crystallization (entropy, specific heat, m
Chabot, B.	70	Solubility of Phlogopite in Basic Aqueous Solutions Under Hydroth
Chai, B.H.T.	169	Synthesis of Rare-Earth Carbonates Under Hydrothermal Conditions
Chakraborty, D.	170	On the Incorporation of Metallic Impurities in Synthetic Quartz S
Chakraverty, B.K.	171	Low-Temperature Specific Heat and Magnetic Susceptibility of Nonm
Challeton, D.	650	Influence of the CaCO3/GeO2 Ratio on Y Sm Lu Ca Fe Ge Garnet Film
Chalyi, V.P.	725	Phase Diagram of Pseudobinary System CdTe-ZnSe (Bridgman)
Chaminade, J.P.	785	Preparation, Optical and Dielectric Properties of Na5W3O9F5 (melt
Champier, G.	857	Scanning Electron Microscopy of Growth Surfaces of Synthetic Quar
Champness, C.H.	876	Czochralski Growth of Tellurium Single Crystals (Te)
	877	Etch Pit Orientation in Czochralski-Grown Tellurium Crystals (Te)
Chan, S.-K.	172	On the Growth of NH4Cl Dendrites at Very Low Supersaturation (sol
Chang, C.E.	173	Computer Simulation of Convection in Floating Zone Melting. I.
	174	Computer Simulation of Convection in Floating Zone Melting. II. C
	175	Thermocapillary Convection in Floating Zone Melting: Influence o
Chang, L.L.	176	Semiconductor Superlattices by MBE and Their Characterization (Ga
Chapman, G.E.	177	Silicide Formation by High-Dose Si+-Ion Implantation of Pd (elect
Charil, J.	745	Properties of Homogeneously Doped ZnSe Single Crystals Obtained b
Charlan, G.B.	180	Crystal Growth and Crystal Chemistry of NiAs-Type Compounds: MnSb
Cheetham, A.K.	781	The Preparation and Crystal Structures of BiReO4 and BiRe2O6 (vap
Chen, H.-Y.	178	The Crystal Structure and Twinning Behavior of Ferric Molybdate,
	891	Crystal Growth and Structure of BiVO4 (Czochralski)
Chen, P.S.	179	Convection Phenomena During the Growth of Sodium Chlorate Crystal
Chen, T.	180	Crystal Growth and Crystal Chemistry of NiAs-Type Compounds: MnSb
Cheng, C.L.	403	Growth of CuInS2 and Its Characterization (vapor transport)
Cherepanova, T.A.	181	Kinetics of Crystallization of Alloys of Multiple Constituents (t
	182	Fast Growth of Ordered AB Crystals: A Monte Carlo Simulation for
	1013	Survey of Monte Carlo Simulations of Crystal Surfaces and Crystal
Cherian, K.A.	748	Overgrowths on Diamond at Atmospheric Pressure (C, flux)
Chernitsyna, M.A.	50	Properties of CuCr2Se4 Crystals Obtained from Solutions in Melts

	51	Crystal Growth of CuCr2Se4
Chernov, A.A.	184	Theoretical Analysis of Equilibrium Adsorption Layers in CVD Syst
Chernov, A.A.(ed.)	183	Growth of Crystals, Volume 11 (review, theory)
Chernov, M.A.	554	Growing Monocrystalline Bars of SiC from the Gas Phase (vapor tra
Chernyshov, A.I.	185	Preparation and Investigation of Crystals of Solid Solutions of t
Cheung, D.T.	803	Liquid Phase Epitaxial Growth of Ga1-xInxSb on GaSb by Stepwise G
Chevalier, B.	137	Czochralski Crystal Growth of Rare Earth Tetraborides (YB4, TbB4,
Chevalier, R.	80	Structure of New Rhenium Apatites Containing Large Anions: Ba10(
	81	Crystal Structure of La3ReO8
	103	Crystal Structure of Double Oxides of Rhenium. II. The Lanthanum
	104	High Pressure Synthesis and Structure of a New Pyrochlore Contain
	1029	Crystal Structure of Li2CaUF8 (flux)
Chevallier, J.	600	Luminescence and Bandgap Studies of High Purity MgxZn1-xTe Ternar
	792	Preparation, Analysis and Properties of High Purity Zn(1-x)MgxTe
Chevrel, R.	186	New Ternary Mo(II)-Compounds InxMo15Se19 Containing Mo6Se8 and Mo
Chevy, A.	187	Growth of Crystalline Slabs of Layered InSe by the Czochralski Me
Chiao, S.M.	188	LPE Growth of Quaternary GayIn1-yAsxP1-x (film)
Chiba, T.	15	Positron Annihilation in ReO3 (vapor transport)
Chinn, S.R.	189	Lasing and Fluorescence in K5NdLi2F10 (flux)
Chkheidze, I.M.	823	Heteroepitaxial Structures of Si on Sapphire, Obtained by Vacuum
Cho, A.Y.	190	Single-Crystal-Aluminum Schottky-Barrier Diodes Prepared by Molec
Chopra, K.L.	868	Electroless Deposition of Semiconductor Films (PbS, PbSe, Pb1-xHg
	869	Electroless Deposition of Epitaxial Pb1-xHgxS Films
Choukroun, S.	191	Germanium-Iodine Equilibrium from 700 to 1000 K (Ge, vapor transp
Christensen, A.N.	192	Crystal Growth of Niobium Nitride and Niobium Carbide Nitride (Nb
Christoffersen, J.	193	The Affinity of Crystal Growth and Dissolution in Aqueous Solutio
Christoffersen, M.R.	193	The Affinity of Crystal Growth and Dissolution in Aqueous Solutio
Christou, A.	194	Crystalline Defects in Solid Phase Epitaxy Si Films Deposited at
Chu, M.	1059	Liquid Phase Growth of HgCdTe Epitaxial Layers
Chu, S.S.	195	Solar Cells from Zone-Refined Metallurgical Silicon (Si, Czochral
Chu, T.L.	195	Solar Cells from Zone-Refined Metallurgical Silicon (Si, Czochral
	196	Purification and Characterization of Metallurgical Silicon (Si, m
Cirafici, S.	197	The Praseodymium-Gallium System from 0 to 50 At% Gallium (melt, P
Clauss, A.	198	Crystallochemical Study of the System Ni3Pb2S2-Ni3Pb2Se2 and Note
Clauws, P.	1023	Optical Properties of V6O13 Single Crystals in the Metallic and t
Clawson, A.R.	199	Control of Substrate Degradation in InP LPE Growth with PH3 Parti
Clink, B.L.	200	Preparation and Characterization of a Europium(II) Bromide-Chlori
Clyne, T.W.	201	Theoretical Solute Redistribution During a Modified Form of Zone
Cockayne, B.	202	A Complexity in the Solidification Behaviour of Molten Y3Al5O12 (
	203	Single Crystal Growth of 12CaO.7Al2O3 (Czochralski)
	1076	Ca12Al14O33: A New Piezoelectric Material (Czochralski)
Coe, I.M.	272	The Role of B2O3 in PbO/PbF2/B2O3 Fluxes (Y3Al5O12, theory)
Coetzer, J.	973	The Crystal Structure of the Solid Electrolyte Ag44I53(C11H30N3)3
Cohen, U.	204	Electrocrystallization of NbO and Hollow NbO2 Needles from Molten
Cole, D.	164	An Improved System for the Bridgman Growth of Crystals with Toxic
Coller, B.A.W.	779	Effect of S-35 Tracer, Particle Size and Imperfection on the Solu
Collin, G.	102	Structure of La2Fe1.76S5 (melt)
	543	Crystal Structure and Physical Properties of CuO0.65VS2 (vapor tr
	544	Preparation, Crystal Structure, and Physical Properties (transpor
Collins, D.M.	576	Design Considerations for Molecular Beam Epitaxy Systems (review)
Collins, R.	624	The Electrical, Optical and Photoconducting Properties of Fe2-xCr
Collomb, A.	205	Hydrothermal Synthesis of Several Mixed Oxides of the A(6+)B2(3+)
Collongues, R.	629	Growth from Skull-Melting of Zirconia-Rare Earth Oxide Crystals (
Constant, G.	291	Single Crystals with Compositions in the Range TiB1.89 - TiB1.96
	1109	Crystal Growth of Gallium Arsenide by Chemical Vapor Deposition f
Cooper, A.S.	621	X-Ray Diffuse Scattering from Alkali, Silver, and Europium beta-A
	791	Superconductivity of Single-Crystal NbGe2
Copley, J.R.D.	459	Local Modes in Rb1-cKc Alloys: A Neutron Scattering Study (Bridg
Corbett, J.M.	129	On the Ordering of Fe Atoms in FexNbS2 (vapor transport)
	1090	Intercalation-Induced Shift of the Absorption Edge in ZrS2 and Hf
Coriell, S.R.	206	Lateral Solute Segregation During Unidirectional Solidification o
Cornut, B.	293	Magnetic, Transport, and Thermal Properties of Ferromagnetic EuB6
Corny, P.	70	Solubility of Phlogopite in Basic Aqueous Solutions Under Hydroth
Cosandey, F.	207	Characterization of (001) Tilt Boundaries Produced by Epitaxial G
Cotton, F.A.	208	La8Ru4O21: A Mixed-Valence Ternary Ruthenium Oxide of a New Hexa
	209	The Crystal Structure of La3Ru3O11: A New Cubic KSbO3 Derivative
Coulet, A.L.	210	Observation by Chemical Attack and X-Ray Topography of Cellular S
Courbion, G.	772	Hydrothermal Growth of the Systems MF-AlF3 and NH4F-FeF3 in Hydro
Cousseins, J.C.	37	The Systems RbF-EuF3 and CsF-EuF3. Study of the Phases RbLnF4 an
	1029	Crystal Structure of Li2CaUF8 (flux)
Cowher, M.E.	211	Low Temperature CVD Garnet Growth (Er3Fe5O12, (Eu,Y)3Fe5O12 and (
Cowlam, N.	896	The Crystal Growth and Atomic Structure of As2(Se,S)3 Compounds (
Cox, D.E.	891	Crystal Growth and Structure of BiVO4 (Czochralski)
Crocker, A.J.	47	Liquidus Measurements in the Pb-Sn-Te System (Pb1-xSnxTe, PbTe, f
Cross, L.E.	108	Polar Twins in Lithium Metasilicate and Metagermanate (Li2SiO3, L
	109	Crystal Growth of Antimony Sulphur Iodide (SbSi, flux)

	141	Crystal Growth and Characterization of Lithium Metasilicate (Li2S
Crouch, P.K.	212	Properties of GaN Grown on Sapphire Substrates (epitaxy)
Csillag, F.J.	1079	Transport Properties of the Systems SnS2-SnI4 and SnS2-I2 (SnS2,
Cueilleron, J.	213	Growth of Single Crystals of Beta Rhombohedral Boron by a Modifie
Cutter, J.R.	214	Growth of Single Crystals of Zinc Selenide from the Vapour Phase
Dabiri, A.E.	61	Observations of the Movement of a Molten Zone in Germanium Under
Dabringhaus, H.	628	Molecular (beam) Processes of Condensation and Evaporation of Alk
Dad Khattak, G.	215	Specific Heat of V6O11 Between 0.4 and 50 K (vapor transport, the
Daguet, C.	447	Crystal Growth and Crystallographic Data of Some Ln2(MoO4)3 Type
Damen, J.P.M.	216	Thin Film Growth of (MnxZn1-x)FeO4 by Liquid Phase Epitaxy on Bri
Damien, D.	153	Americium Ditelluride (AmTe2-x
Daniel, D.R.	583	Liquid Encapsulation Zone-Refining of PbS (zone melting, vapor tr
Darack, S.	31	On the Resistivity of the Magnetically Ordered Ground State of th
Darby, D.B.	126	Nature, Origin and Effect of Dislocations in Epitaxial Semiconduc
Das, B.N.	217	Preparation of V3Ga Single Crystals by Iodine Transport
Dashevsky, M.Ya.	218	Impurity Effects on Low-Angle Boundary Formation in Silicon Singl
Daval, J.	650	Influence of the CaCO3/GeO2 Ratio on Y Sm Lu Ca Fe Ge Garnet Film
Davarashvili, O.I.	668	Investigation of Liquidus Surface in Pb-Sn-Te System by Simplex L
Davey, J.E.	194	Crystalline Defects in Solid Phase Epitaxy Si Films Deposited at
Davey, R.J.	134	Solvent Effects and the Growth Kinetics of Ionic Crystals (CaSO4.
	166	Evaluation of Supersaturation in Crystal Growth from Solution (th
	219	The Role of Dislocations in the Growth of Ammonium Dihydrogen Pho
Davidova, L.B.	128	The Effect of the Solidification Front Shape on the Temmperature
Davis, J.G.	220	Electrical and Optical Properties of Copper Indium Ditelluride Cr
Davis, P.R.	221	The Adsorption of Zr onto W(100) Surfaces
Davison, W.	1063	The Flux Growth of Some New Rare Earth and Iron Group Complex Oxi
	1064	Flux Growth of Crystals of Some Transition Metal Fluorides. Part
Dayan, M.	162	A Simple Method for Growing V3Si Single Crystals (electron beam m
de Brouckere, L.	222	An Improved Method of Stirring for LPE Growth (Sm,Y)3(Fe,Ga)5O12
de Jonge, W.J.M.	77	Crystallographic and Magnetic Structure of RbFeCl3.2D2O and CsFeC
de Jongh, L.J.	106	Magnetic and Neutron Scattering Experiments on the Antiferromagne
de Kock, A.J.R.	223	The Effect of Doping on Microdefect Formation in As-Grown Disloca
De Lange, G.J.	795	Preparation, Structures, and Properties of Niobium Chalcogenide H
De Mattei, R.C.	224	Conditions for Stable Growth of Epitaxial GaP Layers by Molten Sa
	225	Growth Rate Limitations in Electrochemical Crystallization (metal
De Pape, R.	772	Hydrothermal Growth of the Systems MF-AlF3 and NH4F-FeF3 in Hydro
de Pater, C.J.	226	Neutron-Scattering Study of the Incommensurate Phase Transition o
de Saja, A.	227	Growth and Some Properties of Cerium Sulphate Enneahydrate Single
de Saja, J.A.	227	Growth and Some Properties of Cerium Sulphate Enneahydrate Single
Debnam, W.J.	212	Properties of GaN Grown on Sapphire Substrates (epitaxy)
Deguchi, Y.	228	Preparation Method of High Purity Aluminum Single Crystals with L
Dejima, K.	1103	Formation of Amorphous Mn-Bi Films and Their Crystallization Proc
Delapalme, A.	154	Magnetization and Neutron Studies of UTe and USb0.8Te0.2
Delfino, M.	229	Thermal and Dielectric Properties of LiKSO4 and LiCsSO4
	567	Single Crystal Growth and Properties of Deuterated Triglycine Flu
Delong, M.C.	808	A Gas Handling System for Exacting Crystal Growth Preparations (T
Dem'yanets, L.N.	230	Hydrothermal Crystallization of Sodium Zirconogermanates (Na2ZrGe
Dement'ev, Yu.S.	231	Kinetics of Gallium Phosphide Synthesis by Chemical Gas Transport
Demianets, L.N.	232	Hydrothermal Crystallization of Magnetic Oxides (review)
	233	X-Ray Topographic Study of Growth Defects in Hydrothermal Na2CoGe
	234	Hydrothermal Synthesis of Crystals (review)
	256	Na2CoGeO4 Single Crystal Growth and Study of Their Piezoelectric
Denes, G.	235	About SnF2 Stannous Flouride. I. Crystallochemistry of alpha-SnF2
Denis, J.P.	756	Optical Properties of Cs2NaBiCl6
Denisenko, G.A.	458	Growth, Spectroscopy, and Stimulated Emission of Cubic Bi4Ge3O12
DePasquali, G.	474	Structural and Magnetic Properties of "One-Dimensional" Barium Va
Depmeier, W.	236	An Application of the Flux Function Method to a Five-Component Sy
Dernier, P.D.	190	Single-Crystal-Aluminum Schottky-Barrier Diodes Prepared by Molec
	621	X-Ray Diffuse Scattering from Alkali, Silver, and Europium beta-A
Desai, C.C.	237	Flux Growth and Characterization of Fluorite Crystals (CaF2)
Desai, C.F.	110	Growth & Etching of InBi Single Crystals (melt, zone melting)
Desvignes, J.M.	238	Flux Growth of Single Magnetic Sublattice Antiferromagnetic Garne
Deville Cavellin, C.	434	Raman Spectra of ZrS3
Dhalenne, G.	239	Growth of Oriented Bicrystals of Nickel Oxide (NiO, arc image, fl
Dich, N.T.	240	Preparation and Basic Physical Properties of BiTeI Single Crystal
Diehl, R.	241	The Crystal Structure of Triclinic WO3 (sublimation)
Dierssen, G.H.	242	Seeded Growth of Large Single Crystals of CdS from the Vapor Phas
Dillin, D.R.	663	Aging Studies on Hydrous Lutetium Oxide (cubic Lu(OH)3
Dion, M.	317	A New Structural Family M2M'Ta5O15. Crystal Structure of CaTlTa
Dishon, G.	834	State-of-The-Art of Crystal Growth and Nuclear Spectroscopic Eval
Dishovsky, N.	121	Growth of Bi2(SeO4)3 Single Crystals in Silica Gels
	122	Growth of Single Crystals of Ag2SeO4 in Silica Gels
Distler, G.I.	243	On the Growth Mechanism of Sodalite Single Crystals Grown by the
Dittmar, G.	244	Growth of Homogeneous Mixed Crystals with a Large Segregation Ten
Dmitruk, M.V.	245	Spectroscopic Properties of Er3Al5-xGaxO12 Films Obtained by Liqu
Dobrovinskaya, E.R.	246	Profile-Sapphire and Its Structural Perfection (alpha-Al2O3, melt

Dogguy-Smiri, L.	247	Twinning and Crystal Structure of Mn0.75Ga2.17S4 (flux)
Dohler, G.H.	248	Periodic Doping Structure in GaAs (GaAs, Ga1-xAlxAs, molecular be
Doi, A.	249	Improvement of Crystal Composition in Ga1-xAlxAs LPE Layers Grown
Dolinskaya, N.Yu.	1035	Crystallization of Bi-Sb Solid Solutions in Capillaries (zone mel
Domanskii, A.I.	902	Crystal Structure of Scandium Metaphosphate Sc(PO3)3 (solution)
Donakova, A.V.	989	Crystallization Kinetics of Lithium and Strontium Nitrate Crystal
Donald, I.W.	250	Infra-Red Transmitting Materials. Part 1. Crystalline Materials
Dorfman, V.F.	251	Growth Kinetics and Structure of Epitaxial Garnet Layers as Funct
Dorner, B.	672	Dispersion of Acoustic Phonons in Layered SrGa2 (melt, slow cooli
Dovchenko, G.V.	16	Acoustooptic Characteristics of LiBi(MoO4)2 (Czochralski)
Down, M.G.	252	Low Temperature Method for the Growth of Lithium Nitride Single C
Drafall, L.E.	813	Crystal-Chemistry, Crystal-Growth, and Phase-Equilibria of Apatit
Dreeben, A.B.	479	Maximum Stable Zone Length in Float-Zone Growth of Small-Diameter
Drofenik, M.	253	Crystal Growth of Some Alkaline-Earth Rare-Earth Pentaoxometallat
Druminski, M.	254	Dichlorosilane as Silicon Source for Epitaxial Films on Insulator
du Tremolet de Lacheisser	631	Magnetization, Magnetocrystalline Anisotropy, Magnetostriction an
Dubitskii, G.A.	1032	Synthesis of Single Crystals of Cubic Boron Nitride in Systems Co
Duchemin, M.J.	255	Kinetics of Silicon Growth Under Low Hydrogen Pressure (vapor tra
Duderov, N.G.	233	X-Ray Topographic Study of Growth Defects in Hydrothermal Na2CoGe
	256	Na2CoGeO4 Single Crystal Growth and Study of Their Piezoelectric
Dudnik, E.F.	16	Acoustooptic Characteristics of LiBi(MoO4)2 (Czochralski)
Dugua, J.	257	Crystallization of Sodium Perborate from Aqueous Solutions. I.
	258	Crystallization of Sodium Perborate from Aqueous Solutions. II.
Dugue, J.	259	Crystal Structure of Lanthanum Polysulfide LaS2
	1048	Mixed Oxysulfides of Chromium III and Rare Earths (LaCrOS2, CeCrO
Dung, N.-H.	247	Twinning and Crystal Structure of Mn0.75Ga2.17S4 (flux)
Dunning, W.J.	572	The Variability of the Rate of Growth of Adamantane Crystals from
Dupuis, M.M.	532	Semi-Periodic Closed Growth Lamellae on the Crystal-Melt Interfac
Dupuy, M.	98	Transmission Electron Microscopy Observations of Precipitates in
Durand, J.M.	393	Fast Growth in GaAs VPE at Low Temperature and High Partial Press
Durif, A.	142	Crystal Structure of Cu4(PO4)2O (flux)
	260	Structure of Silver Dichromate: Ag2Cr2O7 (solution)
	261	Crystal Structure of the Mercury Nitro-Phosphate Hg4PO4NO3.H2O (m
	608	Crystal Structure of AgH2PO4. Crystallography of AgH2AsO4 (solut
	609	Crystal Structure of a New Variety of Nickel Pyrophosphate: Ni2P2
	986	Crystal Structure of Monophosphate AgCoPO4
Dwight, K.	624	The Electrical, Optical and Photoconducting Properties of Fe2-xCr
Dyer, L.D.	262	Dislocation-Free Czochralski Growth of (110) Silicon Crystals (Si
Dziuba, Z.	341	Crystallization of HgTe Crystals from the Vapour Phase
Eberhardt, A.	371	Two-Phase (alpha/beta)-Brass Bicrystals Produced by a Solid-Solid
	372	Two-Phase (alpha/beta)-Brass Bicrystals Produced by a Solid-Solid
Ebina, A.	45	Growth of ZnSxSe1-x Layers on ZnS and ZnSe Substrates by a Solid-
Ebisuzaki, Y.	263	Growth of Carbon Tetrabromide Crystals for Optical Studies (CBr4,
Eckardt, R.C.	343	Heteroepitaxy of a Deposited Amorphous Germanium Layer on a Silic
Edwards, J.P.N.	376	Dynamic Sublimation Technique for Growing Mercuric Iodide Crystal
Eggimann, P.	290	The Binary Systems Cobalt-Gallium and Nickel-Gallium Compared (Co
Egorov, K.G.	345	Behavior of Boron in Ge Single Crystals Grown by the Czochralski
Egorov, V.L.	151	Preparation and Physical Properties of Boracite Crystals (Chemica
Ehara, S.	264	Photoconduction of Lead Trititanate (PbTi3O7, flux)
Eick, H.A.	200	Preparation and Characterization of a Europium(II) Bromide-Chlori
Eidenson, A.M.	218	Impurity Effects on Low-Angle Boundary Formation in Silicon Singl
Eigermann, W.	265	Solubility Curves in High-Temperature Melts for the Growth of Sin
Finullaev, A.V.	817	Synthesis, Preparation of Single Crystals, and Physicochemical In
Eisner, J.	266	The Physical Properties of TGS Single Crystals, Grown from Aqueou
Ekstrom, T.	267	A Series of Lead Tungsten Bronzes (PbxWO3)
El-Mahallawy, N.A.	268	Effect of Some Thermal Parameters on the Directional Solidificati
Elaatmani, M.	785	Preparation, Optical and Dielectric Properties of Na5W3O9F5 (melt
Ellis, R.J.	360	Silicon Ribbon Growth via the Ribbon-To-Ribbon (RTR) Technique:
Eltoukhy, A.H.	269	Growth and Electrical Properties of Sputter-Deposited Single-Crys
Elwell, D.	224	Conditions for Stable Growth of Epitaxial GaP Layers by Molten Sa
	270	Priorities in the Initial Use of Spacelab for Crystal Growth (vap
	271	Man-Made Gemstones (review)
	272	The Role of B2O3 in PbO/PbF2/B2O3 Fluxes (Y3Al15O12, theory)
Emel'yanova, T.A.	50	Properties of CuCr2Se4 Crystals Obtained from Solutions in Melts
	51	Crystal Growth of CuCr2Se4
Emoto, T.	654	Preparation of Fine Sodium Chloride Crystals (NaCl)
Endo, S.	410	Electrical Properties of p- and n-Type CuInSe2 Single Crystals (m
Endo, Y.	715	The Effects of M(M=Mg, Ca and Ba) Ions on the KCl Flux Growth of
Enoki, T.	273	Specific Heat of a Quasi-Two-Dimensional Antiferromagnet Ni(OH)2,
Entine, G.	1053	Crystal Growth of CdTe for X-Ray Detectors (review)
Erdelyi, G.	90	Connections Between Thermodynamic Quantities and Vacancy and Diff
Ermisch, W.	564	Growth of Pb1-xSnXTe Single Crystals of Graded Composition by the
Erofeev, V.N.	128	The Effect of the Solidification Front Shape on the Temperature
Esaki, L.	176	Semiconductor Superlattices by MBE and Their Characterization (Ga
Eshelman, F.R.	274	Single-Crystal Elastic Constants of Al2Cu (Bridgman, arc melting)
Esin, V.O.	275	Preferred Growth Directions and Perfection of Single Crystals of

Etourneau, J.	137	Czochralski Crystal Growth of Rare Earth Tetraborides (YB4, TbB4,
Ettenberg, M.	728	Growth Effects in the Heteroepitaxy of III-V Compounds (review)
Evans, B.L.	667	The Thermal Expansion of 2H-MoS2 and 2H-WSe2 Between 10 and 320 K
Evianova, N.F.	1047	Crystallization of LiNbO3 from Solution in Borate, Vanadate, and
Eyer, A.	276	A Double-Ellipsoid Mirror Furnace for Zone Crystallization Experi
Fabian, W.	550	Crystal Growth of Triple Nitrites in Gels (K2CuPb(NO2)6, Tl2CuPb(
Fabig, H.	277	Solution Growth and Some Properties of GaP Bulk Crystals
Fadly, M.	445	Flux Growth of Lead Neodymium Chlorapatite (Na2Nd2Pb6(PO4)6Cl, sl
Fagen, E.A.	278	Optical Properties of Zn3P2 (evaporation)
Faile, S.P.	215	Specific Heat of V6O11 Between 0.4 and 50 K (vapor transport, the
	279	Modified Chemical Vapor Growth of Cinnabar (HgS) and GaP in Close
	280	Growth of Uranium Dioxide Crystals Using Tellurium Tetrachloride
	288	Electron Energy-Loss Spectroscopy of TiS2 (vapor transport)
Faktor, M.M.	281	The Modified Entrainment Method and Its Application to the Study
	333	The Liquidus Temperature and Growth-Dissolution Kinetics of Garne
Fal'kevih, E.S.	687	Influence of Growth Conditions on Mirodefect Distribution in Disl
Falck, L.	282	The Crystal Structure of CuTeO4 (hydrothermal)
Falckenberg, R.	283	The Verneuil Process (review)
	284	Powder Feeder for Crystals of Large Diameter Grown by the Verneui
Farag, M.M.	268	Effect of Some Thermal Parameters on the Directional Solidificati
Farnworth, B.	1004	Raman and Infrared Spectra of CdIn2S4 and ZnIn2S4 (vapor transpor
Farrington, G.C.	285	Hydronium Beta" Alumina: A Fast Proton Conductor (0.84Na20.0.84M
	286	Ionic Conductivity in N30+ Beta Alumina (solution)
Farrow, R.F.C.	287	A High Temperature High Purity Source for Metal Beam Epitaxy (Ag,
Fattah, A.	1002	A Model for Macroscopic Czochralski Growth: Theoretical and Expe
	1071	Measurements of the Contact Angle Between Melt and Crystal During
Faust, J.W.,Jr.	369	Thermal Analysis of Solidification in Web-Dendritic Ribbon Growth
Fedorenko, A.I.	501	Growth of Antimony on Bismuth (Sb film, epitaxy)
Fedoroff, M.	792	Preparation, Analysis and Properties of High Purity Zn(1-x)MgxTe
Fedorov, V.A.	46	Structure, Spectroscopy, and Stimulated Emission of Crystals of Y
	231	Kinetics of Gallium Phosphide Synthesis by Chemical Gas Transport
Fedoseev, D.V.	862	Temperature Field Dynamics in the Synthesis of Diamonds (kinetics
Feigelson, R.S.	224	Conditions for Stable Growth of Epitaxial GaP Layers by Molten Sa
	225	Growth Rate Limitations in Electrochemical Crystallization (metal
Feldkamp, L.A.	288	Electron Energy-Loss Spectroscopy of TiS2 (vapor transport)
Feldmann, P.	767	Magnetic Properties of a Single Crystal of NdF3 Between 2 and 4.2
Fenin, V.V.	44	Growing Optical films of Aluminum Yttrium Garnet (RE substituted
Fenner, J.	25	Telluride Halides of Copper with High Ionic Partial Conductivity.
Ferianc, M.	289	Analysis of the Static Method of the Growth of Cadmium Sulphide C
Ferrand, B.	650	Influence of the CaCO3/GeO2 Ratio on Y Sm Lu Ca Fe Ge Garnet Film
Ferretti, A.	891	Crystal Growth and Structure of BiVO4 (Czochralski)
Ferris, S.	53	The Growth of In-Rich Bulk Single Crystals of GaxIn1-xPyAs1-y Via
Feschotte, P.	290	The Binary Systems Cobalt-Gallium and Nickel-Gallium Compared (Co
Feurer, R.	291	Single Crystals with Compositions in the Range TiB1.89 - TiB1.96
Figat, R.A.	865	Preparation and Characterization of Ruthenium Dioxide Crystals (R
Finney, J.L.	123	Monte Carlo Simulation of the Crystal-Melt Interface of a Lennard
Fiorito, G.	292	A Possible Method for the Growth of Homogeneous Mercury Cadmium T
Fischer, O.	186	New Ternary Mo(II)-Compounds InxMo15Se19 Containing Mo6Se8 and Mo
Fisk, Z.	293	Magnetic, Transport, and Thermal Properties of Ferromagnetic EuB6
	791	Superconductivity of Single-Crystal NbGe2
Flanders, D.C.	899	Oriented Crystal Growth on Amorphous Substrates Using Artificial
Flerova, S.A.	294	Change in the Energy Band Structure of Eulytine on Transition to
Fletcher, J.	126	Nature, Origin and Effect of Dislocations in Epitaxial Semiconduc
Florian, M.	788	The Epitaxial Growth of Reactively Sputtered CuI Films on Rocksal
Fluckiger, U.	295	On the Preparation of Pure, Doped and Reduced KNbO3 Single Crysta
Fontcuberta, J.	296	Mechanism of Habit Change of ADP Crystals by Fe3+, Based on Mossb
Forzatti, P.	297	Solid State Reactions to CdTeMoO6 and Its Structural Characteriza
Fossum, J.G.	298	High-Efficiency p+-n-n+ Back-Surface-Field Silicon Solar Cells (S
Foti, G.	996	Grain Size Dependence in a Self-Implanted Silicon Layer on Laser
Fourcaudot, G.	299	Vapor Phase Transport and Crystal Growth of Molybdenum Trioxide a
Fourcroy, P.H.	300	Crystal Structure of the Zinc Iodide ZnI2 (vapor transport)
Fournier, M.-T.	36	Crystal Structure of RbEu3F10 (melt)
Fournier, M.T.	37	The Systems RbF-EuF3 and CsF-EuF3. Study of the Phases RbLnF4 an
Fourquet, J.L.	772	Hydrothermal Growth of the Systems MF-AlF3 and NH4F-FeF3 in Hydro
Fouskova, A.	138	The Growth of Single Crystals of Langbeinites Rb2Cd2(SO4)3, Tl2Cd
Foxon, C.T.	301	Surface Processes Controlling the Growth of GaxIn1-xAs and GaxIn1
Franceschi, E.	197	The Praseodymium-Gallium System from 0 to 50 At% Gallium (melt, P
Franchini-Angela, M.	302	Growth Morphology of Weddellite (CaC2O4.2XH2O, gel)
Frank, F.C.	333	The Liquidus Temperature and Growth-Dissolution Kinetics of Garne
Frank, R.I.	303	Performance of a New High-Intensity Silicon Solar Cell (Si, float
Franke, H.	991	Effects of Interface Kinetics on the Growth Rate of Dendrites (Sn
Franke, V.D.	989	Crystallization Kinetics of Lithium and Strontium Nitrate Crystal
Franke, W.	304	The Morphology of Hydrothermally Grown Strontium Paracelsian (Sr[
Franzosi, P.	305	Epitaxial Growth of Single Crystal Cd1-xZnxS Layers on (111) GaAs
Fredericks, W.J.	579	Simultaneous Diffusion of Calcium and Strontium in KCl Single Cry
Freedman, G.M.	452	Aluminum Redistribution in EFG of Silicon Ribbon (Si)

Freeland, P.E.	68	Dislocation-Free and Low-Dislocation Quartz Prepared by Hydrother
Frenk, D.M.	306	Preparation of Films of IV-VI Compounds in a Quasi-Sealed Volume
Freudenreich, W.	38	Solution Growth of RbAg4I5 Crystals
Frey, M.L.	307	Elastic Constants of Niobium-Zirconium, Hafnium, and Tungsten All
Friedberg, S.A.	464	Uniaxial Ferromagnetism in NiZrF6.6H2O (solution)
Frikkee, E.	106	Magnetic and Neutron Scattering Experiments on the Antiferromagne
Fripp, A.L.	212	Properties of GaN Grown on Sapphire Substrates (epitaxy)
Froidevaux, Y.	555	Structural and Electrical Properties of Layered Transition Metal
Frolov, A.A.	746	Dependence of Degree of Perfection of Single Crystals of FeGe2 on
	861	Growth Characteristics of Crystals of Certain Metallides (alpha-F
Frumar, M.	975	Some Electrical Properties of GeBi2Te4 Single Crystals (Bridgman)
Fujibayashi, K.	852	GaxIn1-xSb Crystals Grown by Liquid Phase Epitaxy (melt)
Fujii, H.	373	Magnetic Properties of Rare Earth Copper Intermetallic Compounds,
	374	Magnetic Properties of Rare Earth Copper Intermetallic Compounds,
	430	Magnetic Properties of (Mn,Cr)P and (Mn,Fe)P Compounds (Mn1-xFexP
Fujimoto, K.	421	Crystallography of the fcc Phase in Uni-Directionally Solidified
Fujinaga, T.	425	Birefringence in CdSiP2 (solution)
Fujioka, T.	534	Evidence for a Universal Law of Dendritic Growth Rates (theory)
Fujita, S.	308	Liquid-Phase Epitaxy of ZnSe from Zn-Ga Solution (film)
	309	Growth of Cubic ZnS, ZnSe and ZnSxSe1-x Single Crystals by Iodine
	938	Growth of High-Purity ZnTe Single Crystals by the Sublimation Tra
Fujita, T.	310	High Temperature Form of Pb2WO5 and Transformation Phenomena to I
Fujiwara, H.	373	Magnetic Properties of Rare Earth Copper Intermetallic Compounds,
	374	Magnetic Properties of Rare Earth Copper Intermetallic Compounds,
Fujiwara, N.	674	Synthesis of Nonstoichiometric Zirconium Carbide Whiskers by Chem
Fukazawa, T.	943	Analysis of Dislocations in Gd3Ga5O12 by a Repeated Etching Techn
Fukuda, T.	311	Growth of LiTaO3 Single Crystals for Saw Device Applications (Czo
Fukui, T.	312	Organometallic VPE Growth of InAs
	313	New Gaseous Impurity and Its Effect on the Purity of LPE GaAs (va
Funakubo, H.	857	Scanning Electron Microscopy of Growth Surfaces of Synthetic Quar
Furmanik, Z.	341	Crystallization of HgTe Crystals from the Vapour Phase
Furuhata, Y.	466	Morphology Control of NdP5O14 Single Crystals Grown from Polyphos
Furukawa, Y.	314	On the Growth Mechanism of Polycrystalline Snow Crystals with a S
	493	Epitaxial Relationships During the Formation of Three-Dimensional
Furukoshi, M.	477	Formation of Defects in Zinc Selenide Crystals Grown from the Mel
Gabbe, D.	1068	Magneto-Optical Properties of KTb3F10 and LiTbF4 Crystals (Czochr
Gabor, T.	242	Seeded Growth of Large Single Crystals of CdS from the Vapor Phas
Gabunia, D.L.	967	Some Peculiarities in the Growth and Properties of Beta-Boron Cry
Gabuniya, D.L.	968	Shape of the Crystallization Front During Electron-Beam Zone Refi
Gaiduk, E.A.	18	Computer Model of Autoepitaxial Growth of Ge Films (theory)
Gallagher, D.A.	315	Raman Scattering Studies in the Solid Electrolytes of the RbAg4I5
Galy, J.	316	The Crystal Structure of Te3Nb2O11
	547	Crystal Structure of Se2TiO6 (slow cooling)
Ganne, M.	317	A New Structural Family M2M'Ta5O15. Crystal Structure of CaTlTa
Gardner, R.N.	318	Controlled Growth of alpha-Fe Single Crystal Whiskers (hydrogen r
Garrard, B.J.	319	Flux Growth of Some Fluoride Crystals Under Reducing Conditions.
	320	Viscosity Measurements in the System Dy2O3-K2O-MoO3 (DyKMo2O8, me
	1063	The Flux Growth of Some New Rare Earth and Iron Group Complex Oxi
Garrett, I.	281	The Modified Entrainment Method and Its Application to the Study
Garrett, J.D.	321	Growth of Large Single Crystals of Heusler Alloys, Ni2MnSn and Ni
Garside, J.	166	Evaluation of Supersaturation in Crystal Growth from Solution (th
	322	Direct Observation of Secondary Nuclei Production (KAl(SO4)2.12H2
	904	The Thermodynamic Driving Force for Crystallization from Solution
Garton, G.	323	Crystal Growth of Some Rare Earth Trifluorides (GdF3, TbF3, DyF3,
	595	Flux Growth of Polymorphic Rare-Earth Disilicates, R2Si2O7 (R = T
	900	Flux Growth and Characterization by X-ray Topography of Rare Eart
Gasanov, G.Sh.	593	alpha-In2Te3 (melt)
Gasparrini, G.	292	A Possible Method for the Growth of Homogeneous Mercury Cadmium T
Gasperin, M.	81	Crystal Structure of La3ReO8
	103	Crystal Structure of Double Oxides of Rhenium. II. The Lanthanum
Gatos, H.C.	437	Liquid-Phase Electroepitaxy: Growth Kinetics (Peltier effect, Ga
	1088	Crystal Growth and Segregation Under Zero Gravity: Ge (Bridgman)
Gauthier, E.	123	Monte Carlo Simulation of the Crystal-Melt Interface of a Lennard
Gavrilova, T.A.	885	Growth of Pb0.8Sn0.2Te Films in a Quasi-Closed Space (vacuum evap
Gazda, S.	771	The Czochralski Growth of Bismuth-Germanium Oxide Single Crystals
Geballe, T.H.	1073	Czochralski Growth of A15 Structure Intermetallic Compounds (V3Si
Gebert, E.	324	The Crystal Structure of Lithium Uranate (Li2UO4, flux)
Gehlig, R.	824	Structural Phase Transition in Mixed Crystals WxMo1-xO3 (melt)
Geil, W.	325	Interaction Between Technological Parameters and Crystal Quality
	326	Effects of Radial and Axial Temperature Gradients on Tensions in
Genequand, P.	836	Pyroelectricity and Related Properties in the Fresnoite Pseudobin
George, J.	327	Growth of Phthalic Anhydride in a Closed Crystal-Vapour System (o
George, V.	328	Growth of Single Crystals of Cubic Zinc Sulphide by Double Decomp
Georgi, C.	843	Epitaxial Layers of CuInSe2 on GaAs
Gerasimenko, V.S.	509	Preparation and Principal Physicochemical Properties of Alkali Me
	726	Phase Equilibria and Properties of Single Crystals of Solid Solut

Gerbaux, X.	329	Infrared Trichroism of Ferroelectric Triglycine Sulphate Thin Fil
Gertner, E.R.	803	Liquid Phase Epitaxial Growth of Ga1-xInxSb on GaSb by Stepwise 3
Gervais, A.	239	Growth of Oriented Bicrystals of Nickel Oxide (NiO, arc image, fl
Ghadekar, S.R.	407	The Growth, X-Ray Studies and Microstructure of Solution-Grown Mi
Ghez, R.	330	Correlation of the Interface Reaction Constant with Elementary Su
	893	Conditions for Constant Growth Rate by LPE from a Cooling, Static
	894	Growth and Dissolution Kinetics of III-V Heterostructures Formed
Ghezzi, C.	305	Epitaxial Growth of Single Crystal Cd1-xZnxS Layers on (111) GaAs
Ghobarkar, H.	304	The Morphology of Hydrothermally Grown Strontium Paracelsian (Sr[
Ghose, S.	331	The Crystal Structure of Synthetic Lautarite, Ca(IO3)2 (flux)
Gibart, P.	332	Vapor Growth of HgCr2Se4 (epitaxy, theory)
Giess, E.A.	333	The Liquidus Temperature and Growth-Dissolution Kinetics of Garne
Giling, L.J.	118	Mechanisms of the Chemical Vapour Deposition of Silicon (Si)
	1015	Surface Morphology of HCl Etched Silicon Wafers (Si)
	1019	The Influence of Adsorption and Step Reconstruction on the Growth
Gill, J.S.	334	Kinetics of Growth of Calcium Sulfate Crystals at Heated Metal Su
Gilles, J.-C.	1034	Preparation, Stability, and Preliminary Crystallographic Study of
Gillessen, K.	603	Growth of InP Crystals by the Synthesis Solute Diffusion Method
Gillson, J.L.	867	Polymorphism in Bi2Sn2O7 (flux)
Gilmer, G.H.	1069	Thermodynamic Properties of Surface Steps (theory)
Ginderow, D.	335	Crystal Structures of Pb4In9S17 and Pb3In6.67S13 (melt)
Girault, B.	336	Synthesis and Characterization of the Ternary Compound ZnGeP2 (me
Gits-Leon, S.	337	Effect of Stirring on Crystalline Quality of Solution Grown Cryst
Givargizov, E.I.	338	Growth of Whiskers by the Vapor-Liquid-Solid Mechanism
Gladkaya, I.S.	1032	Synthesis of Single Crystals of Cubic Boron Nitride in Systems Co
Glass, A.M.	339	Calcium Orthovanadate, Ca3(VO4)2 - A New High-Temperature Ferroel
Glass, H.L.	340	LPE Growth of Lithium Ferrite on Spinel Substrate Crystals (films
Glaunsinger, W.S.	357	Physical Properties of the Quarternary Chalcogenides Cu2(I)B(II)C
Glazer, A.M.	800	The Effect of Impurities on the Piezoelectric Properties of Lithi
Godeny, I.	90	Connections Between Thermodynamic Quantities and Vacancy and Diff
Gokieli, B.	906	Reactive Pulse Plasma Crystallization of Diamond and Diamond-Like
Golacki, Z.	341	Crystallization of HgTe Crystals from the Vapour Phase
	342	Crystallization of HgCdTe Mixed Crystals from the Vapour Phase by
Goldman, A.M.	162	A Simple Method for Growing V3Si Single Crystals (electron beam m
Golecki, I.	343	Heteroepitaxy of a Deposited Amorphous Germanium Layer on a Silic
	537	Solar Furnace Annealing of Amorphous Si Layers (films)
Golenko, V.P.	344	Growth of Zircon Single Crystals from a Solution in a Melt (ZrSiO
Golic, L.	253	Crystal Growth of Some Alkaline-Earth Rare-Earth Pentaoxometallat
Goltz, G.	573	The Preparation of Large Spherical Iron Single Crystals (Fe, stra
Gombia, E.	305	Epitaxial Growth of Single Crystal Cd1-xZnxS Layers on (111) GaAs
Gomi, M.	5	Magnetic and Mossbauer Studies of FeNi2BO5 and FeNiBO4 (ferrites,
Goncharov, L.A.	345	Behavior of Boron in Ge Single Crystals Grown by the Czochralski
	346	Behavior of Be as an Impurity During Crystallization of Ge by the
Gonda, S.	617	Nitrogen Doping into GaAs1-xPx Using Ionized Beam in Molecular Be
	1104	Cathodoluminescence of Impurity-Doped Aluminum Nitride Films Prod
Gonda, T.	347	Morphology of Ice Droxtals Grown from Supercooled Water Droplets
Gondrand, M.	205	Hydrothermal Synthesis of Several Mixed Oxides of the A(6+)B2(3+)
	389	Structure of Monocrystalline WV2O6 at 298 and 383 K (vapor transp
Goodman, C.H.L.	348	Crystal Growth - Theory and Techniques, Volume 2 (review, epitaxy
	349	Subsidiary Electrical Heating for Verneuil Furnaces in the USSR (
Goreaud, M.	350	The Thallium Tungstate Tl2W4O13: A Tunnel Structure Related to t
Gorecki, T.	351	Heat Capacity of Metals Near the Melting Point and the Vacancy Me
Gorochov, O.	543	Crystal Structure and Physical Properties of CuO0.65VS2 (vapor tr
	544	Preparation, Crystal Structure, and Physical Properties (transpor
Gorog, T.	352	Epitaxial Growth of A(III)B(V) Semiconductors from Vapour Phase (
Goryainov, L.A.	57	Analytical Study of the Thermal Transfer Process in an Installati
Gossard, A.C.	353	GaAs/AlAs Layered Films (epitaxy)
	761	Molecular Beam Epitaxy of Ge and Ga1-xAlxAs Ultra Thin Film Super
	762	Crystal Growth Kinetics in (GaAs)n-(AlAs)m Superlattices Deposite
Goto, M.	665	Flux Growth of Bi2WO6 Single Crystal Below the Transformation Tem
Gottschalch, V.	354	Liquid Phase Epitaxial Deposition of GaP on GaAs
Gourmala, M.	299	Vapor Phase Transport and Crystal Growth of Molybdenum Trioxide a
	835	Metal-Insulator Transitions and Phase Diagram of (Ti1-xVx)407: E
Gouskov, A.	187	Growth of Crystalline Slabs of Layered InSe by the Czochralski Me
	336	Synthesis and Characterization of the Ternary Compound ZnGeP2 (me
Grange, G.	532	Semi-Periodic Closed Growth Lamellae on the Crystal-Melt Interfac
Grange, Y.	541	Crystal Growth by the Thermic Screen Translation (TST) Technique:
Greedan, J.E.	321	Growth of Large Single Crystals of Heusler Alloys, Ni2MnSn and Ni
Greene, J.E.	269	Growth and Electrical Properties of Sputter-Deposited Single-Crys
Greenwell, D.W.	7	Congruent (Diffusionless) Vapor Transport (theory, kinetics)
Gregory, R.O.	358	Limitations in Using Kilohertz Radio Frequencies for Float Zone S
Griffin, J.A.	1068	Magneto-Optical Properties of KTb3F10 and LiTbF4 Crystals (Czochr
Griffith, W.T.	355	On the Transition from Faceted to Non-Faceted Growth in Melt-Grow
Grimouille, G.	446	Crystal Growth and Imperfections in Gadolinium Molybdate and Othe
	447	Crystal Growth and Crystallographic Data of Some Ln2(MoO4)3 Type
Grosskreutz, W.	458	Growth, Spectroscopy, and Stimulated Emission of Cubic Bi4Ge3O12

Gruehn, R.	787	New Observations in the Chemical Transport of GeO2. I. Transport
Grun, R.	356	The Growth of beta-Si3N4 Single Crystals
Grushin, A.I.	306	Preparation of Films of IV-VI Compounds in a Quasi-Sealed Volume
Gruska, R.P.	911	Preparation and Properties of Two Indium Antimony Selenides (In2S
Gruttner, A.	186	New Ternary Mo(II)-Compounds InxMo15Se19 Containing Mo6Se8 and Mo
Grzesiak, E.	127	Chemical Synthesis and Crystal Growth of Laser Quality Praseodymi
Gschneidner, K.A.,Jr.	620	Single Crystal Growth by the Horizontal Levitation Zone Melting M
Guedu, H.	836	Pyroelectricity and Related Properties in the Fresnoite Pseudobin
Guen, L.	357	Physical Properties of the Quarternary Chalcogenides Cu2(I)B(II)C
Guillaume, J.C.	600	Luminescence and Bandgap Studies of High Purity MgxZn1-xTe Ternar
	792	Preparation, Analysis and Properties of High Purity Zn(1-x)MgxTe
Guillot, M.	767	Magnetic Properties of a Single Crystal of NdF3 Between 2 and 4.2
Guitel, J.-C.	261	Crystal Structure of the Mercury Nitro-Phosphate Hg4PO4NO3.H2O (m
	608	Crystal Structure of AgH2PO4. Crystallography of AgH2AsO4 (solut
Guitel, J.C.	21	Crystal Structure of beta-KEr3F10 (melt, slow cooling)
	142	Crystal Structure of Cu4(PO4)2O (flux)
	609	Crystal Structure of a New Variety of Nickel Pyrophosphate: Ni2P2
	986	Crystal Structure of Monophosphate AgCoPO4
Guittard, M.	259	Crystal Structure of Lanthanum Polysulfide LaS2
	983	Crystal Structure of the Spinel CdErS4 (slow cooling)
	1048	Mixed Oxysulfides of Chromium III and Rare Earths (LaCrOS2, CeCrO
Gukov, O.Ya.	1003	Heterogeneous Equilibria and Mass Transfer Kinetics in the System
Gupta, B.K.	14	Single Crystal Growth of Stannous Selenide (SnSe, Bridgman)
Gupta, K.P.	358	Limitations in Using Kilohertz Radio Frequencies for Float Zone S
Gurin, V.N.	359	The Preparation of Single Crystals of the Rare Earth Borides by t
Gurtler, R.	546	Properties of Rapidly-Grown RTR Silicon Sheet and Their Effects o
Gurtler, R.W.	360	Silicon Ribbon Growth via the Ribbon-To-Ribbon (RTR) Technique:
Guseinov, G.D.	361	Structure and Growth Peculiarities of Tl-Se-TlInSe2 (TlSe, zone m
Guseinov, G.G.	22	Growth and X-Ray Analysis of Single Crystals of EuSb4S7 and EuSb2
	361	Structure and Growth Peculiarities of Tl-Se-TlInSe2 (TlSe, zone m
	817	Synthesis, Preparation of Single Crystals, and Physicochemical In
Guseinov, M.S.	22	Growth and X-Ray Analysis of Single Crystals of EuSb4S7 and EuSb2
Gustin, D.	1071	Measurements of the Contact Angle Between Melt and Crystal During
Hadni, A.	329	Infrared Trichroism of Ferroelectric Triglycine Sulphate Thin Fil
Hagen, W.	362	Tetragonal Distortion in Heteroepitaxial Layers: Ge On GaAs (vap
	363	In Situ X-Ray Topography of Epitaxial Ge Layers During Growth
Hagenmuller, P.	137	Czochralski Crystal Growth of Rare Earth Tetraborides (YB4, TbB4,
	191	Germanium-Iodine Equilibrium from 700 to 1000 K (Ge, vapor transp
Hager, R.J.	849	Solution Growth of Indium-Doped Silicon (Si)
Haire, R.G.	153	Americium Ditelluride (AmTe2-x
Hall, H.T.,Jr.	918	VPE Growth of AlxGa1-xAs (theory)
	919	Organometallic VPE Growth of AlxGa1-xAs (vapor transport)
Hall, N.J.G.	845	Magnetic Properties of beta-Phase Iron-Germanium (FexGe, slow coo
Hallais, J.	974	Ellipsometric Assessment of (Ga,Al)As/GaAs Epitaxial Layers Durin
Hallett, J.	28	Influence of Environmental Saturation and Electric Field on Growt
Hamakawa, Y.	948	Liquid Phase Epitaxy of AlxGa1-xP on GaP and Its Application to D
	1098	Characterization of the Optical Properties of LPE InxGa1-xAsyP1-y
Hamamura, K.	364	Growth and Morphology of Copper-Nickel Alloy Crystals by Hydrogen
Hamersky, J.	365	Influence of the Source Material and the Source Shape on the Subl
Hamid, S.A.	366	Electron Microscopic Characterization of the Hydrothermal Growth
Handa, V.K.	812	Kinetics of Crystallization of Barium Tungstate from Solutions in
Hanke, G.	1002	A Model for Macroscopic Czochralski Growth: Theoretical and Expe
Hansel, T.	844	Structural and Electrical Properties of CuGaSe2 Thin Films on GaA
Hara, S.	1093	Morphology of Iron Pyrite Crystals (FeS2, vapor transport)
Hara, T.	988	Impurity Effect on the Formation of Terraces in GaAs LPE Growth (
Harano, Y.	367	Measurement of Crystallization of Potassium Bromate from Its Quie
Harbec, J.Y.	434	Raman Spectra of ZrS3
Hari Babu, V.	908	Growth and Dissolution of BaFBr Crystals (flux)
Harmon, T.C.	368	Optically Pumped LPE-Grown Hg1-xCdxTe Lasers (films, flux)
Harrill, M.D.	369	Thermal Analysis of Solidification in Web-Dendritic Ribbon Growth
Harris, J.E.	74	The Effects of Growth Speed on the Compositional Variations in Cr
	75	A Study of Casting in the CdxHg1-xTe System
	164	An Improved System for the Bridgman Growth of Crystals with Toxic
Harrison, H.R.	370	Skull Melter Growth of Magnetite (Fe3O4)
Haschke, J.M.	531	Preparation, Phase Equilibria, and Crystal Chemistry of La, Pr, a
Hashimoto, K.	405	Electrical Properties of TlTe Single Crystals (melt, theory)
Hashimoto, S.	371	Two-Phase (alpha/beta)-Brass Bicrystals Produced by a Solid-Solid
	372	Two-Phase (alpha/beta)-Brass Bicrystals Produced by a Solid-Solid
Hashimoto, T.	632	The Melt Growth and Thermal Expansion of Na2CO3 Crystals (Bridgma
Hashimoto, Y.	373	Magnetic Properties of Rare Earth Copper Intermetallic Compounds,
	374	Magnetic Properties of Rare Earth Copper Intermetallic Compounds,
Hasiguti, R.R.	375	Morphology and Growth Mechanism of Vapor Grown Cd Crystals as Aff
Hassan, M.A.	376	Dynamic Sublimation Technique for Growing Mercuric Iodide Crystal
Hatto, P.	47	Liquidus Measurements in the Pb-Sn-Te System (Pb1-xSnxTe, PbTe, f
Hattori, T.	264	Photoconduction of Lead Trititanate (PbTi3O7, flux)
Haubenreisser, W.	59	A Fluctuation Theory of Normal Crystal Growth for Metallic One-Co

	765	Analytical Theory of Crystal Growth of Two-Component Systems I.
Haupt, H.J.	377	Crystal Structure, Growth, and Optical Properties of Alkali-Triha
Haussuhl, S.	378	Ferroelasticity and Phase Transformation in Rb2Hg(CN)4 Spinel (so
Hawthorne, F.C.	379	The Crystal Structure of Ba2V2O7 (melt)
Hayama, F.	931	Crystal Growth of Co Ferrite on Fine Acicular gamma-Fe2O3 Particl
Hayashi, H.	49	Growth of a Crystal Needle of Si2ON2 by Nitridization in the Si-S
	858	Vapor Phase Growth of Thin Plates of Cr2O3 Single Crystals
Hayashi, I.	518	Second Harmonic Generation in a Sputtered LiNbO3 Film on MgO (epi
Hayashi, S.	380	Preferred Growth Direction of Pure Cu Single Crystals by the Floa
	381	Growth of Cobalt Single Crystals by the Electron Beam Floating-Zo
Heaton, J.L.	1057	GaAs Vapour Phase Epitaxy Using Computer Control
Hebecker, C.	570	Preparation and Crystal Structure of CsAlF4
Heidner, E.	382	Protein Crystallizations (solution, organics)
Heidrich, K.	377	Crystal Structure, Growth, and Optical Properties of Alkali-Triha
Heifets, A.Ya.	978	Kinetic Characteristics of Growth of Y3Al5O12 Single Crystals fro
Heimann, R.B.	304	The Morphology of Hydrothermally Grown Strontium Paracelsian (Sr[
Hein, K.	86	Growth and Physical Properties of ZnSiP2 (vapor transport, soluti
	1086	Partition of Principal Components in the Course of Preparation of
Henkie, Z.	383	Growth of USb2 Single Crystals and Their Structural Perfection (v
Henoc, P.	114	Growth and Characterization of Undoped ZnSe Epitaxial Layers Obta
Henry, R.L.	384	InP Growth and Properties (Czochralski, doped)
Herkert, R.	583	Liquid Encapsulation Zone-Refining of PbS (zone melting, vapor tr
Herman, C.J.	1088	Crystal Growth and Segregation Under Zero Gravity: Ge (Bridgman)
Hermoneit, B.	458	Growth, Spectroscopy, and Stimulated Emission of Cubic Bi4Ge3O12
Herron, L.W.	385	Heat-Flow-Controlled Growth During Li2B4O7 Crystallization (melt)
Hibiya, T.	386	Electrical Properties and Charge Imbalance for Ca,Ge-Substituted
Hidaka, Y.	386	Electrical Properties and Charge Imbalance for Ca,Ge-Substituted
Higashi, I	679	The Hardness of NbB2 Single Crystals (flux, slow cooling)
Hilborn, R.B.,Jr.	369	Thermal Analysis of Solidification in Web-Dendritic Ribbon Growth
Hildisch, L.	277	Solution Growth and Some Properties of GaP Bulk Crystals
Hillel, R.	557	Crystal Structure of the Indium Selenide In2Se3 (vapor transport)
Hillhouse, R.W.A.	387	Plasma Growth of Rutile Crystals and Their Photoelectronic Proper
Hine, S.	875	Epitaxial Thin Iron Films on Antimony (Fe, vapor transport)
Hirai, T.	696	Growth, Morphology and Slip System of alpha-Si3N4 Single Crystal
Hiramatsu, M.	710	Edge Emission in Melt-Grown ZnSexS1-x (Bridgman)
Hirano, H.	311	Growth of LiTaO3 Single Crystals for Saw Device Applications (Czo
Hirao, M.	249	Improvement of Crystal Composition in Ga1-xAlxAs LPE Layers Grown
Hirata, M.	642	Polytypism and Amorphousness in Silicon Whiskers (Si, vapor trans
Hirokawa, Y.	711	Chemical Vapor Deposition of Single-Crystalline ZnO Film with Smo
Hiwatari, Y.	965	Geometrical Analysis of Crystallization of the Soft-Core Model in
Ho, J.H.	388	(111) Cu2O Growth Modes on (111)Cu Surfaces (epitaxy)
	1045	Recent Developments in the Microstructural Characterization of Co
Hoche, H.	105	Interaction Between Successive Monatomic Step Trains of Different
Hocking, M.G.	42	The Vapour Transport of NbB2 and TaB2 (kinetics, theory)
Hockly, M.	126	Nature, Origin and Effect of Dislocations in Epitaxial Semiconduc
Hodeau, J.L.	389	Structure of Monocrystalline V2O6 at 298 and 383 K (vapor transp
	390	The Crystal Structure of V4O7 at 120 degrees K (vapor transport)
	391	Structural Aspects of the Metal-Insulator Transitions in (Ti0.997
Hodorowicz, S.	525	On the Effect of pH of Solution on the Kinetics of Crystallizatio
Hodorowicz, S.A.	392	The Growth Kinetics of Urea Monocrystals from Aqueous Solution (C
Hoekstra, H.R.	324	The Crystal Structure of Lithium Uranate (Li2UO4, flux)
Hogan, L.M.	489	Fivefold Twinned Silicon Crystals Grown in an Al-16 wt.% Si Melt
Hogg, J.H.C.	1058	Indium Polytelluride In2Te5(II) (melt)
Hollan, L.	393	Fast Growth in GaAs VPE at Low Temperature and High Partial Press
Holland, G.E.	620	Single Crystal Growth by the Horizontal Levitation Zone Melting M
Holloway, H.	394	MBE Techniques for IV-VI Optoelectronic Devices (review)
Holmes, D.E.	395	Fiber Optic Probe for Thermal Profiling of Liquids During Crystal
Holt, D.B.	73	Growth of Films of CdS Evaporated in Vacuum onto Silicon (vapor t
Holtz, J.E.	1057	GaAs Vapour Phase Epitaxy Using Computer Control
Holzer, A.	834	State-of-The-Art of Crystal Growth and Nuclear Spectroscopic Eval
Homma, S.	645	Interface Shape Transitions in Czochralski Grown YAG Crystals (Y3
Hong, H.Y.-P.	396	Crystal Structure and Ionic Conductivity of Li24Zn(GeO4)4 and oth
	397	Crystal Structure of K5NdLi2F10 (flux)
Hong, H.Y-P.	189	Lasing and Fluorescence in K5NdLi2F10 (flux)
Honjo, G.	949	Roles of Lattice Fitting in Epitaxy (theory, Ni, Cu, Pd, Pt, Fe,
	950	Techniques for Routine UHV in situ Electron Microscopy of Growth
Honma, S.	644	Interface Shape Transitions in the Czochralski Growth of Dy3Al5O1
	646	The Growth of Dy3Al5O12 (Czochralski)
Hooper, R.M.	115	The Growth of Single Crystals of Some Organic Compounds by the Cz
Hopkins, R.H.	855	Solute Partitioning During Silicon Dendritic Web Growth (Si, theo
Hoppe, R.	135	The First Oxomanganate(II): Na14Mn2O9 = Na14[MnO4]2O
	398	Advances in Preparative Chemistry of Oxides and Fluorides (review
	1082	A Novel Ruthenium Bronze: KRu4O8 (solution)
Horak, J.	240	Preparation and Basic Physical Properties of BiTeI Single Crystal
Hori, M.	9	Epitaxial Growth of Potassium Lithium Niobate Single-Crystal Film
Horig, W.	843	Epitaxial Layers of CuInSe2 on GaAs

Horikoshi, Y.	312	Organometallic VPE Growth of InAs
Horinaka, H.	935	Photoluminescence Properties of CuGaSe2 Grown by Iodine Vapour Tr
Hornstra, J.	399	Determination of the Lattice Constant of Epitaxial Layers of III-
Hornung, J.	433	Influence of the VPE Parameters on the Epitaxial Film Properties
Hoshikawa, K.	1050	Interface Growth Feature and Voids in Sapphire Ribbon Crystals (a
	1051	Dislocations in Sapphire Ribbon Crystals Grown by the Edge-Define
Hosoya, S.	484	Re-Investigation of the Re-Entrant Corner Effect in Twinned Cryst
Hottier, F.	974	Ellipsometric Assessment of (Ga,Al)As/GaAs Epitaxial Layers Durin
Huber, W.	38	Solution Growth of RbAg4I5 Crystals
	39	Layer Perovskites of the (CnH2n+1NH3)2MX4 and NH3(CH2)mNH3MX4 Fam
	569	CdTe Thin Films Grown by Hot Wall Epitaxy
Hull, G.W.	1073	Czochralski Growth of A15 Structure Intermetallic Compounds (V3Si
Hulliger, F.	400	Low Temperature Thermal and Magnetic Properties of CeP and CeAs (
	401	Crystal Structure and Antiferromagnetism of EuSb2 (melt)
Humphreys, C.J.	149	A Theoretical Study of Temperature Distributions During Czochrals
Hurle, D.T.J.(ed.)	66	Crystal Growth (Proceedings of the Third International Summer Sch
Huseynov, B.	598	The Growth of CdAs2 and ZnAs2 Single Crystals from the Vapor Pha
Hussain, A.	402	On the Alkali Metal Tungsten Bronzes, in Particular those of Pota
Huyberts, M.A.H.	216	Thin Film Growth of (MnxZn1-x)Fe04 by Liquid Phase Epitaxy on Bri
Hwang, H.L.	403	Growth of CuInS2 and Its Characterization (vapor transport)
Idler, K.L.	692	Crystallographic Studies and Structural Systematics of the C2/c A
Igaki, K.	647	Vapor Phase Transport and Stoichiometry Control of Cadmium Sulfo-
	648	Vapor Phase Transport and Stoichiometry Control of Cadmium Sulfid
Ii, N.	404	Single Crystal Growth of Akermanite (Ca2MgSi2O7) and Gehlenite (C
	483	Interface Shape and Horizontal Variations of Al and Ga Contents i
	879	Single Crystal Growth of Substituted Yttrium Iron Garnets Y3Fe5-x
Ikari, T.	405	Electrical Properties of TlTe Single Crystals (melt, theory)
Ikeda, T.	582	Dielectric and Optical Properties of K5Nd(MoO4)4 (Czochralski)
Il'in, A.G.	231	Kinetics of Gallium Phosphide Synthesis by Chemical Gas Transport
Illarionov, A.A.	1003	Heterogeneous Equilibria and Mass Transfer Kinetics in the System
Imamura, Y.	951	Quasi-Grown Layers in Liquid Phase Epitaxial Growth (In1-xGaxAs)
Immorlica, A.A.,Jr.	574	Temperature Gradient Measurement in Liquid Epitaxial Growth Syste
Inagaki, M.	675	Growth of Single Crystals of Na3Al2Li3F12 Garnet Under Hydrotherm
Indenbom, V.L.	406	Formation of Stress and Dislocations in Crystal Growth (Si, Ge, G
Ingle, S.G.	407	The Growth, X-Ray Studies and Microstructure of Solution-Grown Mi
Inokuchi, K.	703	Observations of Defects in LPE GaAs Revealed by New Chemical Etch
Inoue, M.	957	Growth of GaAs1-xSbx Crystals by Steady-State Liquid Phase Epitax
Inoue, N.	1052	Growth of Stacking Faults by Bardeen-Herring Mechanism in Czochra
Inoue, T.	380	Preferred Growth Direction of Pure Cu Single Crystals by the Floa
	408	An Evidence of Spiral Growth on the Surface of Melt-Grown KCl Cry
	409	Effects of Growth Rates on the Defect Generation in KCl Crystals
Inoue, Y.	937	In-Process Monitoring and Control of Doping Gas Concentration Dur
Inuishi, Y.	463	Electrical Conductivity and Galvanomagnetic Effects in (SN)x Sing
	938	Growth of High-Purity ZnTe Single Crystals by the Sublimation Tra
	939	Growth by Travelling Heater Method and Characteristic of Undoped
	957	Growth of GaAs1-xSbx Crystals by Steady-State Liquid Phase Epitax
Irie, T.	410	Electrical Properties of p- and n-Type CuInSe2 Single Crystals (m
Irwin, J.C.	1004	Raman and Infrared Spectra of CdIn2S4 and ZnIn2S4 (vapor transpor
Isaacs, T.J.	411	Crystal Growth of Tl3VS4 (Bridgman-Stockbarger)
Isawa, N.	461	Improvements in SSD (synthesis, solute diffusion) Growth of GaP
Iseler, G.W.	412	Optical and Electrical Properties of CdGeAs2 (Bridgman)
Ishibashi, S.	413	Epitaxial Growth of CrO2 on Sapphire in Air (vapor transport)
	414	Epitaxial Growth of Ferromagnetic CrO2 Films in Air
Ishibashi, T.	375	Morphology and Growth Mechanism of Vapor Grown Cd Crystals as Aff
Ishibashi, Y.	632	The Melt Growth and Thermal Expansion of Na2CO3 Crystals (Bridgma
	633	Optical and Dilatometric Studies of KCaCl3 and RbCaCl3 Crystals (
Ishida, M.	415	Epitaxial Growth of Ferroelectric PLZT ((Pb,La)(Zr,Ti)O3) Thin Fi
	828	Liquid Phase Epitaxial Growth of CdTe in the CdTe-CdCl2 System (f
Ishida, T.	615	Epitaxial Growth of ZnS by Zn-S-H2 CVD Method
Ishii, M.	416	Liquid Phase Epitaxial Growth of Zn and S Doped GaAs
	819	Effect of Addition of Vanadium on the Chemical Transport of TiSx,
	943	Analysis of Dislocations in Gd3Ga5O12 by a Repeated Etching Techn
Ishii, T.	1100	Molecular Beam Epitaxy of GaSb and GaSbxAs1-x (films)
Ishikawa, Y.	956	Magnetic Structure of GdCu1-xZnx System (melt)
Ishino, T.	580	Synthesis and Characterization of Europium(II)-Holoborates, Eu2B5
Ishitani, A.	518	Second Harmonic Generation in a Sputtered LiNbO3 Film on MgO (epi
Ishizawa, Y.	964	Fermi Surface Measurements of ZrB2 by the de Haas van Alphen Effe
Ishizuki, H.	417	Growth Mechanism of Multi-Steps on the Surface of Vapour Grown Fe
	418	Growth of Triple Unit Cell Steps on the Surface of Vapour Grown F
	419	Step Growth by Elementary Steps and Multi-Steps on the Surface of
Ismailov, M.Z.	361	Structure and Growth Peculiarities of Tl-Se-TlInSe2 (TlSe, zone m
Itaya, Y.	35	Conditions of LPE Growth for Lattice Matched GaInAsP/InP DH Laser
Ito, H.	422	Growth of p-Type InP Single Crystals by the Temperature Gradient
Ito, K.	420	Preparation of Ge-Si Epitaxial Alloys by Sputtering
	421	Crystallography of the fcc Phase in Uni-Directionally Solidified
	422	Growth of p-Type InP Single Crystals by the Temperature Gradient

Ito, R.	249	Improvement of Crystal Composition in Ga1-xAlxAs LPE Layers Grown
Ito, S.	423	Vapor Growth of Hexagonal GeO2 Needle Crystals
	424	Vapor Growth of Zn2GeO4 Single Crystals
Ito, T.	311	Growth of LiTaO3 Single Crystals for Saw Device Applications (Czo
Itoh, H.	947	Growth Parameters and Crystal Morphology of Vapor-Deposited Niobi
Itoh, N.	425	Birefringence in CdSiP2 (solution)
Itoh, S.	651	Ionized-Cluster Beam Epitaxial Growth of GaP Films on GaP and Si
Ittyachen, M.A.	3	New Gel Method for Growing Large Needles and Single Crystals of L
	519	Growth Kinetics and Habit Modification of Barium Molybdate Single
Ivanova, I.V.	866	Electron Diffraction Study of Phase Transformations of the Compou
Iwai, S.	719	The Crystal Structure of Cs6W11O36
	720	The Crystal Structure of K2W4O13
Iwakoshi, S.	926	Chemical Vapor Growth of Titanium Diboride by a Modified Hot Wire
Iwanaga, H.	426	Crystal Growth in c Direction and Crystallographic Polarity in Zn
	427	Growth Mechanism of Hollow ZnO Crystals from ZnSe. II
	428	Crystal Growth and Sublimation in II-VI Compounds Along their Pol
Iwanov, D.	429	Direct Synthesis and Structure of Epitaxial Cadmium Sulphide Laye
Iwasaki, H.	643	Epitaxial Growth of KNdP4O12 Laser Waveguides (KLaP4O12, solution
	923	Large-Sized InP Single Crystals by the Synthesis, Solute Diffusio
Iwasaki, N.	712	Horizontal Zone-Melting of Indium Antimonide in Vacuum. I. Basi
Iwata, N.	430	Magnetic Properties of (Mn,Cr)P and (Mn,Fe)P Compounds (Mn1-xFexP
Izumi, F.	431	Growth of Anatase (TiO2) Crystals by Chemical Transport Reactions
Izumi, O.	33	Orientation and Temperature Dependence of the Flow Stress in the
Jacobs, K.	354	Liquid Phase Epitaxial Deposition of GaP on GaAs
Jacobsen, C.S.	87	A New Organic Low Temperature Conductor: HMTSF-TNAP (solution, or
Jacquier, B.	756	Optical Properties of Cs2NaBiCl6
Jain, A.K.	14	Single Crystal Growth of Stannous Selenide (SnSe, Bridgman)
Jain, P.C.	432	Effect of Growth Conditions of Polytypism in Cadmium Iodide Cryst
Jakobus, T.	433	Influence of the VPE Parameters on the Epitaxial Film Properties
Jandl, S.	434	Raman Spectra of ZrS3
Jansen, A.K.	435	Morphological Stability Analysis in Chemical Vapour Deposition Pr
	1010	Morphological Stability Analysis in Chemical Vapour Deposition Pr
	1011	Interface Morphology in Chemical Vapour Deposition on Profiled Su
Janssen, E.M.W.	436	Crystal Growth and the Crystal Structures of Two Modifications of
Jastrzebski, L.	437	Liquid-Phase Electroepitaxy: Growth Kinetics (Peltier effect, Ga
Jean-Louis, A.M.	114	Growth and Characterization of Undoped ZnSe Epitaxial Layers Obta
	745	Properties of Homogeneously Doped ZnSe Single Crystals Obtained b
Jedamzik, D.	1002	A Model for Macroscopic Czochralski Growth: Theoretical and Expe
Jeffes, J.H.E.	42	The Vapour Transport of NbB2 and TaB2 (kinetics, theory)
	73	Growth of Films of CdS Evaporated in Vacuum onto Silicon (vapor t
Jeitschko, W.	596	Ternary Lanthanoid-Transition Metal Pnictides with ThCr2Si2-Type
Jeitschko, W.	438	Synthesis and Crystal Structure of Iron Polyphosphide FeP4
Jellinek, F.	814	The Crystal Structure of Niobium Trisulfide, NbS3
Jensen, J.D.	439	Electrical Properties of Pb1-xCdxS Epitaxial Films (vapor transpo
John, M.	86	Growth and Physical Properties of ZnSiP2 (vapor transport, soluti
John, V.	237	Flux Growth and Characterization of Fluorite Crystals (CaF2)
Johnston, D.C.	293	Magnetic, Transport, and Thermal Properties of Ferromagnetic EuB6
	791	Superconductivity of Single-Crystal NbGe2
Joly, J.P.	695	Solution Growth of Sparingly Soluble Single Crystals from Soluble
Jones, A.L.	779	Effect of S-35 Tracer, Particle Size and Imperfection on the Solu
Jones, G.A.	867	Polymorphism in Bi2Sn2O7 (flux)
Jones, G.R.	440	The Growth and Properties of Crystalline Rubidium and Cesium Niob
	1076	Ca12Al14O33: A New Piezoelectric Material (Czochralski)
Jonker, H.D.	1018	Flux Growth of ZnGa2O4 Single Crystals
Jorgensen, J.-E.	441	Growth of Chromium Silicide, Cr3Si, Crystals (float zone melting)
Joshi, M.S.	442	Single Crystal Growth of Ferroelectric Potassium Dihydrogen Ortho
	443	Nucleation in Supersaturated Potassium Dihydrogen Orthophosphate
	444	Growth of Fibrous Sodium Aluminosilicate Crystals in Silica Gel (
Joubert, J.C.	104	High Pressure Synthesis and Structure of a New Pyrochlore Contain
	205	Hydrothermal Synthesis of Several Mixed Oxides of the A(6+)B2(3+)
	389	Structure of Monocrystalline WV2O6 at 298 and 383 K (vapor transp
	527	Crystal Chemistry of Perovskite Phases of the Systems Th0.25NbO3-
	661	Synthesis and Crystal Structure of a New Mixed Oxide "FeV3O8" (Fe
	662	Synthesis and Crystal Structure of a New Mixed Oxide FeV2O6OH0.5:
	1036	Crystal Structure of CuTa2O6 (flux)
Joukoff, B.	445	Flux Growth of Lead Neodymium Chlorapatite (Na2Nd2Pb6(PO4)6Cl, sl
	446	Crystal Growth and Imperfections in Gadolinium Molybdate and Othe
	447	Crystal Growth and Crystallographic Data of Some Ln2(MoO4)3 Type
Joyce, B.A.	301	Surface Processes Controlling the Growth of GaxIn1-xAs and GaxIn1
	686	Temperature Range for Growth of Autoepitaxial GaAs Films by MBE (
Kaczmarek, F.	127	Chemical Synthesis and Crystal Growth of Laser Quality Praseodymi
Kadeckova, S.	1028	Growth Striations in Fe-3wt%Si Single Crystals Grown by Floating
Kaganov, V.Ya.	473	Structural Transitions and Phase Composition of Thin Films of Alu
Kahlweit, M.	172	On the Growth of NH4Cl Dendrites at Very Low Supersaturation (sol
Kainthla, R.C.	868	Electroless Deposition of Semiconductor Films (PbS, PbSe, Pb1-xHg
Kaito, C.	448	Coalescence Growth of Smoke Particles Prepared by a Gas-Evaporati

	881	Growth of Thallium Iodide Crystals in Silica Hydrogels (TlI, Tl2I
Kakinoki, J.	874	Morphology and Structure of ZnS Grown from the Vapour Phase
Kaldis, E.	79	Magnetic Mixed Valent TmSe (melt)
Kaldis, E. (ed.)	449	Current Topics in Materials Science, Volume 2 (Review)
	450	Current Topics in Materials Science, Volume 1 (review)
	946	Vapour Growth and Epitaxy (review)
Kalejs, J.P.	451	Impurity Redistribution in EFG (Si, ribbon, theory, kinetics)
	452	Aluminum Redistribution in EFG of Silicon Ribbon (Si)
Kalinin, V.B.	721	New Ferroelectric Na3Sc2(PO4)3 (melt)
Kalinnikov, V.T.	50	Properties of CuCr2Se4 Crystals Obtained from Solutions in Melts
	51	Crystal Growth of CuCr2Se4
	586	Magnetic Properties of CdCr2Se4-ZnCr2Se4 Single Crystals (melt)
	1037	Photoferromagnetic Effect and Photoconductivity of Cd1-xHgxCr2Se4
Kalisher, M.H.	691	Thin Films of LiNbO3, Doped with Na+ and Co2+ plus Zr4+, Grown by
Kambas, K.	453	Far Infrared Optical Study of alpha-In2Se3 Compound (melt, slow c
Kamegashira, N.	454	Preparation of ThSiO4 Single Crystals by a Vapour Phase Reaction
	455	Growth of Single Crystals of U1-xThxO2 Solid Solutions by Chemica
	674	Synthesis of Nonstoichiometric Zirconium Carbide Whiskers by Chem
Kamigaki, N.	228	Preparation Method of High Purity Aluminum Single Crystals with L
Kamijoh, T.	456	Single Crystal Growth of LiInS2 (directional solidification)
	520	Single Crystal Growth of Li2ZnGe (melt)
Kaminskii, A.A.	46	Structure, Spectroscopy, and Stimulated Emission of Crystals of Y
	457	Growth, Spectroscopic Investigations, and Some New Stimulated Emi
	458	Growth, Spectroscopy, and Stimulated Emission of Cubic Bi4Ge3O12
Kamitakahara, W.A.	459	Local Modes in Rb1-cKc Alloys: A Neutron Scattering Study (Bridg
Kamiya, K.	822	Structure and Stability of Alkali Germanate Crystals and Glasses
Kan, H.	460	Liquid Phase Epitaxial Growth of Lattice-Matched GaP-Ga0.5Al0.5P-
Kanai, W.	716	Growth of NaNd(MoO4)2 Single Crystals from the Ternary System Nd2
Kaneko, K.	461	Improvements in SSD (synthesis, solute diffusion) Growth of GaP
Kaneko, T.	462	Growth Rate of Iron Whiskers (Fe, vapor transport, theory)
Kaneta, K.	5	Magnetic and Mossbauer Studies of FeNi2BO5 and FeNiBO4 (ferrites,
	6	Mossbauer Study of FeMo2S4 (vapor transport)
Kaneto, K.	463	Electrical Conductivity and Galvanomagnetic Effects in (SN)x Sing
Kaplow, R.	303	Performance of a New High-Intensity Silicon Solar Cell (Si, float
Karelin, B.V.	27	Doping of Silicon Crystals with Phosphorus During Growth from the
Kargareteli, L.N.	733	Synthesis of Single Crystal LnPO4 (vapor transport)
Kariya, T.	948	Liquid Phase Epitaxy of AlxGa1-xP on GaP and Its Application to D
	1098	Characterization of the Optical Properties of LPE InxGa1-xAsyP1-y
Karnezos, M.	464	Uniaxial Ferromagnetism in NiZrF6.6H2O (solution)
Karolczak, J.	127	Chemical Synthesis and Crystal Growth of Laser Quality Praseodymi
Karunakaran, C.	465	Effect of Pressure on Electrical Resistivity and Thermoelectric P
Kasano, H.	466	Morphology Control of NdP5O14 Single Crystals Grown from Polyphos
Kashida, S.	467	Ultrasonic Velocities of K2PbCu(NO2)6 Near the Cooperative Jahn-T
Kashihara, Y.	468	Upper and Lower Critical Fields of TaS2(Pyridine)1/2 (vapor trans
Kashiwaya, K.	228	Preparation Method of High Purity Aluminum Single Crystals with L
Kasper, E.	502	Si-MBE: Growth and Sb Doping (Si, film)
Kasuga, M.	426	Crystal Growth in c Direction and Crystallographic Polarity in Zn
Katayama, S.	35	Conditions of LPE Growth for Lattice Matched GaInAsP/InP DH Laser
Katayama, T.	499	Electrical Resistivity in In-Heusler Cu2.0Mn0.94In1.0 Single Crys
Katayama, Y.	883	Molecular Beam and Solid-Phase Epitaxies of Silicon Under Ultra-H
Kato, A.	958	Catalytic Effect of Nickel on the Growth of Zirconium Carbide Whi
	959	Catalytic effects of Various Metals and Refractory Oxides on the
Kato, Y.	469	Chromium Doped GaAs Vapor Phase Epitaxy
	490	Crystal Growth and Assessment of SnxPb1-xTe Mixed Crystals (PbxSn
Katsuno, H.	460	Liquid Phase Epitaxial Growth of Lattice-Matched GaP-Ga0.5Al0.5P-
Katsuta, H.	470	Physico-Chemical Properties of the Non-Stoichiometric VS2 and V5S
Katty, A.	471	Crystal Growth and Characterization of In1.9As0.1Se3 (vapor trans
	544	Preparation, Crystal Structure, and Physical Properties (transpor
Katuski, Z.	127	Chemical Synthesis and Crystal Growth of Laser Quality Praseodymi
Katzer, D.	105	Interaction Between Successive Monatomic Step Trains of Different
Kawabata, A.	9	Epitaxial Growth of Potassium Lithium Niobate Single-Crystal Film
	10	Elastic and Piezoelectric Properties of Potassium Lithium Niobate
	11	Epitaxial Growth of Potassium Lithium Niobate Single-Crystal Film
	711	Chemical Vapor Deposition of Single-Crystalline ZnO Film with Smo
	882	High Rate Epitaxial Growth of ZnO Films on Sapphire by Planar Mag
Kawabuchi, K.	472	Deposition of Crystals from the Plasmas of ZrO2, HfO2, ThO2, and C
Kawada, I.	819	Effect of Addition of Vanadium on the Chemical Transport of TiSx,
Kawada, K.	675	Growth of Single Crystals of Na3Al2Li3F12 Garnet Under Hydrotherm
Kawai, S.	964	Fermi Surface Measurements of ZrB2 by the de Haas van Alphen Effe
	1092	Preparation of TiCx Single Crystal with Homogeneous Compositions
Kawamoto, J.	881	Growth of Thallium Iodide Crystals in Silica Hydrogels (TlI, Tl2I
Kazimirov, V.P.	473	Structural Transitions and Phase Composition of Thin Films of Alu
Kazinets, M.M.	866	Electron Diffraction Study of Phase Transformations of the Compou
Kazumura, M.	970	Variation of Liquid and Solid Compositions During LPE Growth from
Kedves, F.J.	90	Connections Between Thermodynamic Quantities and Vacancy and Diff
Keesom, P.H.	215	Specific Heat of V6O11 Between 0.4 and 50 K (vapor transport, the

Kelber, J.   474   Structural and Magnetic Properties of "One-Dimensional" Barium Va
                           610   Magnetic and Electrical Properties of BaVS3 and BaVxTi1-xS3 (flux
Keller, K.W.   105   Interaction Between Successive Monatomic Step Trains of Different
Kelm, R.W.,Jr.   195   Solar Cells from Zone-Refined Metallurgical Silicon (Si, Czochral
Kennedy, E.F.   343   Heteroepitaxy of a Deposited Amorphous Germanium Layer on a Silic
                           537   Solar Furnace Annealing of Amorphous Si Layers (films)
Kerimova, E.M.   361   Structure and Growth Peculiarities of Tl-Se-TlInSe2 (TlSe, zone m
Kern, R.   542   Physical Methods Used for the Characterization of Modes of Epitax
Kershaw, R.   624   The Electrical, Optical and Photoconducting Properties of Fe2-xCr
Kervalishvili, P.D.   345   Behavior of Boron in Ge Single Crystals Grown by the Czochralski
                           346   Behavior of Be as an Impurity During Crystallization of Ge by the
Khalikov, B.   733   Synthesis of Single Crystal LnPO4 (vapor transport)
Kharchenko, L.Yu.   475   Solubility of KLa(MoO4)2 in Aqueous K2MoO4 Solutions in Hydrother
Khodadad, P.   131   The Compound Zn4(P2S6)3 (melt)
                           132   Crystal Structure of Zn4(P2S6)3 (melt)
                           981   The Compounds Ag7PS6 and Ag7PSe6
Khorvat, A.M.   345   Behavior of Boron in Ge Single Crystals Grown by the Czochralski
Khrjapov, V.T.   590   New Model of LPE Growth: Growth Rate Calculation (theory)
Khrusanova, M.   759   Synthesis and Crystal Growing of La3NbO7
Khuseinov, B.   476   Linear Expansion Anisotropy of CdGa2Se4 (melt)
Khvedelidze, A.G.   967   Some Peculiarities in the Growth and Properties of Beta-Boron Cry
                           968   Shape of the Crystallization Front During Electron-Beam Zone Refi
Kibbel, H.   502   Si-MBE: Growth and Sb Doping (Si, film)
Kihara, K.   1096   Bismuth Tellurides: BiTe and Bi4Te3 (melt)
Kikuma, I.   477   Formation of Defects in Zinc Selenide Crystals Grown from the Mel
Kildal, H.   412   Optical and Electrical Properties of CdGeAs2 (Bridgman)
                           635   Phase Studies, Crystal Growth, and Optical Properties of CdGe(As1
Kim, K.M.   478   Morphological Instability Under Constitutional Supercooling Durin
                           479   Maximum Stable Zone Length in Float-Zone Growth of Small-Diameter
Kimata, M.   1100   Molecular Beam Epitaxy of GaSb and GaSbxAs1-x (films)
                           1101   Molecular Beam Epitaxy of InxGa1-xSb (x = 0 to 1) (GaSb)
Kimura, K.   415   Epitaxial Growth of Ferroelectric PLZT ((Pb,La)(Zr,Ti)O3) Thin Fi
Kimura, S.   410   Electrical Properties of p- and n-Type CuInSe2 Single Crystals (m
                           480   Evaluation of Yttrium Iron Garnet Single Crystals Grown by the Fl
                           483   Interface Shape and Horizontal Variations of Al and Ga Contents i
                           879   Single Crystal Growth of Substituted Yttrium Iron Garnets Y3Fe5-x
                           880   Growth of Mg2TiO4 Single Crystals by the Floating Zone Method
Kino, T.   228   Preparation Method of High Purity Aluminum Single Crystals with L
Kinzel, D.L.   1057   GaAs Vapour Phase Epitaxy Using Computer Control
Kirovskaya, I.A.   185   Preparation and Investigation of Crystals of Solid Solutions of t
Kirsten, P.   86   Growth and Physical Properties of ZnSiP2 (vapor transport, soluti
Kisel', Yu.I.   60   Reaction of Germanium (111) Single Crystals with GeBr4 Vapor (Ge,
Kishi, K.   481   Morphological Change from Copper/alpha-copper-Zinc Whiskers to De
Kishino, K.   35   Conditions of LPE Growth for Lattice Matched GaInAsP/InP DH Laser
Kitagawa, M.   482   Heteroepitaxial Growth of ZnS on GaP by the Close-Spaced Techniqu
                           828   Liquid Phase Epitaxial Growth of CdTe in the CdTe-CdCl2 System (f
Kitamura, K.   480   Evaluation of Yttrium Iron Garnet Single Crystals Grown by the Fl
                           483   Interface Shape and Horizontal Variations of Al and Ga Contents i
                           645   Interface Shape Transitions in Czochralski Grown YAG Crystals (Y3
                           879   Single Crystal Growth of Substituted Yttrium Iron Garnets Y3Fe5-x
                           880   Growth of Mg2TiO4 Single Crystals by the Floating Zone Method
                           954   Growth of FeTiO3 (Ilmenite) Crystals by the Floating-Zone Method
Kitamura, M.   484   Re-Investigation of the Re-Entrant Corner Effect in Twinned Cryst
Klein, P.J.   485   The Influence of Impurities on the Crystallization of Supercooled
Klein, M.L.   850   Polytypism and the Vibrational Properties of PbI2
Klein, M.V.   315   Raman Scattering Studies in the Solid Electrolytes of the RbAg4I5
Klevtsov, P.V.   475   Solubility of KLa(MoO4)2 in Aqueous K2MoO4 Solutions in Hydrother
                           486   Crystal Structure of the Double Molybdate K2Ni(MoO4)2 (slow cooli
Klevtsova, R.F.   486   Crystal Structure of the Double Molybdate K2Ni(MoO4)2 (slow cooli
Klikorka, J.   975   Some Electrical Properties of GeBi2Te4 Single Crystals (Bridgman)
Klinkova, L.A.   52   Crystallization of WO2 from the Gas Phase (vapor transport)
                           487   Transfer of VO2 by Iodine (vapor transport, doped)
Kliya, M.O.   488   Crystals of CTA-DNA: A Growth and Morphological Study (solution)
Knook, B.   1002   A Model for Macroscopic Czochralski Growth: Theoretical and Expe
Kobayashi, K.   489   Fivefold Twinned Silicon Crystals Grown in an Al-16 wt.% Si Melt
                           950   Techniques for Routine UHV in situ Electron Microscopy of Growth
Kobayashi, K.L.I.   490   Crystal Growth and Assessment of SnxPb1-xTe Mixed Crystals (PbxSn
                           883   Molecular Beam and Solid-Phase Epitaxies of Silicon Under Ultra-H
Kobayashi, N.   491   Computational Simulation of the Melt Flow During Czochralski Grow
                           492   Power Required to Form a Floating Zone and the Zone Shape (theory
Kobayashi, T.   313   New Gaseous Impurity and Its Effect on the Purity of LPE GaAs (va
                           314   On the Growth Mechanism of Polycrystalline Snow Crystals with a S
                           493   Epitaxial Relationships During the Formation of Three-Dimensional
                           494   Chemical Vapor Deposition of Scandium Hydride (ScH2, epitaxy, fil
                           715   The Effects of M(M=Mg, Ca and Ba) Ions on the KCl Flux Growth of
                           934   Electron Microscopic Characterization of Cubic BN Prepared by Wat

Kobzareva, S.A.	243	On the Growth Mechanism of Sodalite Single Crystals Grown by the
Kodaira, K.	423	Vapor Growth of Hexagonal GeO2 Needle Crystals
	424	Vapor Growth of Zn2GeO4 Single Crystals
Kodama, H.	431	Growth of Anatase (TiO2) Crystals by Chemical Transport Reactions
Koelsch, M.F.	255	Kinetics of Silicon Growth Under Low Hydrogen Pressure (vapor tra
Koffyberg, F.P.	920	Electrical and Optical Properties of High-Purity p-type Single Cr
Kohra, K.	1052	Growth of Stacking Faults by Bardeen-Herring Mechanism in Czochra
Koizumi, H.	643	Epitaxial Growth of KNdP4O12 Laser Waveguides (KLaP4O12, solution
Kojima, T.	495	Crystal Growth of Mn15Si26 (vapor transport, Bridgman)
Kolar, D.	253	Crystal Growth of Some Alkaline-Earth Rare-Earth Pentaoxometallat
Kolb, E.D.	68	Dislocation-Free and Low-Dislocation Quartz Prepared by Hydrother
	496	Hydrothermal Synthesis of Aluminum Orthophosphate (AlPO4)
Kolesar, D.F.	618	Surface Photovoltage Experiments on SrTiO3 Electrodes (theory)
Kolina, A.V.	94	The Mechanism of Growth of KDP Crystals in the Presence of Impuri
Komatsu, H.	380	Preferred Growth Direction of Pure Cu Single Crystals by the Floa
	381	Growth of Cobalt Single Crystals by the Electron Beam Floating-Zo
	408	An Evidence of Spiral Growth on the Surface of Melt-Grown KCl Cry
	409	Effects of Growth Rates on the Defect Generation in KCl Crystals
Komatsubara, K.F.	490	Crystal Growth and Assessment of SnxPb1-xTe Mixed Crystals (PbxSn
	883	Molecular Beam and Solid-Phase Epitaxies of Silicon Under Ultra-H
Komem, Y.	207	Characterization of (001) Tilt Boundaries Produced by Epitaxial G
Komunjer, L.	601	A New Method to Follow Crystal Growth by Coulter Counter (solutio
Konagai, M.	497	High Efficiency GaAs Thin Film Solar Cells by Peeled Film Technol
Konak, A.R.	498	The Kinetics of Dissolution (theory)
Kondo, K.	499	Electrical Resistivity in In-Heusler Cu2.0Mn0.94In1.0 Single Crys
Kondo, S.	500	LPE Growth of Li(Nb,Ta)O3 Solid-Solution Thin Film Waveguides on
	922	Direct Measurement of Temperature Dependence of Lattice Mismatche
Kondratenko, V.V.	501	Growth of Antimony on Bismuth (Sb film, epitaxy)
Konig, U.	502	Si-MBE: Growth and Sb Doping (Si, film)
Kopylov, Yu.L.	903	Dependence of the Optical Homogeneity of Single Germanosillenite
Korec, J.	503	Model of Silicon Epitaxial Growth in SiCl4-HCl-H2 System Based on
	504	Flow-Graph Model of Epitaxial Growth of A(III)B(V) Compounds from
Kornblit, A.	505	Heat Capacity of EuS near the Ferromagnetic Transition (Bridgman,
Korsukova, M.M.	359	The Preparation of Single Crystals of the Rare Earth Borides by t
Kosak, W.	136	Local Selective Homoepitaxy of Silicon at Reduced Temperatures Us
Kosicka, M.	1108	The Crystal Structure of Lithium Fluorosulphate LiSO3F (solution)
Kostecky, G.	567	Single Crystal Growth and Properties of Deuterated Triglycine Flu
Kostenko, V.I.	588	Improving the Electrical Characteristics of CdS Single Crystals G
Kostiner, E.	29	The Crystal Structure of Cu4(PO4)20 (flux)
Kotani, T.	506	An LPE Growth of Thick InxGa1-xAs
Kotova, Yu.A.	231	Kinetics of Gallium Phosphide Synthesis by Chemical Gas Transport
Kotru, P.N.	507	On the Dissolution of Grown Synthetic Quartz Inside the Autoclave
	508	Flux Growth of Lanthanum Borate, LaBO3
Kovach, S.K.	509	Preparation and Principal Physicochemical Properties of Alkali Me
Kovaleva, I.S.	1037	Photoferromagnetic Effect and Photoconductivity of Cd1-xHgxCr2Se4
Kozuka, H.	522	Wide Silicon Ribbon Crystals (EDFG, Si)
	523	Evaluation of Temperature Distribution of Melt in Silicon Ribbon
Kramer, P.	354	Liquid Phase Epitaxial Deposition of GaP on GaAs
Kravchenko, V.B.	903	Dependence of the Optical Homogeneity of Single Germanosillenite
Krivonosova, A.S.	275	Preferred Growth Directions and Perfection of Single Crystals of
Kroger, F.A.	510	Defects in Single Crystals Prepared at High Temperature (theory,
Krok, J.	722	Formation of Mixed Crystals in the System CuxZn1-xCr2Se4
Krylov, A.S.	511	Control of the Composition of Crystals Grown in Electric and Magn
Kubiak, R.	512	Growth and Morphology of Spherical Crystals of omega-Phase in the
Kubodera, K.	643	Epitaxial Growth of KNdP4O12 Laser Waveguides (KLaP4O12, solution
Kubota, E.	923	Large-Sized InP Single Crystals by the Synthesis, Solute Diffusio
Kubota, M.	676	Unit Cell Step Structures on Vapour-Grown Cr2S3 Single Crystals
Kucha, V.V.	903	Dependence of the Optical Homogeneity of Single Germanosillenite
Kucharczyk, D.	513	Accurate X-ray Determination of the Lattice Parameters and the Th
Kuhn, G.	690	Growth and Electrical Properties of Epitaxial CuInS2 Thin Films o
	843	Epitaxial Layers of CuInSe2 on GaAs
	844	Structural and Electrical Properties of CuGaSe2 Thin Films on GaA
Kuhne, H.	514	Temperature Dependence of the Degree of Doping in Silicon Epitaxi
	515	Temperature Dependence of Degree of Doping in Growth of Epitaxial
Kuiper, A.E.T.	516	Thin Film Deposition From Beams of Ionized Atoms and Clusters (Ag
Kukkonen, C.A.	288	Electron Energy-Loss Spectroscopy of TiS2 (vapor transport)
Kumagawa, M.	517	Observations of Microfacets Near Irregularly Remelted Surfaces in
Kumashiro, Y.	680	Single-Crystal Growth of Tantalum Diboride (TaB2, zone melting, s
Kunomura, K.	518	Second Harmonic Generation in a Sputtered LiNbO3 Film on MgO (epi
Kunte, N.S.	465	Effect of Pressure on Electrical Resistivity and Thermoelectric P
Kunzel, H.	377	Crystal Structure, Growth, and Optical Properties of Alkali-Triha
Kurbanov, K.R.	4	Effect of Crucible Shape on a Crystallization Front (Bridgman, Ki
Kurbanov, Kh.M.	733	Synthesis of Single Crystal LnPO4 (vapor transport)
Kurbanov, T.Kh.	817	Synthesis, Preparation of Single Crystals, and Physicochemical In
Kurien, K.V.	519	Growth Kinetics and Habit Modification of Barium Molybdate Single
Kuriyama, K.	456	Single Crystal Growth of LiInS2 (directional solidification)

	520	Single Crystal Growth of Li2ZnGe (melt)
Kuriyama, M.	521	Crystal Perfection in Czochralski Grown Nickel Single Crystals (N
Kuroda, E.	522	Wide Silicon Ribbon Crystals (EDFG, Si)
	523	Evaluation of Temperature Distribution of Melt in Silicon Ribbon
	524	Growth and Characterization of Silicon Ribbon Crystals Grown with
Kuznetsov, D.A.	525	On the Effect of pH of Solution on the Kinetics of Crystallizatio
Kuznetsov, G.F.	526	Real Structure of Heteroepitaxial Zinc Oxide Films Grown on Sapph
Kuznetsov, V.A.	592	Hydrothermal Crystallization in the System BaO-GeO2-Na2O-H2O (bar
Kuznetsov, V.G.	587	Structure and Luminescence Properties of alpha-CsNd(PO3)4 Crystal
Kyoto, M.	63	Mass Spectrometric Study on Chemical Transport of VO2
L'Haridon, P.	597	Crystal Structure of beta-Eu2(II)SiO4 (melt)
La Vecchia, A.	540	Epitaxial Growth of Antimony Electrodeposits on Oriented Single C
Labbe, Ph.	350	The Thallium Tungstate Tl2W4O13: A Tunnel Structure Related to t
Labeau, M.	389	Structure of Monocrystalline WV2O6 at 298 and 383 K (vapor transp
	527	Crystal Chemistry of Perovskite Phases of the Systems Th0.25NbO3-
Lacmann, R.	991	Effects of Interface Kinetics on the Growth Rate of Dendrites (Sn
Ladd, A.J.C.	528	Structure of the Lennard-Jones (100) Crystal-Liquid Interface (th
Lagerstedt, O.	649	Properties of VPE-Grown GaN Doped with Al and Some Iron-Group Met
Lagowski, J.	437	Liquid-Phase Electroepitaxy: Growth Kinetics (Peltier effect, Ga
Lam, S.-T.	529	Surface Charge and Ledge Dynamics in Cadmium Sulfide (CdS, evapor
Lampert, B.	530	Anisotropy of Round Evaporation Spirals on Rocksalt Surfaces (NaC
Lance-Gomez, E.T.	531	Preparation, Phase Equilibria, and Crystal Chemistry of La, Pr, a
Lander, G.H.	154	Magnetization and Neutron Studies of UTe and USb0.8Te0.2
Landers, R.	532	Semi-Periodic Closed Growth Lamellae on the Crystal-Melt Interfac
Lang, J.	533	Nitrides--Structures and Crystal Growth (review, vapor transport,
Langer, J.S.	534	Evidence for a Universal Law of Dendritic Growth Rates (theory)
Langlois, W.E.	535	Effect of the Buoyancy Parameter on Czochralski Bulk Flow in Garn
	536	Digital Simulation of Czochralski Bulk Flow in Microgravity (Al2O
Lanir, M.	1059	Liquid Phase Growth of HgCdTe Epitaxial Layers
LaPlaca, S.J.	865	Preparation and Characterization of Ruthenium Dioxide Crystals (R
Larson, M.A.	322	Direct Observation of Secondary Nuclei Production (KAl(SO4)2.12H2
Latypov, Z.M.	101	Liquidus of the Ternary System Pb-Tl-Te
Lau, S.S.	177	Silicide Formation by High-Dose Si+-Ion Implantation of Pd (elect
	343	Heteroepitaxy of a Deposited Amorphous Germanium Layer on a Silic
	537	Solar Furnace Annealing of Amorphous Si Layers (films)
Laudan, H.	611	Czochralski Growth of Barium Hexaaluminate Single Crystals (BaAl1
Laudise, R.A.	68	Dislocation-Free and Low-Dislocation Quartz Prepared by Hydrother
	496	Hydrothermal Synthesis of Aluminum Orthophosphate (AlPO4)
	538	Are Icicles Single Crystals? (H2O)
Laugier, A.	600	Luminescence and Bandgap Studies of High Purity MgxZn1-xTe Ternar
Launay, J.-C.	124	Preparation by Vapor Transport of Monocrystals of AlF3, FeF3, and
Launay, J.C.	191	Germanium-Iodine Equilibrium from 700 to 1000 K (Ge, vapor transp
Laurent, Y.	533	Nitrides--Structures and Crystal Growth (review, vapor transport,
	597	Crystal Structure of beta-Eu2(II)SiO4 (melt)
Lavina, A.	539	Electron Microscopy Study of Some Molybdenum Oxide Crystals (MoO3
Lazzari, M.	540	Epitaxial Growth of Antimony Electrodeposits on Oriented Single C
Le Flem, G.	825	The Crystal Structure of a New High -Nd- Concentration Laser Mate
Le Gal, H.	541	Crystal Growth by the Thermic Screen Translation (TST) Technique:
Le Gall, H.	238	Flux Growth of Single Magnetic Sublattice Antiferromagnetic Garne
	767	Magnetic Properties of a Single Crystal of NdF3 Between 2 and 4.2
Le Lay, G.	542	Physical Methods Used for the Characterization of Modes of Epitax
Le Marechal, L.	774	Solubility of AlPO4 in Hydrothermal Solutions of H3PO4
Le Marouille, J.Y.	235	About SnF2 Stannous Flouride. I. Crystallochemistry of alpha-SnF2
Le Nagard, N.	543	Crystal Structure and Physical Properties of CuO0.65VS2 (vapor tr
	544	Preparation, Crystal Structure, and Physical Properties (transpor
Le Traon, F.	757	New Thiochlorides of Molybdenum II: Mo6Cl10Y (Y = S, Se, Te): S
Lebesque, J.V.	106	Magnetic and Neutron Scattering Experiments on the Antiferromagne
Leblanc, M.	772	Hydrothermal Growth of the Systems MF-AlF3 and NH4F-FeF3 in Hydro
Lebugle, A.	1109	Crystal Growth of Gallium Arsenide by Chemical Vapor Deposition f
Lefaucheux, F.	233	X-Ray Topographic Study of Growth Defects in Hydrothermal Na2CoGe
	337	Effect of Stirring on Crystalline Quality of Solution Grown Cryst
	545	Crystal Quality of Gel Grown CaHPO4.2H2O Crystals
Lefever, R.A.	175	Thermocapillary Convection in Floating Zone Melting: Influence o
	179	Convection Phenomena During the Growth of Sodium Chlorate Crystal
	863	Influence of Magnetic Field on Vertical Bridgman-Stockbarger Grow
Legge, R.	546	Properties of Rapidly-Grown RTR Silicon Sheet and Their Effects o
Legros, J.-P.	547	Crystal Structure of Se2TiO6 (slow cooling)
Lehmann, G.	64	Semiconductor Single Crystals Obtained by the Traveling Solvent M
Leiss, M.	577	Epitaxy of NdAl3(BO3)4 For Thin Film Minature Lasers (flux)
	578	Epitaxial Layers of the Laser Material Nd(Ga,Cr)3(BO3)4 (flux)
Lejcek, P.	548	The Growth of Nickel Single Crystals by Recrystallization Method
	887	The Growth of Thin Bismuth Single Crystals with High Resistivity
Lemaignan, C.	163	In-Situ Observation of Solidification Interfaces with a Very High
	549	Unidirectional Solidification of Irregular Silver-Germanium Eutec
Lempicki, A.	189	Lasing and Fluorescence in K5NdLi2F10 (flux)
Lendvay, E.	352	Epitaxial Growth of A(III)B(V) Semiconductors from Vapour Phase (

Lent, B.	202	A Complexity in the Solidification Behaviour of Molten Y3A15012 (
	203	Single Crystal Growth of 12CaO.7Al2O3 (Czochralski)
	1076	Ca12Al14O33: A New Piezoelectric Material (Czochralski)
Lentz, A.	550	Crystal Growth of Triple Nitrites in Gels (K2CuPb(NO2)6, Tl2CuPb(
Leonov, P.A.	345	Behavior of Boron in Ge Single Crystals Grown by the Czochralski
Leonyuk, N.I.	551	Neodymium Incorporation into YAl-Borate Crystals in Preparation f
Leroux, G.	447	Crystal Growth and Crystallographic Data of Some Ln2(MoO4)3 Type
Lesk, I.A.	360	Silicon Ribbon Growth via the Ribbon-To-Ribbon (RTR) Technique:
Leskela, M.	1008	Scandium Hydrogenselenite (Sc(HSeO3)3, evaporation, solution)
LeTraon, A.	757	New Thiochlorides of Molybdenum II: Mo6Cl10Y (Y = S, Se, Te): S
Leu, C.Y.	403	Growth of CuInS2 and Its Characterization (vapor transport)
Leu, Y.-T.	552	Preparation and Properties of Bulk GaxIn1-xAs
Leung, S.Y.	1068	Magneto-Optical Properties of KTb3F10 and LiTbF4 Crystals (Czochr
Leung, W.H.	553	Nitrilotri(Methylenephosphonic Acid) Adsorption on Barium Sulfate
Levin, V.I.	554	Growing Monocrystalline Bars of SiC from the Gas Phase (vapor tra
Levshin, V.A.	1037	Photoferromagnetic Effect and Photoconductivity of Cd1-xHgxCr2Se4
Levy, F.	555	Structural and Electrical Properties of Layered Transition Metal
Leycuras, C.	767	Magnetic Properties of a Single Crystal of NdF3 Between 2 and 4.2
Li, P.W.	556	Epitaxial Growth of CdS on CuInSe2
Liaw, J.H.W.	340	LPE Growth of Lithium Ferrite on Spinel Substrate Crystals (films
Lichtensteiger, M.	1088	Crystal Growth and Segregation Under Zero Gravity: Ge (Bridgman)
Likforman, A.	557	Crystal Structure of the Indium Selenide In2Se3 (vapor transport)
Lilley, P.	558	Mass Transport Processes Associated with the Epitaxial Growth of
Lillicrap, B.J.	559	Orientation Dependence of l-Alanine Incorporation in TGS Crystals
	560	Growth and Electrical Properties of Deuterated TGS Produced by th
Lilov, S.K.	561	Study of Temperature Distribution in High-Temperature Furnaces fo
	562	Study of Silicon Carbide Epitaxial Growth Kinetics in the SiC-C S
Lim, T.C.	691	Thin Films of LiNbO3, Doped with Na+ and Co2+ plus Zr4+, Grown by
Lindqvist, O.	282	The Crystal Structure of CuTeO4 (hydrothermal)
	316	The Crystal Structure of Te3Nb2O11
Lingart, Yu.K.	563	Melting and Crystallization of Corundum (alpha-Al2O3)
Link, R.	64	Semiconductor Single Crystals Obtained by the Traveling Solvent M
	564	Growth of Pb1-xSnxTe Single Crystals of Graded Composition by the
Linz, A.	1068	Magneto-Optical Properties of KTb3F10 and LiTbF4 Crystals (Czochr
Lipko, H.A.	565	Crystallisation of Some Ferrite Materials from Lead Flux with Add
Littlejohn, M.A.	89	Liquid Phase Epitaxy of II-IV-V2, Chalcopyrite Compounds (ZnSnAs2
Litvinov, L.A.	246	Profile-Sapphire and Its Structural Perfection (alpha-Al2O3, melt
Litvinova, I.Yu.	687	Influence of Growth Conditions on Mirodefect Distribution in Disl
Lobachev, A.N.	233	X-Ray Topographic Study of Growth Defects in Hydrothermal Na2CoGe
	234	Hydrothermal Synthesis of Crystals (review)
	243	On the Growth Mechanism of Sodalite Single Crystals Grown by the
	256	Na2CoGeO4 Single Crystal Growth and Study of Their Piezoelectric
	777	Synthesis and Hydrothermal Growth of the Ferroelectric SbNbO4, Sb
Locke, K.E.	321	Growth of Large Single Crystals of Heusler Alloys, Ni2MnSn and Ni
Loeksmanto, W.	766	Crystal Chemistry of a Tellurate IV of Indium III: In2Te3O9
Loginova, R.V.	18	Computer Model of Autoepitaxial Growth of Ge Films (theory)
Loiacono, G.	339	Calcium Orthovanadate, Ca3(VO4)2 - A New High-Temperature Ferroel
Loiacono, G.M.	229	Thermal and Dielectric Properties of LiKSO4 and LiCsSO4
	566	Crystal Growth from Solution Using Cylindrical Seeds (TGS, TBFB)
	567	Single Crystal Growth and Properties of Deuterated Triglycine Flu
Longinotti, L.D.	31	On the Resistivity of the Magnetically Ordered Ground State of th
Lonnee, J.E.	307	Elastic Constants of Niobium-Zirconium, Hafnium, and Tungsten All
Lopez-Otero, A.	568	Hot Wall Epitaxy (II-VI compounds, IV-VI compounds, III-V compoun
	569	CdTe Thin Films Grown by Hot Wall Epitaxy
Losch, R.	570	Preparation and Crystal Structure of CsAlF4
Loshchenov, V.B.	587	Structure and Luminescence Properties of alpha-CsNd(PO3)4 Crystal
Losiak, P.	240	Preparation and Basic Physical Properties of BiTeI Single Crystal
Louisy, A.	571	Physical Properties and Lithium Intercalates of CrPS4 (slow cooli
Lubetkin, S.D.	572	The Variability of the Rate of Growth of Adamantane Crystals from
Lubitz, K.	573	The Preparation of Large Spherical Iron Single Crystals (Fe, stra
Lucas, J.	235	About SnF2 Stannous Flouride. I. Crystallochemistry of alpha-SnF2
Luck, J.	583	Liquid Encapsulation Zone-Refining of PbS (zone melting, vapor tr
Ludington, B.W.	574	Temperature Gradient Measurement in Liquid Epitaxial Growth Syste
Lum, W.Y.	199	Control of Substrate Degradation in InP LPE Growth with PH3 Parti
Lundager Madsen, H.E.	575	Localized Adsorption in One Layer on a Crystal-Solution Interface
Lundgren, L.	30	Properties of Fe2P Crystals Prepared from a Liquid Copper Medium
Luscher, P.E.	576	Design Considerations for Molecular Beam Epitaxy Systems (review)
Lutovich, A.S.	19	The Mechanism of Silicon Epitaxial Layer Growth from Ion-Molecula
Lutz, F.	577	Epitaxy of NdAl3(BO3)4 For Thin Film Minature Lasers (flux)
	578	Epitaxial Layers of the Laser Material Nd(Ga,Cr)3(BO3)4 (flux)
Lyons, M.H.	281	The Modified Entrainment Method and Its Application to the Study
Machi, Y.	669	Epitaxial Growth of ZnSe by Vapor Flow Method
	670	Preparation and Optical Properties of ZnSe Epitaxial Layers by a
Machida, H.	579	Simultaneous Diffusion of Calcium and Strontium in KCl Single Cry
Machida, K.	580	Synthesis and Characterization of Europium(II)-Holoborates, Eu2B5
Madhava, M.R.	581	Elastic Constants and Stability of bcc In-Tl Alloys (Bridgman)

Maeda, M.	582	Dielectric and Optical Properties of K5Nd(MoO4)4 (Czochralski)
Maekawa, S.	1102	Molecular Beam Epitaxy of ZnSexTe1-x(film)
Magari, S.	472	Deposition of Crystals from the Plasmas of ZrO2, HfO2, ThO2 and C
Maier, H.	583	Liquid Encapsulation Zone-Refining of PbS (zone melting, vapor tr
Majni, G.	584	Solid Phase Epitaxial Growth of Si Through Al Film
	585	Growth Kinetics of (111) Si Through an Al Layer by Solid Phase Ep
Makhotkin, V.E.	586	Magnetic Properties of CdCr2Se4-ZnCr2Se4 Single Crystals (melt)
	1037	Photoferromagnetic Effect and Photoconductivity of Cd1-xHgxCr2Se4
Maki, M.	522	Wide Silicon Ribbon Crystals (EDFG, Si)
	523	Evaluation of Temperature Distribution of Melt in Silicon Ribbon
	524	Growth and Characterization of Silicon Ribbon Crystals Grown with
Makita, Y.	617	Nitrogen Doping into GaAs1-xPx Using Ionized Beam in Molecular Be
	1102	Molecular Beam Epitaxy of ZnSexTe1-x(film)
Makovetskaya, L.A.	119	Preparation and Properties of Single Crystals of Solid Solutions
Makowski, J.	341	Crystallization of HgTe Crystals from the Vapour Phase
	342	Crystallization of HgCdTe Mixed Crystals from the Vapour Phase by
Makram, H.	445	Flux Growth of Lead Neodymium Chlorapatite (Na2Nd2Pb6(PO4)6Cl, sl
Maksimova, S.I.	587	Structure and Luminescence Properties of alpha-CsNd(PO3)4 Crystal
Malikov, V.Ya.	588	Improving the Electrical Characteristics of CdS Single Crystals G
Malinin, A.J.	589	New Model of LPE Growth: Layer Thickness Calculation (theory, ki
	590	New Model of LPE Growth: Growth Rate Calculation (theory)
Malinin, A.Yu.	591	Theory of Crystallization During Epitaxial Growth from a Solution
Malinovskii, Yu.A.	592	Hydrothermal Crystallization in the System BaO-GeO2-Na2O-H2O (bar
Malmejac, Y.	549	Unidirectional Solidification of Irregular Silver-Germanium Eutec
Mamedov, A.S.	593	alpha-In2Te3 (melt)
Mamedov, K.P.	593	alpha-In2Te3 (melt)
Maqsood, A.	594	Relative Supersaturation and Supercooling in Fluxed Melt Systems
	595	Flux Growth of Polymorphic Rare-Earth Disilicates, R2Si2O7 (R = T
	1061	The Flux Growth of Some Rare-Earth and Iron Group Complex Oxides
	1064	Flux Growth of Crystals of Some Transition Metal Fluorides. Part
Marchand, R.	533	Nitrides--Structures and Crystal Growth (review, vapor transport,
	596	Ternary Lanthanoid-Transition Metal Pnictides with ThCr2Si2-Type
	597	Crystal Structure of beta-Eu2(II)SiO4 (melt)
Marchee, W.G.J.	193	The Affinity of Crystal Growth and Dissolution in Aqueous Solutio
Marenkin, S.F.	598	The Growth of CdAs2 and ZnAs2 Single Crystals from the Vapor Pha
	599	Growth of CdP4 Single Crystals (evaporation, vapor transport)
Marezio, M.	390	The Crystal Structure of V4O7 at 120 degrees K (vapor transport)
	391	Structural Aspects of the Metal-Insulator Transitions in (Ti0.997
	661	Synthesis and Crystal Structure of a New Mixed Oxide "FeV3O8" (Fe
	662	Synthesis and Crystal Structure of a New Mixed Oxide FeV2O6OH0.5:
	1036	Crystal Structure of CuTa2O6 (flux)
Marine, J.	600	Luminescence and Bandgap Studies of High Purity MgxZn1-xTe Ternar
Mark, W.	282	The Crystal Structure of CuTeO4 (hydrothermal)
Markovic, M.	601	A New Method to Follow Crystal Growth by Coulter Counter (solutio
Marnier, G.	602	Growth Conditions for Sodium-Titanium Bronze Crystals in NaBO2 (f
Marshall, A.J.	603	Growth of InP Crystals by the Synthesis Solute Diffusion Method
Martin, E.P.	604	Application of a Heat Pipe to Czochralski Growth. Part I. Growth
Martin, M.S.	133	Crystal Growth and EPR Study of Cr+-Doped MgTe Single Crystals (s
Marumo, F.	719	The Crystal Structure of Cs6W11O36
	720	The Crystal Structure of K2W4O13
Maruyama, S.	605	Surface Morphology of GaAs Layers Grown by Liquid-Phase Epitaxy (
Marychev, V.V.	1035	Crystallization of Bi-Sb Solid Solutions in Capillaries (zone mel
Mashchenko, V.G.	107	Growth of Nickel-Zinc Ferrite Single Crystals from BaO-B2O3-ZnO-N
Mashio, T.	264	Photoconduction of Lead Trititanate (PbTi3O7, flux)
Maslov, V.N.	606	Reproductive Epitaxy
Maslyak, N.T.	306	Preparation of Films of IV-VI Compounds in a Quasi-Sealed Volume
Mason, K.N.	607	Growth and Characterization of Transition Metal Silicides (chemic
Masse, G.	157	Growth of Single Crystals of Cinnabar (alpha-HgS) at Rather Low T
Masse, R.	261	Crystal Structure of the Mercury Nitro-Phosphate Hg4PO4NO3.H2O (m
	608	Crystal Structure of AgH2PO4. Crystallography of AgH2AsO4 (solut
	609	Crystal Structure of a New Variety of Nickel Pyrophosphate: Ni2P2
	986	Crystal Structure of Monophosphate AgCoPO4
Massenet, O.	474	Structural and Magnetic Properties of "One-Dimensional" Barium Va
	610	Magnetic and Electrical Properties of BaVS3 and BaVxTi1-xS3 (flux
Masumoto, K.	681	Vapor-Phase Growth of Epitaxial Ga1-xInxSb Using Open-Tube Flow S
Mateika, D.	71	Preparation of Barium Lead Hexa-Aluminate Single Crystal Layers b
	611	Czochralski Growth of Barium Hexaaluminate Single Crystals (BaAl1
Mathur, P.C.	612	Effect of High-Temperature Annealing in Cadmium on the Electrical
Matsubara, K.	944	New Developments in Ionized-Cluster Beam and Reactive Ionized-Clu
	945	Ionized-Cluster Beam Deposition and Epitaxy (review)
Matsubara, T.	518	Second Harmonic Generation in a Sputtered LiNbO3 Film on MgO (epi
Matsuda, H.	965	Geometrical Analysis of Crystallization of the Soft-Core Model in
Matsuda, M.	522	Wide Silicon Ribbon Crystals (EDFG, Si)
	523	Evaluation of Temperature Distribution of Melt in Silicon Ribbon
	524	Growth and Characterization of Silicon Ribbon Crystals Grown with
Matsuda, N.	613	Vapour-Phase Epitaxial Growth of ZnS on GaP

Matsui, J.	614	X-Ray Study of LEC-Grown InP Crystals (Czochralski, theory)
	856	Impurity Effect on Grown-In Dislocation Density of InP and GaAs C
Matsumoto, T.	615	Epitaxial Growth of ZnS by Zn-S-H2 CVD Method
	931	Crystal Growth of Co Ferrite on Fine Acicular gamma-Fe2O3 Particl
	1096	Bismuth Tellurides: BiTe and Bi4Te3 (melt)
Matsumura, S.	311	Growth of LiTaO3 Single Crystals for Saw Device Applications (Czo
Matsunami, H.	415	Epitaxial Growth of Ferroelectric PLZT ((Pb,La)(Zr,Ti)O3) Thin Fi
	616	Heteroepitaxial Growth of beta-SiC on Silicon Substrate Using SiC
	699	Growth and Morphology of 6H-SiC Epitaxial Layers by CVD (film)
Matsuo, M.	830	Ferroelectricity in (N(CH3)4)2ZnCl4 (evaporation)
	831	Ferroelectricity in (N(CH3)4)2CoCl4
Matsushima, Y.	617	Nitrogen Doping into GaAs1-xPx Using Ionized Beam in Molecular Be
	1100	Molecular Beam Epitaxy of GaSb and GaSbxAs1-x (films)
Matsushita, T.	423	Vapor Growth of Hexagonal GeO2 Needle Crystals
	424	Vapor Growth of Zn2GeO4 Single Crystals
Mattern, N.	798	On the Chemical Transport of Some Niobium Oxides with TeCl4 (II)
Matteson, S.	177	Silicide Formation by High-Dose Si+-Ion Implantation of Pd (elect
Matveev, S.I.	344	Growth of Zircon Single Crystals from a Solution in a Melt (ZrSiO
Mauersberger, P.	377	Crystal Structure, Growth, and Optical Properties of Alkali-Triha
Maunaye, M.	533	Nitrides--Structures and Crystal Growth (review, vapor transport,
Maurin, G.	26	Electrocrystallization. Fundamental Aspects (review, theory)
Maurin, M.	766	Crystal Chemistry of a Tellurate IV of Indium III: In2Te3O9
Mavlonov, Sh.	476	Linear Expansion Anisotropy of CdGa2Se4 (melt)
Mavroides, J.G.	618	Surface Photovoltage Experiments on SrTiO3 Electrodes (theory)
Mayer, J.W.	177	Silicide Formation by High-Dose Si+-Ion Implantation of Pd (elect
	343	Heteroepitaxy of a Deposited Amorphous Germanium Layer on a Silic
	996	Grain Size Dependence in a Self-Implanted Silicon Layer on Laser
McCollum, B.	189	Lasing and Fluorescence in K5NdLi2F10 (flux)
McCollum, B.C.	397	Crystal Structure of K5NdLi2F10 (flux)
McDaniel, C.L.	639	Alkali Ion Exchange Reactions with RbAlSiO4: A New Metastable Pol
McKelvy, M.J.	737	A New Technique for Preparation of Copper Single Crystals from th
McKinley, A.	619	An Investigation of Thin Silver Films on Cleaved Silicon Surfaces
McLellan, R.B.	470	Physico-Chemical Properties of the Non-Stoichiometric VS2 and V5S
McMasters, O.D.	620	Single Crystal Growth by the Horizontal Levitation Zone Melting M
McMillan, P.W.	250	Infra-Red Transmitting Materials. Part 1. Crystalline Materials
McQuillan, B.	76	The Synthesis of the First Stage Graphite Salt C8(+)AsF6(-) and I
McWhan, D.B.	621	X-Ray Diffuse Scattering from Alkali, Silver, and Europium beta-A
McWilliams, G.E.	199	Control of Substrate Degradation in InP LPE Growth with PH3 Parti
Mehrotra, K.	622	Polytypism in Vapor Grown Crystals of Cadmium Bromide (CdBr2)
	870	Growth and Polytypism in Vapor-Grown Mixed Cadmium Halide Crystal
Meier, D.L.	464	Uniaxial Ferromagnetism in NiZrF6.6H2O (solution)
Melikhov, I.V.	623	Crystallization of Salts from Supersaturated Solutions: Diffusio
Melnikov, O.K.	243	On the Growth Mechanism of Sodalite Single Crystals Grown by the
Menge, G.	1044	AlSiP3, A Compound with a Novel Wurtzite-Pyrite Intergrowth Struc
Menyuk, N.	412	Optical and Electrical Properties of CdGeAs2 (Bridgman)
Merchant, P.	624	The Electrical, Optical and Photoconducting Properties of Fe2-xCr
Mercier, J.	299	Vapor Phase Transport and Crystal Growth of Molybdenum Trioxide a
	610	Magnetic and Electrical Properties of BaVS3 and BaVxTi1-xS3 (flux
Merenda, P.	754	The Growth of GaxIn1-xAs on (100) InP by Liquid-Phase Epitaxy (fl
	907	Bridgman and Czochralski Growth of beta-Phase NaxV2O5
Merkulova, A.I.	625	Preparation of Single Crystals of the Magnetic Semiconductor CdCr
Mermant, G.	626	Preparation, Crystal Growth and Characterization of Mixed Bromide
Metselaar, R.	1016	Liquid Phase Epitaxial Growth of Copper Ferrite Films (CuFe2O4)
	1017	Liquid Phase Epitaxial Growth of Lithium Ferrite-Aluminate Films
	1018	Flux Growth of ZnGa2O4 Single Crystals
Meuller, M.H.	474	Structural and Magnetic Properties of "One-Dimensional" Barium Va
Meyer, H.J.	627	Linear Growth Rate of Calcite in Aqueous Solution (CaCO3, kinetic
	628	Molecular (beam) Processes of Condensation and Evaporation of Alk
Michalski, A.	906	Reactive Pulse Plasma Crystallization of Diamond and Diamond-Like
Michel, D.	629	Growth from Skull-Melting of Zirconia-Rare Earth Oxide Crystals (
	630	Ceramic Eutectics in the Systems ZrO2-Ln2O3 (Ln = Lanthanide): U
Michelutti, B.	631	Magnetization, Magnetocrystalline Anisotropy, Magnetostriction an
Midgley, D.	900	Flux Growth and Characterization by X-ray Topography of Rare Eart
Midorikawa, M.	632	The Melt Growth and Thermal Expansion of Na2CO3 Crystals (Bridgma
	633	Optical and Dilatometric Studies of KCaCl3 and RbCaCl3 Crystals (
Mihama, K.	821	Crystal Structure and Habit of Fine Metal Particles Formed by Gas
Mihara, M.	988	Impurity Effect on the Formation of Terraces in GaAs LPE Growth (
Mihelcic, M.	1071	Measurements of the Contact Angle Between Melt and Crystal During
Miki, H.	852	GaxIn1-xSb Crystals Grown by Liquid Phase Epitaxy (melt)
Mikkelsen, J.C.,Jr.	180	Crystal Growth and Crystal Chemistry of NiAs-Type Compounds: MnSb
	634	Bridgman-Stockbarger Crystal Growth of Li2Ti3O7
	635	Phase Studies, Crystal Growth, and Optical Properties of CdGe(As1
Mil'vidskii, M.G.	636	Creation of Defects During the Growth of Semiconductor Single Cry
Miller, B.I.	552	Preparation and Properties of Bulk GaxIn1-xAs
Miller, D.C.	637	The Strain Configuration Within the Facet of Czochralski Grown In
	638	The Effect of Melt Flow Phenomena on the Perfection of Czochralsk

245

Miller, R.F.	782	The Growth and Structure of Epitaxial Films of the Rare-Earth Dih
Milligan, W.O.	663	Aging Studies on Hydrous Lutetium Oxide (cubic Lu(OH)3)
	664	Cerium Trihydroxide (Ce(OH)3, gel)
Mimoto, H.	308	Liquid-Phase Epitaxy of ZnSe from Zn-Ga Solution (film)
	309	Growth of Cubic ZnS, ZnSe and ZnSxSe1-x Single Crystals by Iodine
Minadjinov, M.S.	590	New Model of LPE Growth: Growth Rate Calculation (theory)
Minari, F.	768	Growth of High Perfection Ag-Sn Crystals (in Ag solvent)
Mininzon, Yu.M.	777	Synthesis and Hydrothermal Growth of the Ferroelectric SbNbO4, Sb
Minor, D.B.	639	Alkali Ion Exchange Reactions with RbAlSiO4: A New Metastable Pol
Mirtskhulava, A.A.	823	Heteroepitaxial Structures of Si on Sapphire, Obtained by Vacuum
Misawa, S.	1104	Cathodoluminescence of Impurity-Doped Aluminum Nitride Films Prod
Mischgofsky, F.H.	39	Layer Perovskites of the (CnH2n+1NH3)2MX4 and NH3(CH2)mNH3MX4 Fam
	640	Observations on Step Bunching for the Layer Perovskite (C3H7NH3)2
	641	Face Stability and Growth Rate Variations of the Layer Perovskite
Misiuk, A.	383	Growth of USb2 Single Crystals and Their Structural Perfection (v
Mitsui, S.	666	Growth Temperature Dependence in Molecular Beam Epitaxy of Galliu
Miyamoto, Y.	642	Polytypism and Amorphousness in Silicon Whiskers (Si, vapor trans
Miyauchi, T.	935	Photoluminescence Properties of CuGaSe2 Grown by Iodine Vapour Tr
Miyazawa, S.	500	LPE Growth of Li(Nb,Ta)O3 Solid-Solution Thin Film Waveguides on
	643	Epitaxial Growth of KNdP4O12 Laser Waveguides (KLaP4O12, solution
	677	Flux Growth of LiNdP4O12 Single Crystals
	678	Top-Seeded Flux Growth of LiNdP4O12 Single Crystals
Miyazawa, Y.	644	Interface Shape Transitions in the Czochralski Growth of Dy3Al5O1
	645	Interface Shape Transitions in Czochralski Grown YAG Crystals (Y3
	646	The Growth of Dy3Al5O12 (Czochralski)
Mizetskaya, I.B.	725	Phase Diagram of Pseudobinary System CdTe-ZnSe (Bridgman)
Mizuno, H.	927	Single Crystal Growth of Zirconium Carbide by a Modified Hot-Fila
	928	Single Crystal Growth of Titanium Carbide from the Vapor by a Mod
Mochizuki, K.	647	Vapor Phase Transport and Stoichiometry Control of Cadmium Sulfo-
	648	Vapor Phase Transport and Stoichiometry Control of Cadmium Sulfid
Moiseenko, M.M.	994	Fractionation of Stable Isotopes and Impurities During Zone Recry
Molotkov, A.P.	1110	On the Theory of Normal Growth of Crystals in Binary Systems (kin
Monemar, B.	649	Properties of VPE-Grown GaN Doped with Al and Some Iron-Group Met
Monier, J.C.	350	The Thallium Tungstate Tl2W4O13: A Tunnel Structure Related to t
Moon, R.L.	188	LPE Growth of Quaternary GayIn1-yAsxP1-x (film)
Moret, J.	282	The Crystal Structure of CuTeO4 (hydrothermal)
	766	Crystal Chemistry of a Tellurate IV of Indium III: In2Te3O9
Morgret, R.	1068	Magneto-Optical Properties of KTb3F10 and LiTbF4 Crystals (Czochr
Mori, M.	644	Interface Shape Transitions in the Czochralski Growth of Dy3Al5O1
Mori, Y.	469	Chromium Doped GaAs Vapor Phase Epitaxy
	480	Evaluation of Yttrium Iron Garnet Single Crystals Grown by the Fl
	645	Interface Shape Transitions in Czochralski Grown YAG Crystals (Y3
	646	The Growth of Dy3Al5O12 (Czochralski)
	653	A Thermodynamic Factor Influencing the Growth Rate and Purity of
Moriceau, H.	650	Influence of the CaCO3/GeO2 Ratio on Y Sm Lu Ca Fe Ge Garnet Film
Morimoto, K.	651	Ionized-Cluster Beam Epitaxial Growth of GaP Films on GaP and Si
Morin, D.	987	Phase Diagram and Crystal Growth of Lead Phosphovanadates (Pb3(Px
Moriya, K.	652	Observation of Growth Defects in Synthetic Quartz Crystals by Lig
Morizane, K.	469	Chromium Doped GaAs Vapor Phase Epitaxy
	653	A Thermodynamic Factor Influencing the Growth Rate and Purity of
Morozov, N.N.	151	Preparation and Physical Properties of Boracite Crystals (Chemica
Morrish, A.H.	845	Magnetic Properties of beta-Phase Iron-Germanium (FexGe, slow coo
Morrison, J.A.	850	Polytypism and the Vibrational Properties of PbI2
	851	Low Temperature Thermal Properties of PbI2 (Bridgman)
Motegi, H.	654	Preparation of Fine Sodium Chloride Crystals (NaCl)
Motojima, S.	655	A Quick Survey Method for the Study of CVD Conditions (TiN, vapor
	656	Chemical Vapor Growth of LaB6 Whiskers and Crystals Having a Shar
	926	Chemical Vapor Growth of Titanium Diboride by a Modified Hot Wire
	927	Single Crystal Growth of Zirconium Carbide by a Modified Hot-Fila
	928	Single Crystal Growth of Titanium Carbide from the Vapor by a Mod
	929	On the Relations Between the Number of Gold Drops and TiP Whisker
Moysenko, A.E.	1057	GaAs Vapour Phase Epitaxy Using Computer Control
Moze, O.	65	Low Temperature Conductivity of Gd2(SO4)3 8H2O (crystal) and Dy2Ti
Mroczkowski, S.	169	Synthesis of Rare-Earth Carbonates Under Hydrothermal Conditions
Muamatsu, K.	310	High Temperature Form of Pb2WO5 and Transformation Phenomena to I
Mukai, S.	617	Nitrogen Doping into GaAs1-xPx Using Ionized Beam in Molecular Be
Muller-Buschbaum, Hk.	657	A New Preparation Technique in Solid State Chemistry. I. High Te
	658	A New Preparative Technique in Solid State Chemistry. II. High Pr
Muller-Krumbhaar, H.	659	Growth Spirals in a Chemical Potential Varying in Space or Time (
	660	Kinetics of Crystal Growth. Microscopic and Phenomenological The
Muller-Vogt, G.	265	Solubility Curves in High-Temperature Melts for the Growth of Sin
Muller, G.	99	GaSb and InSb Crystals Grown by a Vertical and Horizontal Travell
	842	Czochralski Growth of Li3N Crystals
Muller, J.	577	Epitaxy of NdAl3(BO3)4 For Thin Film Minature Lasers (flux)
	661	Synthesis and Crystal Structure of a New Mixed Oxide "FeV3O8" (Fe
	662	Synthesis and Crystal Structure of a New Mixed Oxide FeV2O6OH0.5:

Muller, W.	160	Single Crystal and Film Growth of Actinide Pnictides by Chemical
	161	Preparation and Single-Crystal Growth of Protactinium Arsenides b
	912	Crystal Growth of Actinide Dioxides by Chemical Transport
Mullica, D.F.	663	Aging Studies on Hydrous Lutetium Oxide (cubic Lu(OH)3)
	664	Cerium Trihydroxide (Ce(OH)3, gel)
Mullin, J.B.(ed.)	66	Crystal Growth (Proceedings of the Third International Summer Sch
Mullin, J.W.	905	A Method for the Determination of Precipitation Induction Periods
Munir, Z.A.	529	Surface Charge and Ledge Dynamics in Cadmium Sulfide (CdS, evapor
Murakami, M.	881	Growth of Thallium Iodide Crystals in Silica Hydrogels (TlI, Tl2I
Murakami, Y.	882	High Rate Epitaxial Growth of ZnO Films on Sapphire by Planar Mag
Muramatsu, K.	264	Photoconduction of Lead Trititanate (PbTi3O7, flux)
	665	Flux Growth of Bi2WO6 Single Crystal Below the Transformation Tem
Muranaka, S.	63	Mass Spectrometric Study on Chemical Transport of VO2
Murotani, T.	666	Growth Temperature Dependence in Molecular Beam Epitaxy of Galliu
Murray, R.	667	The Thermal Expansion of 2H-MoS2 and 2H-WSe2 Between 10 and 320 K
Muszynski, Z.	668	Investigation of Liquidus Surface in Pb-Sn-Te System by Simplex L
Mutaftschiev, B.	532	Semi-Periodic Closed Growth Lamellae on the Crystal-Melt Interfac
Mutsukura, N.	669	Epitaxial Growth of ZnSe by Vapor Flow Method
	670	Preparation and Optical Properties of ZnSe Epitaxial Layers by a
Muzyka, D.R.	671	Control of Structure and Properties. Controlling Microstructures
Myer, H.	672	Dispersion of Acoustic Phonons in Layered SrGa2 (melt, slow cooli
Nagai, H.	673	A Simple Analysis of Vapor Phase Growth - Citing an Instance of G
Nagata, F.	934	Electron Microscopic Characterization of Cubic BN Prepared by Wat
Nagel, E.	254	Dichlorosilane as Silicon Source for Epitaxial Films on Insulator
Naito, K.	455	Growth of Single Crystals of U1-xThxO2 Solid Solutions by Chemica
	674	Synthesis of Nonstoichiometric Zirconium Carbide Whiskers by Chem
Naito, M.	730	Anisotropy of Upper Critical Field in Superconducting 2H-NbS2 (th
Naka, S.	675	Growth of Single Crystals of Na3Al2Li3F12 Garnet Under Hydrotherm
Nakada, I.	418	Growth of Triple Unit Cell Steps on the Surface of Vapour Grown F
	419	Step Growth by Elementary Steps and Multi-Steps on the Surface of
	676	Unit Cell Step Structures on Vapour-Grown Cr2S3 Single Crystals
	1066	Preparation of Some Chalcogenide Spinel Single Crystals and Their
Nakagawa, M.	718	Growth of alpha-Ag2S Whiskers in a VLS System (vapor transport)
Nakahira, M.	819	Effect of Addition of Vanadium on the Chemical Transport of TiSx,
Nakai, S.	506	An LPE Growth of Thick InxGa1-xAs
Nakamura, E.	654	Preparation of Fine Sodium Chloride Crystals (NaCl)
Nakamura, S.	731	Raman Scattering in 3R-NbS2 (vapor transport, theory)
Nakano, J.	677	Flux Growth of LiNdP4O12 Single Crystals
	678	Top-Seeded Flux Growth of LiNdP4O12 Single Crystals
Nakano, K.	679	The Hardness of NbB2 Single Crystals (flux, slow cooling)
	680	Single-Crystal Growth of Tantalum Diboride (TaB2, zone melting, s
Nakatani, I.	681	Vapor-Phase Growth of Epitaxial Ga1-xInxSb Using Open-Tube Flow S
Nakau, T.	425	Birefringence in CdSiP2 (solution)
Nalbandyan, A.G.	682	Enantiomorphism of alpha-LiIO3 Single Crystals (solution)
Nalbandyan, H.G.	682	Enantiomorphism of alpha-LiIO3 Single Crystals (solution)
Namikawa, T.	413	Epitaxial Growth of CrO2 on Sapphire in Air (vapor transport)
	414	Epitaxial Growth of Ferromagnetic CrO2 Films in Air
Nancollas, G.H.	334	Kinetics of Growth of Calcium Sulfate Crystals at Heated Metal Su
	553	Nitrilotri(Methylenephosphonic Acid) Adsorption on Barium Sulfate
	683	The Growth of Crystals in Solution (review, theory, alkaline eart
Nanev, C.	429	Direct Synthesis and Structure of Epitaxial Cadmium Sulphide Laye
Nanjo, J.	1093	Morphology of Iron Pyrite Crystals (FeS2, vapor transport)
Naramoto, H.	684	Growth of Niobium Single Crystals by Pulling from a Melt on a Ped
Narang, R.S.	115	The Growth of Single Crystals of Some Organic Compounds by the Cz
Narayan, J.	1106	Electrical and Structural Characteristics of Laser-Induced Epitax
Narayanan Kutty, T.R.	1039	Effect of Mineralizers on the Growth Morphology of Cubic Thulium
Nassau, J.	685	The History and Present Status of Synthetic Diamond (C, review, d
Nassau, K.	685	The History and Present Status of Synthetic Diamond (C, review, d
Neave, J.H.	686	Temperature Range for Growth of Autoepitaxial GaAs Films by MBE (
Neimark, K.N.	687	Influence of Growth Conditions on Mirodefect Distribution in Disl
Nenasheva, S.N.	688	Ag5GaS4 -- A New Phase in the System Ag-Ga-S (AgGaS2)
Nenow, D.	689	Appearance of Non-Singular Surfaces on Vapour-Grown Ice Crystals
Neumann, H.	690	Growth and Electrical Properties of Epitaxial CuInS2 Thin Films o
	843	Epitaxial Layers of CuInSe2 on GaAs
	844	Structural and Electrical Properties of CuGaSe2 Thin Films on GaA
Neurgaonkar, R.R.	691	Thin Films of LiNbO3, Doped with Na+ and Co2+ plus Zr4+, Grown by
Nevskii, O.B.	591	Theory of Crystallization During Epitaxial Growth from a Solution
Nevsky, O.B.	589	New Model of LPE Growth: Layer Thickness Calculation (theory, ki
	590	New Model of LPE Growth: Growth Rate Calculation (theory)
Newnham, R.E.	108	Polar Twins in Lithium Metasilicate and Metagermanate (Li2SiO3, L
Ng, H.N.	692	Crystallographic Studies and Structural Systematics of the C2/c A
Nguyen, V.D.	610	Magnetic and Electrical Properties of BaVS3 and BaVxTi1-xS3 (flux
Nicholson, D.	164	An Improved System for the Bridgman Growth of Crystals with Toxic
Nicolau, I.F.	693	Solution Growth of Sparingly Soluble Single Crystals from Soluble
	694	Solution Growth of Sparingly Soluble Single Crystals from Soluble
	695	Solution Growth of Sparingly Soluble Single Crystals from Soluble

Nicolet, M-A.	537	Solar Furnace Annealing of Amorphous Si Layers (films)
Niftiev, G.M.	593	alpha-In2Te3 (melt)
Nihira, H.	702	Mechanisms of Chemical Vapor Deposition of Silicon (Si, epitaxy)
Niihara, K.	696	Growth, Morphology and Slip System of alpha-Si3N4 Single Crystal
Nikanorov, S.P.	359	The Preparation of Single Crystals of the Rare Earth Borides by t
Nikiforov, K.G.	780	Preparation and Certain Electrical Properties of CdCr2S4 Single C
Niklewski, T.	513	Accurate X-ray Determination of the Lattice Parameters and the Th
Nikolic, P.M.	697	Some Optical Properties of GeS2 Single Crystals (Bridgman)
Nishida, A.	468	Upper and Lower Critical Fields of TaS2(Pyridine)1/2 (vapor trans
Nishida, I.	495	Crystal Growth of Mn15Si26 (vapor transport, Bridgman)
Nishijima, M.	1103	Formation of Amorphous Mn-Bi Films and Their Crystallization Proc
Nishinaga, T.	698	Studies of LPE Ripple Based on Morphological Stability Theory
	713	Si Contamination in Epitaxial Boron Monophosphide (BP)
Nishino, S.	616	Heteroepitaxial Growth of beta-SiC on Silicon Substrate Using SiC
	699	Growth and Morphology of 6H-SiC Epitaxial Layers by CVD (film)
Nishino, T.	1098	Characterization of the Optical Properties of LPE InxGa1-xAsyP1-y
Nishio, M.	700	Vapor Growth of Zinc Telluride by Closed Tube Method (ZnTe, epita
Nishiyama, G.	858	Vapor Phase Growth of Thin Plates of Cr2O3 Single Crystals
Nishizawa, J.	701	Aspects of Silicon Epitaxy (review, Si)
	702	Mechanisms of Chemical Vapor Deposition of Silicon (Si, epitaxy)
	703	Observations of Defects in LPE GaAs Revealed by New Chemical Etch
	705	Surface Morphology of GaAs Grown by Vapor Phase Epitaxy
Nitsche, R.	148	Sublimation Growth and X-Ray Topographic Characterization of CdTe
	276	A Double-Ellipsoid Mirror Furnace for Zone Crystallization Experi
Nittono, O.	426	Crystal Growth in c Direction and Crystallographic Polarity in Zn
	704	An X-Ray Double-Crystal Method Utilizing Non-Parallel Setting for
Noghinov, A.L.	590	New Model of LPE Growth: Growth Rate Calculation (theory)
Noguchi, T.	308	Liquid-Phase Epitaxy of ZnSe from Zn-Ga Solution (film)
	309	Growth of Cubic ZnS, ZnSe and ZnSxSe1-x Single Crystals by Iodine
Nolang, B.I.	236	An Application of the Flux Function Method to a Five-Component Sy
Nomura, S.	5	Magnetic and Mossbauer Studies of FeNi2BO5 and FeNiBO4 (ferrites,
	1001	Crystal Growth, and Magnetic and Mossbauer Studies of Sr(Fe0.766W
Nonomura, Y.	705	Surface Morphology of GaAs Grown by Vapor Phase Epitaxy
Nord, A.G.	706	Structural Studies of the Solid Electrolyte High-LiTa3O8 (electro
Nordblad, P.	30	Properties of Fe2P Crystals Prepared from a Liquid Copper Medium
Normua, S.	1093	Morphology of Iron Pyrite Crystals (FeS2, vapor transport)
Nosyrev, N.A.	230	Hydrothermal Crystallization of Sodium Zirconogermanates (Na2ZrGe
Notzel, N.	564	Growth of Pb1-xSnxTe Single Crystals of Graded Composition by the
Novitskaya, G.N.	725	Phase Diagram of Pseudobinary System CdTe-ZnSe (Bridgman)
Nowak, Z.	707	Heteroepitaxial Homogeneous CdxHg1-xTe Films (vapor transport)
O'Hara, C.	800	The Effect of Impurities on the Piezoelectric Properties of Lithi
	1076	Ca12Al14O33: A New Piezoelectric Material (Czochralski)
Oda, H.	955	Growth and Properties of Strontium Cobaltate Single Crystals (SrC
Odile, J.P.	471	Crystal Growth and Characterization of In1.9As0.1Se3 (vapor trans
	708	Crystal Growth and Characterization of the Transition-Metal Phosp
Oeder, R.	709	Growth and Properties of Pb(WO4)1-x(MoO4)x Mixed Crystals (Czochr
	848	Experiments on Surface Tension Driven Flow in Floating Zone Melti
Ogawa, H.	700	Vapor Growth of Zinc Telluride by Closed Tube Method (ZnTe, epita
Ogawa, T.	652	Observation of Growth Defects in Synthetic Quartz Crystals by Lig
	937	In-Process Monitoring and Control of Doping Gas Concentration Dur
	965	Geometrical Analysis of Crystallization of the Soft-Core Model in
Ogita, N.	965	Geometrical Analysis of Crystallization of the Soft-Core Model in
Ohishi, M.	710	Edge Emission in Melt-Grown ZnSexS1-x (Bridgman)
Ohmori, K.	710	Edge Emission in Melt-Grown ZnSexS1-x (Bridgman)
Ohnishi, S.	711	Chemical Vapor Deposition of Single-Crystalline ZnO Film with Sno
	882	High Rate Epitaxial Growth of ZnO Films on Sapphire by Planar Mag
Ohno, I.	712	Horizontal Zone-Melting of Indium Antimonide in Vacuum. I. Basi
Ohsawa, J.	713	Si Contamination in Epitaxial Boron Monophosphide (BP)
Ohta, K.	455	Growth of Single Crystals of U1-xThxO2 Solid Solutions by Chemica
Ohta, M.	714	Observations of Etch Pits on As-Grown Faces of Brushite Crystals
Ohtake, H.	884	Crystal Growth of Boron Monophosphide Using a B2H6-PH3-H2 System
Oishi, S.	715	The Effects of M(M=Mg, Ca and Ba) Ions on the KCl Flux Growth of
	716	Growth of NaNd(MoO4)2 Single Crystals from the Ternary System Nd2
Oka, K.	717	Growth of Single Crystals of Lead Monoxide (PbO, flux)
Okabe, T.	718	Growth of alpha-Ag2S Whiskers in a VLS System (vapor transport)
Okada, K.	719	The Crystal Structure of Cs6W11O36
	720	The Crystal Structure of K2W4O13
Okamoto, T.	373	Magnetic Properties of Rare Earth Copper Intermetallic Compounds,
	374	Magnetic Properties of Rare Earth Copper Intermetallic Compounds,
	430	Magnetic Properties of (Mn,Cr)P and (Mn,Fe)P Compounds (Mn1-xFexP
Okonenko, S.A.	721	New Ferroelectric Na3Sc2(PO4)3 (melt)
Okonska-Kozlowska, I.	722	Formation of Mixed Crystals in the System CuxZn1-xCr2Se4
Okuda, H.	858	Vapor Phase Growth of Thin Plates of Cr2O3 Single Crystals
	859	Growth of Cr2O3 Crystals by the Decomposition of a K2Cr2O7 Melt (
	860	Flux Growth of Cr2O3 Crystals from a K2Cr2O7-K2CrO4-K2B4O7 Mixtur
Okuda, T.	710	Edge Emission in Melt-Grown ZnSexS1-x (Bridgman)

Okunev, Yu.A.	723	Relation Between Growth Conditions and Dislocation Structure in G
Okuno, Y.	703	Observations of Defects in LPE GaAs Revealed by New Chemical Etch
	705	Surface Morphology of GaAs Grown by Vapor Phase Epitaxy
Okuyama, F.	724	Recarburization of Monocrystalline Ditungsten Carbide as Revealed
Oleinik, G.S.	725	Phase Diagram of Pseudobinary System CdTe-ZnSe (Bridgman)
Olekseyuk, I.D.	726	Phase Equilibria and Properties of Single Crystals of Solid Solut
Oliver, J.D.	664	Cerium Trihydroxide (Ce(OH)3, gel)
Olsen, G.H.	727	Single-Crystal Growth of Mixed (La, Eu, Y, Ce, Ba, Cs) Hexaboride
	728	Growth Effects in the Heteroepitaxy of III-V Compounds (review)
	729	Crystal Growth and Properties of Binary, Ternary and Quaternary (
Olson, B.	865	Preparation and Characterization of Ruthenium Dioxide Crystals (R
Olson, J.M.	808	A Gas Handling System for Exacting Crystal Growth Preparations (T
Onabe, K.	730	Anisotropy of Upper Critical Field in Superconducting 2H-NbS2 (th
Onari, S.	731	Raman Scattering in 3R-NbS2 (vapor transport, theory)
Ono, A.	431	Growth of Anatase (TiO2) Crystals by Chemical Transport Reactions
Ono, S.	381	Growth of Cobalt Single Crystals by the Electron Beam Floating-Zo
Oota, K.	367	Measurement of Crystallization of Potassium Bromate from Its Quie
Oota, T.	732	Synthesis of Potassium Hexatitanate Fibers by the Hydrothermal De
Opilski, A.	100	Crystal Growth of the Para- and Antiferroelectric Phases of Deute
Oppermann, H.	789	Preparation of Single V305 Crystals by Chemical Transport Reactio
	797	On the Chemical Transport of Some Niobium Oxides with TeCl4. I. C
	798	On the Chemical Transport of Some Niobium Oxides with TeCl4 (II)
Orlov, P.B.	345	Behavior of Boron in Ge Single Crystals Grown by the Czochralski
Orlovskii, V.P.	151	Preparation and Physical Properties of Boracite Crystals (Chemica
	733	Synthesis of Single Crystal LnPO4 (vapor transport)
Ortiz, C.	539	Electron Microscopy Study of Some Molybdenum Oxide Crystals (MoO3
Osborne, W.N.	566	Crystal Growth from Solution Using Cylindrical Seeds (TGS, TBFB)
	567	Single Crystal Growth and Properties of Deuterated Triglycine Flu
Oseroff, S.	293	Magnetic, Transport, and Thermal Properties of Ferromagnetic EuB6
Osica, V.D.	488	Crystals of CTA-DNA: A Growth and Morphological Study (solution)
Osiko, V.V.	20	Synthesis and Crystal Growth of Refractory Materials by RF Meltin
	46	Structure, Spectroscopy, and Stimulated Emission of Crystals of Y
	457	Growth, Spectroscopic Investigations, and Some New Stimulated Emi
Ossart, P.	987	Phase Diagram and Crystal Growth of Lead Phosphovanadates (Pb3(Px
Ostorero, J.	445	Flux Growth of Lead Neodymium Chlorapatite (Na2Nd2Pb6(PO4)6Cl, sl
Ota, Y.	734	n-Type Doping Techniques in Silicon Molecular Beam Epitaxy by Sim
Otooni, M.A.	735	A New Technique for the Preparation of Copper Single Crystals (Cu
	736	A New Technique for Preparation of Thin Single Crystal Films of C
	737	A New Technique for Preparation of Copper Single Crystals from th
Otsubo, M.	852	GaxIn1-xSb Crystals Grown by Liquid Phase Epitaxy (melt)
Ott, H.R.	79	Magnetic Mixed Valent TmSe (melt)
	400	Low Temperature Thermal and Magnetic Properties of CeP and CeAs (
Ottaviani, G.	584	Solid Phase Epitaxial Growth of Si Through Al Film
	585	Growth Kinetics of (111) Si Through an Al Layer by Solid Phase Ep
Otto, H.H.	738	Crystal Data and Crystal Growth of PbGe307 (melt)
	739	Crystal Data and Crystal Growth of Pb5Ge07 (melt)
Ouvrard, G.	571	Physical Properties and Lithium Intercalates of CrPS4 (slow cooli
Oyama, Y.	703	Observations of Defects in LPE GaAs Revealed by New Chemical Etch
Ozaki, K.	655	A Quick Survey Method for the Study of CVD Conditions (TiN, vapor
Ozaki, T.	654	Preparation of Fine Sodium Chloride Crystals (NaCl)
Padalko, A.G.	776	Preparation and Properties of Monocrystalline Films of CdSb (melt
Pajaczkowska, A.	740	Physicochemical Properties and Crystal Growth of A(II)B(VI)-MnB(V
	769	Dissociation Processes and Crystal Growth of Gadolinium Gallium G
	770	On the Preparation of Pseudobrookite Single Crystals by Chemical
Pak, K.	698	Studies of LPE Ripple Based on Morphological Stability Theory
Pakhomov, V.I.	151	Preparation and Physical Properties of Boracite Crystals (Chemica
Palkina, K.K.	587	Structure and Luminescence Properties of alpha-CsNd(PO3)4 Crystal
Palvadeau, P.	1031	Vanadium (III) Oxychloride: Magnetic, Optical and Electrical Pro
Pandya, D.K.	868	Electroless Deposition of Semiconductor Films (PbS, PbSe, Pb1-xHg
	869	Electroless Deposition of Epitaxial Pb1-xHgxS Films
Pannetier, J.	235	About SnF2 Stannous Flouride. I. Crystallochemistry of alpha-SnF2
Paoletti, A.(ed.)	741	Physics of Magnetic Garnets (review)
Paorici, C.	742	Interrelation Between Interface and Vapour Transport Kinetics in
	743	Crystal Growth and Properties of CuGaxIn1-xSe2 Chalcopyrite Compo
	744	A Temperature Variation Method for the Growth of Chalcopyrite Cry
Papadopoulo, A.C.	745	Properties of Homogeneously Doped ZnSe Single Crystals Obtained b
Papushina, T.I.	746	Dependence of Degree of Perfection of Single Crystals of FeGe2 on
Pardo, M.-P.	247	Twinning and Crystal Structure of Mn0.75Ga2.17S4 (flux)
Parent, C.	825	The Crystal Structure of a New High -Nd- Concentration Laser Mate
Parke, A.W.	619	An Investigation of Thin Silver Films on Cleaved Silicon Surfaces
Parker, H.S.	890	Crystal Growth of the Solid Electrolyte (C5H5NH)2Cu5Br7 From the
Pashkova, A.V.	551	Neodymium Incorporation into YAl-Borate Crystals in Preparation f
Passig, G.	72	A New Substrate Holder for Liquid Phase Epitaxy
Passoni, D.	292	A Possible Method for the Growth of Homogeneous Mercury Cadmium T
Pastol, J.L.	792	Preparation, Analysis and Properties of High Purity Zn(1-x)MgxTe
Pastor, J.M.	227	Growth and Some Properties of Cerium Sulphate Enneahydrate Single

Pateau, L.	987	Phase Diagram and Crystal Growth of Lead Phosphovanadates (Pb3(Px
Patel, A.R.	747	Growth and Characterization of Mixed Sr:Ca Tartrates (Sr1-xCazC4H
	748	Overgrowths on Diamond at Atmospheric Pressure (C, flux)
	749	Modified Gel Technique to Grow Single Crystals of KClO4
	750	Crystal Growth of Potassium Perchlorate in Gelatin Gels (KClO4)
	751	Nucleation and Growth of KClO4 Single Crystals in Silica Gels
	752	Gel Growth and Perfection of Orthorhombic Potassium Perchlorate S
Patel, H.B.	12	Growth of Tungstenite Single Crystals by Direct Vapour Transport
Patel, J.R.	68	Dislocation-Free and Low-Dislocation Quartz Prepared by Hydrother
Patel, S.M.	328	Growth of Single Crystals of Cubic Zinc Sulphide by Double Decomp
Paterson, L.	65	Low Temperature Conductivity of Gd2(SO4)38H2O (crystal) and Dy2Ti
Pausch, H.	657	A New Preparation Technique in Solid State Chemistry. I. High Te
	658	A New Preparation Technique in Solid State Chemistry. II. High Pr
Pavlovska, A.	753	Interface Roughness of Vapor-Grown Adamantine Crystals (C10H16)
Pearce, G.	376	Dynamic Sublimation Technique for Growing Mercuric Iodide Crystal
Pearsall, T.P.	754	The Growth of GaxIn1-xAs on (100) InP by Liquid-Phase Epitaxy (fl
Pech, S.	755	A New Technique for Determining the Kinetics of Crystal Growth fr
Pelissier, J.	163	In-Situ Observation of Solidification Interfaces with a Very High
Pelle, F.	756	Optical Properties of Cs2NaBiCl6
Peraldo Bicelli, L.	540	Epitaxial Growth of Antimony Electrodeposits on Oriented Single C
Perez Y Jorba, M.	629	Growth from Skull-Melting of Zirconia-Rare Earth Oxide Crystals (
	630	Ceramic Eutectics in the Systems ZrO2-Ln2O3 (Ln = Lanthanide): U
Perrier de la Bathie, R.	631	Magnetization, Magnetocrystalline Anisotropy, Magnetostriction an
Perrin, C.	757	New Thiochlorides of Molybdenum II: Mo6Cl10Y (Y = S, Se, Te): S
Perrot, P.	120	Crystal Structure of Ba2Fe6O11 (flux)
Perry, P.B.	758	The Optical Absorption Edge of Single-Crystal AlN Prepared by a C
Peshev, P.	759	Synthesis and Crystal Growing of La3NbO7
	760	On the Preparation of Nickel Ferrite Single Crystals by Chemical
	770	On the Preparation of Pseudobrookite Single Crystals by Chemical
Peskin, V.F.	777	Synthesis and Hydrothermal Growth of the Ferroelectric SbNbO4, Sb
Peters, D.	690	Growth and Electrical Properties of Epitaxial CuInS2 Thin Films o
Peterson, S.W.	324	The Crystal Structure of Lithium Uranate (Li2UO4, flux)
Petroff, P.M.	761	Molecular Beam Epitaxy of Ge and Ga1-xAlxAs Ultra Thin Film Super
	762	Crystal Growth Kinetics in (GaAs)n-(AlAs)m Superlattices Deposite
Petrushinina, S.A.	251	Growth Kinetics and Structure of Epitaxial Garnet Layers as Funct
Pettenpaul, E.	1043	Preparation of Pure and Doped Silicon Carbide by Pyrolysis of Sil
Pfeiffer, H.	763	Basic Consideration for an Analytic Theory of Two-Component Cryst
	764	On the Analytic Treatment of the Growth of Rough Interfaces (theo
	765	Analytical Theory of Crystal Growth of Two-Component Systems I.
Pfister, J.C.	98	Transmission Electron Microscopy Observations of Precipitates in
Philippot, E.	282	The Crystal Structure of CuTeO4 (hydrothermal)
	766	Crystal Chemistry of a Tellurate IV of Indium III: In2Te3O9
Picard, J.	767	Magnetic Properties of a Single Crystal of NdF3 Between 2 and 4.2
Picaud, B.	36	Crystal Structure of RbEu3F10 (melt)
	37	The Systems RbF-EuF3 and CsF-EuF3. Study of the Phases RbLnF4 an
Pichaud, B.	768	Growth of High Perfection Ag-Sn Crystals (in Ag solvent)
Pickhardt, V.Y.	898	Impurity Dopant Incorporation and Diffusion During Molecular Beam
Piekarczyk, W.	565	Crystallisation of Some Ferrite Materials from Lead Flux with Add
	769	Dissociation Processes and Crystal Growth of Gadolinium Gallium G
	770	On the Preparation of Pseudobrookite Single Crystals by Chemical
	771	The Czochralski Growth of Bismuth-Germanium Oxide Single Crystals
Pink, H.	1004	Raman and Infrared Spectra of CdIn2S4 and ZnIn2S4 (vapor transpor
Piotrowksi, T.	707	Heteroepitaxial Homogeneous CdxHg1-xTe Films (vapor transport)
Piotrowski, J.	707	Heteroepitaxial Homogeneous CdxHg1-xTe Films (vapor transport)
Pisarevsky, Yu.V.	256	Na2CoGeO4 Single Crystal Growth and Study of Their Piezoelectric
Pishchik, V.V.	246	Profile-Sapphire and Its Structural Perfection (alpha-Al2O3, melt
Plet, F.	772	Hydrothermal Growth of the Systems HF-AlF3 and NH4F-FeF3 in Hydro
Ploog, K.	248	Periodic Doping Structure in GaAs (GaAs, Ga1-xAlxAs, molecular be
Plovnick, R.H.	556	Epitaxial Growth of CdS on CuInSe2
Pobedimskaya, E.A.	592	Hydrothermal Crystallization in the System BaO-GeO2-Na2O-H2O (bar
Podolinski, V.V.	773	Investigation of Structural Formation and Growth Kinetics of Orga
Poignant, H.	774	Solubility of AlPO4 in Hydrothermal Solutions of H3PO4
Poindessault, R.	775	Preparation of Photoluminescent and Conductive ZnSe:I Crystals by
Ponomarev, V.F.	776	Preparation and Properties of Monocrystalline Films of CdSb (melt
Popolitov, V.I.	777	Synthesis and Hydrothermal Growth of the Ferroelectric SbNbO4, Sb
Popov, A.I.	997	Variations of Impurity Distribution in the Modified Czochralski M
	998	Growth and Properties of Silicon Crystals With Automatic Diameter
Popov, V.P.	60	Reaction of Germanium (111) Single Crystals with GeBr4 Vapor (Ge,
	778	Anisotropy of Ion Transport in Stoichiometric Rutile Single Cryst
Popovic, Z.V.	697	Some Optical Properties of GeS2 Single Crystals (Bridgman)
Potemski, R.M.	892	An Automated System for the Growth of Multilayered Structures in
Potschke, J.	485	The Influence of Impurities on the Crystallization of Supercooled
Pouchard, M.	124	Preparation by Vapor Transport of Monocrystals of AlF3, FeF3, and
	191	Germanium-Iodine Equilibrium from 700 to 1000 K (Ge, vapor transp
Pougnet, A.M.	447	Crystal Growth and Crystallographic Data of Some Ln2(MoO4)3 Type
Pouillard, G.	120	Crystal Structure of Ba2Fe6O11 (flux)

250

Pouilly, G.	836	Pyroelectricity and Related Properties in the Fresnoite Pseudobin
Povzner, A.A.	861	Growth Characteristics of Crystals of Certain Metallides (alpha-F
Powell, J.L.	779	Effect of S-35 Tracer, Particle Size and Imperfection on the Solu
Pradhananga, R.	871	An Inexpensive Zone Melting Furnace for Purification and Crystal
Prasad, B.	871	An Inexpensive Zone Melting Furnace for Purification and Crystal
Predel, B.	25	Telluride Halides of Copper with High Ionic Partial Conductivity.
Preisser, F.	848	Experiments on Surface Tension Driven Flow in Floating Zone Melti
Premachandran, S.K.	327	Growth of Phthalic Anhydride in a Closed Crystal-Vapour System (o
Primont, J.	626	Preparation, Crystal Growth and Characterization of Mixed Bromide
Primot, J.	987	Phase Diagram and Crystal Growth of Lead Phosphovanadates (Pb3(Px
Prodan, A.	129	On the Ordering of Fe Atoms in FexNbS2 (vapor transport)
Prokhorov, A.M.	20	Synthesis and Crystal Growth of Refractory Materials by RF Meltin
	245	Spectroscopic Properties of Er3Al5-xGaxO12 Films Obtained by Liqu
Protasova, V.I.	475	Solubility of KLa(MoO4)2 in Aqueous K2MoO4 Solutions in Hydrother
Pulham, R.J.	252	Low Temperature Method for the Growth of Lithium Nitride Single C
Queisser, H.J.	363	In Situ X-Ray Topography of Epitaxial Ge Layers During Growth
Quelch, M.J.T.	74	The Effects of Growth Speed on the Compositional Variations in Cr
	75	A Study of Casting in the CdxHg1-xTe System
Queyroux, F.	1034	Preparation, Stability, and Preliminary Crystallographic Study of
Rabenau, A.	25	Telluride Halides of Copper with High Ionic Partial Conductivity.
Radautsan, S.I.	625	Preparation of Single Crystals of the Magnetic Semiconductor CdCr
	780	Preparation and Certain Electrical Properties of CdCr2S4 Single C
Rae Smith, A.R.	781	The Preparation and Crystal Structures of BiReO4 and BiRe2O6 (vap
Rahman Khan, M.S.	782	The Growth and Structure of Epitaxial Films of the Rare-Earth Dih
Randles, M.H.	783	Liquid Phase Epitaxial Growth of Magnetic Garnets (review)
Rao, A.V.	749	Modified Gel Technique to Grow Single Crystals of KClO4
	750	Crystal Growth of Potassium Perchlorate in Gelatin Gels (KClO4)
Rao, K.K.	784	Crystal Growth and Selective Etching Studies on Sodium Chlorate C
Rao, N.	1039	Effect of Mineralizers on the Growth Morphology of Cubic Thulium
Rasmussen, S.E.	441	Growth of Chromium Silicide, Cr3Si, Crystals (float zone melting)
Raveau, B.	350	The Thallium Tungstate Tl2W4O13: A Tunnel Structure Related to t
Ravez, J.	785	Preparation, Optical and Dielectric Properties of Na5W3O9F5 (melt
Ravishankar, P.S.	786	The Stage II Transient in Off-Eutectic Solidification (theory)
Ravitz, S.F.	1	Rate Equations for Dislocation-Free and Dislocation-Assisted Grow
Ray, I.	912	Crystal Growth of Actinide Dioxides by Chemical Transport
Rebizant, J.	161	Preparation and Single-Crystal Growth of Protactinium Arsenides b
Reddy, K.N.	12	Growth of Tungstenite Single Crystals by Direct Vapour Transport
Redlich, W.	787	New Observations in the Chemical Transport of GeO2. I. Transport
Reiche, P.	457	Growth, Spectroscopic Investigations, and Some New Stimulated Emi
Reichelt, K.	530	Anisotropy of Round Evaporation Spirals on Rocksalt Surfaces (NaC
	788	The Epitaxial Growth of Reactively Sputtered CuI Films on Rocksal
Reichelt, W.	789	Preparation of Single V3O5 Crystals by Chemical Transport Reactio
Reidinger, F.	790	Single Crystal Preparation and Characterization of the Group VA M
Reilly, J.J.	790	Single Crystal Preparation and Characterization of the Group VA M
Reimer, H.-H.	172	On the Growth of NH4Cl Dendrites at Very Low Supersaturation (sol
Reis, A.H.,Jr.	324	The Crystal Structure of Lithium Uranate (Li2UO4, flux)
Reis, A.H.Jr.	474	Structural and Magnetic Properties of "One-Dimensional" Barium Va
Remeika, J.P.	621	X-Ray Diffuse Scattering from Alkali, Silver, and Europium beta-A
	791	Superconductivity of Single-Crystal NbGe2
Repinskii, S.M.	60	Reaction of Germanium (111) Single Crystals with GeBr4 Vapor (Ge,
Revcolevschi, A.	239	Growth of Oriented Bicrystals of Nickel Oxide (NiO, arc image, fl
Revel, G.	792	Preparation, Analysis and Properties of High Purity Zn(1-x)MgxTe
Rhodes, C.A.	369	Thermal Analysis of Solidification in Web-Dendritic Ribbon Growth
Riabtsev, N.G.	668	Investigation of Liquidus Surface in Pb-Sn-Te System by Simplex L
Ricard, J.	793	Evaluation of Verneuil Sapphire as a Substrate for Silicon on an
Rice, C.E.	208	La8Ru4O21: A Mixed-Valence Ternary Ruthenium Oxide of a New Hexa
	209	The Crystal Structure of La3Ru3O11: A New Cubic KSbO3 Derivative
Richards, E.A.	794	Crystal Growth of Bi3B5O12 From its Stoichiometric Melt (kinetics
Richardson, M.W.	236	An Application of the Flux Function Method to a Five-Component Sy
Richter-Van Leeuwen, G.K.	39	Layer Perovskites of the (CnH2n+1NH3)2MX4 and NH3(CH2)mNH3MX4 Fam
Rijnsdorp, J.	795	Preparation, Structures, and Properties of Niobium Chalcogenide H
Riley, N.	796	Effects of Rotation on Czochralski Crystal Growth (theory)
Rimini, E.	996	Grain Size Dependence in a Self-Implanted Silicon Layer on Laser
Ristic, R.I.	219	The Role of Dislocations in the Growth of Ammonium Dihydrogen Pho
Ritschel, M.	797	On the Chemical Transport of Some Niobium Oxides with TeCl4. I. C
	798	On the Chemical Transport of Some Niobium Oxides with TeCl4 (II)
Rivet, J.	300	Crystal Structure of the Zinc Iodide ZnI2 (vapor transport)
Rivolta, B.	540	Epitaxial Growth of Antimony Electrodeposits on Oriented Single C
Robert, M.C.	233	X-Ray Topographic Study of Growth Defects in Hydrothermal Na2CoGe
	337	Effect of Stirring on Crystalline Quality of Solution Grown Cryst
	545	Crystal Quality of Gel Grown CaHPO4.2H2O Crystals
Robertson, A.S.	76	The Synthesis of the First Stage Graphite Salt C8(+)AsF6(-) and I
Robertson, D.S.	440	The Growth and Properties of Crystalline Rubidium and Cesium Niob
	799	The Preparation and Properties of Single Crystal Copper Phosphide
	800	The Effect of Impurities on the Piezoelectric Properties of Lithi
	801	The Cadmium Oxide-Tungsten Oxide Phase System and Growth of Cadmi

Robertson, J.M.	71	Preparation of Barium Lead Hexa-Aluminate Single Crystal Layers b
	216	Thin Film Growth of (MnxZn1-x)FeO4 by Liquid Phase Epitaxy on Bri
	802	Liquid Phase Epitaxy of Garnets (review)
Rode, J.R.	803	Liquid Phase Epitaxial Growth of Ga1-xInxSb on GaSb by Stepwise G
Rodier, N.	132	Crystal Structure of Zn4(P2S6)3 (melt)
Rodriguez, R.	296	Mechanism of Habit Change of ADP Crystals by Fe3+, Based on Mossb
	804	Unstable Growth of ADP Crystals (NH4H2PO4, theory, solution, kine
Rogers, T.	805	Some Thermodynamic Considerations for the Growth of Layered Dicha
Rogl, P.	806	The Crystal Structure of ThBC
Roland, G.W.	411	Crystal Growth of Tl3VS4 (Bridgman-Stockbarger)
Romanenko, V.N.	511	Control of the Composition of Crystals Grown in Electric and Magn
Romanowski, Z.	906	Reactive Pulse Plasma Crystallization of Diamond and Diamond-Like
Romeo, N.	743	Crystal Growth and Properties of CuGaxIn1-xSe2 Chalcopyrite Compo
Rommeluere, J.F.	600	Luminescence and Bandgap Studies of High Purity MgxZn1-xTe Ternar
	792	Preparation, Analysis and Properties of High Purity Zn(1-x)MgxTe
Rooijmans, C.J.M. (ed.)	807	Crystals for Magnetic Applications, Vol. 1 (garnets, review, hydr
Rosenberger, F.	7	Congruent (Diffusionless) Vapor Transport (theory, kinetics)
	808	A Gas Handling System for Exacting Crystal Growth Preparations (T
	809	Low-Stress Physical Vapor Growth (PVT) (KCl, KBr, KCl:KCN, KCN:KO
	1074	On Diffusive-Advective Interfacial Mass Transfer (theory, solutio
Rossnick, M.	358	Limitations in Using Kilohertz Radio Frequencies for Float Zone S
Roth, R.S.	639	Alkali Ion Exchange Reactions with RbAlSiO4: A New Metastable Pol
Roth, W.L.	286	Ionic Conductivity in N30+ Beta Alumina (solution)
Rouaux, Y.	630	Ceramic Eutectics in the Systems ZrO2-Ln2O3 (Ln = Lanthanide): U
Rouchaud, J.C.	792	Preparation, Analysis and Properties of High Purity Zn(1-x)MgxTe
Roux, M.T.	21	Crystal Structure of beta-KEr3F10 (melt, slow cooling)
Rouxel, J.	1031	Vanadium (III) Oxychloride: Magnetic, Optical and Electrical Pro
Roy, B.N.	810	Kinetics of Crystal Growth of Calcium Tungstate from Solutions in
	811	Kinetics of Crystal Growth of Strontium Tungstate from Solutions
	812	Kinetics of Crystallization of Barium Tungstate from Solutions in
Roy, D.M.	813	Crystal-Chemistry, Crystal-Growth, and Phase-Equilibria of Apatit
Roy, R.	813	Crystal-Chemistry, Crystal-Growth, and Phase-Equilibria of Apatit
Rozin, K.M.	978	Kinetic Characteristics of Growth of Y3Al5O12 Single Crystals fro
Rubin, J.J.	552	Preparation and Properties of Bulk GaxIn1-xAs
Rull, F.	227	Growth and Some Properties of Cerium Sulphate Enneahydrate Single
Rundqvist, S.	30	Properties of Fe2P Crystals Prepared from a Liquid Copper Medium
Runsdorp, J.	814	The Crystal Structure of Niobium Trisulfide, NbS3
Rupp, L.W.,Jr.	31	On the Resistivity of the Magnetically Ordered Ground State of th
Ruppel, D.	578	Epitaxial Layers of the Laser Material Nd(Ga,Cr)3(BO3)4 (flux)
Rusaikin, M.P.	184	Theoretical Analysis of Equilibrium Adsorption Layers in CVD Syst
Rusche, C.	192	Crystal Growth of Niobium Nitride and Niobium Carbide Nitride (Nb
Rusek, A.	906	Reactive Pulse Plasma Crystallization of Diamond and Diamond-Like
Russell, G.J.	815	The Growth of CdS in Sealed Silica Capsules (vapor transport)
	816	Vapour Growth and Defect Characterisation of Large Single Crystal
Rustamov, P.G.	22	Growth and X-Ray Analysis of Single Crystals of EuSb4S7 and EuSb2
	817	Synthesis, Preparation of Single Crystals, and Physicochemical In
Rustamov, V.D.	361	Structure and Growth Peculiarities of Tl-Se-TlInSe2 (TlSe, zone m
Rutz, R.F.	758	The Optical Absorption Edge of Single-Crystal AlN Prepared by a C
Ryskin, G.Ya.	23	Influence of Isotopic Composition on the Density and Molecular Vo
Rzajeva, L.A.	361	Structure and Growth Peculiarities of Tl-Se-TlInSe2 (TlSe, zone m
Sabinina, I.V.	885	Growth of Pb0.8Sn0.2Te Films in a Quasi-Closed Space (vacuum evap
Sachs, E.M.	818	In-Situ Carbon-Silicon Ribbon Thermocouple (Si, equipment)
Sadowski, J.	707	Heteroepitaxial Homogeneous CdxHg1-xTe Films (vapor transport)
Saeki, M.	819	Effect of Addition of Vanadium on the Chemical Transport of TiSx,
Saidov, M.S.	820	Growing Single Crystals of Silicon-Germanium Solid Solutions by E
Saito, H.	732	Synthesis of Potassium Hexatitanate Fibers by the Hydrothermal De
	1094	Vapor Phase Growth of Alumina Whiskers by Hydrolysis of Aluminum
Saito, Y.	499	Electrical Resistivity in In-Heusler Cu2.0Mn0.94In1.0 Single Crys
	821	Crystal Structure and Habit of Fine Metal Particles Formed by Gas
	881	Growth of Thallium Iodide Crystals in Silica Hydrogels (TlI, Tl2I
Sakata, T.	495	Crystal Growth of Mn15Si26 (vapor transport, Bridgman)
Sakiyama, K.	582	Dielectric and Optical Properties of K5Nd(MoO4)4 (Czochralski)
Sakka, S.	822	Structure and Stability of Alkali Germanate Crystals and Glasses
Sakudo, T.	717	Growth of Single Crystals of Lead Monoxide (PbO, flux)
Sakuma, E.	680	Single-Crystal Growth of Tantalum Diboride (TaB2, zone melting, s
Sakvarelidze, L.G.	823	Heteroepitaxial Structures of Si on Sapphire, Obtained by Vacuum
Salje, E.	241	The Crystal Structure of Triclinic WO3 (sublimation)
	824	Structural Phase Transition in Mixed Crystals WxMo1-xO3 (melt)
Salmon, R.	825	The Crystal Structure of a New High -Nd- Concentration Laser Mate
Samiev, S.Kh.	599	Growth of CdP4 Single Crystals (evaporation, vapor transport)
Samojlovich, Yu.A.	826	Dynamics of Non-Equilibrium Crystallization of an Alloy (theory,
Sangwall, K.	827	Selective Etching of Cesium Iodide Crystals and Comparison of Gro
Sanygin, V.N.	776	Preparation and Properties of Monocrystalline Films of CdSb (melt
Saraie, J.	482	Heteroepitaxial Growth of ZnS on GaP by the Close-Spaced Techniqu
	828	Liquid Phase Epitaxial Growth of CdTe in the CdTe-CdCl2 System (f
Sarkisov, S.E.	457	Growth, Spectroscopic Investigations, and Some New Stimulated Emi

	458	Growth, Spectroscopy, and Stimulated Emission of Cubic Bi4Ge3O12
Sasaki, A.	951	Quasi-Grown Layers in Liquid Phase Epitaxial Growth (In1-xGaxAs)
	952	Composition Latching Phenomenon and Lattice Mismatch Effects in L
Sato, K.	499	Electrical Resistivity in In-Heusler Cu2.0Mn0.94In1.0 Single Crys
	1065	Solution Growth of ZnS, ZnSe, CdS and Their Mixed Compounds Using
Satou, M.	414	Epitaxial Growth of Ferromagnetic CrO2 Films in Air
Saunders, G.A.	581	Elastic Constants and Stability of bcc In-Tl Alloys (Bridgman)
Savage, A.	761	Molecular Beam Epitaxy of Ge and Ga1-xAlxAs Ultra Thin Film Super
	762	Crystal Growth Kinetics in (GaAs)n-(AlAs)m Superlattices Deposite
Savitsky, E.M.	829	Growth of Single Crystals of High Melting Metal Alloys and Compou
Sawada, S.	830	Ferroelectricity in (N(CH3)4)2ZnCl4 (evaporation)
	831	Ferroelectricity in (N(CH3)4)2CoCl4
Sberveglieri, G.	743	Crystal Growth and Properties of CuGaxIn1-xSe2 Chalcopyrite Compo
Schaefer, R.J.	832	The Validity of Steady-State Dendrite Growth Models (theory)
	833	Reply to Comment of R. Trivedi on "The Validity of Steady-State D
Scharmann, A.	709	Growth and Properties of Pb(WO4)1-x(MoO4)x Mixed Crystals (Czochr
	847	Some Evidence for the Existence and Magnitude of a Critical Maran
	848	Experiments on Surface Tension Driven Flow in Floating Zone Melti
Schaub, B.	600	Luminescence and Bandgap Studies of High Purity MgxZn1-xTe Ternar
Scheiber, H.,Jr.	552	Preparation and Properties of Bulk GaxIn1-xAs
Schieber, M.	834	State-of-The-Art of Crystal Growth and Nuclear Spectroscopic Eval
Schleich, D.	1031	Vanadium (III) Oxychloride: Magnetic, Optical and Electrical Pro
Schleich, D.M.	571	Physical Properties and Lithium Intercalates of CrPS4 (slow cooli
Schlenker, C.	835	Metal-Insulator Transitions and Phase Diagram of (Ti1-xVx)4O7: E
Schluckebier, G.	25	Telluride Halides of Copper with High Ionic Partial Conductivity.
Schmelczer, R.	401	Crystal Structure and Antiferromagnetism of EuSb2 (melt)
Schmid, H.	236	An Application of the Flux Function Method to a Five-Component Sy
	836	Pyroelectricity and Related Properties in the Fresnoite Pseudobin
	837	Gas Phase Synthesis of Epitaxial Layers of Nickel-Chlorine Boraci
Schmidt, W.	838	Dislocation Propagation in Czochralski Grown Gadolinium Gallium G
Schmit, J.L.	839	LPE Growth of Hg0.60Cd0.40Te from Te-Rich Solution (film, vapor t
Schmugge, K.	325	Interaction Between Technological Parameters and Crystal Quality
	326	Effects of Radial and Axial Temperature Gradients on Tensions in
Schneider, H.A.	86	Growth and Physical Properties of ZnSiP2 (vapor transport, soluti
Schobinger-Papamantellos,	840	On the Magnetic Structure and Magnetic Phase Transitions of Tb5Ge
Schonherr, E.	841	Measurements of Growth Parameters During the Crystal Growth from
	842	Czochralski Growth of Li3N Crystals
Schoolar, R.B.	439	Electrical Properties of Pb1-xCdxS Epitaxial Films (vapor transpo
Schoonman, J.	1012	Luminescence and Ionic Conductivity of Bariumfluorochloride (BaFC
Schouten, W.J.	516	Thin Film Deposition From Beams of Ionized Atoms and Clusters (Ag
Schreiber, H.,Jr.	54	Melt and Solution Growth of Bulk Single Crystals of Quaternary II
Schujko, A.	479	Maximum Stable Zone Length in Float-Zone Growth of Small-Diameter
Schultze, D.	457	Growth, Spectroscopic Investigations, and Some New Stimulated Emi
	458	Growth, Spectroscopy, and Stimulated Emission of Cubic Bi4Ge3O12
Schulz, U.	1086	Partition of Principal Components in the Course of Preparation of
Schumann, B.	690	Growth and Electrical Properties of Epitaxial CuInS2 Thin Films o
	843	Epitaxial Layers of CuInSe2 on GaAs
	844	Structural and Electrical Properties of CuGaSe2 Thin Films on GaA
Schurer, P.J.	845	Magnetic Properties of beta-Phase Iron-Germanium (FexGe, slow coo
Schurmann, E.	846	Progress of Temperature and Concentration in the Region of Solidi
Schuthof, J.	40	Preparation of Pure Hydroxyapatite Single Crystals by Hydrotherma
Schwabe, D.	709	Growth and Properties of Pb(WO4)1-x(MoO4)x Mixed Crystals (Czochr
	847	Some Evidence for the Existence and Magnitude of a Critical Maran
	848	Experiments on Surface Tension Driven Flow in Floating Zone Melti
Schwanebeck, J.	162	A Simple Method for Growing V3Si Single Crystals (electron beam m
Schwuttke, G.H.	1099	Analysis of Vapor-Deposited Silicon Carbide Films on Silicon Ribb
Scifres, D.R.	152	Integrated Optical Devices Fabricated by MBE (GaAs, Ga1-xAlxAs, r
Scott, W.	849	Solution Growth of Indium-Doped Silicon (Si)
Sealy, B.J.	156	The Thermal Oxidation of GaAs
Sears, W.M.	850	Polytypism and the Vibrational Properties of PbI2
	851	Low Temperature Thermal Properties of PbI2 (Bridgman)
Sedgwick, T.O.	211	Low Temperature CVD Garnet Growth (Er3Fe5O12 and (Eu, Y)3Fe5O12 and (
Seeber, B.	186	New Ternary Mo(II)-Compounds InxMo15Se19 Containing Mo6Se8 and Mo
Segawa, K.	852	GaxIn1-xSb Crystals Grown by Liquid Phase Epitaxy (melt)
Sehgal, H.K.	869	Electroless Deposition of Epitaxial Pb1-xHgxS Films
Seibert, G.	853	Spacelab and Material Processing Facilities and Experiments (revi
	854	Spacelab and Past and Future Crystal Growth Experiments in Space
Seidensticker, R.G.	855	Solute Partitioning During Silicon Dendritic Web Growth (Si, theo
Sekerka, R.F.	206	Lateral Solute Segregation During Unidirectional Solidification o
	534	Evidence for a Universal Law of Dendritic Growth Rates (theory)
Seki, Y.	614	X-Ray Study of LEC-Grown InP Crystals (Czochralski, theory)
	856	Impurity Effect on Grown-In Dislocation Density of InP and GaAs C
Sekiguchi, Y.	857	Scanning Electron Microscopy of Growth Surfaces of Synthetic Quar
Sekiya, T.	858	Vapor Phase Growth of Thin Plates of Cr2O3 Single Crystals
	859	Growth of Cr2O3 Crystals by the Decomposition of a K2Cr2O7 Melt (
	860	Flux Growth of Cr2O3 Crystals from a K2Cr2O7-K2CrO4-K2B4O7 Mixtur

253

Selin, V.V.	116	Growth of Si Strips by Stepanov's Method
	723	Relation Between Growth Conditions and Dislocation Structure in G
Semenov, V.N.	861	Growth Characteristics of Crystals of Certain Metallides (alpha-F
Semenova-Tyan-Shanskaya,	862	Temperature Field Dynamics in the Synthesis of Diamonds (kinetics
Semiletov, S.A.	526	Real Structure of Heteroepitaxial Zinc Oxide Films Grown on Sapph
Semrad, E.E.	509	Preparation and Principal Physicochemical Properties of Alkali Me
Sen, S.	863	Influence of Magnetic Field on Vertical Bridgman-Stockbarger Grow
	864	Non-Constant Distribution Coefficients for Directionally Solidifi
Sergent, M.	186	New Ternary Mo(II)-Compounds InxMo15Se19 Containing Mo6Se8 and Mo
	757	New Thiochlorides of Molybdenum II: Mo6Cl10Y (Y = S, Se, Te): S
Sethi, B.R.	612	Effect of High-Temperature Annealing in Cadmium on the Electrical
Sevast'yanov, M.A.	984	Epitaxial Growth of BaTiO3 Vacuum Condensates on SrTiO3 Crystals
Shabaltai, A.A.	46	Structure, Spectroscopy, and Stimulated Emission of Crystals of Y
Shabunina, G.G.	1037	Photoferromagnetic Effect and Photoconductivity of Cd1-xHgxCr2Se4
Shafer, M.W.	865	Preparation and Characterization of Ruthenium Dioxide Crystals (R
Shafizade, R.B.	866	Electron Diffraction Study of Phase Transformations of the Compou
Shamah, A.M.	896	The Crystal Growth and Atomic Structure of As2(Se, S)3 Compounds (
Shannette, G.W.	307	Elastic Constants of Niobium-Zirconium, Hafnium, and Tungsten All
Shannon, R.D.	867	Polymorphism in Bi2Sn2O7 (flux)
Sharkhatunyan, R.O.	682	Enantiomorphism of alpha-LiIO3 Single Crystals (solution)
Sharma, N.C.	868	Electroless Deposition of Semiconductor Films (PbS, PbSe, Pb1-xHg
	869	Electroless Deposition of Epitaxial Pb1-xHgxS Films
Sharma, O.P.	612	Effect of High-Temperature Annealing in Cadmium on the Electrical
Sharma, S.D.	870	Growth and Polytypism in Vapor-Grown Mixed Cadmium Halide Crystal
Sharon, M.	871	An Inexpensive Zone Melting Furnace for Purification and Crystal
Shaulov, A.	229	Thermal and Dielectric Properties of LiKSO4 and LiCsSO4
Shaw, N.	560	Growth and Electrical Properties of Deuterated TGS Produced by th
Shdanov, A.V.	128	The Effect of the Solidification Front Shape on the Temmperature
Sheikhet, E.G.	687	Influence of Growth Conditions on Mirodefect Distribution in Disl
Shenderovich, I.L.	998	Growth and Properties of Silicon Crystals With Automatic Diameter
Shepelev, Yu.F.	902	Crystal Structure of Scandium Metaphosphate Sc(PO3)3 (solution)
Sherry, E.G.	872	The Structure of K2Er(NO3)5 (flux)
Sherwood, J.N.	115	The Growth of Single Crystals of Some Organic Compounds by the Cz
Shevchenko, V.Y.	598	The Growth of CdAs2 and ZnAs2 Single Crystals from the Vapor Pha
Shevchenko, V.Ya.	599	Growth of CdP4 Single Crystals (evaporation, vapor transport)
Shevelev, V.V.	873	Capture of Migrating Adatoms by Point Capture Centres for Differe
Shibata, N.	426	Crystal Growth in c Direction and Crystallographic Polarity in Zn
	427	Growth Mechanism of Hollow ZnO Crystals from ZnSe. II
	428	Crystal Growth and Sublimation in II-VI Compounds Along their Pol
Shichiri, T.	874	Morphology and Structure of ZnS Grown from the Vapour Phase
Shick, L.K.	638	The Effect of Melt Flow Phenomena on the Perfection of Czochralsk
Shigematsu, T.	875	Epitaxial Thin Iron Films on Antimony (Fe, vapor transport)
Shih-chun, C.	1107	Magneto-Optical Properties of Some Rare Earth Iron Garnets (flux,
Shih, I.	876	Czochralski Growth of Tellurium Single Crystals (Te)
	877	Etch Pit Orientation in Czochralski-Grown Tellurium Crystals (Te)
Shilo, I.	983	Crystal Structure of the Spinel CdErS4 (slow cooling)
Shimada, S.	424	Vapor Growth of Zn2GeO4 Single Crystals
Shimanoe, T.	666	Growth Temperature Dependence in Molecular Beam Epitaxy of Galliu
Shimaoka, G.	878	Epitaxial Growth of Some II-VI Compound Films Evaporated Onto Ele
Shimazu, M.	264	Photoconduction of Lead Trititanate (PbTi3O7, flux)
Shimizu, S.	704	An X-Ray Double-Crystal Method Utilizing Non-Parallel Setting for
Shimony, Y.	95	Crystal Growth and Phase Transitions in the NbxCr1-xO2 (x greater
Shin, S.H.	1059	Liquid Phase Growth of HgCdTe Epitaxial Layers
Shindo, I.	404	Single Crystal Growth of Akermanite (Ca2MgSi2O7) and Gehlenite (C
	480	Evaluation of Yttrium Iron Garnet Single Crystals Grown by the Fl
	483	Interface Shape and Horizontal Variations of Al and Ga Contents i
	879	Single Crystal Growth of Substituted Yttrium Iron Garnets Y3Fe5-x
	880	Growth of Mg2TiO4 Single Crystals by the Floating Zone Method
	955	Growth and Properties of Strontium Cobaltate Single Crystals (SrC
Shinozaki, S.S.	288	Electron Energy-Loss Spectroscopy of TiS2 (vapor transport)
Shiojiri, M.	881	Growth of Thallium Iodide Crystals in Silica Hydrogels (TlI, Tl2I
Shiokawa, J.	580	Synthesis and Characterization of Europium(II)-Holoborates, Eu2B5
Shiosaki, T.	9	Epitaxial Growth of Potassium Lithium Niobate Single-Crystal Film
	11	Epitaxial Growth of Potassium Lithium Niobate Single-Crystal Film
	711	Chemical Vapor Deposition of Single-Crystalline ZnO Film with Smo
	882	High Rate Epitaxial Growth of ZnO Films on Sapphire by Planar Mag
Shirafuji, J.	939	Growth by Travelling Heater Method and Characteristic of Undoped
	957	Growth of GaAs1-xSbx Crystals by Steady-State Liquid Phase Epitax
Shiraki, Y.	883	Molecular Beam and Solid-Phase Epitaxies of Silicon Under Ultra-H
Shiroishi, Y.	830	Ferroelectricity in (N(CH3)4)2ZnCl4 (evaporation)
	831	Ferroelectricity in (N(CH3)4)2CoCl4
Shkavro, A.G.	473	Structural Transitions and Phase Composition of Thin Films of Alu
Shlichta, P.J.	179	Convection Phenomena During the Growth of Sodium Chlorate Crystal
Shoemaker, G.L.	29	The Crystal Structure of Cu4(PO4)20 (flux)
Shohno, K.	884	Crystal Growth of Boron Monophosphide Using a B2H6-PH3-H2 System
Shotov, A.P.	668	Investigation of Liquidus Surface in Pb-Sn-Te System by Simplex L

Shperun, V.M.	306	Preparation of Films of IV-VI Compounds in a Quasi-Sealed Volume
Shtrikman, H.	960	Non-Seeded Growth of Large Single Pb1-xSnxTe Crystals on a Quartz
	961	Growth Study of Large Non-Seeded Pb1-xSnxTe Single Crystals (vapo
	962	Improved Nucleation and a Planar Interface in LPE Grown Pb1-xSnxT
Shul'man, S.G.	359	The Preparation of Single Crystals of the Rare Earth Borides by t
Shupegin, M.L.	251	Growth Kinetics and Structure of Epitaxial Garnet Layers as Funct
Shvaiko-Shvaikovskii, V.E	778	Anisotropy of Ion Transport in Stoichiometric Rutile Single Cryst
Sidorov, Yu.G.	885	Growth of Pb0.8Sn0.2Te Films in a Quasi-Closed Space (vacuum evap
Siegel, W.	86	Growth and Physical Properties of ZnSiP2 (vapor transport, soluti
Sienko, M.J.	171	Low-Temperature Specific Heat and Magnetic Susceptibility of Nonm
Siffert, P.	886	New Methods of Making Silicon Solar Cells (review)
Sil'vestrova, I.M.	256	Na2CoGeO4 Single Crystal Growth and Study of Their Piezoelectric
Sima, V.	548	The Growth of Nickel Single Crystals by Recrystallization Method
	887	The Growth of Thin Bismuth Single Crystals with High Resistivity
Simon, B.	257	Crystallization of Sodium Perborate from Aqueous Solutions. I.
	258	Crystallization of Sodium Perborate from Aqueous Solutions. II.
	888	Kinetics of Growth and Dissolution of Sodium Chlorate in Diffusio
Since, J.J.	610	Magnetic and Electrical Properties of BaVS3 and BaVxTi1-xS3 (flux
Singh, G.	889	The Occurrence of High Period Polytype of SiC at the Interface of
Singh, S.R.	889	The Occurrence of High Period Polytype of SiC at the Interface of
Sinyakov, I.V.	688	Ag5GaS4 -- A New Phase in the System Ag-Ga-S (AgGaS2)
Sinyakova, E.F.	688	Ag5GaS4 -- A New Phase in the System Ag-Ga-S (AgGaS2)
Sirdeshmukh, D.B.	784	Crystal Growth and Selective Etching Studies on Sodium Chlorate C
Skarstad, P.M.	890	Crystal Growth of the Solid Electrolyte (C5H5NH)2Cu5Br7 From the
Skrebkova, E.D.	52	Crystallization of WO2 from the Gas Phase (vapor transport)
	487	Transfer of VO2 by Iodine (vapor transport, doped)
Skwarcz, J.	914	Optical and Magneto-Optical Properties of Sm0.55Tb0.45FeO3 (flux)
Sleight, A.W.	867	Polymorphism in Bi2Sn2O7 (flux)
	891	Crystal Growth and Structure of BiVO4 (Czochralski)
Slesarev, V.N.	1032	Synthesis of Single Crystals of Cubic Boron Nitride in Systems Co
Slivka, V.Yu.	509	Preparation and Principal Physicochemical Properties of Alkali Me
Small, M.B.	892	An Automated System for the Growth of Multilayered Structures in
	893	Conditions for Constant Growth Rate by LPE from a Cooling, Static
	894	Growth and Dissolution Kinetics of III-V Heterostructures Formed
Smirnov, A.E.	827	Selective Etching of Cesium Iodide Crystals and Comparison of Gro
Smirnov, I.A.	359	The Preparation of Single Crystals of the Rare Earth Borides by t
Smirnova, G.F.	119	Preparation and Properties of Single Crystals of Solid Solutions
Smith, B.A.	895	Some Aspects of the Design and Construction of the Cold Crucible
	896	The Crystal Growth and Atomic Structure of As2(Se,S)3 Compounds (
Smith, D.L.	897	Nonstoichiometry and Carrier Concentration Control in MBE of Comp
	898	Impurity Dopant Incorporation and Diffusion During Molecular Beam
Smith, H.I.	899	Oriented Crystal Growth on Amorphous Substrates Using Artificial
Smith, J.F.	274	Single-Crystal Elastic Constants of Al2Cu (Bridgman, arc melting)
Smith, S.H.	900	Flux Growth and Characterization by X-ray Topography of Rare Eart
	901	The Growth of LaAlO3 Crystals Enriched in Isotope 170 (flux)
	1063	The Flux Growth of Some New Rare Earth and Iron Group Complex Oxi
Smith, W.A.	229	Thermal and Dielectric Properties of LiKSO4 and LiCsSO4
Smolin, Yu.I.	902	Crystal Structure of Scandium Metaphosphate Sc(PO3)3 (solution)
Snowball, G.	799	The Preparation and Properties of Single Crystal Copper Phosphide
Sobol', A.A.	46	Structure, Spectroscopy, and Stimulated Emission of Crystals of Y
Sobolev, A.T.	903	Dependence of the Optical Homogeneity of Single Germanosillenite
Sohnel, O.	904	The Thermodynamic Driving Force for Crystallization from Solution
	905	A Method for the Determination of Precipitation Induction Periods
Sokolov, E.B.	231	Kinetics of Gallium Phosphide Synthesis by Chemical Gas Transport
Sokolowska, A.	906	Reactive Pulse Plasma Crystallization of Diamond and Diamond-Like
Sokolowski, M.	906	Reactive Pulse Plasma Crystallization of Diamond and Diamond-Like
Sol, N.	907	Bridgman and Czochralski Growth of beta-Phase NaxV2O5
Soled, S.	471	Crystal Growth and Characterization of In1.9As0.1Se3 (vapor trans
	708	Crystal Growth and Characterization of the Transition-Metal Phosp
Solonichnyi, Ya.V.	306	Preparation of Films of IV-VI Compounds in a Quasi-Sealed Volume
Somaiah, K.	908	Growth and Dissolution of BaFBr Crystals (flux)
Sood, A.K.	909	Metastable Pb1-xCdxS Epitaxial Films. I. Growth and Physical Pr
	910	Metastable Pb1-xCdxS Epitaxial Films. II. Electrical Properties
Sopori, B.	546	Properties of Rapidly-Grown RTR Silicon Sheet and Their Effects o
Spear, K.E.	109	Crystal Growth of Antimony Sulphur Iodide (SbSI, flux)
	141	Crystal Growth and Characterization of Lithium Metasilicate (Li2S
Spiesser, M.	911	Preparation and Properties of Two Indium Antimony Selenides (In2S
Spirlet, J.C.	160	Single Crystal and Film Growth of Actinide Pnictides by Chemical
	161	Preparation and Single-Crystal Growth of Protactinium Arsenides b
	912	Crystal Growth of Actinide Dioxides by Chemical Transport
Spirlet, M.R.	913	Dodecatungstophosphoric Acid-21-Water by Neutron Diffraction (eva
Spitsyn, V.I.	994	Fractionation of Stable Isotopes and Impurities During Zone Recry
Spyridelis, J.	453	Far Infrared Optical Study of alpha-In2Se3 Compound (melt, slow c
Srivastava, O.N.	8	Electronic Bandgap Measurements of SnS2 Polytypes (vapor transpor
Stacy, W.T.	92	X-Ray Topographic Study of Growth Sector Strains in KDP Single Cr
	223	The Effect of Doping on Microdefect Formation in As-Grown Disloca

Stager, C.V.   321   Growth of Large Single Crystals of Heusler Alloys, Ni2MnSn and Ni
Staples, E.J.   691   Thin Films of LiNbO3, Doped with Na+ and Co2+ plus Zr4+, Grown by
Startsev, S.A.   128   The Effect of the Solidification Front Shape on the Temmperature
Stasyuk, N.P.   509   Preparation and Principal Physicochemical Properties of Alkali Me
Stefanovich, S.Yu.   721   New Ferroelectric Na3Sc2(PO4)3 (melt)
Stepanov, N.N.   359   The Preparation of Single Crystals of the Rare Earth Borides by t
Stepniak, A.   914   Optical and Magneto-Optical Properties of Sm0.55Tb0.45FeO3 (flux)
Stewart, A.M.   855   Solute Partitioning During Silicon Dendritic Web Growth (Si, theo
Steyn, J.B.   73   Growth of Films of CdS Evaporated in Vacuum onto Silicon (vapor t
Stillinger, F.H.   915   Study of Melting and Freezing in the Gaussian Core Model by Molec
Stoenner, R.W.   790   Single Crystal Preparation and Characterization of the Group VA M
Stoika, T.M.   726   Phase Equilibria and Properties of Single Crystals of Solid Solut
Stolpakova, T.M.   16   Acoustooptic Characteristics of LiBi(MoO4)2 (Czochralski)
Stone, J.   916   Nd:Y2O3 Single-Crystal Fiber Laser: Room-Temperature cw Operatio
Stout, J.W.   917   Crystalline Vanadium (II) Fluoride, VF2. Preparation, Structure,
Stoyanova, V.   689   Appearance of Non-Singular Surfaces on Vapour-Grown Ice Crystals
Stringfellow, G.B.   918   VPE Growth of AlxGa1-xAs (theory)
                      919   Organometallic VPE Growth of AlxGa1-xAs (vapor transport)
Strobel, P,   920   Electrical and Optical Properties of High-Purity p-type Single Cr
Stroz, K.   56   Growth and Phase Transitions in Single Crystals of (KxNa1-x)NbO3
Stucky, G.   474   Structural and Magnetic Properties of "One-Dimensional" Barium Va
Stutius, W.   921   Organometallic Vapor Deposition of Epitaxial ZnSe Films on GaAs S
Subbarao, S.N.   911   Preparation and Properties of Two Indium Antimony Selenides (In2S
Suematsu, Y.   35   Conditions of LPE Growth for Lattice Matched GaInAsP/InP DH Laser
Suery, M.   372   Two-Phase (alpha/beta)-Brass Bicrystals Produced by a Solid-Solid
Sueur, G.   80   Structure of New Rhenium Apatites Containing Large Anions: Ba10(
Sugii, K.   500   LPE Growth of Li(Nb,Ta)O3 Solid-Solution Thin Film Waveguides on
                      922   Direct Measurement of Temperature Dependence of Lattice Mismatche
                      923   Large-Sized InP Single Crystals by the Synthesis, Solute Diffusio
Sugimoto, M.   497   High Efficiency GaAs Thin Film Solar Cells by Peeled Film Technol
                      924   Magnetic Spinel Single Crystals by Bridgman Technique (review)
Sugita, Y.   937   In-Process Monitoring and Control of Doping Gas Concentration Dur
Sugitani, Y.   1097 Flux Growth of Double Oxides of Niobium and Rare-Earth Elements (
Sugiura, T.   925   Low Temperature Growth of GaP LPE Layers from Indium Solvent (flu
Sugiyama, K.   655   A Quick Survey Method for the Study of CVD Conditions (TiN, vapor
                      656   Chemical Vapor Growth of LaB6 Whiskers and Crystals Having a Shar
                      926   Chemical Vapor Growth of Titanium Diboride by a Modified Hot Wire
                      927   Single Crystal Growth of Zirconium Carbide by a Modified Hot-Fila
                      928   Single Crystal Growth of Titanium Carbide from the Vapor by a Mod
                      929   On the Relations Between the Number of Gold Drops and TiP Whisker
Suib, S.L.   930   Gel Growth of Single Crystals of Some Rubidium and Cesium Tin Hal
Sukegawa, T.   460   Liquid Phase Epitaxial Growth of Lattice-Matched GaP-Ga0.5Al0.5P-
                      925   Low Temperature Growth of GaP LPE Layers from Indium Solvent (flu
Sumiya, K.   931   Crystal Growth of Co Ferrite on Fine Acicular gamma-Fe2O3 Particl
Sun, C.Y.   403   Growth of CuInS2 and Its Characterization (vapor transport)
Sunagawa, I.   484   Re-Investigation of the Re-Entrant Corner Effect in Twinned Cryst
                      932   Vapor Growth and Epitaxy of Minerals and Synthetic Crystals (revi
                      933   Modes of Vibrations in Step Trains: Rhythmical Bunching (theory)
Surbeck, H.   38   Solution Growth of RbAg4I5 Crystals
Suryanarayana, S.   465   Effect of Pressure on Electrical Resistivity and Thermoelectric P
Susa, K.   934   Electron Microscopic Characterization of Cubic BN Prepared by Wat
Susaki, M.   935   Photoluminescence Properties of CuGaSe2 Grown by Iodine Vapour Tr
Susaki, W.   416   Liquid Phase Epitaxial Growth of Zn and S Doped GaAs
Sutherland, H.H.   1058 Indium Polytelluride In2Te5(II) (melt)
Suzuki, K.   386   Electrical Properties and Charge Imbalance for Ca,Ge-Substituted
                      470   Physico-Chemical Properties of the Non-Stoichiometric VS2 and V5S
Suzuki, T.   936   Asymmetric 180 degree Bloch Walls in Fe Single Crystal Films (the
                      937   In-Process Monitoring and Control of Doping Gas Concentration Dur
Suzuki, Y.   1100 Molecular Beam Epitaxy of GaSb and GaSbxAs1-x (films)
Sweet, D.A.   796   Effects of Rotation on Czochralski Crystal Growth (theory)
Swiggard, E.M.   384   InP Growth and Properties (Czochralski, doped)
Swirkowics, M.   771   The Czochralski Growth of Bismuth-Germanium Oxide Single Crystals
Sysoev, L.A.   588   Improving the Electrical Characteristics of CdS Single Crystals G
Szymanski, M.   127   Chemical Synthesis and Crystal Growth of Laser Quality Praseodymi
Tadano, H.   703   Observations of Defects in LPE GaAs Revealed by New Chemical Etch
Taguchi, T.   938   Growth of High-Purity ZnTe Single Crystals by the Sublimation Tra
                      939   Growth by Travelling Heater Method and Characteristic of Undoped
Tairov, Yu.M.   554   Growing Monocrystalline Bars of SiC from the Gas Phase (vapor tra
                      561   Study of Temperature Distribution in High-Temperature Furnaces fo
                      562   Study of Silicon Carbide Epitaxial Growth Kinetics in the SiC-C S
                      940   Growth Kinetics of Epitaxial Layers of Silicon Carbide Obtained b
                      941   Investigation of Growth Processes of Ingots of Silicon Carbide Si
                      942   Investigations of Kinetic and Thermal Conditions of Silicon Carbi
Tajima, K.   956   Magnetic Structure of GdCu1-xZnx System (melt)
Takada, T.   63   Mass Spectrometric Study on Chemical Transport of VO2
                      875   Epitaxial Thin Iron Films on Antimony (Fe, vapor transport)

Takagi, K.	943	Analysis of Dislocations in Gd3Ga5O12 by a Repeated Etching Techn
Takagi, T.	944	New Developments in Ionized-Cluster Beam and Reactive Ionized-Clu
	945	Ionized-Cluster Beam Deposition and Epitaxy (review)
Takagi, Y.	632	The Melt Growth and Thermal Expansion of Na2CO3 Crystals (Bridgma
	633	Optical and Dilatometric Studies of KCaCl3 and RbCaCl3 Crystals (
Takahashi, K.	497	High Efficiency GaAs Thin Film Solar Cells by Peeled Film Technol
Takahashi, K.(ed.)	946	Vapour Growth and Epitaxy (review)
Takahashi, T.	45	Growth of ZnSxSe1-x Layers on ZnS and ZnSe Substrates by a Solid-
	947	Growth Parameters and Crystal Morphology of Vapor-Deposited Niobi
Takahashi, Y.	655	A Quick Survey Method for the Study of CVD Conditions (TiN, vapor
	656	Chemical Vapor Growth of LaB6 Whiskers and Crystals Having a Shar
	926	Chemical Vapor Growth of Titanium Diboride by a Modified Hot Wire
	927	Single Crystal Growth of Zirconium Carbide by a Modified Hot-Fila
	928	Single Crystal Growth of Titanium Carbide from the Vapor by a Mod
	929	On the Relations Between the Number of Gold Drops and TiP Whisker
Takakura, H.	948	Liquid Phase Epitaxy of AlxGa1-xP on GaP and Its Application to D
	1098	Characterization of the Optical Properties of LPE InxGa1-xAsyP1-y
Takamizawa, H.	480	Evaluation of Yttrium Iron Garnet Single Crystals Grown by the Fl
Takaoka, H.	944	New Developments in Ionized-Cluster Beam and Reactive Ionized-Clu
	1052	Growth of Stacking Faults by Bardeen-Herring Mechanism in Czochra
Takase, T.	1101	Molecular Beam Epitaxy of InxGa1-xSb (x = 0 to 1) (GaSb)
Takashige, M.	830	Ferroelectricity in (N(CH3)4)2ZnCl4 (evaporation)
	831	Ferroelectricity in (N(CH3)4)2CoCl4
Takayanagi, K.	949	Roles of Lattice Fitting in Epitaxy (theory, Ni, Cu, Pd, Pt, Fe,
	950	Techniques for Routine UHV in situ Electron Microscopy of Growth
Takebe, H.	309	Growth of Cubic ZnS, ZnSe and ZnSxSe1-x Single Crystals by Iodine
Takeda, Y.	675	Growth of Single Crystals of Na3Al2Li3F12 Garnet Under Hydrotherm
	951	Quasi-Grown Layers in Liquid Phase Epitaxial Growth (In1-xGaxAs)
	952	Composition Latching Phenomenon and Lattice Mismatch Effects in L
Takei, H.	494	Chemical Vapor Deposition of Scandium Hydride (ScH2, epitaxy, fil
	953	Growth of Fayalite (Fe2SiO4) Single Crystals by the Floating Zone
	954	Growth of FeTiO3 (Ilmenite) Crystals by the Floating-Zone Method
	955	Growth and Properties of Strontium Cobaltate Single Crystals (SrC
Takei, K.	956	Magnetic Structure of GdCu1-xZnx System (melt)
Takenaka, N.	957	Growth of GaAs1-xSbx Crystals by Steady-State Liquid Phase Epitax
Takenouchi, K.	364	Growth and Morphology of Copper-Nickel Alloy Crystals by Hydrogen
Takigawa, M.	929	On the Relations Between the Number of Gold Drops and TiP Whisker
Talwar, P.L.	612	Effect of High-Temperature Annealing in Cadmium on the Electrical
Tamari, N.	958	Catalytic Effect of Nickel on the Growth of Zirconium Carbide Whi
	959	Catalytic effects of Various Metals and Refractory Oxides on the
	960	Non-Seeded Growth of Large Single Pb1-xSnxTe Crystals on a Quartz
	961	Growth Study of Large Non-Seeded Pb1-xSnxTe Single Crystals (vapo
	962	Improved Nucleation and a Planar Interface in LPE Grown Pb1-xSnxT
Tanaka, A.	925	Low Temperature Growth of GaP LPE Layers from Indium Solvent (flu
Tanaka, H.	963	Isotope Effects on the Central Peak Phenomena in KD3(SeO3)2 (solu
Tanaka, S.	730	Anisotropy of Upper Critical Field in Superconducting 2H-NbS2 (th
	1000	Optical Phonon Modes and Localized Effective Charges of Transitio
Tanaka, T.	415	Epitaxial Growth of Ferroelectric PLZT ((Pb,La)(Zr,Ti)O3) Thin Fi
	416	Liquid Phase Epitaxial Growth of Zn and S Doped GaAs
	482	Heteroepitaxial Growth of ZnS on GaP by the Close-Spaced Techniqu
	616	Heteroepitaxial Growth of beta-SiC on Silicon Substrate Using SiC
	699	Growth and Morphology of 6H-SiC Epitaxial Layers by CVD (film)
	828	Liquid Phase Epitaxial Growth of CdTe in the CdTe-CdCl2 System (f
	964	Fermi Surface Measurements of ZrB2 by the de Haas van Alphen Effe
	1092	Preparation of TiCx Single Crystal with Homogeneous Compositions
Tanemura, M.	965	Geometrical Analysis of Crystallization of the Soft-Core Model in
Taniguchi, S.	934	Electron Microscopic Characterization of Cubic BN Prepared by Wat
Tanner, B.K.	900	Flux Growth and Characterization by X-ray Topography of Rare Eart
Taranets, V.A.	940	Growth Kinetics of Epitaxial Layers of Silicon Carbide Obtained b
Tareen, J.A.K.	1039	Effect of Mineralizers on the Growth Morphology of Cubic Thulium
Tarricone, L.	743	Crystal Growth and Properties of CuGaxIn1-xSe2 Chalcopyrite Compo
Tatarchenko, V.A.	128	The Effect of the Solidification Front Shape on the Temperature
	966	Preparation of Silicon Plates by Pulling from a Molten Bath Accor
Tatarintsev, V.M.	20	Synthesis and Crystal Growth of Refractory Materials by RF Meltin
Tate, I.	715	The Effects of M(M=Mg, Ca and Ba) Ions on the KCl Flux Growth of
	716	Growth of NaNd(MoO4)2 Single Crystals from the Ternary System Nd2
Tatsuzaki, I.	963	Isotope Effects on the Central Peak Phenomena in KD3(SeO3)2 (solu
Tavadze, F.N.	967	Some Peculiarities in the Growth and Properties of Beta-Boron Cry
	968	Shape of the Crystallization Front During Electron-Beam Zone Refi
Tejada, J.	296	Mechanism of Habit Change of ADP Crystals by Fe3+, Based on Mossb
	804	Unstable Growth of ADP Crystals (NH4H2PO4, theory, solution, kine
Telfer, J.R.	801	The Cadmium Oxide-Tungsten Oxide Phase System and Growth of Cadmi
Temkin, D.E.	873	Capture of Migrating Adatoms by Point Capture Centres for Differe
	969	Region of Metastability During Normal Growth in a Binary Alloy (k
Tempei, A.	843	Epitaxial Layers of CuInSe2 on GaAs
Tempel, A.	690	Growth and Electrical Properties of Epitaxial CuInS2 Thin Films o

	844	Structural and Electrical Properties of CuGaSe2 Thin Films on GaA
Tennant, W.E.	803	Liquid Phase Epitaxial Growth of Ga1-xInxSb on GaSb by Stepwise G
Teramoto, I.	970	Variation of Liquid and Solid Compositions During LPE Growth from
Terashima, K.	971	Low-Defect InSb Crystal Growth by InN Doping (Czochralski)
Ternisien D'Ouville, T.	600	Luminescence and Bandgap Studies of High Purity MgxZn1-xTe Ternar
Terukov, E.I.	789	Preparation of Single V3O5 Crystals by Chemical Transport Reactio
Teske, C.L.	972	Growth and X-Ray Study of Cu2SrGeS4 and Cu2BaGeS4 (flux)
Tezlevan, V.E.	625	Preparation of Single Crystals of the Magnetic Semiconductor CdCr
	780	Preparation and Certain Electrical Properties of CdCr2S4 Single C
Thackeray, M.M.	973	The Crystal Structure of the Solid Electrolyte Ag44I53(C11H30N3)3
Theeten, J.B.	974	Ellipsometric Assessment of (Ga,Al)As/GaAs Epitaxial Layers Durin
Thiel, F.A.	53	The Growth of In-Rich Bulk Single Crystals of GaxIn1-xPyAs1-y Via
	54	Melt and Solution Growth of Bulk Single Crystals of Quaternary II
	552	Preparation and Properties of Bulk GaxIn1-xAs
Thomas, D.	120	Crystal Structure of Ba2Fe6O11 (flux)
Thomas, G.E.	516	Thin Film Deposition From Beams of Ionized Atoms and Clusters (Ag
Thomas, J.O.	706	Structural Studies of the Solid Electrolyte High-LiTa3O8 (electro
Thoni, W.	79	Magnetic Mixed Valent TmSe (melt)
Tichy, L.	975	Some Electrical Properties of GeBi2Te4 Single Crystals (Bridgman)
Tie-lian, K.	1107	Magneto-Optical Properties of Some Rare Earth Iron Garnets (flux,
Tieghi, G.	297	Solid State Reactions to CdTeMoO6 and Its Structural Characteriza
Tiller, W.A.	992	Interface Morphology During Crystallization - I. Single Filament
	993	Interface Morphology During Crystallization - II. Single Filament
Tilley, R.J.D.	267	A Series of Lead Tungsten Bronzes (PbxWO3)
Timofeeva, V.A.	93	Some Problems of the Crystallization Kinetics and Growth Mechanis
	159	Phase Equilibria in Solvent-Garnet Systems (flux, Y3Al5O12, Y3Fe5
	976	Rost Kristallov iz Rastvorov-Rasplavov (Crystal Growth from High-
	977	Growth of Bulk Yttrium Iron Garnet Crystals from High Temperature
Timofeeva, V.S.	978	Kinetic Characteristics of Growth of Y3Al5O12 Single Crystals fro
Timoshechkin, M.I.	46	Structure, Spectroscopy, and Stimulated Emission of Crystals of Y
	245	Spectroscopic Properties of Er3Al5-xGaxO12 Films Obtained by Liqu
	457	Growth, Spectroscopic Investigations, and Some New Stimulated Emi
Tippman, H.	837	Gas Phase Synthesis of Epitaxial Layers of Nickel-Chlorine Boraci
Tippmann, H.	836	Pyroelectricity and Related Properties in the Fresnoite Pseudobin
Titchmarsh, J.M.	126	Nature, Origin and Effect of Dislocations in Epitaxial Semiconduc
Titov, A.V.	979	X-Ray and Dielectric Investigations of Pb2CrO5 (melt)
Tiwari, G.P.	980	Correlation of Diffusion Behaviour with the Entropy of Fusion (th
Toffoli, P.	131	The Compound Zn4(P2S6)3 (melt)
	132	Crystal Structure of Zn4(P2S6)3 (melt)
	981	The Compounds Ag7PS6 and Ag7PSe6
Tolksdorf, W.	982	Crystal Growth of Magnetic Garnets from High-Temperature Solution
Tomas, A.	983	Crystal Structure of the Spinel CdErS4 (slow cooling)
Tomashik, V.N.	725	Phase Diagram of Pseudobinary System CdTe-ZnSe (Bridgman)
Tomashpol'skii, Yu.Ya.	984	Epitaxial Growth of BaTiO3 Vacuum Condensates on SrTiO3 Crystals
Toncheva, L.T.	985	The Bending of Silicon Wafers by Thin Polycrystalline Silicon Fil
Tordjman, I.	261	Crystal Structure of the Mercury Nitro-Phosphate Hg4PO4NO3.H2O (m
	986	Crystal Structure of Monophosphate AgCoPO4
Torres, J.	987	Phase Diagram and Crystal Growth of Lead Phosphovanadates (Pb3(Px
Toshev, A.	760	On the Preparation of Nickel Ferrite Single Crystals by Chemical
	770	On the Preparation of Pseudobrookite Single Crystals by Chemical
Toth, L.	162	A Simple Method for Growing V3Si Single Crystals (electron beam m
Toudic, Y.	774	Solubility of AlPO4 in Hydrothermal Solutions of H3PO4
Tournoux, M.	317	A New Structural Family M2"M'Ta5O15. Crystal Structure of CaTlTa
Toyoda, N.	988	Impurity Effect on the Formation of Terraces in GaAs LPE Growth (
Trainist, T.P.	473	Structural Transitions and Phase Composition of Thin Films of Alu
Travadzhyan, M.G.	554	Growing Monocrystalline Bars of SiC from the Gas Phase (vapor tra
Treivus, E.B.	392	The Growth Kinetics of Urea Monocrystals from Aqueous Solution (C
	989	Crystallization Kinetics of Lithium and Strontium Nitrate Crystal
Trigunayat, G.C.	432	Effect of Growth Conditions of Polytypism in Cadmium Iodide Cryst
Triodina, N.S.	243	On the Growth Mechanism of Sodalite Single Crystals Grown by the
Trivedi, R.	990	Morphological Stability of a Solid Particle Growing from a Binary
	991	Effects of Interface Kinetics on the Growth Rate of Dendrites (Sn
	992	Interface Morphology During Crystallization - I. Single Filament,
	993	Interface Morphology During Crystallization - II. Single Filament
Trivedi, S.B.	111	Growth and Cleavage of Se-Te Alloy Crystals (Bridgman)
	112	Growth of Hollow Se-Te Whisker Crystals (theory)
Troitskii, O.A.	994	Fractionation of Stable Isotopes and Impurities During Zone Recry
Trottier, D.	1029	Crystal Structure of Li2CaUF8 (flux)
Trunov, N.N.	995	Simulation of the Growth of a Metallic Face-Centered Cubic Lattic
Tsagareishvili, G.V.	968	Shape of the Crystallization Front During Electron-Beam Zone Refi
Tseng, W.	194	Crystalline Defects in Solid Phase Epitaxy Si Films Deposited at
Tseng, W.F.	343	Heteroepitaxy of a Deposited Amorphous Germanium Layer on a Silic
	537	Solar Furnace Annealing of Amorphous Si Layers (films)
	996	Grain Size Dependence in a Self-Implanted Silicon Layer on Laser
Tsevtkov, V.F.	561	Study of Temperature Distribution in High-Temperature Furnaces fo
Tsiganov, A.D.	994	Fractionation of Stable Isotopes and Impurities During Zone Recry

Tsuda, N.	15	Positron Annihilation in ReO3 (vapor transport)
Tsuji, S.	415	Epitaxial Growth of Ferroelectric PLZT ((Pb,La)(Zr,Ti)O3) Thin Fi
Tsujikawa, I.	273	Specific Heat of a Quasi-Two-Dimensional Antiferromagnet Ni(OH)2,
Tsunashima, A.	424	Vapor Growth of Zn2GeO4 Single Crystals
Tsuo, Y.H.	229	Thermal and Dielectric Properties of LiKSO4 and LiCsSO4
Tsurkan, V.V.	50	Properties of CuCr2Se4 Crystals Obtained from Solutions in Melts
Tsutsumi, M.	714	Observations of Etch Pits on As-Grown Faces of Brushite Crystals
Tsvetkov, V.F.	554	Growing Monocrystalline Bars of SiC from the Gas Phase (vapor tra
	562	Study of Silicon Carbide Epitaxial Growth Kinetics in the SiC-C S
	940	Growth Kinetics of Epitaxial Layers of Silicon Carbide Obtained b
	941	Investigation of Growth Processes of Ingots of Silicon Carbide Si
	942	Investigations of Kinetic and Thermal Conditions of Silicon Carbi
Tu, C.C.	403	Growth of CuInS2 and Its Characterization (vapor transport)
Turovskii, B.M.	997	Variations of Impurity Distribution in the Modified Czochralski M
	998	Growth and Properties of Silicon Crystals With Automatic Diameter
Ubbelohde,A.R.	999	The Molten State of Matter. Melting and Crystal Structure (melt,
Uchida, S.	1000	Optical Phonon Modes and Localized Effective Charges of Transitio
Uchino, K.	6	Mossbauer Study of FeMo2S4 (vapor transport)
	1001	Crystal Growth, and Magnetic and Mossbauer Studies of Sr(Fe0.766W
Uchiyama, S.	698	Studies of LPE Ripple Based on Morphological Stability Theory
	713	Si Contamination in Epitaxial Boron Monophosphide (BP)
Ueda, A.	965	Geometrical Analysis of Crystallization of the Soft-Core Model in
Uehara, S.	500	LPE Growth of Li(Nb,Ta)O3 Solid-Solution Thin Film Waveguides on
Uelhoff, W.	1002	A Model for Macroscopic Czochralski Growth: Theoretical and Expe
	1071	Measurements of the Contact Angle Between Melt and Crystal During
Uemura, C.	1095	InP Single Crystal Growth by the Synthesis, Solute Diffusion Meth
Ueno, S.	714	Observations of Etch Pits on As-Grown Faces of Brushite Crystals
Ugai, Ya.A.	1003	Heterogeneous Equilibria and Mass Transfer Kinetics in the System
Ukhlinov, G.A.	1035	Crystallization of Bi-Sb Solid Solutions in Capillaries (zone mel
Umarov, B.S.	476	Linear Expansion Anisotropy of CdGa2Se4 (melt)
Umerov, R.S.	820	Growing Single Crystals of Silicon-Germanium Solid Solutions by E
Unger, W.K.	1004	Raman and Infrared Spectra of CdIn2S4 and ZnIn2S4 (vapor transpor
Unoki, H.	717	Growth of Single Crystals of Lead Monoxide (PbO, flux)
Ura, M.	937	In-Process Monitoring and Control of Doping Gas Concentration Dur
Urusovskaya, A.A.	827	Selective Etching of Cesium Iodide Crystals and Comparison of Gro
Ushio, M.	1005	Single Crystal Growth of Alexandrite in a V2O5 Flux (BeAl2O4)
Uyeda, R.	821	Crystal Structure and Habit of Fine Metal Particles Formed by Gas
	1006	Growth of Polyhedral Metal Crystallites in Inactive Gas (Mg, revi
Vaidya, S.N.	465	Effect of Pressure on Electrical Resistivity and Thermoelectric P
	1007	Theory of Melting, Vacancy Model
Valentino, A.J.	638	The Effect of Melt Flow Phenomena on the Perfection of Czochralsk
Valkonen, J.	1008	Scandium Hydrogenselenite (Sc(HSeO3)3, evaporation, solution)
van de Wijgert, W.M.	223	The Effect of Doping on Microdefect Formation in As-Grown Disloca
van den Berg, P.J.	40	Preparation of Pure Hydroxyapatite Single Crystals by Hydrotherma
van den Brekel, C.H.J.	435	Morphological Stability Analysis in Chemical Vapour Deposition Pr
	1009	Morphological Stability Analysis of Growth from the Vapour (theor
	1010	Morphological Stability Analysis in Chemical Vapour Deposition Pr
	1011	Interface Morphology in Chemical Vapour Deposition on Profiled Su
van den Heuvel, G.P.M.	1012	Luminescence and Ionic Conductivity of Bariumfluorochloride (BaFC
van der Berg, G.J.	1002	A Model for Macroscopic Czochralski Growth: Theoretical and Expe
van der Eerden, J.P.	182	Fast Growth of Ordered AB Crystals: A Monte Carlo Simulation for
	1013	Survey of Monte Carlo Simulations of Crystal Surfaces and Crystal
	1020	Monte Carlo Simulation of a (111) Diamond Face Around the Rougnen
van der Hart, A.W.A.	1002	A Model for Macroscopic Czochralski Growth: Theoretical and Expe
van der Hoek, B.	105	Interaction Between Successive Monatomic Step Trains of Different
van der Leeden, G.A.	196	Purification and Characterization of Metallurgical Silicon (Si, m
van der Linden, W.H.	40	Preparation of Pure Hydroxyapatite Single Crystals by Hydrotherma
van der Merwe, J.H.	1014	The Role of Lattice Misfit in Epitaxy (review, theory)
van der Putte, P.	1015	Surface Morphology of HCl Etched Silicon Wafers (Si)
van der Straten, P.J.M.	1016	Liquid Phase Epitaxial Growth of Copper Ferrite Films (CuFe2O4)
van der Straten, P.J.M.	1017	Liquid Phase Epitaxial Growth of Lithium Ferrite-Aluminate Films
van der Straten, P.J.M.	1018	Flux Growth of ZnGa2O4 Single Crystals
van Dijk, C.	226	Neutron-Scattering Study of the Incommensurate Phase Transition o
Van Enckevort, W.J.P.	1015	Surface Morphology of HCl Etched Silicon Wafers (Si)
	1019	The Influence of Adsorption and Step Reconstruction on the Growth
	1020	Monte Carlo Simulation of a (111) Diamond Face Around the Rougnen
van Erk, W.	1021	The Growth Kinetics of Garnet Liquid Phase Epitaxy Using Horizont
	1022	A Solubility Model for Rare-Earth Iron Garnets in a PbO/B2O3 Solu
Van Hove, W.	1023	Optical Properties of V6O13 Single Crystals in the Metallic and t
van Leeuwen, C.	1024	On the Driving Force for Crystallization: The Growth Affinity (t
	1025	On the Presentation of Growth Curves for Growth from Solution (th
Van Loo, F.J.J.	78	Crystallography of Aligned Fe-Al Eutectoid (FeAl, FeAl2)
van Nam, N.	843	Epitaxial Layers of CuInSe2 on GaAs
	844	Structural and Electrical Properties of CuGaSe2 Thin Films on GaA
Van Rosmalen, G.M.	193	The Affinity of Crystal Growth and Dissolution in Aqueous Solutio
Van Run, A.M.J.G.	1026	Computation of Striated Impurity Distributions in Melt-Grown Crys

van Suchtelen, J.	1027	Coarsening of Eutectic Structures During and After Unidirectional
van Vlimmeren, Q.A.G.	77	Crystallographic and Magnetic Structure of RbFeCl3.2D2O and CsFeC
Van Wormer, K.A.	1038	A New Technique for Simulating Two-Dimensional Nucleation Models
Vanderwyck, A.H.B.	1059	Liquid Phase Growth of HgCdTe Epitaxial Layers
Vanek, P.	1028	Growth Striations in Fe-3wt%Si Single Crystals Grown by Floating
Vasek, P.	887	The Growth of Thin Bismuth Single Crystals with High Resistivity
Vassilev, I.S.	985	The Bending of Silicon Wafers by Thin Polycrystalline Silicon Fil
Vaughan, W.R.	129	On the Ordering of Fe Atoms in FexNbS2 (vapor transport)
Vedrine, A.	36	Crystal Structure of RbEu3F10 (melt)
	37	The Systems RbF-EuF3 and CsF-EuF3. Study of the Phases RbLnF4 an
	1029	Crystal Structure of Li2CaUF8 (flux)
Venables, J.A.	1030	Kinetic Studies of Nucleation and Growth at Surfaces (review)
Venevtsev, Yu.N.	155	Ferroelectric (Pb,Ba)5Ge3O11 Crystals (flux)
	721	New Ferroelectric Na3Sc2(PO4)3 (melt)
	979	X-Ray and Dielectric Investigations of Pb2CrO5 (melt)
Venien, J.P.	1031	Vanadium (III) Oxychloride: Magnetic, Optical and Electrical Pro
Venkateswara Rao, A.	751	Nucleation and Growth of KClO4 Single Crystals in Silica Gels
	752	Gel Growth and Perfection of Orthorhombic Potassium Perchlorate S
Vennik, J.	1023	Optical Properties of V6O13 Single Crystals in the Metallic and t
Verbaere, A.	317	A New Structural Family M2"M'Ta5O15. Crystal Structure of CaTlTa
Vereshchagin, L.F.	1032	Synthesis of Single Crystals of Cubic Boron Nitride in Systems Co
Verscharen, W.	1033	X-Ray and Neutron Diffraction of Na2MnFeF7: A Variety of Trigona
Veselago, V.G.	50	Properties of CuCr2Se4 Crystals Obtained from Solutions in Melts
	586	Magnetic Properties of CdCr2Se4-ZnCr2Se4 Single Crystals (melt)
	1037	Photoferromagnetic Effect and Photoconductivity of Cd1-xHgxCr2Se4
Vetter, G.	1034	Preparation, Stability, and Preliminary Crystallographic Study of
Vettier, C.	621	X-Ray Diffuse Scattering from Alkali, Silver, and Europium beta-A
Viala, J.C.	213	Growth of Single Crystals of Beta Rhombohedral Boron by a Modifie
Viechnicki, D.	167	Melt Growth of Nd:Y3Al5O12(Nd:YAG) Using the Heat Exchange Method
Vigdorovich, V.N.	1035	Crystallization of Bi-Sb Solid Solutions in Capillaries (zone mel
Vignes-Adler, M.	755	A New Technique for Determining the Kinetics of Crystal Growth fr
Vijayakumar, V.	465	Effect of Pressure on Electrical Resistivity and Thermoelectric P
Vincent, H.	1036	Crystal Structure of CuTa2O6 (flux)
Vinogradova, G.I.	1037	Photoferromagnetic Effect and Photoconductivity of Cd1-xHgxCr2Se4
Viola, M.S.	1038	A New Technique for Simulating Two-Dimensional Nucleation Models
Vishnyakov, E.M.	4	Effect of Crucible Shape on a Crystallization Front (Bridgman, Ki
Viswanathan, K.	824	Structural Phase Transition in Mixed Crystals WxMo1-xO3 (melt)
Viswanathiah, M.N.	1039	Effect of Mineralizers on the Growth Morphology of Cubic Thulium
Vitt, B.	709	Growth and Properties of Pb(WO4)1-x(MoO4)x Mixed Crystals (Czochr
Vlasse, M.	825	The Crystal Structure of a New High -Nd- Concentration Laser Mate
Vogt, O.	154	Magnetization and Neutron Studies of UTe and USb0.8Te0.2
Vohl, P.	1040	Synthesis and Crystal Growth of CdGeP2 (melt)
	1041	Synthesis and Crystal Growth of CdGeP2 (vapor transport)
Vollenkle, H.	83	The Crystal Structure of the Tetragermanate K2Ba[Ge4O9]2
von Allmen, M.	537	Solar Furnace Annealing of Amorphous Si Layers (films)
	1042	Dynamics of Laser-Induced Formation of Palladium Silicide (Pd2Si)
von Molnar, S.	293	Magnetic, Transport, and Thermal Properties of Ferromagnetic EuB6
von Muench, W.	1043	Preparation of Pure and Doped Silicon Carbide by Pyrolysis of Sil
von Schnering, H.G.	1044	AlSiP3, A Compound with a Novel Wurtzite-Pyrite Intergrowth Struc
Vook, R.W.	388	(111) Cu2O Growth Modes on (111)Cu Surfaces (epitaxy)
	1045	Recent Developments in the Microstructural Characterization of Co
Voron'ko, Yu.K.	46	Structure, Spectroscopy, and Stimulated Emission of Crystals of Y
Voronkov, V.V.	1046	Mass Transfer at the Surface of a Crystal Near to its Boundary wi
Voronkova, V.I.	1047	Crystallization of LiNbO3 from Solution in Borate, Vanadate, and
Voroshilov, Yu.V.	509	Preparation and Principal Physicochemical Properties of Alkali Me
Vovan, T.	1048	Mixed Oxysulfides of Chromium III and Rare Earths (LaCrOS2, CeCrO
Vrolijk, J.W.G.A.	78	Crystallography of Aligned Fe-Al Eutectoid (FeAl, FeAl2)
	1049	Crystallography of Aligned Cu-In Eutectoid (Cu-Cu7In3)
Wachter, P.	79	Magnetic Mixed Valent TmSe (melt)
Wada, K.	1050	Interface Growth Feature and Voids in Sapphire Ribbon Crystals (a
	1051	Dislocations in Sapphire Ribbon Crystals Grown by the Edge-Define
	1052	Growth of Stacking Faults by Bardeen-Herring Mechanism in Czochra
Wagner, R.J.	343	Heteroepitaxy of a Deposited Amorphous Germanium Layer on a Silic
Wagner, S.	220	Electrical and Optical Properties of Copper Indium Ditelluride Cr
Waintal, A.	631	Magnetization, Magnetocrystalline Anisotropy, Magnetostriction an
Wald, F.V.	452	Aluminum Redistribution in EFG of Silicon Ribbon (Si)
	1053	Crystal Growth of CdTe for X-Ray Detectors (review)
Walker, P.J.	323	Crystal Growth of Some Rare Earth Trifluorides (GdF3, TbF3, DyF3,
	901	The Growth of LaAlO3 Crystals Enriched in Isotope 170 (flux)
	1054	Crystal Growth of SrCl2 and Solid Solutions of SrCl2-PrCl3 and Sr
	1055	Crystal Growth of K2CuF4 and Some Solid Solutions of K2CuF4-K2ZnF
	1056	Crystal Growth of Sodium Sulphide, Na2S (hygroscopic, Bridgman)
Walline, R.E.	1057	GaAs Vapour Phase Epitaxy Using Computer Control
Walpole, J.N.	394	MBE Techniques for IV-VI Optoelectronic Devices (review)
Walsh, W.M.,Jr.	31	On the Resistivity of the Magnetically Ordered Ground State of th
Walton, P.D.	1058	Indium Polytelluride In2Te5(II) (melt)

Wan, C.	331	The Crystal Structure of Synthetic Lautarite, Ca(IO3)2 (flux)
Wang, C.C.	1059	Liquid Phase Growth of HgCdTe Epitaxial Layers
Wani, P.A.	13	Growth Conditions and Crystal Structure Parameters of Layer Compo
Wanklyn, B.	594	Relative Supersaturation and Supercooling in Fluxed Melt Systems
Wanklyn, B.M.	319	Flux Growth of Some Fluoride Crystals Under Reducing Conditions.
	320	Viscosity Measurements in the System Dy2O3-K2O-MoO3 (DyKMo2O8, me
	508	Flux Growth of Lanthanum Borate, LaBO3
	595	Flux Growth of Polymorphic Rare-Earth Disilicates, R2Si2O7 (R = T
	1060	Effects of Modifying Starting Compositions for Flux Growth (DyPO4
	1061	The Flux Growth of Some Rare-Earth and Iron Group Complex Oxides
	1062	Flux Growth of Crystals of RKMo2O8, R2MoO6 and R6MoO12 in the Sys
	1063	The Flux Growth of Some New Rare Earth and Iron Group Complex Oxi
	1064	Flux Growth of Crystals of Some Transition Metal Fluorides. Part
Warasteh, M.	198	Crystallochemical Study of the System Ni3Pb2S2-Ni3Pb2Se2 and Note
Washiyama, M.	1065	Solution Growth of ZnS, ZnSe, CdS and Their Mixed Compounds Using
Watanabe, A.	665	Flux Growth of Bi2WO6 Single Crystal Below the Transformation Tem
Watanabe, H.	614	X-Ray Study of LEC-Grown InP Crystals (Czochralski, theory)
	651	Ionized-Cluster Beam Epitaxial Growth of GaP Films on GaP and Si
	856	Impurity Effect on Grown-In Dislocation Density of InP and GaAs C
	955	Growth and Properties of Strontium Cobaltate Single Crystals (SrC
Watanabe, N.	956	Magnetic Structure of GdCu1-xZnx System (melt)
Watanabe, T.	1066	Preparation of Some Chalcogenide Spinel Single Crystals and Their
Watatani, S.	931	Crystal Growth of Co Ferrite on Fine Acicular gamma-Fe2O3 Particl
Webb, G.W.	1067	Cold Working Nb3Al in the bcc Structure and then Converting to th
Webber, H.	799	The Preparation and Properties of Single Crystal Copper Phosphide
Weber, K.	198	Crystallochemical Study of the System Ni3Pb2S2-Ni3Pb2Se2 and Note
Weber, M.J.	1068	Magneto-Optical Properties of KTb3F10 and LiTbF4 Crystals (Czochr
Weber, T.A.	915	Study of Melting and Freezing in the Gaussian Core Model by Molec
Wedeking, G.	846	Progress of Temperature and Concentration in the Region of Solidi
Weeks, J.D.	1069	Thermodynamic Properties of Surface Steps (theory)
Weiss, A.	672	Dispersion of Acoustic Phonons in Layered SrGa2 (melt, slow cooli
Weiss, H.	1070	Materials Behaviour in Low Gravity Environment (melt, solution, v
Weiss, R.	838	Dislocation Propagation in Czochralski Grown Gadolinium Gallium G
Weller, P.F.	930	Gel Growth of Single Crystals of Some Rubidium and Cesium Tin Hal
Welter, J.-M.	1072	Properties and Synthesis of Nb-H Interstitial Alloys (electron be
Welz, F.	982	Crystal Growth of Magnetic Garnets from High-Temperature Solution
Wendl, W.	265	Solubility Curves in High-Temperature Melts for the Growth of Sin
Wenzl, H.	1002	A Model for Macroscopic Czochralski Growth: Theoretical and Expe
	1071	Measurements of the Contact Angle Between Melt and Crystal During
	1072	Properties and Synthesis of Nb-H Interstitial Alloys (electron be
Wernick, J.H.	1073	Czochralski Growth of A15 Structure Intermetallic Compounds (V3Si
Westman, S.	43	Spherical Monocrystals for X-Ray Work Obtained by Plasma Remeltin
Westphal, G.H.	809	Low-Stress Physical Vapor Growth (PVT) (KCl, KBr, KCl:KCN, KCN:KO
	1074	On Diffusive-Advective Interfacial Mass Transfer (theory, solutio
Weyher, J.	1075	Some Notes on the Growth Kinetics and Morphology of VLS Silicon C
Whatmore, R.W.	1076	Ca12Al14O33: A New Piezoelectric Material (Czochralski)
Whitehouse, C.R.	1077	Structural Studies of Compounds in the Series GaSxSe(1-x) (0 less
	1078	Optimization of Conditions for the Growth of Gallium Selenide and
Wiart, R.	26	Electrocrystallization. Fundamental Aspects (review, theory)
Wieczorek, C.	254	Dichlorosilane as Silicon Source for Epitaxial Films on Insulator
Wiedemeier, H.	1079	Transport Properties of the Systems SnS2-SnI4 and SnS2-I2 (SnS2,
Wiegers, G.A.	436	Crystal Growth and the Crystal Structures of Two Modifications of
	795	Preparation, Structures, and Properties of Niobium Chalcogenide H
Wiegmann, W.	761	Molecular Beam Epitaxy of Ge and Ga1-xAlxAs Ultra Thin Film Super
	762	Crystal Growth Kinetics in (GaAs)n-(AlAs)m Superlattices Deposite
Wilcox, W.R.	175	Thermocapillary Convection in Floating Zone Melting: Influence o
	179	Convection Phenomena During the Growth of Sodium Chlorate Crystal
	786	The Stage II Transient in Off-Eutectic Solidification (theory)
	863	Influence of Magnetic Field on Vertical Bridgman-Stockbarger Grow
	864	Non-Constant Distribution Coefficients for Directionally Solidifi
	1080	Non-Constant Distribution Coefficients from Experimental Data: T
Wilcox, W.R.(ed.)	1081	Preparation and Properties of Solid State Materials, Vol. 4, Morp
Wilhelm, M.	1082	A Novel Ruthenium Bronze: KRu4O8 (solution)
Wilkes, J.G.	165	On the Cooling Rates of Large-Diameter Silicon Crystals (Si, theo
Williams, G.M.	287	A High Temperature High Purity Source for Metal Beam Epitaxy (Ag,
Williams, R.H.	619	An Investigation of Thin Silver Films on Cleaved Silicon Surfaces
Willig, A.	544	Preparation, Crystal Structure, and Physical Properties (transpor
Wilson, L.O.	1083	On Interpreting a Quantity in the Bruton, Prim and Slichter Equat
	1084	A New Look at the Burton, Prim, and Slichter Model of Segregation
Winckler, E.	842	Czochralski Growth of Li3N Crystals
Windgassen, N.R.	1085	Growing and Polishing Large Thallium Single Crystals (Tl, melt)
Winkler, K.	82	Growth of ZnSiP2 from Gas Phase Under Non-Stoichiometric Conditio
	86	Growth and Physical Properties of ZnSiP2 (vapor transport, soluti
	1086	Partition of Principal Components in the Course of Preparation of
Winzer, A.	1087	Quantum-Mechanical Relations in the Calculation of Heats of Fusio
Witt, A.F.	437	Liquid-Phase Electroepitaxy: Growth Kinetics (Peltier effect, Ga

	604	Application of a Heat Pipe to Czochralski Growth. Part I. Growth
	1088	Crystal Growth and Segregation Under Zero Gravity: Ge (Bridgman)
Wittmer, M.	1042	Dynamics of Laser-Induced Formation of Palladium Silicide (Pd2Si)
Wold, A.	357	Physical Properties of the Quarternary Chalcogenides Cu2(I)B(II)C
	471	Crystal Growth and Characterization of In1.9As0.1Se3 (vapor trans
	624	The Electrical, Optical and Photoconducting Properties of Fe2-xCr
	708	Crystal Growth and Characterization of the Transition-Metal Phosp
	911	Preparation and Properties of Two Indium Antimony Selenides (In2S
	920	Electrical and Optical Properties of High-Purity p-type Single Cr
Wolff, L.R.	78	Crystallography of Aligned Fe-Al Eutectoid (FeAl, FeAl2)
	1049	Crystallography of Aligned Cu-In Eutectoid (Cu-Cu7In3)
Wondre, F.R.	1062	Flux Growth of Crystals of RKMo208, R2MoO6 and R6MoO12 in the Sys
	1063	The Flux Growth of Some New Rare Earth and Iron Group Complex Oxi
	1064	Flux Growth of Crystals of Some Transition Metal Fluorides. Part
Wood, J.D.C.	559	Orientation Dependence of 1-Alanine Incorporation in TGS Crystals
	560	Growth and Electrical Properties of Deuterated TGS Produced by th
Wood, R.F.	1106	Electrical and Structural Characteristics of Laser-Induced Epitax
Wood, V.M.	560	Growth and Electrical Properties of Deuterated TGS Produced by th
Woodcock, L.V.	140	A Molecular Dynamics Study of Surface Melting (theory, vapor tran
	528	Structure of the Lennard-Jones (100) Crystal-Liquid Interface (th
Woods, J.	214	Growth of Single Crystals of Zinc Selenide from the Vapour Phase
	387	Plasma Growth of Rutile Crystals and Their Photoelectronic Proper
	815	The Growth of CdS in Sealed Silica Capsules (vapor transport)
	816	Vapour Growth and Defect Characterisation of Large Single Crystal
Wu, K.	909	Metastable Pb1-xCdxS Epitaxial Films. I. Growth and Physical Pr
	910	Matastable Pb1-xCdxS Epitaxial Films. II. Electrical Properties
Wunderlich, B.	97	Crystallization During Polymerization of Lithium Dihydrogen Phosp
Wunderlich, B. (ed.)	1089	Selected Papers on Polymer Crystallization (Review)
Yacobi, B.G.	1090	Intercalation-Induced Shift of the Absorption Edge in ZrS2 and Hf
Yaeger, I.	1091	Crystal Growth and Magnetic Properties of NdFeTiO5 (flux)
Yagi, K.	949	Roles of Lattice Fitting in Epitaxy (theory, Ni, Cu, Pd, Pt, Fe,
	950	Techniques for Routine UHV in situ Electron Microscopy of Growth
Yagi, T.	963	Isotope Effects on the Central Peak Phenomena in KD3(SeO3)2 (solu
Yajima, F.	1092	Preparation of TiCx Single Crystal with Homogeneous Compositions
Yamada, E.	700	Vapor Growth of Zinc Telluride by Closed Tube Method (ZnTe, epita
Yamada, I.	944	New Developments in Ionized-Cluster Beam and Reactive Ionized-Clu
	945	Ionized-Cluster Beam Deposition and Epitaxy (review)
Yamada, S.	1093	Morphology of Iron Pyrite Crystals (FeS2, vapor transport)
Yamada, T.	49	Growth of a Crystal Needle of Si2ON2 by Nitridation in the Si-S
	677	Flux Growth of LiNdP4O12 Single Crystals
	678	Top-Seeded Flux Growth of LiNdP4O12 Single Crystals
	1103	Formation of Amorphous Mn-Bi Films and Their Crystallization Proc
Yamaguchi, A.	506	An LPE Growth of Thick InxGa1-xAs
Yamaguchi, T.	427	Growth Mechanism of Hollow ZnO Crystals from ZnSe. II
	428	Crystal Growth and Sublimation in II-VI Compounds Along their Pol
	947	Growth Parameters and Crystal Morphology of Vapor-Deposited Niobi
Yamai, I.	732	Synthesis of Potassium Hexatitanate Fibers by the Hydrothermal De
	1094	Vapor Phase Growth of Alumina Whiskers by Hydrolysis of Aluminum
Yamamoto, A.	830	Ferroelectricity in (N(CH3)4)2ZnCl4 (evaporation)
	831	Ferroelectricity in (N(CH3)4)2CoCl4
	1095	InP Single Crystal Growth by the Synthesis, Solute Diffusion Meth
Yamamoto, M.	463	Electrical Conductivity and Galvanomagnetic Effects in (SN)x Sing
	948	Liquid Phase Epitaxy of AlxGa1-xP on GaP and Its Application to D
Yamamoto, N.	935	Photoluminescence Properties of CuGaSe2 Grown by Iodine Vapour Tr
Yamana, K.	1096	Bismuth Tellurides: BiTe and Bi4Te3 (melt)
Yamanaka, H.	970	Variation of Liquid and Solid Compositions During LPE Growth from
Yamasaki, Y.	1097	Flux Growth of Double Oxides of Niobium and Rare-Earth Elements (
Yamazaki, T.	347	Morphology of Ice Droxtals Grown from Supercooled Water Droplets
Yamazoe, Y.	1098	Characterization of the Optical Properties of LPE InxGa1-xAsyP1-y
Yanagisawa, K.	320	Viscosity Measurements in the System Dy2O3-K2O-MoO3 (DyKMo2O8, me
	1064	Flux Growth of Crystals of Some Transition Metal Fluorides. Part
Yang, K.H.	1099	Analysis of Vapor-Deposited Silicon Carbide Films on Silicon Ribb
Yano, M.	1100	Molecular Beam Epitaxy of GaSb and GaSbxAs1-x (films)
	1101	Molecular Beam Epitaxy of InxGa1-xSb (x = 0 to 1) (GaSb)
Yanovskii, V.K.	1047	Crystallization of LiNbO3 from Solution in Borate, Vanadate, and
Yao, T.	1102	Molecular Beam Epitaxy of ZnSexTe1-x(film)
Yaron, G.	834	State-of-The-Art of Crystal Growth and Nuclear Spectroscopic Eval
Yatsuya, S.	821	Crystal Structure and Habit of Fine Metal Particles Formed by Gas
Yoneda, N.	424	Vapor Growth of Zn2GeO4 Single Crystals
Yoneyama, T.	712	Horizontal Zone-Melting of Indium Antimonide in Vacuum. I. Basi
Yoo, H.I.	196	Purification and Characterization of Metallurgical Silicon (Si, m
Yoshida, K.	1103	Formation of Amorphous Mn-Bi Films and Their Crystallization Proc
Yoshida, S.	1104	Cathodoluminescence of Impurity-Doped Aluminum Nitride Films Prod
Yoshiie, T.	428	Crystal Growth and Sublimation in II-VI Compounds Along their Pol
Yoshino, K.	463	Electrical Conductivity and Galvanomagnetic Effects in (SN)x Sing
Yoshioka, H.	468	Upper and Lower Critical Fields of TaS2(Pyridine)1/2 (vapor trans

Young, I.M.	800	The Effect of Impurities on the Piezoelectric Properties of Lithi
	801	The Cadmium Oxide-Tungsten Oxide Phase System and Growth of Cadmi
	1105	The Central Region of the Calcium Oxide-Gallium Oxide System (CaG
Young, R.T.	1106	Electrical and Structural Characteristics of Laser-Induced Epitax
Yu-chuan, H.	1107	Magneto-Optical Properties of Some Rare Earth Iron Garnets (flux,
Yumoto, H.	375	Morphology and Growth Mechanism of Vapor Grown Cd Crystals as Aff
Yvon, K.	186	New Ternary Mo(II)-Compounds InxMo15Se19 Containing Mo6Se8 and Mo
Zaichko, V.V.	116	Growth of Si Strips by Stepanov's Method
Zak, Z.	1108	The Crystal Structure of Lithium Fluorosulphate LiSO3F (solution)
Zamerowski, T.J.	729	Crystal Growth and Properties of Binary, Ternary and Quaternary (
Zanotti, L.	743	Crystal Growth and Properties of CuGaxIn1-xSe2 Chalcopyrite Compo
	744	A Temperature Variation Method for the Growth of Chalcopyrite Cry
Zaouk, A.	1109	Crystal Growth of Gallium Arsenide by Chemical Vapor Deposition f
Zelenev, Yu.V.	59	A Fluctuation Theory of Normal Crystal Growth for Metallic One-Co
	1110	On the Theory of Normal Growth of Crystals in Binary Systems (kin
Zeleva, G.M.	185	Preparation and Investigation of Crystals of Solid Solutions of t
Zemel, J.N.	909	Metastable Pb1-xCdxS Epitaxial Films. I. Growth and Physical Pr
	910	Matastable Pb1-xCdxS Epitaxial Films. II. Electrical Properties
Zharikov, E.V.	46	Structure, Spectroscopy, and Stimulated Emission of Crystals of Y
	457	Growth, Spectroscopic Investigations, and Some New Stimulated Emi
Zhekov, V.I.	245	Spectroscopic Properties of Er3Al5-xGaxO12 Films Obtained by Liqu
Zimmermann, H.	276	A Double-Ellipsoid Mirror Furnace for Zone Crystallization Experi
Zizic, B.	219	The Role of Dislocations in the Growth of Ammonium Dihydrogen Pho
Zuccalli, G.	744	A Temperature Variation Method for the Growth of Chalcopyrite Cry